# The Chandra X-ray Observatory

Exploring the high energy universe

AAS | IOP Astronomy



## About the program:

AAS-IOP Astronomy ebooks is the official book program of the American Astronomical Society (AAS), and aims to share in depth the most fascinating areas of astronomy, astrophysics, solar physics and planetary science. The program includes publications in the following topics:

- GALAXIES AND COSMOLOGY
- INTERSTELLAR MATTER AND THE LOCAL UNIVERSE
- STARS AND STELLAR PHYSICS
- EDUCATION, OUTREACH AND HERITAGE
- HIGH-ENERGY PHENOMENA AND FUNDAMENTAL PHYSICS
- THE SUN AND THE HELIOSPHERE
- THE SOLAR SYSTEM, EXOPLANETS, AND ASTROBIOLOGY
- INSTRUMENTATION, SOFTWARE, LABORATORY ASTROPHYSICS AND DATA

Books in the program range in level from short introductory texts on fast-moving areas, graduate and upper-level undergraduate textbooks, research monographs, and practical handbooks.

For a complete list of published and forthcoming titles, please visit iopscience.org/books/aas.

## About the American Astronomical Society

The American Astronomical Society (aas.org), established 1899, is the major organization of professional astronomers in North America. The membership (~7,000) also includes physicists, mathematicians, geologists, engineers and others whose research interests lie within the broad spectrum of subjects now comprising the contemporary astronomical sciences. The mission of the Society is to enhance and share humanity's scientific understanding of the universe.

# The Chandra X-ray Observatory

Exploring the high energy universe

**Edited by**
**Belinda Wilkes and Wallace Tucker**
*Center for Astrophysics | Harvard & Smithsonian, 60 Garden St., Cambridge, MA 02138, USA*

**IOP** Publishing, Bristol, UK

Media content for this book is available from https://doi.org/10.1088/978-0-7503-2163-1.

ISBN 978-0-7503-2163-1 (ebook)
ISBN 978-0-7503-2161-7 (print)
ISBN 978-0-7503-2164-8 (myPrint)
ISBN 978-0-7503-2162-4 (mobi)

DOI 10.1088/2514-3433/ab43dc

Version: 20191201

AAS–IOP Astronomy
ISSN 2514-3433 (online)
ISSN 2515-141X (print)

British Library Cataloguing-in-Publication Data: A catalogue record for this book is available from the British Library.

Published by IOP Publishing, wholly owned by The Institute of Physics, London

IOP Publishing, Temple Circus, Temple Way, Bristol, BS1 6HG, UK

US Office: IOP Publishing, Inc., 190 North Independence Mall West, Suite 601, Philadelphia, PA 19106, USA

*For all who have dedicated years, and in many cases decades, to transforming the* Chandra *X-ray Observatory from a dream into reality and using it to explore the high-energy universe.*

# Contents

# Preface

The development of X-ray telescopes, culminating in the *Chandra* X-ray Observatory, is one of the great success stories of modern science and technology. Less than six decades ago, Riccardo Giacconi and his colleagues discovered the first extrasolar X-source. Only 37 years later—Riccardo always insisted it could have been much sooner!—*Chandra* was launched. As the first, and to date only, X-ray telescope capable of producing subarcsecond images, a critical feature for multi-wavelength investigations with optical, infrared, and radio telescopes, *Chandra* revolutionized the field overnight. This was made clear with the discovery of the long-sought neutron star in the center of the Cassiopeia A supernova remnant in *Chandra*'s first-light image.

Twenty years after launch, *Chandra* continues to make discoveries, and it has firmly established its role as one of the most versatile and powerful tools for exploring the universe and the fundamental laws that govern our existence. *Chandra* gives us a spectacular view of the hot, high-energy regions of the universe, reaching down to faint sources that produce only one photon every 13 days, 10 orders of magnitude fainter than the brightest X-ray source, Sco X-1.

*Chandra* observations have probed the geometry of spacetime around a black hole, unveiled the role of accreting supermassive black holes in influencing the evolution of the most massive galaxies, provided some of the best evidence yet for the existence of dark matter, and independently confirmed the presence of dark energy. *Chandra* has tracked the dispersal of heavy elements by supernovae and has measured the flaring rates of young Sun-like stars with implications for the formation of planets. With *Chandra* observations, it is also possible to do research in fundamental physics, by testing the basic principles in domains not accessible on Earth.

This book is intended to be an introduction to a fertile and exciting field of research, as well as a progress report that captures the depth and breadth of the advances and discoveries made in first two decades of *Chandra*.

With this goal in mind, we have assembled contributions from a "stellar" team of scientists. They include Jeremy J. Drake, who discusses X-rays from stars and planetary systems; Patrick Slane on supernovae and their remnants; Michael A. Nowak and Dom J. Walton on X-ray binaries; Giuseppina Fabbiano on X-rays from galaxies; Aneta Siemiginowska and Francesca Civano on supermassive black holes; Paul Nulsen and Brian McNamara on groups and clusters of galaxies; and finally, Steven W. Allen and Adam B. Mantz on galaxy cluster cosmology. Rafaele D'Abrusco and Rafael Martinez-Galarza were coauthors with one of us (B.W.) of an overview of *Chandra* and its operations, and in the prologue, Harvey Tananbaum and Martin C. Weisskopf provide a unique perspective from two scientists who have, between them, 85 years of experience with *Chandra*, most of the time in project leadership positions.

# Acknowledgments

We are extremely grateful to the scientists, engineers and staff of the *Chandra* X-ray Center and the NASA Project Office at Marshall Space Flight Center, without whose expertise and dedication the work described in this book would not have been possible. We are also deeply indebted to the many people who supported us as we prepared the manuscript. These include our IoP editor Leigh Jenkins, for her continuous support and patient responses to our many questions throughout this long process; Peter Edmonds for reviewing all of the figures and obtaining permissions for reproduction as needed; Evan Tingle for working with the various authors on electronic figures; and Chris Ingle at IoP who, in consultation with Evan, prepared interactive electronic figures of various kinds. In addition to various reviews by the co-chapter-authors, all chapters were reviewed by experts in each topic: Dan Schwartz, Jonathan McDowell, Antonella Fruscione, Randall Smith, Vinay Kashyap, Dan Patnaude, Jack Steiner, Dong-Woo Kim, Martin Elvis, Scott Randall, and Bill Forman, all of whom we wish to thank.

R.A., F.C., J.J.D., G.F., R.M.G., P.E.J.N., A.S., P.S., H.T., W.K.T., and B.J.W. acknowledge the support of NASA contract NAS8-03060.

## Chapter 4

Thanks to M. Corcoran, W. Dunn, P. Edmonds, V. Kashyap, B. Snios, W. Tucker, and B. Wargelin for helpful comments, corrections, and suggestions.

## Chapter 6

M.A.N. would like to thank the numerous colleagues who have been influential in driving his thinking about many of these topics, especially Jörn Wilms, Katja Pottschmidt, Christopher Reynolds, Julia Lee, Daryl Haggard, and Sera Markoff. M.A.N. would also like to thank the help of the many young scientists who he had the privilege to work with when they were graduate students and/or postdocs working with *Chandra* observations and who have continued to make significant contributions to the science areas discussed above, especially Lia Corrales, Victoria Grinberg, Sebastian Heinz, Li Ji, Jon Miller, Joseph Neilsen, and Rudy Wijnands. M.A.N. thanks Jack Steiner for giving feedback on a draft of the manuscript. Finally, M.A.N. would like to give special thanks to the High Energy Transmission Grating Spectrometer team at the Massachusetts Institute of Technology *Chandra* X-ray Science Center: Glenn Allen, John Davis, Dan Dewey, David Heunemoerder, John Houck, Katherine Flanagan, Herman Marshall, Michael Noble, Norbert Schulz, Michael Wise, and especially Claude Canizares, the Principal Investigator of the HETGS. D.J.W. is supported by STFC through an Ernest Rutherford fellowship and would like to thank Felix Fürst, Matt Middleton, Matteo Bachetti, Murray Brightman, Marianne Heida, and Ciro Pinto for their continued collaboration, as well as Andy Fabian, Fiona Harrison, and Daniel Stern for their excellent mentorship over the years.

## Chapter 7

A large part of this chapter was written at the Aspen Center for Physics (ACP) and benefited from discussions with other visitors at the Center. The ACP is supported by National Science Foundation grant PHY-1607611. We thank Martin Elvis, Alessandro Paggi, and Andreas Zezas for input and comments on the manuscript.

## Chapter 8

We thank F. Fornasini, M. Jones, M. Elvis, G. Fabbiano, B. Snios, R. Hickox, and B. Wilkes for comments that improved our chapter. We are grateful to J. Comerford, S. Marchesi, H.R. Russell, M. Sobolewska, B. Snios, G. Lanzuisi, and F. Vito for providing updated figures, and to S. LaMassa and R. Canning for sharing unpublished flux sensitivity curves for the Stripe82 and CATS surveys. We also thank all of the authors that allowed the use of their proprietary graphics in this chapter.

## Chapter 9

We thank Scott Randall for providing comments, and Helen Russell and Julie Hlavacek-Larrondo for providing figures. We also wish to thank Elke Roediger and Alex Sheardown for creating the movie associated with Figure 9.13.

## Chapter 10

Parts of this discussion were adapted from "Cosmological Parameters from Observations of Galaxy Clusters," co-written with Gus Evrard and published in the *Annual Review of Astronomy & Astrophysics* (Allen et al. 2011). We are grateful to many colleagues for their valuable insights and contributions to the work presented here, including Mark Allen, Doug Applegate, Camille Avestruz, Nick Battaglia, Lucie Baumount, Brad Benson, Lindsey Bleem, Stefano Borgani, Marusa Bradač Patricia Burchat, David Burke, Rebecca Canning, Doug Clowe, Harald Ebeling, Steven Ehlert, Andy Fabian, Daniel Gruen, Pat Henry, Ricardo Herbonnet, Stefan Hilbert, Julie Hlavacek-Larrondo, Ashley King, Patrick Kelly, Andrey Kravtsov, Anja von der Linden, Michael McDonald, Glenn Morris, Justin Myles, Daisuke Nagai, Emil Noordeh, Saul Perlmutter, David Rapetti, Eduardo Rozo, Eli Rykoff, Robert Schmidt, Tim Schrabback, Neelima Sehgal, Sara Shandera, Aurora Simionescu, Ondrej Urban, Alexey Vikhlinin, Risa Wechsler, Norbert Werner, Adam Wright, and Irina Zhuravleva. S.W.A. is supported in part by the US Department of Energy under contract No. DE-AC02-76SF00515. A.B.M. is supported by the Kavli Institute for Particle Astrophysics and Cosmology at Stanford. S.W.A. acknowledges the hospitality of the Institute of Astronomy, Cambridge, UK, during the completion of this review.

## Chapter 11

I would like to thank those who have provided information and figures for the various projects described in this chapter, in particular Paul Nandra (*Athena*),

Alexey Vikhlinin (Lynx), Grant Tremblay (Lynx), Peter Predehl (eROSITA), Mikail Pavlinsky (SRG), Paul Ray (STROBE-X), Randall Smith (Arcus), Richard Mushotzky (AXIS), Maxim Markevitch (AXIS), Laura Brenneman (HEX-P), Rob Petre (XRISM), and Makoto Tashiro (XRISM).

## Reference

Allen, S. W., Evrard, A. E., & Mantz, A. B. 2011, ARAA, 49, 409

# Contributors biographies

## Harvey Tananbaum

Dr. Harvey Tananbaum is a Senior Astrophysicist at the Smithsonian Astrophysical Observatory (SAO) in Cambridge, Massachusetts. He received his BA in mathematics and physics from Yale University in 1964, and his PhD in physics from MIT in 1968. His thesis research studied a mysterious, highly variable cosmic X-ray source. Later, observations with Uhuru—the first satellite dedicated to X-ray astronomy—along with ground-based optical and radio data were instrumental in showing that this source, Cygnus X-1, was powered by matter falling into a black hole. Dr. Tananbaum directed SAO's High Energy Astrophysics Division from 1981 through 1993. He served as Project Scientist for Uhuru; as Scientific Program Manager for the HEAO-2/Einstein mission, the first imaging telescope for extra-solar X-ray astronomy; and as Principal Investigator and Director of the Einstein Data Center. In 1976, he and Riccardo Giacconi led the team which proposed to NASA to initiate the study and design of a large, long-lived X-ray telescope that was launched in 1999, as the *Chandra* X-Ray Observatory. He organized and led the team which was selected competitively in 1991 to develop and operate the entity now known as the *Chandra* X-ray Center, for which he served as Director from 1991–2014. Dr. Tananbaum received the NASA Exceptional Scientific Achievement Medal, the NASA Public Service Award, and the NASA Medal for Outstanding Leadership. In 2004, he and Dr. Martin Weisskopf were awarded the Bruno Rossi prize of the High Energy Astrophysics Division of the American Astronomical Society, for "...vision, dedication, and leadership in the development, testing, and operation of the *Chandra* X-ray Observatory." In 2005, Dr. Tananbaum was elected to the National Academy of Sciences.

## Martin Weisskopf

Dr. Martin C. Weisskopf is Project Scientist for NASA's *Chandra* X-ray Observatory, Principal Investigator of the Imaging X-ray Polarimetry Explorer (IXPE), and Chief Scientist for X-ray Astronomy at Marshall Space Flight Center (MSFC), where he began his NASA career in 1977. Weisskopf was previously an assistant professor at Columbia University and performed many pioneering experiments in X-ray astronomy—particularly in X-ray polarimetry. He earned a bachelor's degree in physics from Oberlin College and a doctorate in physics from Brandeis University. Weisskopf is author or co-author of over 350 publications—including refereed journal articles, book articles, monographs and papers in conference proceedings and has received numerous honors—including the Rossi Prize of the High Energy Astrophysics Division of the American Astronomical Society (shared with Dr. H. Tananbaum). He is also a Fellow of both the American Physical Society and the SPIE.

## Wallace Tucker

Dr. Wallace Tucker served as the science spokesperson for the *Chandra* X-ray Center at the Center for Astrophysics | Harvard & Smithsonian for 20 years. In that capacity he worked closely with and contributed written material for the *Chandra* Communications & Public Engagement program, which has received recognition for its projects and digital content through two international communication awards and numerous US digital awards. Tucker received a BS in Mathematics and a MS in Physics from the University of Oklahoma, and PhD in Physics from the University of California, San Diego in 1966. He was a research associate at Cornell University, and an Assistant Professor of Space Science at Rice University, before moving to American Science & Engineering (AS&E) to become the head of the theoretical astrophysics group. From 1972 until 1996 he worked as a consultant for AS&E and held positions at the Smithsonian Astrophysical Observatory, United States International University, University of California, Irvine, and University of California, San Diego before joining the *Chandra* X-ray Center. He is a member of the American Astronomical Society, the American Association for the Advancement of Science, and the International Astronomical Union. He is the author or co-author of six books on astronomy for the general reader including *The X-ray Universe* with Riccardo Giacconi, and two textbooks, *Radiative Processes in Astrophysics* and *Heath Physical Science.* He has authored or co-authored numerous scientific and popular articles on astrophysics. He has also written three prize-winning stage plays, one of which was published in an anthology of American Indian plays, and is active in land conservation, having co-founded the Fallbrook Land Conservancy, which has preserved more than 2,000 acres of open space.

## Belinda Wilkes

Dr. Belinda Wilkes is a Senior Astrophysicist at the Center for Astrophysics | Harvard & Smithsonian. She has served as Director of the *Chandra* X-ray Center, which operates the *Chandra* X-ray Observatory on contract with NASA, since 2014. Dr. Wilkes received her BSc in Astronomy and Physics from the University of St. Andrews, Scotland in 1978 and her PhD in Astronomy from the University of Cambridge, England, in 1982. She spent two years at the University of Arizona's Steward Observatory as a NATO postdoctoral fellow, and moved to CfA's High Energy Astrophysics Division in 1984. She is a Fellow of the Royal Astronomical Society and of the Cambridge Philosophical Society, and a member of the American Astronomical Society, the International Astronomical Union, the American Physical Society and the American Association for the Advancement of Science. She has received numerous Smithsonian Institution awards including the Exceptional Accomplishment Awards and NASA Group Achievement Awards, and a NASA MSFC Director's Commendation. In 2018 she was elected an Honorary Fellow of Jesus College, Cambridge University, England. Her research involves X-ray and multi-wavelength studies of active galaxies. She is author or co-author of over 460 publications, including refereed science papers, book chapters, papers in conference proceedings, abstracts, white papers, author or editor of several books, and of science articles in the public media.

## Raffaele D'Abrusco

Dr. Raffaele D'Abrusco is a staff astrophysicist at the Center for Astrophysics | Harvard & Smithsonian and is responsible for the archive operations team of the Chandra X-ray Observatory. He works on classification of high-energy sources, evolution of galaxies through the spatial distribution of globular clusters and the application of data mining tools to large/complex astronomical datasets. He was educated at the University of Naples "Federico II" (Italy) as both an undergraduate (2004) and postgraduate (2007). Before joining the CXC, he has held postdoctoral appointments at the University of Naples, University of Padua and the High Energy Astrophysics Division at the CfA.

## Rafael Martínez-Galarza

Dr. Rafael Martínez-Galarza is an astrophysicist at the Smithsonian Astrophysical Observatory (SAO) and the deputy end-to-end scientist for the *Chandra* X-ray Center (CXC) Data Systems, where he has had a significant participation in the release of the *Chandra* Source Catalog 2.0. His research focuses on multi-wavelength studies of star-forming galaxies, and on the use of machine learning techniques for the exploration of large astronomical datasets. Before joining the CXC, he was a postdoctoral fellow at SAO and a teaching fellow at Harvard University's Institute for Applied Computational Science, where he taught stochastic optimization and machine learning. He has worked as a calibration scientist for the Mid-Infrared Instrument for the *James Webb Space Telescope* and is currently a member of the LSST science collaboration in transients and variable stars. He holds a PhD in astronomy from Leiden University in the Netherlands.

## Jeremy Drake

Dr. Jeremy Drake is a Senior Astrophysicist at the Smithsonian Astrophysical Observatory of the Harvard-Smithsonian Center for Astrophysics in Cambridge, Massachusetts, USA. He graduated with a DPhil degree in 1989 from the University of Oxford, and was subsequently awarded a NATO Postdoctoral Fellowship to work at the University of Texas at Austin, USA, on high-resolution spectroscopy of stars. A subsequent move to the University of California, Berkely, saw Drake move higher up in the stellar atmosphere to study stellar coronae using what was to be the newly-launched NASA *Extreme Ultraviolet Explorer* (EUVE). This provided a natural platform to move to higher energies still, and at the end of 1995 he moved to SAO to work on the *Advanced X-ray Astrophysics Facility*, subsequently re-named the *Chandra X-ray Observatory.* He studies the high-energy aspects of stellar physics and how they impact star and planet formation, stellar evolution, and planetary radiation environments.

## Patrick Slane

Dr. Patrick Slane is a Senior Astrophysicist at the Center for Astrophysics | Harvard & Smithsonian. His research centers on high energy astrophysics, with particular concentration on supernova remnants, young neutron stars, and pulsar winds. He completed his undergraduate studies at the University of Wisconsin—Whitewater, a masters degree in mathematics at the University of Wisconsin—Milwaukee, and his PhD in physics at the University of Wisconsin—Madison. He has served the scientific community as a member of the Executive Committees for the High Energy Astrophysics Division of the AAS and the Division of Astrophysics of the APS. He has been the recipient of numerous NASA Group Achievement awards and Smithsonian Superior Accomplishment awards, and is an elected Fellow of the APS. He is currently the Assistant Director for Science at the *Chandra* X-ray Center where he also leads the Science Mission Planning Team for NASA's *Chandra* X-ray Observatory.

## Michael Nowak

Dr. Michael Nowak is a research professor in physics at Washington University in St. Louis. He studied physics at the Massachusetts Institute of Technology (MIT; SB 1987) and Stanford University (MS 1988, PhD 1992), and held postdoctoral positions at the Canadian Institute of Theoretical Astrophysics and the University of Colorado, before becoming a research scientist for the *Chandra* X-ray Observatory at MIT (2001–2018). He studies the high energy astrophysics of stellar mass black holes and neutron stars in our own galaxy, as well as supermassive black holes in the centers of our own and other galaxies. He has been especially interested in using X-ray variability and spectroscopy to probe relativistic effects in these systems.

## Dominic Walton

Dr. Dominic Walton is a senior research fellow in astrophysics at the University of Cambridge, UK. He has spent most of his career working on a wide variety of different accreting systems, including Galactic X-ray binaries, ultraluminous X-ray sources and active galactic nuclei, focusing primarily on the study of their emission in the X-ray band by utilizing a variety of different space-based facilities. Educated at Durham University (undergraduate, 2008) and the University of Cambridge (PhD, 2012), he has also previously held research positions at the California Institute of Technology, working with the *NuSTAR* X-ray observatory, and NASA's Jet Propulsion Laboratory, having won a NASA postdoctoral research fellowship. He is currently supported by STFC after being awarded an Ernest Rutherford fellowship.

## Giuseppina Fabbiano

Dr. Giuseppina (Pepi) Fabbiano is a Senior Astrophysicist at the Smithsonian Astrophysical Observatory (Center for Astrophysics | Harvard & Smithsonian, Cambridge MA), which she joined in 1975. A native of Italy, she has worked in X-ray Astronomy since obtaining her Doctorate in Physics from the University of Palermo (Italy) in 1973. She has been instrumental in establishing the field of X-ray studies of galaxies. Over the years, she has studied the different components of the X-ray emission of the Milky Way and external galaxies, including neutron stars and black holes in binary systems, supernova remnants, the hot interstellar medium and nuclear massive black holes. She uses X-ray observations, together with data throughout the observable spectrum, to investigate the joint evolution of galaxies and black holes in the universe. She is the author of over 270 refereed papers in major scientific journals, including two invited reviews in the Annual Review of Astronomy and Astrophysics, and several reviews and book contributions. Throughout her career, Fabbiano has been involved in the operations and data management of major NASA X-ray observatories, and is presently the Head of the *Chandra* X-ray Observatory Data Systems Division. She has served in several national and international scientific and data management committees. She is a member of the executive committee of the International Virtual Observatory Alliance, an organization whose purpose is to foster standards for data interoperability throughout astronomy. She is also a member of the board of the Aspen Center for Physics.

## Aneta Siemiginowska

Dr. Aneta Siemiginowska is a Senior Astrophysicist in the High Energy Astrophysics Division of the Center for Astrophysics | Harvard and Smithsonian, and a member of the Science Data System team at the Chandra X-ray Center. She studied at the University of Warsaw (MS 1985), and Nicolaus Copernicus Astronomical Center in Warsaw (PhD 1991). She has worked in both theoretical and observational aspects of X-ray astronomy with interests in extragalactic radio sources, quasars, powerful jets and statistical methods. She has discovered several hundred kiloparsec long relativistic X-ray jets associated with distant quasars and initiated early studies of young radio sources in X-rays with *Chandra*. In addition to X-ray research, she is also a founding member of the International CHASC AstroStatistics Collaboration promoting communication and algorithm development between astrophysicists and statisticians. She served as Chair of the American Astronomical Society Working Group on Astroinformatics and Astrostatistics (2013-2019), promoting awareness of the applications of advanced computer science, statistics and allied branches of applied mathematics to further the goals of astronomical and astrophysical research.

## Francesca Civano

Dr. Francesca Civano is an Astrophysicist at the Center for Astrophysics | Harvard & Smithsonian, where she is the Deputy Manager for Data Processing at NASA's *Chandra* X-ray Observatory and also researches black holes and galaxies. She obtained her PhD at the Bologna University in 2007. She was a postdoctoral fellow at the Harvard-Smithsonian center for Astrophysics from 2007–2012, a research associate at Dartmouth College and Yale University from 2012–2016. Dr. Civano is the PI of the *Chandra* COSMOS Legacy survey, a 2.8 million seconds program to observe the Cosmic Evolution Survey field with *Chandra* and learn about black hole and galaxy co-evolution. She also led a 3.1 million seconds survey of the same COSMOS field with NASA's *NuSTAR*, and was just awarded 585k seconds to observe the North Ecliptic Pole field. Her research interests include the interplay and evolution of galaxies and the active supermassive black holes at their center (and not!) using data from all the electromagnetic spectrum. Dr. Civano is deeply involved in the second release of the *Chandra* Source Catalog.

## Paul Nulsen

Dr. Paul Nulsen is a Senior Astrophysicist at the Smithsonian Astrophysical Observatory in Cambridge, Massachusetts, where he manages operations for the *Chandra* High Resolution Camera. His research centres on galaxy groups and clusters, particularly X-ray observations and theory of the physical processes affecting their hot atmospheres. Following a BSc University of Western Australia (1975) and PhD at the University of Cambridge (1980), he has held positions at the Institute of Astronomy, Cambridge, the Australian National University, the University of Wollongong and the Harvard-Smithsonian Center for Astrophysics.

## Brian McNamara

Dr. Brian McNamara is Professor, Department Chair, and University Research Chair in Physics & Astronomy at the University of Waterloo, in Ontario, Canada. He is an Affiliate of the Perimeter Institute for Theoretical Physics, former visiting member of the Harvard-Smithsonian Center for Astrophysics, and former Director of the Guelph-Waterloo Physics Institute. After receiving a PhD at the University of Virginia in 1991, McNamara took a postdoctoral fellowship at the Kapteyn Laboratory in Groningen, The Netherlands. From 1993 to 2000 he was a staff member at the *Chandra* X-ray Center and Harvard-Smithsonian Center for Astrophysics. From 2000 to 2006, McNamara was a professor of Physics & Astronomy at Ohio University. Since 2006 he has taught physics and astronomy at the University of Waterloo, where he remains today. McNamara studies galaxies and clusters of galaxies. He is interested in how they form and evolve under the influence of powerful radio jets launched by supermassive black holes. His most recent work involves making measurements with the earth-orbiting *Chandra* X-ray Observatory and the Atacama Large Millimeter Array, which is the most powerful telescope in existence. He is a former member of the Hitomi X-ray Observatory Science Team and is current team member of its successor, the XRISM X-ray observatory, which is planned for launch from Japan in 2022

## Steven Allen

Dr. Steve Allen is a Professor of Physics at Stanford University and the SLAC National Accelerator Laboratory. His research interests span the astrophysics of galaxy clusters and their use as cosmological probes, utilizing a broad array of multi-wavelength observations. Educated in the UK, he received his PhD in 1995 from the University of Cambridge, where he also held a PPARC Postdoctoral Fellowship and a Royal Society University Research

Fellowship. He joined Stanford in 2005 and was co-recipient of the 2008 Bruno Rossi Prize of the American Astronomical Society.

## Adam Mantz

Dr. Adam Mantz is a Research Scientist in Physics at Stanford University. His research centers on the cosmology and astrophysics of clusters of galaxies, particularly using X-ray and radio observations of the intracluster medium, and the development of robust statistical techniques for astrophysical applications. He received his PhD from Stanford in 2009, and held a NASA Postdoctoral Fellowship at Goddard Space Flight Center, followed by a postdoctoral position at the University of Chicago.

# Prologue

**Harvey Tananbaum and Martin C Weisskopf**

We share some recollections of important milestones—highlights and challenges, setbacks and successes—as seen through our eyes as the Smithsonian Astrophysical Observatory (SAO) team lead/initial Director of the *Chandra* X-ray Center and the NASA Marshall Space Flight Center (MSFC) Project Scientist. Of necessity, the process of compressing the first 23 years of *Chandra* history (from proposal through launch) into a brief chapter requires a very subjective selection of events to include.

## The Early Years (1976–1981)

In 1976 April, several SAO scientists submitted an unsolicited proposal to NASA to study a 1.2 m diameter X-ray telescope national space observatory. Riccardo Giacconi was the Principal Investigator (PI) for that proposal, with Harvey Tananbaum as co-PI, and Paul Gorenstein, Rick Harnden, Pat Henry, Ed Kellogg, Steve Murray, Herb Schnopper, and Leon van Speybroeck as co-Investigators. Many of the key parameters for this proposed mission were first laid out in a 1963 "white paper" for the long-term development of X-ray astronomy, also led by Riccardo Giacconi, based on the 1962 discovery of the first extrasolar X-ray source (Sco X-1) and the all-sky X-ray background. Key parameters for the 1976 proposal were a long life (~10 years), excellent angular resolution (~0.5″), and high sensitivity (1.2 m diameter outer mirror, effective area over 1000 $cm^2$ for wavelengths longer than 4 Å, and high detection efficiency). The broadly stated science objectives included studies of stellar structure and evolution, large-scale galactic phenomena, active galaxies, rich clusters of galaxies, and cosmology.

Even though the *High Energy Astronomy Observatory* telescope mission (*HEAO-B*, later nicknamed *Einstein*) had not yet launched, the SAO team was confident about its potential discoveries while being concerned about its relatively short lifetime (nominally limited to one year by the onboard propellant). Based on preliminary discussions with NASA HQ, the team believed that they would be receptive to a proposal to initiate technology work and studies for a more powerful, longer life successor mission. Within a year, the NASA Associate Administrator for Space Science, Noel Hinners, approved the request. Anticipating a natural transition of the

doi:10.1088/2514-3433/ab43dcch0

work force at MSFC that was managing the *Large Space Telescope* (*LST*—later renamed the *Hubble Space Telescope*, *HST*) and *HEAO-B*, NASA HQ assigned the management of the new study to MSFC, with an endorsement of this selection from SAO. Martin Weisskopf left Columbia University and joined MSFC in 1977 as the first (and to date the only) Project Scientist for the mission, which was soon named the *Advanced X-ray Astrophysics Facility* (*AXAF*). The suggestion for the name came from Noel Hinners, who recommended, for political expediency, that we not use the word "telescope" following the just-completed Congressional approval to start the LST.

From the start, the *AXAF* study was organized with MSFC responsible for overall management and initial system engineering studies and SAO providing scientific and technical support to MSFC and the Project Scientist. Our perspective was, and still is, that this teaming agreement provided *AXAF* with end-to-end leadership where science considerations were integrated with engineering, schedule, and cost across all areas of the project. Quoting *AXAF* Program Manager, Fred Wojtalik: "This culture [the *AXAF* team] is easier to work with. I'm not sure what it is, but they (the scientists) have helped us tremendously, when we need to make trades and try to make ends meet. They are much more members of a team" (Tucker & Tucker 2001, pp. 62–63).

NASA also appointed an *AXAF* Science Working Group (SWG) to assist in establishing the science requirements and to provide other guidance during the early phase of the mission. This group, chaired by R. Giacconi with M. Weisskopf as vice-chair, first met in December 1977. The members were E. Boldt, S. Bowyer, G. Clark, A. Davidsen, G. Garmire, W. Kraushaar, R. Novick, M. Oda, A. Opp, K. Pounds, S. Shulman, H. Tananbaum, J. Trümper, and A. Walker, representing most of the X-ray astronomy groups around the world. The SWG documented their findings in a report published as NASA TM-78285 in May 1980, providing important inputs to the *Astronomy and Astrophysics for the 1980s* Decadal Survey. The SWG was disbanded prior to the 1983 Announcement of Opportunity to propose for *AXAF* instruments, telescope scientist, and interdisciplinary scientists.

Since the 1960s, the US astronomy community has met at roughly 10 year intervals under the auspices of the National Academy of Sciences to review the state of the field and to recommend priorities for ground- and space-based astronomy for the coming decade (hence the commonly used Decadal Survey moniker). The Decadal Survey for the 1980s, chaired by George Field and undoubtedly influenced by results from the *Einstein* mission, unanimously recommended, as its highest priority, a major new program, "*An Advanced Astrophysics Facility* (*AXAF*)" operated as a permanent national observatory in space (National Research Council 1982, p. 133). The Report went on to say: "Because of its power and versatility, *AXAF* will profoundly influence and enhance the development of nearly all areas of Galactic and extragalactic astronomy" (National Research Council 1982, p. 135). The Committee also recommended institutional arrangements akin to *HST*'s Science Institute to manage the science operations "and to facilitate the participation of the science community in the acquisition and interpretation of X-ray observations" (National Research Council 1982, p. 135). It is important to note that the thinking and language here show that less than two decades after the initial discoveries, X-ray astronomy was seen as being important to all astronomers and not simply to the expert practitioners of the field.

## Technology Development, Mission Studies, Selection of Science Instruments and Prime Contractor (1981–1989)

The years from 1981 to 1989 were very active. Highlights included numerous trade studies such as how to configure the mission with multiple focal plane instruments—should the optics be moved to place the image on the detector of choice for a given science observation or should one move the detectors to place them at the best focus; should the pointing be fixed on a target or should the observatory be "dithered" to minimize detector damage due to bright sources, as well as impacts due to loss of flux which falls within or crosses over gaps between chips or node boundaries within a chip; and should the observatory be serviced using the *Space Station* or the shuttle (as was done for *HST*). The most important technical study during this period involved the construction and X-ray testing of the Technology Mirror Assembly (TMA) shown in Figure 0.1. The TMA was built to achieve the *AXAF* imaging requirements, of order 10× smaller radius for the 50% and 90% encircled energy (EE) than either *Einstein* or *ROSAT* (*Röntgensatellit*, led by West Germany with US and UK participation, and launched in 1990). The TMA was sized to be a two-third version of the innermost paraboloid–hyperboloid pair of the baseline *AXAF* design, with a 0.41 m diameter and 6 m focal length to allow for X-ray testing at MSFC's X-ray facility that had been used for *Einstein Observatory* calibration.

It is worth noting that, initially, there were three commercial manufacturers interested in the *AXAF* optics. All three were funded to provide a TMA. Ultimately, only one of these, the Perkin Elmer Corporation, delivered a completed TMA. After

**Figure 0.1.** The technology mirror assembly. (Courtesy of NASA.)

a rocky start that required repolishing due to a sign error and lack of metrology for the midfrequency spatial regime, Perkin Elmer delivered a TMA that demonstrated 90% EE within a 1″ radius, more than satisfying *AXAF* requirements. Technically speaking, because of the scaling of its size, tolerances on several of the TMA parameters were more stringent than those for *AXAF*; thus, the successful testing of the TMA verified to us the ability to build the *AXAF* optics. However, we would subsequently be challenged to address scale-up issues.

The Announcement of Opportunity for instruments, a telescope scientist (TS), and a team of interdisciplinary scientists (IDSs) was released in 1983, with selections awarded in 1985. At the time, six instruments were chosen: the *AXAF* (later Advanced) CCD Imaging Spectrometer (ACIS), G. Garmire, PI; The High Resolution Camera (HRC), S. Murray, PI; The Low Energy Transmission Grating Spectrometer (LETGS), A. Brinkman PI; The High Energy Transmission Grating (HETG) and the Focal Plane Crystal Spectrometer (FPCS), C. Canizares, PI; and the X-ray Spectrometer (XRS), S. Holt, PI. In addition, L. Van Speybroeck was selected as the telescope scientist (TS), and a number of IDSs completed the selection: A. Wilson, A. Fabian, J. Linsky, R. Giacconi, and R. Mushotzky. Including the head of the *AXAF* Mission Support Team at SAO, H. Tananbaum, a new Science Working Group was established, with M. Weisskopf as the Chair. The team is shown in Figure 0.2. The FPCS was removed upon the recommendation of the SWG in 1988 as part of a cost-saving action, and the XRS was subsequently moved to the *AXAF-S* mission (see more below).

This time period also saw the inclusion of industry contractors performing Phase B studies and then responding to a Request for Proposal (RFP) for flight development

**Figure 0.2.** Members of the second *AXAF* Science Working Group (SWG), formed in 1985. From left to right: A. Wilson (UMD), A. Fabian (IoA, Cambridge, UK), J. Linsky (JILA, NIST), H. Tananbaum (SAO), A. Bunner (NASA, ex-officio), S. Holt (GSFC), M.C. Weisskopf (MSFC, chair), R. Giacconi (STScI), A. Brinkman (SRON, The Netherlands), S. Murray (SAO), G. Garmire (PSU), L. Van Speybroeck (SAO), C. Canizares (MIT), and R. Mushotzky (GSFC). The picture was taken at MSFC, in front of the Space Science Laboratory (Building 4481, which no longer exists). (Courtesy of NASA.)

in 1988. Two teams, one led by Lockheed (the prime contractor for *HST*) and the other by TRW (the prime contractor for *Einstein*) responded to the RFP. To ensure that both teams would be positioned to work with the only company that had successfully produced a TMA, the proposers were instructed that the optical elements would be provided by Perkin Elmer although assembling these elements into a telescope assembly would be part of their proposals. After a several-month-long source evaluation, TRW and its team were awarded the prime contract.

## Politics, a Breakthrough Accomplishment, and a New Mission Configuration (1983–1992)

As part of the effort to move *AXAF* forward, our team engaged in "politics" with fellow scientists, NASA HQ, the Office of Management and Budget (OMB), and Congress. In late 1983, two highly respected infrared astronomers met with the NASA Astrophysics Division Director, Charlie Pellerin, advocating the reprioritization of the recommendations of the 1980 Decadal Survey. They proposed moving the Shuttle Infrared Telescope Facility (SIRTF) to the top of the list for free-flying observatories. In 1984 October, George Field wrote to Charlie Pellerin reconfirming the priorities established by the Astronomy Survey Committee. Pellerin then met with SIRTF and *AXAF* leaders to explain that *AXAF* was ready to move into preliminary design (Phase B) while SIRTF was not, so he would continue to follow the guidance provided by the Decadal Survey.

Endeavoring to find a path forward for both missions, Pellerin and his deputy George Newton organized a meeting of senior astronomers. Pellerin asked Martin Harwit, then at Cornell and author of the popular book *Cosmic Discovery*, to chair the discussion. The group identified 10 of the most significant problems in astrophysics along with the telescopes and instruments required to address these questions. This approach tied together infrared (SIRTF), optical-UV (*HST*), X-ray (*AXAF*), and gamma-ray (GRO) observatories. With the subsequent suggestion of a name by George Field, the Great Observatories concept was formulated. Led by Martin Harwit, who was assisted by Dr. Valerie Neal, the team generated a booklet with simple sketches and direct wording to explain how the Great Observatories together would explore the universe. Figure 0.3, adapted from the booklet (Harwit & Neal 1986), displays the electromagnetic spectrum, a temperature scale, sketches of the four observatories, and notional images of objects they might study. No longer could opponents argue that the already approved *HST* and GRO missions meant that there was no need for *AXAF* or SIRTF.

Although *AXAF* had been ranked as the top priority NASA astrophysics mission by the 1980 Decadal Survey, major science initiatives at NASA also included candidate planetary, solar-terrestrial, and Earth science missions. Each discipline advocated for their own priorities; however, *HST* delays and overruns were working against *AXAF*. In 1985, the Office of Space Science and Applications (OSSA) ranked TOPEX with its ocean measurements as its top choice, and followed that in 1986 with the International Solar-Terrestrial Physics Program as its next priority. The situation evolved dramatically in spring 1987, when Lennard Fisk was named to

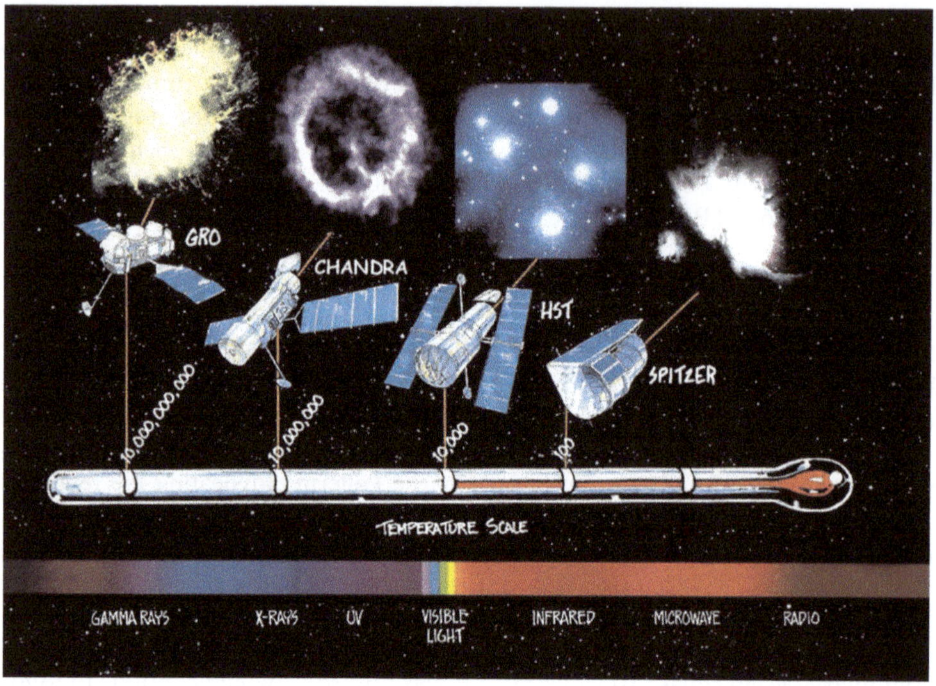


**Figure 0.3.** Figure from the Great Observatories brochure (adapted from the booklet, Harwit & Neal 1986), which shows the bands comprising the electromagnetic spectrum with the corresponding temperature scale for thermal emission, sketches of the four Great Observatories, and notional images of objects they might study. (Courtesy of NASA.)

lead OSSA. Harvey Tananbaum together with Harvard astronomer Jonathan Grindlay met with Fisk. They discussed the Great Observatories concept indicating, for example, how *AXAF* would study the hot gas in clusters, filling the space between the galaxies seen in the optical band. They also described how *AXAF* was a simpler observatory than *HST* with more relaxed pointing and stability requirements through its counting of individual X-ray photons rather than integrating over longer intervals; how the key technology for the X-ray optics was already well advanced via the TMA program; and how the *AXAF* operations would be much simpler. Many of these same points were subsequently emphasized to Fisk's Space and Earth Sciences Advisory Committee at its 1987 May 27–29 meeting, following which *AXAF* was selected as the top priority OSSA new start for FY89.

NASA then submitted a request for a formal start on *AXAF* to the OMB. OMB turned down the request as well as a subsequent appeal by NASA—most likely because of fiscal concerns. NASA decided to expend a "silver bullet" and said it would appeal the OMB decision directly to President Ronald Reagan. The appeal was to be in the form of a single chart (pre-PowerPoint) explaining why the President should overrule the OMB. The HQ *AXAF* Program Manager, Arthur Fuchs, called both Tananbaum and Weisskopf to ask for suggestions. They provided the following assessment: science arguments would not be effective, nor would possible technology spin-offs and education opportunities; the most promising direction should focus on global leadership. The Soviet Union, labeled the Evil Empire by President Reagan, had an active X-ray astronomy program on board the *Mir* space station. Japan, with its very strong economy and the yen overpowering the dollar, also had an impressive X-ray program. Tananbaum suggested a chart with three flags—U.S., Soviet Union, and Japan—and words to the effect: Who will lead the world in X-ray astronomy? Fuchs took the suggestions to Pellerin and Fisk. Notwithstanding its out-of-the-box character, they drew up such a chart for NASA Administrator James Fletcher to take to the White House. Before seeing President Reagan, Fletcher was asked by Reagan's inner circle of advisors what he intended to present. Upon seeing the chart, they exclaimed something like: "Why didn't you tell us this program was so important? It's back in the budget." Of course, we do not know what President Reagan might have done, but the approach bore fruit for *AXAF*.

In parallel with these efforts, the *AXAF* science leaders also organized letter-writing campaigns such as that led by Tananbaum and MIT's George Clark in 1983 urging the OSSA Associate Administrator to start on Phase B industry studies and to solicit instrument proposals for *AXAF*. Subsequently, drawing on the Great Observatories initiative, the *AXAF* scientists developed *AXAF*-focused booklets, brochures, fact sheets, and the like. The *AXAF* brochure even won a prize. After the new start was included in the NASA FY89 budget request, the NASA/*AXAF* industry team began to focus on the Congressional Authorization and Appropriations Subcommittees with jurisdiction over NASA. Particularly strong working relationships were formed with Marty Kress, who led the Senate Authorization staff, and Nancy Ramsey from Senator John Kerry's office; Bill Smith, who led the House Authorization staff; and Jeff Lawrence from Congressman Bill Green's office, who was minority lead, and Dick Malow, who headed staff for the House Appropriations Subcommittee for Veterans Affairs, Housing and Urban Development, and Independent Agencies chaired by Congressman Edward Boland. Another key player in these interactions was Hank Steenstra, who headed TRW's Washington office, knew just about everyone, and had a down-to-earth, likable demeanor, which facilitated initial credibility and acceptance by the D.C. "players" for those of us who arrived as "outsiders."

Malow played a central role in subsequent developments. While the *AXAF* team had established credibility in their discussions, continuing issues with *HST* raised concerns for him, as did input from someone challenging our ability to produce mirrors to the *AXAF* specifications. With the question of *AXAF*'s future on the line, a letter-writing campaign from scientists to Congress seemed to help, with Congressman Boland eventually telling Len Fisk: "You've got to do something about those goddamn letters" (Tucker & Tucker 2001, p. 107).

Ultimately, Malow and Boland worked out a compromise with Fisk, Pellerin, and the *AXAF* team. The aptly named "mirror challenge" provided the necessary funding for us to build the largest pair of *AXAF* mirrors in three years and to demonstrate their performance with an X-ray test by 1991 October 1. If the challenge was met, the full program with its full funding would proceed; if the challenge was not accomplished, the program would be canceled.

With the challenge accepted, work on the largest pair of mirrors at Hughes Danbury (Perkin Elmer's large optics division had been purchased by Hughes) proceeded pretty much on schedule for almost two years. However, in the fall of 1990, polishing cycles no longer showed the anticipated improvements when the mirrors were measured with visible light metrology stations designed to check circularity, axial slope, and surface smoothness on various scales. Following a few months with essentially no progress, MSFC, SAO, NASA HQ, and TRW experts convened with the Hughes Danbury team in 1991 January. We decided to stop the polishing in order to determine what was wrong, even though that would mean falling even further behind schedule. Our assessment (perhaps more of a guess) was that it would take two months to diagnose and fix the problems. The integrated team eventually found binding of cables on one metrology station, slipping of a reference bar on another, and vibrations introduced by a large fan 100 ft away from a third. With the problems identified, fixes were developed, tested, and successfully implemented. Shortly before the work was scheduled to resume, Art Napolitano, the Hughes Danbury program manager, took his senior team members off the job for a one-day retreat to brainstorm how to fit the remaining six months of work into the available two months of schedule. They decided to work around the clock seven days a week with stepped-up computer control of the polishing process. More importantly, they cut down the number of time-consuming visits to the metrology stations by assuming that the polishing would now proceed as planned and less frequent checks would be needed. Amazingly, they actually finished two weeks ahead of their schedule after being four months behind! Len Fisk approved an additional, short smoothing run, which met a goal for imaging quality at 6 keV nearly as sharp as that at 1 keV. The mirror pair was shipped to Eastman Kodak (EK) where the 2 pieces were mounted to a pair of adjustable fixtures.

While the optics were being polished, a dedicated team at MSFC undertook the substantial expansion and extension of the test facility first built to calibrate the *Einstein* telescope and then to test the TMA. MSFC considered this such a high priority that the team working on the test facility met with the Center Director Jack Lee at 7:00 every Thursday morning to review progress. X-rays generated by slamming high-energy electrons into various targets would be transmitted through a 1700 ft-long pipe with diameter increasing from the source to the large vacuum chamber housing the mirrors. The pipe was evacuated so that X-rays could traverse the distance to the mirror without being absorbed, while the long distance ensured a nearly parallel beam of X-rays arriving at the mirror. Of course, everything also had to be superclean to avoid contaminating the mirror surfaces. Figure 0.4 shows an aerial view of the facility.

With the mirrors in place and aligned, the chamber was pumped down over Labor Day weekend 1991. One-dimensional scans across the mirror revealed a puzzling behavior different from the aligned single peaks expected for the set of scans. Was it possible that the mirror had been polished to the wrong shape? Engineers and scientists from SAO, MSFC, and EK spent several nerve-wracking days sorting through possible explanations. Ultimately, they concluded that gravity had caused the mirrors to sag ever so slightly, and the resultant ovalization had

**Figure 0.4.** Aerial view for the X-ray Calibration facility used to calibrate the *AXAF* optics. The source chamber is toward the upper left, the 1700 ft-long pipe is in the center, and the building housing the vacuum chamber and the team working area is at the right. (Courtesy of NASA.)

distorted the image. The cause was traced to separate errors in two of the ray-trace codes used to account for gravity effects along with incompleteness of a third code. The EK engineers computed the force required to offset the gravity effects, and then designed and implemented a set of springs to apply the extra push. When the chamber was pumped down again and the X-rays were turned on, the desired images were seen. Figure 0.5 shows the 2D image generated from the full set of test data with the image having an FWHM of 0.19″, a factor of 2.5 better than the performance required by the mirror challenge. A very important by-product of the challenge was the tremendous team building among the industry leads, NASA, SAO, and the SWG, which was so important for what would follow.

In 1992 January, Charlie Pellerin informed the *AXAF* team that new NASA budgets would no longer support the projected cost of *AXAF*; neither the peak-year funding levels nor the overall (run-out) costs could be accommodated. *AXAF* would need to be descoped to avoid cancellation. This devastating news precipitated six months of discussions, trade studies, negotiations, and at times, pessimism and hard feelings. The team considered more than 20 different combinations of mirror pairs and instrument complements, while struggling to find an acceptable balance between

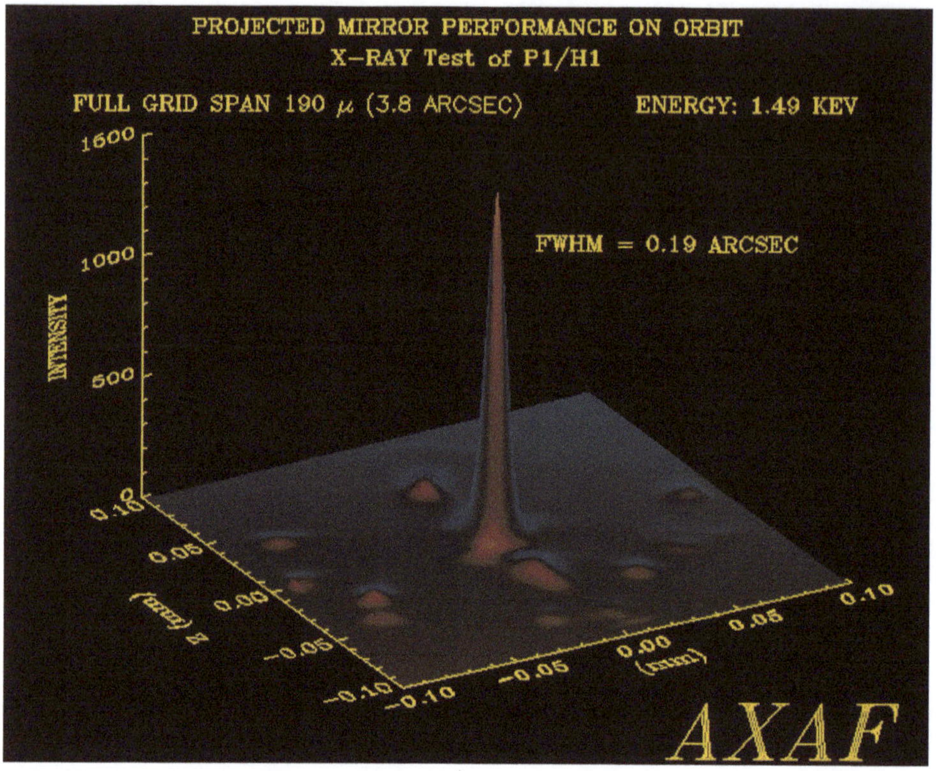


**Figure 0.5.** The two-dimensional image reconstructed from the X-ray exposures scanning across the largest pair of *AXAF* mirrors in 1991 September. The central peak (FWHM of 0.19″) was 2.5 times sharper than required by the "mirror challenge." (Courtesy of NASA.)

science return and concomitant cost. The fantastic rescue of the *HST* mission with astronauts installing "corrective lenses" for its flawed mirror was offset by the realization that servicing missions likely meant total life-cycle costs several times higher than the cost to build and launch a mission. Following a bruising battle, the *AXAF* team had to accept the fact that their mission would not be serviceable. NASA moved to ensure that decision by proposing to place *AXAF* in a high-Earth orbit (unreachable by the shuttle). A high orbit would mean a substantially less massive payload, leading us to focus on two smaller *AXAF* missions—the one for high orbit was called *AXAF-I* with 1″ angular resolution and a still to be determined area accompanied by the two imaging instruments and gratings. The other for low-Earth orbit was called *AXAF-S* with angular resolution closer to 1′ and the X-ray spectrometer (calorimeter) instrument at the focus. After about a year of work, the *AXAF-S* mission was canceled by the US Senate.

For the *AXAF-I* (later simply *AXAF* and then post-launch *Chandra*) concept, the two primary issues were how much mass could be lifted to high-Earth orbit and how much would the revised payload cost. Pellerin held firm to the position that only the innermost and outermost pairs of mirrors could be accommodated, resulting in approximately one-third of the original *AXAF* effective area and undetermined cost savings. A breakthrough of sorts came when the SAO/MSFC team showed that flying only the largest mirror pair, already polished and tested, would save less than 10% of the projected cost, so at most $150M. Still deadlocked, Tananbaum and MIT's Claude Canizares met with Fisk and Pellerin in 1992 April. Fisk reconfirmed the cost constraints and again ruled out servicing. Based on the arguments presented by the scientists, he agreed that at least three pairs of mirrors would be included, and if the mass limits permitted, then four of the original six pairs would comprise the telescope for the high-Earth orbit *AXAF* mission. An innovative approach by TRW to utilize much more graphite–epoxy composite material in the structure cleared the way for the fourth pair of mirrors with acceptable mass reserves. Based on analyses and tests by several team members, Martin Weisskopf had already championed a switch to iridium for the mirror coating to improve the higher energy response. As a result, the reconfigured mission achieves about two-thirds of the original collecting area across the full energy band. Moreover, the high orbit provides viewing efficiency as high as 70%, or a factor of ~1.5 times more than the shuttle-compatible low-Earth orbit. Thus, except for rapid flaring events, the net photon collection for the reconfigured mission is very close to what had been projected for the original *AXAF*. Moreover, there are substantial advantages to the new orbit which we did not fully appreciate until later. No longer transitioning from day to night 15 times every 24 hr limits accompanying power, thermal, and mechanical swings and stresses. So, once the orbit was achieved and the various mechanisms and doors were exercised successfully, a long *AXAF* lifetime became feasible even without servicing.

## Building and Preparing to Launch *AXAF* (1990–1999)

Recognizing the importance of having the *AXAF* Science Center in place while the observatory was being developed and calibrated, NASA issued a Request for Proposals in May of 1990. Ultimately, a team led by SAO in collaboration with MIT and TRW was selected to develop and operate the Science Center. In 1996, as part of a NASA restructuring at the direction of NASA Administrator Dan Goldin, the Operations Control Center, which was to be based at MSFC, was reassigned to the *AXAF* Science Center in Cambridge, MA. Considering the strong interaction between science and observatory operations, this was a wise decision resulting in smoother operations from pre-launch end-to-end tests through the years on orbit as the Observatory has aged, faster responses to Target of Opportunity and Director's Discretionary Time requests when needed, and reduced cost.

In 1992, 15 years after we began the *AXAF* project, we were finally in a position to complete the design and start building! An important ingredient was an unwritten pledge between the *AXAF* Project at MSFC and NASA HQ that the Project would receive the funds we required on the schedule we had set out. Not having to replan the mission each year was a godsend. Of course, there were a number of bumps in the road along the way. For example, during the alignment and assembly of the eight mirror shells into a single telescope at EK in 1996, an unexpected shift was detected in the alignment (and therefore the focus) of one of the shells with respect to the first pair, which had been aligned and bonded to begin the assembly. The shift was of order 1/3″ and within specifications, but substantially larger than what was expected. Because the cause was unknown, the team decided we had to understand the origin and to be confident that the measured results were accurate. Ultimately, we determined that 40 W fluorescent lights in the vertical assembly tower had heated the support structure and then the surrounding air by a fraction of a degree. The resultant small change in the air density changed its index of refraction, which in turn introduced a tiny change in the path of the laser beam used for the alignment. The solution was to make the final alignment measurements and bond the elements with the lights off.

During the same period of time, MIT's Lincoln Laboratory was developing the flight charge-coupled devices (CCDs) for the ACIS instrument. Relatively speaking, they had no difficulty producing front-side illuminated (FI) chips for *AXAF*, but the back-side illuminated (BI) chips, with higher quantum efficiency for lower energy X-rays, were much more of a challenge. Ultimately, three good BI chips were produced, but the best-performing chip was lost in testing. The two others were placed into the six-chip linear array serving as a readout for the HETG, with the better BI chip at the prime focus for this linear array. The rationale was that the moderately worse energy resolution (compared to the FI devices) would not really impact the high-resolution grating spectroscopy and on-axis images over an 8′ field of view for this one CCD would be very useful as a low-energy supplement to the larger ACIS-I 4-chip (all FI) array. The perseverance of the ACIS team in producing BI chips and the placement of this BI CCD at the prime focus turned out to be fortuitous. Early after launch, we detected degradation of the charge transfer efficiency, and thereby the energy resolution, of the FI chips due to bombardment by protons focused by the X-ray optics with the ACIS detector at the prime focus during passage through the radiation belts. The BI chips were far less sensitive to proton hits and sustained much less damage. Once we understood what was happening, the *Chandra* detectors were routinely moved out of the focal plane during passage through the radiation belts.

From fall 1996 through spring 1997, *AXAF* underwent an extensive end-to-end calibration at Marshall's X-Ray Calibration Facility (XRCF) following a precalibration rehearsal using the TMA and flight-like gratings during the summer of 1996. During the period of time when the High Resolution Mirror Assembly (the *AXAF* optics) was present, 3170 X-ray measurements were taken. In addition to the flight focal plane instruments, the calibration also employed a number of nonflight detectors. A thorough technical overview of the calibration may be found in Weisskopf & O'Dell (1997) and O'Dell & Weisskopf (1998) and references therein. One of the most compelling aspects of this activity which, with few exceptions (such as requiring us to leave the facility in light of a tornado warning), operated 24 hr per day, seven days a week, was the camaraderie that developed among the 12 organizations and about 100 people involved on a day-to-day basis.

Following calibration, the integration of the observatory and subsequent verification as we approached launch went as smoothly as one might expect for such a complex mission, meaning that significant additional challenges were encountered. Hiccups worth noting were delays with the integration and test software at TRW, as well as several electrical part alerts requiring examination and in a few cases replacement of components or full electronic boards, ultimately resulting in a launch delay of several months. During a thermal vacuum test of the full observatory in 1998 June, the ACIS door failed to open. Electrical current is used to heat a wax actuator, which expands and turns a piston/shaft system to open the door. When the door failed to open, a fail-safe actuator burst as intended to prevent further damage. Examination of the mechanism at the vendor, Lockheed Martin, was inconclusive as to whether icing or some other problem caused the O-ring seal on the door to stick or whether the mechanism had failed prior to the thermal vacuum test, due to inadvertent transmission of the command to open when the ACIS was first integrated to the spacecraft at ambient pressure. Without a clear-cut explanation, the failed mechanism was replaced with modifications to enable a more gradual opening of the door which would preclude bursting of the fail-safe disk if sticking occurred on orbit. We all held our breath when the time came to open the ACIS door on orbit as we describe below.

In April 1999, a classified Air Force mission experienced the failure of an Inertial Upper Stage (IUS), similar to the one which *AXAF* would also use to achieve high-Earth orbit. Analysis indicated that cables between the two stages of the IUS failed to separate, due to the cables being taped too close to their ends. This problem was remedied by rewrapping the tape on the *AXAF* IUS cables. Despite a total launch delay of 11 months, the ultimate cost growth was only ~2.5% of the $1.5B cost established in 1992 to develop the new version of *AXAF*.

In 1998, NASA announced a naming contest, with the prize, provided by TRW, being an all-expenses paid trip to the launch. The *AXAF* name contest attracted 6000 entries from 50 states and 61 countries, with many imaginative and intriguing entries comprising the suggested name and a short essay supporting the choice. The winning name, *Chandra*, was decided by a panel that included prominent scientists, a space industry executive, and nationally recognized science reporters. Two winners, Tyrel Johnson, a high school student at the time, and Jatila van der Veen, a high school physics and astronomy teacher, were selected (see Figure 0.6). *Chandra* was the nickname for the Nobel-Prize-winning, Indian–American, astrophysicist Subramanyan Chandrasekhar, whose work on late stages of stellar evolution introduced the requirement for something besides white dwarf end-states for stars much more massive than the Sun. This work foreshadowed the discovery of neutron stars and black holes, which are so central to much of our work in X-ray astronomy.

## Launch and First Light (July–August 1999)

The *Chandra* launch was scheduled for just after midnight, and thus early in the morning on Tuesday, 1999 July 20. Notable celebrities in attendance included First Lady Hillary Clinton; composer–singer Judy Collins, who wrote and performed an original song in honor of shuttle commander Eileen Collins; the 1999 FIFA World

**Figure 0.6.** From right to left: the two winners of the *AXAF* naming contest, Tyrel Johnson and Jatila van der Veen, with Mrs. Chandrasekhar at the launch. (Courtesy of NASA.)

Cup Champion U.S. women's soccer team; and the actor Fabio. Many of the notables were present, at least in part, to note the milestone of having the first female commander for a shuttle mission. For his part, Fabio had received an invitation to attend as a guest of Mission Specialist Cady Coleman, although the invitation was actually extended by Cady's fellow crew members without her knowledge as a tension breaker of the sort often employed by the astronauts.

The countdown proceeded flawlessly until an indicator showed a possible fuel leak, and the launch was aborted less than 10 s before liftoff. Subsequent analysis showed that the reading had been spurious, but of course, no one knew that when the liftoff was canceled. Because the abort occurred before the main engines had ignited, there would only be a 48 hr delay before a second attempt to launch could proceed. At a morning weather briefing on Wednesday, the lead meteorologist reported a zero percent chance of weather impacts for the launch. On the bus ride to the viewing bleachers that evening, we saw regular light flashes, which one of us, eager to see *Chandra* launch after 23 years, reported as likely coming from a coastal lighthouse which was not observed two nights earlier. Alas, the flashes were due to lightning, which had not been present on the earlier evening. Even with an extension of the launch window, *Chandra* could not lift off because lightning persisted within 10 miles of the launchpad for nearly an hour after the original launch time.

The third time was the charm following another 24 hr wait. Although many of the distinguished guests had left, our team and most of our families stayed on and we were treated to a spectacular launch at 12:31 AM (EDT) on 2019 July 23. Figure 0.7 shows *Chandra* in the shuttle bay, and Figure 0.8 shows liftoff, including a video of the actual launch (also available online).[1] The shuttle flight itself was not without challenges. A few seconds after liftoff, a short circuit took out computers controlling

[1] http://chandra.harvard.edu/resources/animations/sts93.html

**Figure 0.7.** The *Chandra X-ray Observatory* in *Columbia*'s cargo bay. The IUS is marked by the letters "USA." (Courtesy of NASA.)

**Figure 0.8.** Liftoff of STS 93 on 1999 July 23 with *Chandra* and its IUS on board. (Courtesy of NASA.) The video shows a movie of the launch. Additional materials available online at http://iopscience.iop.org/book/978-0-7503-2163-1.

two of the three engines, but Commander Collins decided to continue the flight using backup computers. When the shuttle reached its parking orbit, it was a few miles short of the targeted altitude. Later analysis showed that a small patching plug had been blown out of a hydrogen tank and a bit of fuel was lost, leading to the lower altitude. This leak was not a problem for *Chandra* given the capabilities of the IUS boosters and the engines built into the spacecraft for achieving the operational orbit, but these issues did lead to a subsequent grounding of the Space Transportation System lasting several months to check out and rework the entire shuttle fleet.

After the shuttle achieved its orbit, the payload bay doors were opened and, about 8 hr after launch, *Chandra* was deployed by Mission Specialist Cady Coleman (Figure 0.9, which includes a video of the deployment). The shuttle then backed away, and shortly afterwards, the IUS was fired—each of the two stages operated flawlessly, boosting *Chandra* to an intermediate orbit of around 200 × 30,000 miles. While most of the launch spectators had already dispersed, the *Chandra* family were glued to NASA TV and cell phones to hear the good news about the deployment and IUS success. The shuttle crew had other responsibilities once *Chandra* was deployed, but they found time to take the crew photo shown in Figure 0.10 with a *Chandra* banner in the background.

Of course, none of us were yet sure that *Chandra* would be a success. Over the next two weeks as we watched (and helped) from the *Chandra* Operations Control Center (OCC) in Cambridge, MA, the spacecraft engines were fired five times to boost *Chandra* to its highly elliptical, initial working orbit of approximately

**Figure 0.9.** *Chandra* deployment from shuttle *Columbia*. (Courtesy of NASA.) The video shows a movie of the deployment. Additional materials available online at http://iopscience.iop.org/book/978-0-7503-2163-1.

**Figure 0.10.** *Columbia*'s crew. Bottom row: Mission Commander Eileen Collins and Mission Specialist Michael Tognini. Top row: Mission Specialist Steven Hawley, Pilot Jeff Ashby, and Mission Specialist Catherine (Cady) Coleman. (Courtesy of NASA.)

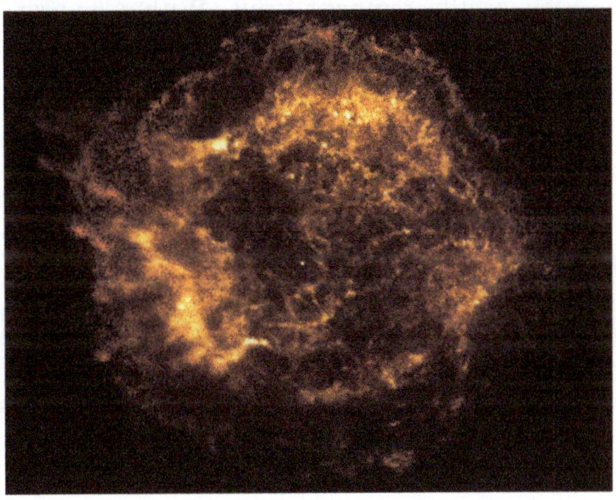

**Figure 0.11.** First *Chandra* image of the supernova remnant (SNR) Cassiopeia A (also available online[2]). The image is 6′ across, and the SNR about 10 light years in diameter. The point source at the center of the remnant is the neutron star formed by the supernova explosion, seen for the first time in this observation. (Courtesy of NASA/CXC/SAO.)

6000 × 86,500 miles (9700 × 139,000 km). The day after the fifth firing was also special, as it marked a flawless opening of the ACIS door, which had failed in the thermal vacuum test at TRW 14 months earlier. On August 12, almost three weeks after launch, the aft and then the front contamination doors on the mirror assembly were opened for the first time. Over the next hour or so, the observatory continued to point stably under gyro control, then the aspect camera locked on stars, and the first

[2] http://chandra.harvard.edu/photo/1999/0237/

X-ray source was detected by the ACIS instrument (Weisskopf et al. 2006). That source was about 3″ in size and was located around 3′–4′ off axis, just about the size expected for a point source that far off axis and without the on-orbit tweaking of the telescope focus or image correction for aspect drift during the observation, per feedback from Telescope Scientist Leon Van Speybroeck. Project Scientist Martin C. Weisskopf, with concurrence of the team present, nicknamed the source Leon X-1. We all left that day confident (perhaps knowing) for the first time that *Chandra* was going to be a great success.

This belief was confirmed on 1999 August 19, when *Chandra* pointed at the supernova remnant Cassiopeia A, a previously known and relatively bright X-ray source, and obtained our official "First Light" image. The image (Figure 0.11) showed much greater detail than any previous X-ray image of Cas A. The real-time data on the computer screen at the OCC showed a distinct point source at the center of the remnant. This was the previously undetectable neutron star formed at the time of the supernova explosion, some 300+ years earlier. The neutron star is so hot that most of its radiation is seen as X-rays, making it too faint to be detected with existing radio and optical telescopes. Previous X-ray telescopes lacked the angular resolution required to separate this relatively faint source from the nearby debris seen as the remnant itself. With *Chandra*'s superb angular resolution and sensitivity, the central source was detected and precisely located using less than 1 hr of exposure time.

## References

Harwit, M., & Neal, V. 1986, The Great Observatories for Space Astrophysics (NASA-CR-176754)

National Research Council, 1982, Astronomy and Astrophysics for the 1980's, Volume 1: Report of the Astronomy Survey Committee (Washington, DC: The National Academies)

O'Dell, S. L., & Weisskopf, M. C. 1998, Proc. SPIE, 3444, 2-18

Tucker, W., & Tucker, K. 2001, Revealing the Universe: The Making of the Chandra X-ray Observatory (Cambridge, MA: Harvard Univ. Press)

Weisskopf, M. C., & O'Dell, S. L. 1997, Proc. SPIE, 3113, 2-17

Weisskopf, M. C., Aldcroft, T. L., Cameron, R. A., et al. 2006, ApJ, 637, 682

# Chapter 1

# Introduction

**Wallace H Tucker**

## 1.1 Exploring the High-energy Universe

X-ray astronomy was born with the space age. Earth's atmosphere is an efficient absorber of X-rays, so the observation of cosmic X-rays had to await the ability to launch detectors above the atmosphere.

In 1949, a payload developed by Herbert Friedman and colleagues at the Naval Research Laboratory (NRL) was launched aboard a V2 rocket and carried to an altitude of 150 km, where it detected X-rays coming from the solar corona (Friedman et al. 1951). It was more than a decade after this discovery when, on 1962 June 18, Riccardo Giacconi and his colleagues at American Science & Engineering (AS&E) used a greatly improved Geiger counter detector aboard an Aerobee rocket to detect the first extrasolar X-ray source, Sco X-1, as well as a diffuse X-ray background (Giacconi et al. 1962).

The discovery of Sco X-1 was confirmed in subsequent rocket flights a few months later by the AS&E group and by the NRL group, who also detected X-rays from the Crab Nebula. The AS&E and NRL teams were soon joined by other X-ray astronomy groups, including those at Lawrence Livermore Laboratories, Lockheed, MIT, NASA's Goddard Space Flight Center, Rice University, University of California, San Diego, and the University of Leicester, who used detectors carried aloft by rockets and balloons to discover more than 30 cosmic X-ray sources by the end of the decade. Six of these sources were associated with supernova remnants, two with extragalactic sources, and the rest with bizarre starlike X-ray sources. With the discovery of pulsars in 1967 by Jocelyn Bell and Anthony Hewish, it was suspected that X-ray stars were also neutron stars, either rapidly rotating ones, like the Crab Nebula pulsar, or fueled by the accretion of plasma from a close binary companion. It was suggested that some of these sources might also harbor black holes.

X-ray astronomy was transformed with the launch of the *Uhuru* X-ray satellite in December of 1970. After taking into account times when the Sun was too bright or

doi:10.1088/2514-3433/ab43dcch1

other efficiency factors, *Uhuru* could accumulate approximately 50,000 s of good observing time per day. In one week, it had accumulated more data than all the previous rocket flights in the history of X-ray astronomy. Further, it was possible to observe a single source for a prolonged period of time, which was impossible with rockets or balloons.

The consequences of the long observing time, combined with the contribution of other groups who made specially designed observations with rocket- and balloon-borne payloads, as well as critical research by observers using optical and radio telescopes, led to the solution to the riddle of the nature of X-ray stars. They were determined to be slowly rotating neutron stars or black holes in binary systems. The observed powerful X-radiation is produced by the gravitational accretion of matter from a companion star onto the neutron star or black hole. The identification of X-ray stars as accreting neutron stars and black holes established that accretion is an important source of power in the universe, from the formation of disks around very young stars to quasars.

In addition to demonstrating that X-ray observations provide insight into the nature of the densest states of matter in the universe, *Uhuru* also showed that extremely low-density matter in clusters of galaxies can also be a strong source of X-ray emission. To fully understand the implications of this discovery, another advance in technology was required. This was provided by the *Einstein Observatory*, which was launched on 1978 November 13. With an angular half-power diameter (HPD) of 10″ and a collecting area two orders of magnitude greater than that employed to study solar X-rays, *Einstein* brought the X-ray sky into focus. X-ray images of supernova remnants and clusters of galaxies, and the detection of faint sources at a level 100,000 times weaker than Sco X-1, convincingly demonstrated the value of an X-ray telescope and opened the way for the development of the *Chandra X-ray Observatory* (Giacconi & Gursky 1974; Giacconi et al. 1979; Tucker & Giacconi 1985; Giacconi & Tananbaum 1980; Giacconi 2008).

*Chandra* operates from 80 eV to 10 keV with unique capabilities for producing subarcsecond X-ray images, locating X-ray sources to high precision, detecting extremely faint sources, and obtaining high-resolution spectra of selected cosmic phenomena. These qualities have established *Chandra* as a versatile and powerful tool for exploring the hot, high-energy regions of the universe. *Chandra* is part of a larger context where, for the first time, subarcsecond imaging of many cosmic sources is available across a wide band of wavelengths. The synergy with NASA's *Hubble* and *Spitzer Space Telescopes*, as well as large ground-based optical, millimeter, and radio telescopes, and the *XMM-Newton* and *NuStar* X-ray observatories, is providing a more complete view of the physical processes at work in the universe.

In this book, we present an overview of how observations using *Chandra*, sometimes alone, but often in conjunction with other telescopes, have deepened our understanding of a rich diversity of topics, from the atmospheres of nearby planets to the event horizons of black holes to galaxy clusters that span millions of light years, and supermassive black holes (SMBHs) in distant quasars.

## 1.2 The *Chandra X-ray Observatory*

The *Chandra X-ray Observatory* was launched on 1999 July 23 by NASA's Space Shuttle *Columbia*, with Eileen Collins commanding. *Chandra* was boosted to a high-Earth elliptical 63.5 hr orbit with an apogee ≈140,000 km and a perigee ≈16,000 km.

*Chandra* consists of three main elements: (1) a telescope containing the High Resolution Mirror Assembly (HRMA), two X-ray transmission gratings that can be inserted into the X-ray path, and a 10 m-long optical bench; (2) a science instrument module (SIM) that holds two focal-plane cameras—the Advanced CCD Imaging Spectrometer (ACIS) and the High Resolution Camera (HRC)—along with mechanisms to adjust their position and focus; and (3) a spacecraft module that provides electrical power, communications, and attitude control. The HRMA, consisting of four nested pairs of cylindrical, grazing-incidence glass-ceramic mirrors coated with iridium, represents a major advance in X-ray imaging capability by providing subarcsecond imaging.

After 20 years, the condition of the *Chandra* observatory remains excellent, with no known limitations that preclude a mission of 25 years, perhaps longer. The extended long observing baseline with stable and well-calibrated instruments, enables temporal studies over timescales from milliseconds to years.

The *Chandra* X-ray Center (CXC) is managed for NASA by the Smithsonian Astrophysical Observatory (SAO), which is part of the Center for Astrophysics |Harvard & Smithsonian located in Cambridge, MA. The CXC operates all aspects of the *Chandra* mission, with NASA's Marshall Space Flight Center (MSFC) providing overall project management and project science oversight. CXC's management includes operating the observatory from the Operations Control Center (OCC) in Burlington, MA, soliciting observing proposals, conducting mission planning and operations, and collecting, distributing, and archiving the data.

Approximately 90% of *Chandra*'s observing time is made available for science proposals submitted by the worldwide astronomical community. The selected proposals generally represent a wide variety of targets, ranging from solar system objects to distant quasars, and a large range of observing times, from snapshots taking only a few kiloseconds (ks), to so-called Visionary projects, which typically will use several megaseconds (Ms) over the course of a project.

An extensive network of bilateral agreements with a large number of specific missions and observatories has enabled coordinated, multiwavelength observations with *Hubble*, *XMM-Newton*, and *Spitzer*, as well as an extensive network of ground-based optical and radio telescopes.

The *Chandra* Data Archive (CDA), the ultimate repository of all data collected by the *Chandra* mission, has proved to be a valuable research tool for studying source variability, given the long baseline provided by the extended mission. The CDA provides access to all *Chandra* observations (17,500 as of 2019 January), calibration data, and internal operational data since the start of the mission. Approximately 14 TB data have been downloaded yearly from a total of 72 countries.

The *Chandra* source catalog contains 315,000 distinct sources, with astrometry, photometry, spectral properties, and variability. Since the beginning of the mission, an average of 14 TB data have been downloaded yearly from the archive from a total of 72 countries.

For a more detailed description of *Chandra* and its operations, see Chapter 2.

## 1.3 Mechanisms for the Production and Absorption of X-Rays in a Cosmic Setting

Chapter 3 summarizes the radiation processes that can produce and absorb X-rays in a cosmic setting. After presenting some basic concepts of classical and quantum theory, specific processes are discussed. These include cyclotron and synchrotron radiation, electron scattering, black body radiation, bremsstrahlung, radiative and dielectronic recombination, and line emission following collisional excitation of ions by electrons.

The production of X-rays of energy $\varepsilon_{\rm keV}$ by the synchrotron process requires electrons of energy $\gamma mc^2$ moving in a magnetic field $B$ such that $B\gamma^2 \sim 10^{11}\varepsilon_{\rm keV}$ Gauss. For example, the observed ~3 keV X-radiation from supernova shock waves with $B \sim 10^{-4}$ Gauss requires electrons with $\gamma \sim 5 \times 10^7$ corresponding to 25 TeV. These extremely energetic electrons are far from equilibrium and typically have a power-law distribution of momenta.

Cosmic X-radiation can also be produced when low-energy photons are scattered up to X-ray energies by high-energy electrons. This can occur in a hot corona above accretion disks or when cosmic microwave background photons scatter off the relativistic electrons that produce synchrotron radio emission in giant radio sources.

Electron scattering plays a key role in limiting the luminosity generated by the accretion onto neutron stars and black holes. For spherically symmetric infall, the radiation force acting on the infalling matter becomes strong enough to offset the inward pull of gravity at a critical luminosity called the Eddington luminosity:

$$L_{\rm Edd} = 4\pi G m_{\rm p} c/\sigma_T = 1.3 \times 10^{38} M_{\rm BH}/M_\odot \text{ erg s}^{-1}. \tag{1.1}$$

Radiation from a plasma[1] of temperature $T$ has a spectrum that peaks at photon energy $\varepsilon \sim kT \sim (T/10^7 K)$ keV, so plasmas with temperatures ~1 MK–100 MK will produce 100 eV–10 keV X-rays. Temperatures of this magnitude are found in stellar coronae, on the surfaces of neutron stars, in optically thick accretion disks around neutron stars and black holes, plasma heated by supernova shock waves, and intergalactic plasma in clusters of galaxies.

Optically thin hot plasmas with $T$ ranging from a few MK to a few 10 MK in stellar coronas and supernova remnants produce a rich spectrum of X-ray line emission from cosmically abundant elements such as oxygen, silicon, and iron. Beyond a few 10 MK, most of the ions are fully ionized (iron ions are an important exception), and bremsstrahlung radiation dominates. Even at the higher

[1] In the pages that follow, the reader may encounter the terms "hot gas" or "gas" in a context that, strictly speaking, refers to a plasma.

temperatures, X-ray line emission plays a role as a diagnostic for determining the abundances of the elements and for accurate estimates of the temperature of the hot plasma.

The analysis of X-ray line emission requires the calculation of the ionization equilibrium for the various ionization stages for a number of elements. For many settings, such as stellar coronae, supernova remnants, and galaxy clusters, the ionization equilibria are determined by a balance between collisional ionization and a combination of radiative and dielectronic recombination. In other cases, such as X-ray binaries and active galactic nuclei (AGNs), photoionization dominates the ionization process.

In addition to the plasma around the source, trace elements, such as oxygen, in the interstellar medium can absorb X-rays on their way to Earth by photoionization. The strong energy dependence of photoionization, roughly $\propto \varepsilon^{-3}$, provides a tool for the study of the abundances of the elements, and in the case for galactic sources, making estimates of the distances to those sources.

Absorption lines can also provide information about absorbing material in the vicinity of the source, e.g., winds flowing away from black holes, and in the search for baryonic matter in vast filaments of intergalactic plasma.

Another important by-product of X-ray absorption is the presence of K-fluorescence lines from iron atoms and ions in the vicinity of black holes. These lines can be used to probe material within a few gravitational radii of the event horizon of black holes.

## 1.4 Stars, Planets, and Solar System Objects

Research in cosmology and extragalactic astrophysics depends on knowledge of how stars form and evolve, yet most star formation occurs in dense regions of dust and gas that absorb optical radiation. X-rays, however, can penetrate the dust and gas, so X-ray observations of star-forming regions are critical for understanding this process. Likewise, the search for habitable planets requires knowledge of the high-energy environment of planets. Progress in both of these areas was hampered by the intrinsic weakness of the X-ray fluxes involved. *Chandra*'s sensitivity and ability to separate individual stars in confused regions was a game-changer. Coupled with infrared and optical observations, *Chandra* data may identify young stars, with and without protoplanetary disks, as well as investigate the process of the triggering of star formation by winds and radiation from massive stars. These and other advances are discussed in Chapter 4, along with highlights of *Chandra*'s observations of solar system objects, planetary nebulae, and isolated white dwarfs.

The general processes for producing X-rays from solar system objects are charge exchange emission, elastic scattering of solar X-ray photons, X-ray fluorescence following inner-shell ionization by solar X-rays, and bremsstrahlung and collisionally excited line emission caused by impact of energetic electrons and ions. Several solar system objects produce X-rays by more than one of these mechanisms.

Charge exchange (CX) X-ray emission occurs when a highly ionized heavy ion encounters neutral material. The ion captures an electron from the neutral material, supergiants have led to a factor 3–14 reduction in the estimated mass-loss rates for these stars.

The X-ray emission from most massive O stars is likely due to interacting shock waves embedded in powerful winds. *Chandra* grating spectra provide rich diagnostics of plasma conditions and allow detailed modeling of the effects of magnetic channeling and other details. *Chandra* has also detected diffuse X-ray emission around O stars from ~10 MK plasma at levels in the range from $10^{33}$–$2 \times 10^{35}$ erg $s^{-1}$. In most cases, the diffuse plasma appears to be generated by stellar winds from massive stars colliding with other winds, the most dramatic example being $\eta$ Carinae.

*Chandra* has also provided new insights into the nature of planetary nebulae. The *Chandra* Planetary Nebula Survey has established that ~50% of planetary nebulae harbor X-ray point sources, while ~30% display emission from hot bubbles generated via shocks and wind interactions. The high frequency of X-ray sources with a hard-X-ray excess that are associated with the central stars points to the frequent presence of binary companions that are likely responsible for nonspherical morphologies.

The end result of the planetary nebula phase is a white dwarf. *Chandra* has observed white dwarf stars in all their phases of activity, from nascent stars in planetary nebulae (PNe) to single white dwarfs on the cooling track, and binary white dwarfs in cataclysmic variables and nova explosions.

If the white dwarf has a nearby companion star, the strong gravity of a white dwarf can accrete gas from the companion. If enough material, mostly in the form of hydrogen plasma, accumulates on the surface of the white dwarf, thermonuclear fusion reactions can occur and intensify, culminating in a nova outburst that can be observed for a period of months to years as the material expands into space.

*Chandra* observed two particularly dramatic nova outbursts, RS Ophiuchi and V745 Sco, which exploded on 2006 February 12 and 2014 February, respectively. The data collected for these novae, especially those using the HETG spectrometer, were used to develop detailed models for the outburst and its aftermath.

## 1.5 Supernovae and Their Remnants

Chapter 5 provides a review of the impact that *Chandra* observations have had on our understanding of many phenomena related to supernovae and supernova remnants: the nature of the progenitor stars, the supernova mechanism, and the evolution of supernova remnants. Apart from being intrinsically interesting for being some of the most powerful and extreme events in nature, the study of supernovae and their remnants is key to understand their role in regulating galactic chemistry, temperature, and pressure conditions in the interstellar medium; producing cosmic rays; triggering star formation; and acting as distance indicators that probe the structure and evolution of the universe.

Two distinct types of supernovae have been identified (a third type, the so-called pair-instability supernova, which is hypothesized to occur in extremely massive ~200 $M_{\odot}$ stars and may have been observed in a few instances, is not discussed in Chapter 5). Core-collapse supernovae are produced by the collapse of the core of a massive star (initial mass $\gtrsim 10\ M_{\odot}$) and leave behind a collapsed remnant in the form of a neutron star or, for stars with initial masses of 50–100 $M_{\odot}$ or more, a black hole. Thermonuclear supernovae are produced by white dwarf stars that have been pushed over their limiting stable mass, either through the accretion of matter from a companion star or merger with another white dwarf.

Both types of supernova release $\sim 10^{51}$ ergs of kinetic energy into the interstellar medium and send shock waves rumbling through interstellar space at speeds of thousands of kilometers per second, leaving shells of multimillion-degree plasma in their wake. They are also the primary mechanism for enriching the host galaxy with elements heavier than carbon, albeit with different concentrations of elements. Core-collapse supernovae enrich their host galaxy with oxygen, magnesium, silicon, etc., whereas thermonuclear supernovae are distinguished by the high concentration of iron they produce.

The interaction of supernova shock waves with the surrounding gas produces radiation that is sensitive to the conditions there. For example, mass loss by the pre-supernova star for hundreds and even thousands of years before the explosion can create a cloud of gas around the star that will light up in X-rays when the supernova shock wave hits it. The intensity of this X-radiation depends on the square of the density, making it a good estimator of the density of the ambient medium. This information, together with the time evolution of the X-ray emission, provides insight into the mass-loss history of the progenitor systems in the late stages leading up to the events.

The emission from a supernova remnant can come from several distinct regions. The outermost region is immediately behind the forward shock wave, where circumstellar or interstellar material is swept up and heated, and electrons can be accelerated to relativistic energies. Inside this region, supernova ejecta are heated by the reverse shock wave. With its combination of arcsecond spatial resolution and high spectral resolution, *Chandra* has made it possible for the first time to separate these two regions. Observations of the outer region reveal the properties of the forward shock, which contains information about the kinetic energy released by the supernova, the properties of the interstellar medium, and the process of accelerating electrons to energies of a few gigaelectronvolts, and by implication, protons to much higher energies. Observations of the reverse shock probe the supernova ejecta, its total mass, composition, and possibly the geometry of the explosion.

In addition to the forward and reverse shock waves produced in supernova remnants, core-collapse supernovae leave behind a collapsed remnant in the form of a neutron star or black hole. If the collapsed remnant is a neutron star, conservation of angular momentum and magnetic flux in the collapse process imply that the neutron star will be a rapidly rotating and highly magnetized object called a pulsar. The combination of rapid rotation and intense magnetic fields will create enormously strong electromagnetic forces that generate ultrahigh-speed polar and equatorial outflows. These outflows, called pulsar wind nebulae, can radiate strongly from radio through gamma-ray energies for thousands of years to create spectacular nebulae such as the Crab Nebula. One of *Chandra*'s outstanding accomplishments has been to resolve the Crab Nebula pulsar wind nebula in unprecedented detail and to discover dozens of other much fainter pulsar wind nebulae.

## 1.6 X-Ray Binaries

Chapter 6 discusses some highlights from *Chandra*'s observations of X-ray binaries. The X-ray emission from supernova shock waves and pulsar wind nebulae will fade away after several thousand years, but if the remnant neutron star or black hole is part of a binary star system, it may become a bright X-ray source once again as it accretes matter from its companion star. X-ray binaries are the brightest X-ray sources in the Galaxy, so they were the first extrasolar X-ray sources discovered and, because of the unique window they provide into the study of matter under extreme physical conditions, continue to be prime targets for X-ray missions.

*Chandra* brings three unique attributes to the study of X-ray binaries: spatial resolution on an arcsecond scale, spectral resolution $E/\delta E \geqslant 1000$, and the ability to observe variations of flux of many orders of magnitude (in principle, 10 orders of magnitude!) from highly variable objects such the black hole X-ray binary V404 Cyg, and Cir X-1, the youngest known neutron star in the Galaxy. These capabilities have enabled studies of the physics of black holes and neutron stars, the accretion process, winds generated in the accretion process, and the interstellar medium both in the disk and halo of the Galaxy.

The accretion flow around a black hole is a complicated system of inflowing hot plasma in an accretion disk, outflowing plasma in the form of winds and/or jets, and a hot corona. Over the last decade, evidence has accumulated for an empirical "fundamental plane" for weakly accreting black holes that links mass and radio and X-ray luminosity over eight orders of magnitude ranging from stellar-mass black holes to SMBHs with masses $\sim 10^9 \, M_\odot$. The ability of *Chandra* to obtain background-free spectra down to low flux levels has become crucial in developing this model.

A fundamental difference between neutron stars and black holes is that neutron stars have a surface. Studies of the X-ray emission from this surface provide a window to the interior of the neutron star and thereby information on the composition of the densest observable matter in the universe. *Chandra*'s spectral resolution has been important for determining any residual heating of the neutron star due to low-level accretion, which can complicate the assessment of cooling processes. The present data indicate that some systems are consistent with slow neutrino cooling processes, but most systems need fast neutrino emission mechanisms to cool the core to the observed temperatures.

In contrast to the weakly accreting systems, there are many compact X-ray sources in the Galaxy that are accreting at a rate $\gtrsim 10^{-4} L_{\rm EDD}$. One of the primary scientific results of the *Chandra* mission has been the demonstration that the Fe K-$\alpha$ lines in X-ray binaries containing black holes are broadened due to gravitational redshifting, and the modeling of these features to constrain the spin of black holes. In particular, it has been determined that the Cygnus X-1 black hole is spinning at more than 90% of its maximum rate.

High-resolution spectra also reveal information about gas flowing away from the black hole. This flow takes two basic forms: jets and winds. In general, the black hole systems alternate between a wind and a jet-like outflow, though there are instances of both occurring together.

*Chandra* HETG spectroscopy of high-mass X-ray binaries shows orbital phase-dependent features that can be attributed to absorption by the stellar wind of the massive companion star. These observations show, e.g., that the stellar winds of the companion stars are composed of hot, low-density regions and cool clumps of higher density. On a larger scale, high-resolution spectroscopic studies suggest that the cold, warm, and hot phases of the interstellar medium are 89%, 8%, and 3% by mass, respectively, in the plane of the Galaxy. Observations along sight lines at high Galactic latitude provide information relevant to models of the nature and distribution of gas above the Galactic plane and in the extended corona of the Galaxy.

It has been known since the *Einstein* and *Rosat* missions that scattering by interstellar gas can produce halos in the images of Galactic X-ray binaries. In *Chandra* images, it is possible to resolve the effects of dust clouds on a scale of a few arcseconds. Because the path length is longer for X-rays that initially travel at small angles to our line of sight before being scattered back into our field of view, time delays can be expected and are in fact observed. A spectacular example of this is from *Chandra* observations of the neutron star X-ray binary Cir X-1, which shows four separate rings, or scattering echoes, due to scattering off of four intervening dust clouds (see Figure 6.10).

The superb spatial resolution of *Chandra* has also enabled advances in the study of X-ray binary systems outside the Galaxy. Two particularly noteworthy classes are the ultraluminous X-ray sources (ULXs) and merging neutron star binary sources, which were discovered as electromagnetic counterparts of gravitational wave sources.

ULXs are so named because their luminosities exceed $10^{39}$ erg s$^{-1}$ $\sim L_{\rm Edd}$ for accretion onto a 10 $M_\odot$ black hole, assuming isotropic emission. Their extreme luminosities stimulated a debate as to whether their implied luminosities are due to super-Eddington accretion onto otherwise normal X-ray binaries, or whether they represent a population of intermediate-mass black holes with masses $\sim 10^4$–$10^5$ $M_\odot$. *Chandra* observations have helped establish a spatial connection between ULXs and regions of recent star formation. This association, together with spectra obtained at energies >10 keV by the *NuStar* observatory, implies that on average, ULXs are young systems that are likely neutron stars or stellar-mass black holes that are accreting matter in an unusual, super-Eddington manner. The discovery of periodic X-ray pulsations in some of the sources shows that these sources, and perhaps all ULXs, are accreting neutron stars. The possibility remains that some are also black holes, and in some extreme cases, intermediate-mass black holes (see also the discussion of ULXs in Chapter 7).

The discovery by the LIGO and Virgo detectors of the gravitational wave source GW 170817 with its electromagnetic counterpart GRB 170817A detected in gamma-rays (*Fermi, INTEGRAL*) and optical (*Swopes Sky Survey*) convincingly demonstrated that this event was produced by the merger of two neutron stars. It also ushered in a new era of multimessenger astrophysics, in which both gravitational waves and photons provide complementary views of the same source. The LIGO detection implies that the merger formed an object with a mass of 2.7 $M_{\odot}$. Observations at optical and infrared wavelengths unveiled the onset and evolution of a radioactivity-powered transient, known as a kilonova. *Chandra*'s first X-ray detection, nine days after the event, and subsequent monitoring along with observations at radio wavelengths, revealed a delayed nonthermal emission component that continued to rise for 100 days after the event, then leveled off for a few months before decaying. Models for the X-ray emission include jets with a fast inner core and a slower outer sheath, or cocoon models where a jet is interacting with ejecta from the initial outburst.

## 1.7 X-Rays from Galaxies

As discussed in detail in Chapter 7, *Chandra*'s high spatial resolution, together with the energy resolution of the ACIS detector, has enabled the detection and study of the three major components of X-ray emission from galaxies: X-ray binaries that compose the predominant component of the X-ray emission of normal galaxies at energies greater than about 2 keV; a hot gaseous component that is prevalent at energies below about 1 keV; and high-energy activity produced by an SMBH in the center of a galaxy.

*Chandra* has detected populations of point-like sources in all galaxies within 20–30 Mpc. The luminosity, spectra, and variability of these sources are consistent with Galactic low-mass and high-mass X-ray binaries. Because all X-ray binaries in a given galaxy are at the same distance, their absolute luminosity can be determined, X-ray luminosity functions can be derived, and their properties correlated with their parent stellar populations.

In galaxies with active star formation, the X-ray binary population is dominated by young, massive binary systems in which a neutron star or black hole is accreting matter from a young, high-mass companion. The donor stars have relatively short lifetimes, $\sim 10^7$ years, and the number of high-mass X-ray binaries in these systems is proportional to the star formation rate in the parent galaxy, as expected. In galaxies with very high star formation rates, the X-ray luminosity function extends to luminosities $\sim 10^{40}$ erg s$^{-1}$, including significant numbers of ULXs, supporting the idea that most ULXs are super-Eddington accretors onto neutron stars or black holes.

Low-mass X-ray binaries are of course also present in all galaxies and dominate the X-ray luminosity in galaxies with low star formation rates. Formulas developed for estimating the relative importance of the two populations show that for a galaxy with $M_* \approx 10^{11}\ M_{\odot}$, X-ray emission from low-mass X-ray binaries will dominate for star formation rates less than about 6–10 $M_{\odot}$ yr$^{-1}$. Above this value, which scales roughly with the stellar mass of the Galaxy, high-mass X-ray binaries are the primary contributor.

The second major component of X-ray emission from a galaxy is from the hot interstellar medium. For star-forming galaxies, the enhanced star formation rate leads to a high rate of supernova explosions and associated heating of the interstellar medium. In some cases (e.g., M82 and NGC 253), the supernova heating rate may be so high as to produce a galactic wind of matter flowing out of the Galaxy. In many, if not all, galaxies, the interactions between galaxies and mergers are the triggers of enhanced star formation. *Chandra* images of three dramatic examples of this (NGC 4485 and NGC 4490, the Antennae galaxies, and NGC 6240) are shown in Chapter 7.

A surprising discovery with the *Einstein Observatory* was the detection of halos of hot plasma associated with elliptical galaxies. These and similar galaxies have an old stellar population and exhibit little or no ongoing star formation, so it was not expected that they would contain hot plasma. With *Chandra*, it became possible to study the hot halos and separate out the various contributions, especially that of the low-mass X-ray binaries, and get an accurate estimate of the luminosity, mass, and temperature of the gaseous component. With these data, it is possible to estimate the amount of dark matter needed to bind the hot plasma to the Galaxy, which is consistent with the estimates obtained by other techniques.

With *Chandra*, it has become possible to explore a wide range of X-ray emission from the nuclei of galaxies, the associated questions of the full range of nuclear black hole masses, and the importance of nuclear activity across galaxy types and cosmic time. Numerous low-luminosity nuclear sources have been detected, from the $4 \times 10^6\ M_{\odot}$ SMBH at the center of the Galaxy to the $\sim 10^{9-10}\ M_{\odot}$ SMBHs in giant elliptical galaxies.

When the SMBHs are accreting gas from their host galaxy, this process can produce a significant fraction of the total energy output from a galaxy. This energy output takes the form of radiation and mechanical energy associated with jets and winds, which can dramatically alter the appearance of the Galaxy, carving large cavities in the interstellar medium. *Chandra* observations of nearby galaxies provide a new perspective on the interaction of the SMBH activity with the host galaxy. For example, they indicate that the standard model of an optically thick torus surrounding the SMBH and screening out radiation perpendicular to the axis of the torus may need some modification to include a more complex, porous torus.

## 1.8 Supermassive Black Holes and Active Galactic Nuclei

The Event Horizon Telescope image of the shadow of the SMBH in the giant elliptical galaxy M87 has provided the best evidence to date for SMBHs. This detection was more of a confirmation than a surprise, as many types of data have indicated for perhaps two decades that most galaxies harbor an SMBH and that accretion of gas into the black hole is the only possible energy source for the observed powerful radiation from the active galactic nuclei, or AGNs. The term AGN is used in reference to a loosely defined class of galaxies in which a compact region, or nucleus, at the center of the Galaxy has a much higher luminosity than normal (some definitions require an AGN luminosity greater than the rest of the parent galaxy) over at least some portion of the electromagnetic spectrum with characteristics indicating that the luminosity is not produced by stars. Theoretically, accretion onto SMBHs was posited as the explanation for the high luminosities observed in connection with the nuclei of active galaxies (AGNs) some five decades ago.

Chapter 8 presents a summary of *Chandra*'s many contributions to understanding SMBH-related phenomena.

Three physical parameters are needed to study the details of black hole accretion: black hole mass $M_{\rm bh}$, the accretion rate $\dot{M}$, and the black hole spin angular momentum $J$. The accretion rate is usually defined in terms of $\dot{M}_{\rm Edd}$, the accretion rate corresponding to the Eddington luminosity, $L_{\rm Edd}$:

$$\dot{M}_{\rm Edd} = \frac{L_{\rm Edd}}{\eta c^2} = \frac{2 \times 10^{-8}(M_{\rm bh}/M_\odot)}{\eta} M_\odot \,{\rm yr}^{-1}, \tag{1.2}$$

where $\eta$ describes the efficiency of converting gravitational energy into radiation and is typically assumed to take on values between 0.06, corresponding to a nonrotating black hole, and 0.4 for the maximum angular momentum:

$$J_{\rm max} = M_{\rm bh} r_g c, \tag{1.3}$$

where

$$r_g = \frac{GM_{\rm bh}}{c^2} \tag{1.4}$$

is the gravitational radius.

Three main regimes of accretion flow can be identified: (1) $\dot{M}/\dot{M}_{\rm Edd} \ll 0.0001$, (2) $0.0001 \lesssim \dot{M}/\dot{M}_{\rm Edd} \lesssim 0.1$, and (3) $0.01 \lesssim \dot{M}/\dot{M}_{\rm Edd}$.

Sgr A*, the SMBH at the center of the Galaxy, is an example of a weakly accreting black hole, with $\dot{M} \sim 10^{-9}$ $\dot{M}_{\rm Edd}$. The M87 black hole, with an estimated SMBH mass $\approx 4 \times 10^9$ $M_\odot$ and accretion rate of $\dot{M} \sim 0.003$–$0.01$ $M_\odot$ yr$^{-1}$ $\sim 0.0002$–$0.004$ $\dot{M}_{\rm Edd}$, is in the intermediate range, and quasars are at the high end of the range.

The standard model for the structure of an AGN consists of several common components: an SMBH, an accretion flow into the SMBH, a hot corona around the SMBH, outflows from the vicinity of the SMBH in the form of winds, jets that can extend from the vicinity of the AGN to well outside the confines of the parent galaxy, clouds farther out that are emitting intense optical line radiation, a hot ionization cone, and a dusty, high-density, asymmetrically distributed, obscuring medium usually described as a torus. X-rays from AGNs originate in the hot corona, a hot wind, or a beamed jet. An estimable model for the wide variety of AGN types (e.g., one showing absorption or not) is that it all depends on one's point of view: obscured and unobscured AGNs are postulated to be intrinsically the same objects seen from different angles with respect to the torus. This model, while it has its appeal, also has its critics, who point out luminosity-dependent and possible age-dependent factors needed to explain the relative numbers of obscured and unobscured AGNs.

Closer in to the SMBH, *Chandra* observations of microlensing by stars in galaxies acting as gravitational lenses for background quasars have shown that the X-ray emission region is more compact than the optical-UV emission region, and is located at distances $<10r_g$ from the SMBH.

Relativistic jets from AGNs were first discovered with radio telescopes in the 1950s, and X-ray jets were detected in M87 and Centaurus A with the *Einstein Observatory* in the 1980s, but it was not until *Chandra* that detailed studies of X-ray jets were possible. To date, X-ray jets have been detected from more than 100 sources. One of the most spectacular of these is the Pictor A radio galaxy, which displays an X-ray jet more than 190 kpc in length. *Chandra* observations show that the jets are interacting strongly with the surrounding gas and likely play a role in a feedback cycle. In this process, discussed in more detail in Section 1.9 below, accretion onto an SMBH triggers the formation of a jet which interacts with the accretion flow, shutting it off, which in turn shuts off the jet, and the cycle begins again.

Such a scenario suggests that the growth of SMBHs and their host galaxies might be closely interconnected. A fundamental constraint on all theories modeling the interplay of SMBH growth and galaxy evolution is the fraction and type of galaxies that host actively accreting SMBHs for which X-ray emission is the most reliable signature.

Among the most valuable information for evaluating theoretical models for the evolution of SMBHs are the various surveys carried out by both *Chandra* and *XMM-Newton*, in coordination with large optical surveys. X-ray observations provide the most efficient selection of AGNs, as they are less affected by obscuration and not normally contaminated by emission from their host galaxies. Two conclusions from these surveys are (1) AGN number densities peak at redshift $z \sim 2$–3 with luminous sources peaking earlier, and (2) the space density of X-ray-selected, high-redshift AGNs is much higher than the one measured from optical surveys alone.

Several plausible mechanisms have been identified for the evolution of SMBHs: (1) major mergers between two roughly equal-sized galaxies, (2) more gradual, so-called secular processes such as accretion of filaments or clumps of gas from galactic halos, or minor mergers driving accretion onto a centrally located SMBH, and (3) cosmological accretion of baryons from dark matter filaments. Evidence is accumulating for a picture in which all scenarios may play important roles at different stages in the coevolution of galaxies and SMBHs.

Notable exceptions to the coevolution trend are the detection by *Chandra* and *XMM-Newton* of high-redshift $z > 3$ AGNs with $M_{\rm bh} > 10^9\ M_{\odot}$ already in place and accompanied by vigorous star formation. Apparently at least some of the most massive SMBHs grew faster than the stellar population of their host galaxies and have not shut down star formation in the process. The detection of the first SMBH remains a task for the future, form in the words of Edwin Hubble, "With increasing distance, our knowledge fades, and fades rapidly··· we measure shadows, and we search among ghostly errors of measurement for landmarks that are scarcely more substantial. The search will continue."

Finally, by way of acknowledging one of the most remarkable careers in the history of astrophysics, note that the problem of the origin of X-ray background radiation, discovered by Riccardo Giacconi and his colleagues in the rocket flight of 1962, was effectively laid to rest by *Chandra* observations, which conclusively established that the main contributors to X-ray background radiation are AGNs. One of the principal investigators on this research, carried out 40 years after the original discovery, was Riccardo Giacconi.

## 1.9 Groups and Clusters of Galaxies

Chapter 9 focuses on the dynamics of the intergalactic hot plasma associated with groups and clusters of galaxies, hereinafter referred to simply as the intracluster medium or ICM. *Chandra*'s unique subarcsecond angular resolution, combined with its capability for spatially resolved spectroscopy, has led to major advances in the study of the ICM, revealing a wealth of structure that has profound implications for understanding the evolution of galaxies, groups and clusters.

X-ray observations have established that the dominant baryonic component of clusters is hot plasma with a temperature 10–100 MK, and a mass as high as $10^{14}$–$10^{15}$ $M_{\odot}$. The mass of the hot plasma in a cluster is roughly 6–10 times that of the mass in stars. The dark matter component has a density ~6 times that of the baryonic matter, so dark matter is primarily responsible for gravitationally binding the hot plasma to the cluster.

The hot plasma in clusters is diffuse, with most of the gas at densities $\sim 10^{-3}$ $\text{cm}^{-3}$. At cluster plasma temperatures, the primary cooling mechanism is thermal bremsstrahlung (see Chapter 3), and the cooling time of the plasma is ~a Hubble time. The plasma therefore retains a "fossil-like" record over hundreds of millions of years, tracing "cosmic violence" which has involved enormous injections of energy powered by outbursts from central SMBHs and high-velocity collisions between galaxies and subclusters.

The role of outbursts from the central SMBH became clear from radio observations and with *Einstein* and *Rosat* observations, which showed that the cooling time of the cluster plasma in the central regions of many clusters is much less than the Hubble time. The high prevalence of short central cooling times together with the lack of any evidence for a sizable component of cooled and cooling gas suggests that heating and cooling rates are linked together in a negative feedback loop. This occurs naturally if cooled or cooling gas fuels a central AGN that provides the power to inhibit further cooling. In simplified terms, the AGN feedback cycle can be summarized as follows: (1) accreting plasma falls toward an SMBH and is heated, converting gravitational potential energy to radiation including X-rays; (2) jets are launched from the SMBH (BH spin is likely important here), reheating the radiating plasma to prevent runaway cooling and pushing aside infalling plasma; (3) with the plasma supply diminished, the jets turn off, and the SMBH returns to an inactive state; (4) accretion resumes and the cycle starts over.

*Chandra* observations of galaxy clusters have demonstrated that X-ray cavities are ubiquitous around the lobes of radio sources hosted by central galaxies in clusters. Simple modeling had suggested that expanding radio lobes would be surrounded by shocked plasma, but this was not found in many cases.

Among the surprising conclusions based on *Chandra* observations of clusters is that radio-mechanical feedback heats gently. Despite ongoing and powerful AGN feedback, the data show that cluster atmospheres have remarkably similar density and temperature profiles. Detailed studies of clusters such as Virgo and Perseus have shown that X-ray bubbles are inflated slowly, driving weak shocks and sound waves, not strong shocks.

Contact discontinuities, shock fronts, and other merger-related structures in clusters were among *Chandra*'s earliest discoveries. In the cluster A2142, *Chandra*'s spatial and spectral resolution showed that the structure was not a shock front with a pressure jump, but rather a contact discontinuity with equal pressures on either side of the discontinuity, indicating subsonic motion, much like a cold front in a weather system. These structures are associated with a sloshing motion in the cluster atmosphere. Merging subclusters with off-center impacts can cause the plasma in the cluster to slosh or oscillate about the cluster potential minimum. These features, seen in both cool core and noncool core clusters, are likely to persist for several gigayears and are commonly seen in deep cluster images.

Some clusters do exhibit shocks, the most spectacular being the merger shock detected from the Bullet Cluster. The discontinuity exhibits both a sharp density edge and temperature jump consistent with a shock front with a Mach number ~2–3.

The shock waves created by outbursts from SMBHs and mergers can accelerate electrons to relativistic energies, where they emit radio synchrotron emission. Over time, the electrons lose their energy, and the radio emission fades. However, observations with *Chandra* and radio telescopes tuned to low frequencies have demonstrated that mergers, through shock waves and turbulence generated by the shock waves, can reaccelerate the electrons and revive the radio sources to produce vast structures called radio relics. A spectacular example is the "Toothbrush," a radio relic about 2 Mpc long produced by a shock in the merging cluster RX J0603.3 +4214.

*Chandra*'s high-resolution imaging of shocks and cold fronts sets limits on thermal conduction. The sharpness of some suggests that the pre- and post-contact regions are thermodynamically isolated. This indicates that thermal conduction is suppressed by magnetic field draping and stretching along the contact region.

Bubbling and mergers create turbulent atmospheres. Early attempts to measure turbulence spectroscopically with *XMM-Newton* were hampered by the low resolution. Nevertheless, they indicated turbulent velocities in clusters of only a few hundred kilometers per second. A novel method to probe turbulence uses surface brightness fluctuations in deep *Chandra* observations of nearby bright clusters, including Virgo and Perseus. The level of turbulent heating is given as $Q_{\mathrm{turb}} \sim \rho V_{\mathrm{t}}/l$, where $\rho$ is the mass density of the plasma, $V_{\mathrm{t}}$ is the turbulent velocity, and $l$ is the injection scale. When plasma motions are subsonic, the root mean square density amplitude perturbations are correlated with velocity perturbations. From surface brightness fluctuations, plasma density fluctuations can be inferred, with the implication that the rate of turbulent heating can balance radiative cooling in the Perseus and Virgo clusters.

AGN feedback also plays a role in the removal of metal-enriched gas from galaxies and the redistribution of these metals in the intracluster plasma. *Chandra* observations of the spatial distribution of metal-rich plasma in a number of galaxy clusters provide evidence that metals have been carried beyond the extent of the inner cavities in all of these clusters, suggesting that this is a common and long-lasting effect sustained over multiple outburst cycles.

Ram-pressure stripping is another process that can enrich the intracluster medium with metals. In massive clusters with a dense intracluster medium, the ram pressure induced by the motion of a galaxy through the core can strip the enriched plasma from the Galaxy. *Chandra* has captured several dramatic images of ram-pressure-stripped tails in galaxy clusters.

## 1.10 Galaxy Cluster Cosmology

The role played by galaxy clusters in modern cosmological research is discussed in Chapter 10. The birth of galaxy cluster cosmology can be traced back to Fritz Zwicky's 1933 paper describing the discovery of dark matter using measurements of the luminosities and velocities of galaxies in the Coma Cluster. *Uhuru* established that galaxy clusters are in general strong X-ray sources, but not until the *Einstein Observatory* and *EXOSAT* was it determined that the baryonic matter content of galaxy clusters is dominated by a hot (10–100 MK) intracluster plasma that radiates primarily at X-ray energies. This hot plasma is confined by the gravity of dark matter. The coupling of these results with big bang nucleosynthesis models showed that the parameter $\Omega_{\mathrm{m}} \sim 0.3\text{–}0.4$, where $\Omega_{\mathrm{m}}$ is the ratio of observed mass density to the critical mass density needed to close the universe. The discovery of cosmic acceleration using observations of Type Ia supernovae showed that the total mass-energy density $\Omega \approx 1$, so an additional mass-energy component, $\Omega_\Lambda$, generally referred to as dark energy, is needed to explain the observations. This plus previous work on the formation of galaxies and clusters of galaxies has led to the $\Lambda$-cold dark matter model ($\Lambda$CDM) in which cosmic structures grow hierarchically under the influence of slowly moving or cold dark matter particles, with small objects collapsing under their self-gravity first, and merging to form larger and more massive objects.

Observations with *Chandra* have enabled rapid advances in the precision and accuracy of measurements of the total cluster masses. A fraction $f_{\mathrm{gas}} \approx 0.17$ of that mass resides in baryons, and $\Omega_m \approx 0.3$.

Measurements of the baryon fraction as a function of redshift have provided the first direct and independent confirmation of the Type Ia supernova results, as well as constraints on the dark energy equation of state parameter, $w = p/\rho$, where $p$ is the dark energy pressure and $\rho$ is its density. The result is consistent with a value of $w = -1$, which corresponds to dark energy being described by a cosmological constant term in general relativity.

The ($\Lambda$CDM) cosmological model predicts that massive galaxy clusters were built up as smaller groups and clusters collided and merged on a timescale governed by the details of the cosmological model. The number of massive clusters formed as a function of mass and time depends sensitively on the initial conditions, as reflected in the spectrum of the initial perturbations in density, as well as the relative amounts of dark matter and dark energy, and the dark energy equation of state. Two studies, using many megaseconds of *Chandra* observations, provided the first robust demonstration that dark energy has slowed the growth of cosmic structure.

*Chandra* observations of galaxy clusters have provided stringent tests of some of the key assumptions of the ΛCDM model, viz. that the dark matter is composed on nonbaryonic particles that move at nonrelativistic speeds, neither emit nor absorb detectable electromagnetic radiation, and interact with other dark matter particles and baryonic matter only via gravity. These assumptions imply, among other things, that as galaxy clusters merge, the collisionless dark matter and X-ray-emitting plasma should become separated as the plasma experiences ram pressure and is slowed. The Bullet Cluster provides a dramatic example of this effect. *Chandra* measures the dominant baryonic mass component, namely the hot intracluster plasma, and gravitational lensing measurements from ground-based observatories and *Hubble* trace the total (dark + baryonic) matter distribution. A clear separation of the two mass components was observed, confirming that dark matter particles interact very weakly with each other with a dark matter self-interaction cross section per unit mass $< 1.25\ \mathrm{cm^2\ g^{-1}}$. This result was confirmed by *Chandra* and *Hubble* observations of the cluster MACS J0025.4−1222. *Chandra* observations of the density profiles of massive, dynamically relaxed galaxy clusters match the predictions of the ΛCDM model, providing additional support for the model.

For dynamically relaxed clusters, combining the measurements of the X-ray intensity and the Sunyaev–Zel'dovich effect can yield the distance to the cluster independent of redshift. *Chandra* data on clusters have been used in two studies to obtain values for the Hubble constant, $H_0$, ranging from 69–77 km $\mathrm{s^{-1}\ Mpc^{-1}}$, with an uncertainty of about 12%.

*Chandra* data on the growth of clusters and the resulting mass function have been used to test nonstandard gravity models that have been proposed as alternatives to the existence of dark energy. In all cases, the results are consistent with general relativity. Cluster X-ray data, in combination with data from the cosmic microwave background, optical observations of baryon acoustic oscillation, and Type Ia supernovae, have helped to improve the limit on neutrino masses to $\sum m_\nu < 0.48$ eV.

Galaxy clusters, as the most massive virialized objects in the universe, are uniquely sensitive to the presence of non-Gaussianity in the primordial density field, the detection of which would open up new possibilities in the quest to understand the physics of inflation. Tests using *Chandra* cluster data have so far failed to detect any evidence of non-Gaussianity.

## 1.11 Future Missions

Finally, in Chapter 11, we take a look at some of the new X-ray astronomy missions in the planning and development stages. These missions will provide expanded sky coverage and push to deeper flux limits and to higher spectral resolution.

## References

Friedman, H., Lichtman, S. W., & Byram, E. T. 1951, PhRv, 83, 1025
Giacconi, R., Branduardi, G., Briel, U., et al. 1979, ApJ, 230, 540
Giacconi, R., Gursky, H., Paolini, F. R., & Rossi, B. B. 1962, PhRvL, 9, 439
Giacconi, R., & Tananbaum, H. 1980, Sci., 209, 865
Giacconi, R., & Gursky, H. 1974, X-ray Astronomy (Dordrecht: Reidel)
Giacconi, R. 2008, Secrets of the Hoary Deep (Baltimore, MD: Johns Hopkins Univ. Press)
Tucker, W., & Giacconi, R. 1985, The X-ray Universe (Cambridge, MA: Harvard Univ. Press)

AAS | IOP Astronomy

The Chandra X-ray Observatory
Exploring the high energy universe
Belinda Wilkes and Wallace Tucker

# Chapter 2

# *Chandra*[1] X-ray Observatory Overview

**Belinda J Wilkes, Raffaele D'Abrusco and Rafael Martínez-Galarza**

## 2.1 Description of the *Chandra* X-Ray Observatory "(*Chandra*)"

### 2.1.1 Launch and Orbit

*Chandra* was launched from Cape Canaveral, FL, aboard the shuttle *Columbia* on 1999 July 23. *Columbia* was commanded by Eileen Collins, the first female commander of a shuttle mission. The Mission Specialist, Cady Coleman, prepared and released *Chandra* soon after the shuttle arrived safely in low-Earth orbit. Command of *Chandra* was then transitioned to the Operations Control Center (OCC) in Cambridge, MA. Over the next two weeks, the prime contractor, Northrop Grumman, controlled and managed the firing of *Chandra*'s own engines to boost it from the initial shuttle orbit, a few miles below expectations (see Chapter 0), into its final orbit. Last-minute updates of the firing sequence were required given the initially low orbit. *Chandra* was successfully maneuvered into its final orbit, with a period of 63.5 hr and an initial apogee of ~140,000 km, ~a third of the distance to the Moon. This orbit is such that *Chandra* spends 16–18 hr within Earth's radiation belts as it approaches its perigee, which was initially at an altitude of ~16,000 km. The orbit has significantly precessed during the past 20 years, both in terms of ellipticity and angle to Earth's axis, while maintaining a stable period.

Northrop Grumman continued to control and manage the turning on of the spacecraft over the next few weeks, leading to the first on-sky image. This random patch of sky happened to include an X-ray source, which was promptly named Leon X-1, for the *Chandra* mirror scientist, Dr. Leon Van Speybroeck. *Chandra* was then pointed to a distant (redshift $z \sim 0.653$) radio-loud quasar, PKS 0637–752, to focus the optics. Because quasars have a bright nucleus that outshines the rest of the Galaxy, this target was expected to be a point source. However, extended X-ray emission was detected on the west side (Figure 2.1). Focusing activities continued

[1] http://cxc.harvard.edu

doi:10.1088/2514-3433/ab43dcch2

**Figure 2.1.** *Chandra* image of the first targeted source, a radio-loud quasar, PKS 0637–752, at a redshift $z = 0.653$, used to focus the telescope optics. (Courtesy of NASA/CXC/SAO.)

using the central point source, the nucleus of the quasar, while scientists and engineers attempted to work out what feature/problem in the telescope/instrument system could be causing the extension. They quickly ruled out all possibilities and concluded, instead, that this observation showed the first major *Chandra* discovery, a 100 kpc-long X-ray jet coincident with the known radio jet on the west side of the source (Figure 2.1; Schwartz et al. 2000). This unexpected discovery opened a new window into the study of relativistic jets which has resulted in major advances in our understanding and continues to be a very active field of study for *Chandra* today (Chapter 9).

The official first-light picture, taken on 1999 August 19, was of the supernova remnant (SNR) Cassiopeia A (Cas A), the remnant of a star that exploded about 300 years ago (see Figure 0.11). Cas A was known from *ROSAT* observations to be extended and include structure, and *Chandra*'s first image met all expectations. A press release was issued on 1999 August[2] 26, and the first paper reporting *Chandra*'s results on Cas A was published in 2000 (Hwang et al. 2000). Cas A has been observed multiple times over *Chandra*'s lifetime, and the scientific discoveries from these rich data have been reported in many papers (Chapter 5).

### 2.1.2 The Spacecraft

The key to *Chandra*'s great advance in angular resolution—subarcsecond FWHM ($\lesssim 0.5''$)—is its High-Resolution Mirror Assembly (HRMA), a 10 m focal length

[2] http://chandra.harvard.edu/photo/1999/0237/

telescope consisting of four nested pairs of cylindrical, grazing-incidence, glass-ceramic mirrors, coated with iridium to enhance their reflectivity at X-ray wavelengths.

*Chandra*'s highly eccentric orbit originally made possible continuous observations of up to ~175 ks, although as the spacecraft ages and the thermal insulation degrades, the need to control the temperature of multiple spacecraft subsystems limits most individual observation lengths in practice. The observing efficiency, averaging ~70% since the start of the mission, is limited primarily by the need to protect the instruments from particles, especially protons, during *Chandra*'s passages through Earth's radiation belts.

The *Chandra X-ray Observatory*[3] (Figure 2.2) consists of three main elements: a telescope containing the HRMA (Section 2.1.3), two X-ray transmission gratings that can be inserted into the X-ray path, and a 10 m-long optical bench; a spacecraft module that provides electrical power, communications, and attitude control; and a science instrument module (SIM) that holds two focal-plane cameras—the Advanced CCD Imaging Spectrometer (ACIS, Section 2.1.4) and the High Resolution Camera (HRC, Section 2.1.5)—along with mechanisms to adjust their position and focus. *Chandra* is 13.8 m in length, extends 19 m across the solar panels, and has a mass of 4800 kg.

A system of gyroscopes, reaction wheels, reference lights, and a CCD-based star camera enables *Chandra* to maneuver between targets and point stably while also providing data for accurately determining the sky positions of observed objects. The

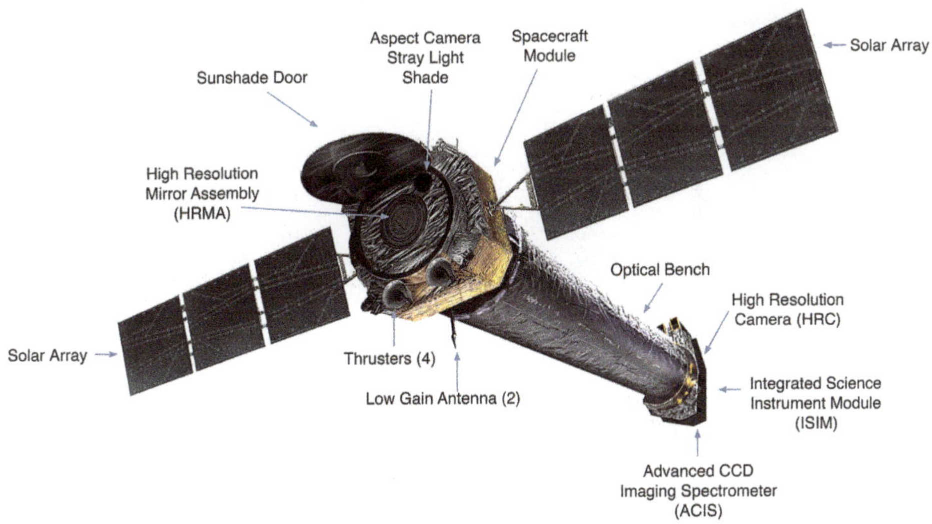


**Figure 2.2.** Schematic diagram of the *Chandra* spacecraft with the main components labeled. The interactive version of this figure is an unlabeled version in which labels may be displayed or removed on demand. (Courtesy of NASA/CXC/SAO & J.Vaughan.) Additional materials available online at http://iopscience.iop.org/book/978-0-7503-2163-1.

[3] http://chandra.si.edu/graphics/resources/illustrations/craft_lable.jpg

blurring of images due to pointing uncertainty is $< 0.15''$, negligibly affecting the resolution of *Chandra*'s mirrors. Absolute positions can be determined to $\leqslant 0.9''$ for 90% of sources, and relative positions to $\sim 0.2''$, providing an unrivaled capability for X-ray source localization.

### 2.1.3 High Resolution Mirror Assembly (HRMA)

The *Chandra* X-ray telescope consists of four pairs of concentric thin-walled, grazing-incidence, Wolter Type-I mirrors. The front mirror of each pair is a paraboloid and the back a hyperboloid. The eight mirrors were fabricated from Zerodur glass, polished, and coated with iridium on a binding layer of chromium. Progressing inwards from the outer pair of mirrors, the pairs are numbered 1, 3, 4, and 6. The numbering reflects the fact that the original design called for six mirror pairs, but pairs 2 and 5 were eliminated to reduce mission costs. The diameters of the pairs of mirrors range from ~1.23–0.65 m for numbers 1–6, respectively. The smallest focuses the highest energy X-rays, and the effective area is largest at lower X-ray energies; see Figure 2.3. For detailed information, please refer to the Proposers' Observatory Guide, Chapter 4.[4]

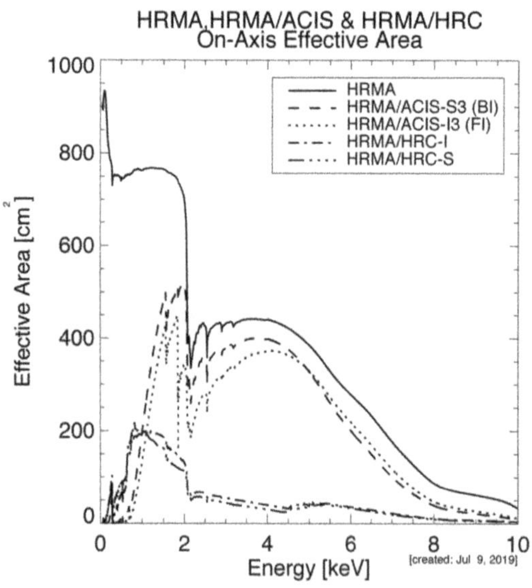


**Figure 2.3.** The HRMA, HRMA/ACIS, and HRMA/HRC effective areas versus X-ray energy. The HRMA/ACIS effective area is the product of the HRMA effective area and the quantum efficiency (QE) of ACIS-I3 (front illuminated) or ACIS-S3 (back illuminated). The HRMA/HRC effective area is the product of the HRMA effective area and the QE of HRC-I or HRC-S at the aimpoints, including the effect of the UV/Ion Shields (UVIS). The figure is reproduced from *Chandra* Cycle 21 Proposers' Observatory Guide, Chapter 4,[5] Figure 4.4. (Courtesy of NASA/CXC.)

[4] http://cxc.cfa.harvard.edu/proposer/POG/html/chap4.html#tth_chAp4

[5] http://cxc.cfa.harvard.edu/proposer/POG/html/chap4.html

### 2.1.4 The Advanced CCD Imaging Spectrometer, ACIS

ACIS contains two arrays of 1024 ×1024 pixel CCDs that provide position, energy, and time information for each detected X-ray photon. The imaging array, ACIS-I (2 × 2 front-illuminated CCDs), has a wide field-of-view (FoV, ~17′ square) and is optimized for spectrally resolved, high-resolution ($\lesssim$0.5″) imaging. The spectroscopy array, ACIS-S (1 × 6 CCDs, two are back illuminated: S3 at the telescope focus, and S0), when used in conjunction with the High Energy Transmission Grating (HETG), provides high-resolution spectroscopy with a resolving power ($E/\Delta E$) up to 1000 (0.4–8 keV energy band).

The four-dimensional ACIS data sets provide X-ray images, low-resolution spectra ($E/\Delta E \sim$ 10–20, ACIS-I, on-axis, and $E/\Delta E \sim$ 15–40, ACIS-S, on-axis), and light curves for any source or region in the image. X-ray detectors such as this provided the first multidimensional data sets. The spatial and spectral dimensions are a lower resolution version of the integral field units (IFUs) which are now in standard use at visible-light telescopes to study an extended source's spectrum as a function of position. The 3D nature of the *Chandra* data sets is illustrated in Figure 2.4, in which the 1 Ms data set of the SNR Cas A, including ~$3 \times 10^8$ photons between 0.5 and 10 keV, is displayed from three angles, and as an interactive figure that rotates through 360°. This extremely rich data set provides unprecedented details about the spatial distribution and spectral properties of individual heavy elements ejected by the supernova.

### 2.1.5 The High Resolution Camera, HRC

The HRC employs two microchannel plate (MCP) detectors, one for wide-field imaging (HRC-I), the other serving as a readout for the Low Energy Transmission Grating

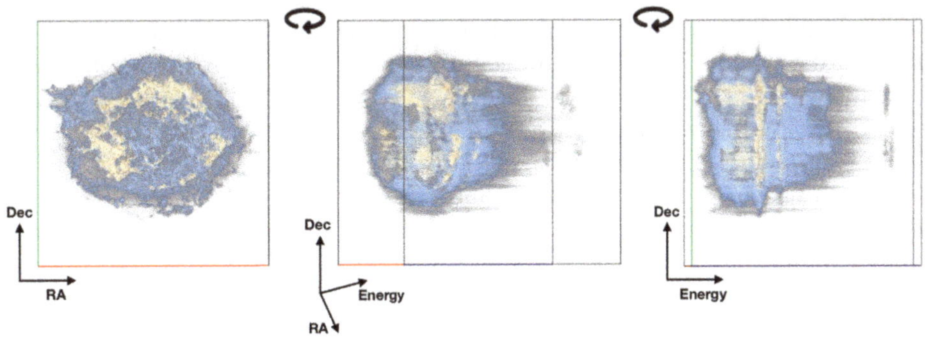


**Figure 2.4.** A rendition of three of the four dimensions of *Chandra* ACIS data (R.A., decl., and energy, but no timing) for the Cas A 1 Ms data set. Three different viewing angles of the same cube are shown (face on, side, and edge on). The interactive version of the figure rotates the view through 360°. The colors (blue, yellow, and red) indicate increasing X-ray photon density, i.e., X-ray brightness. Figures and animation courtesy of Fabio Acero.[6] Additional materials available online at http://iopscience.iop.org/book/978-0-7503-2163-1.

[6] https://github.com/facero/IFU/

(LETG; HRC-S). The HRC detectors, in certain operating modes, have a time resolution as fast as 16 $\mu$s. The HRC-I's spatial resolution, 0.4″, and FoV, 30′, are the highest available on the *Chandra X-ray Observatory*. When operated with the HRC's spectral array (HRC-S), the LETG provides spectral resolution > 1000 at low (0.08–0.2 keV) energies and moderate resolution over the remainder of the *Chandra* energy band.

Each MCP consists of a 10 cm (4 inch) square cluster of 69 million, tiny lead-oxide glass tubes that are about 10 $\mu$m in diameter (one-eighth the thickness of a human hair) and 1.2 mm (1/20 an inch) long. The tubes have a special coating that causes electrons to be released when the tubes are struck by X-rays. These electrons are accelerated down the tube by a high voltage, releasing more electrons as they bounce off the sides of the tube. By the time they leave the end of the tube, they have created a cloud of 30 million electrons. A crossed grid of wires detects this electronic signal and allows the position of the original X-ray to be determined with high precision.

### 2.1.6 Transmission Gratings: HETG, LETG

HETG is the High-Energy Transmission Grating (Canizares et al. 2005). In operation with the HRMA and a focal-plane imager, the complete instrument is referred to as the HETGS—the High-Energy Transmission Grating Spectrometer. The HETGS provides high-resolution spectra (with $E/\Delta E$ up to 1000) between 0.4 keV and 10.0 keV for point and slightly extended (few arcsecond) sources.

The HETG consists of two sets of gratings, each with a different period. One set, the Medium Energy Grating (MEG), intercepts rays from the outer HRMA shells and is optimized for medium energies. The second set, the High Energy Grating (HEG), intercepts rays from the two inner shells and is optimized for high energies. Both gratings are mounted on a single support structure which is swung into the X-ray beam, and therefore are used concurrently. The two sets of gratings are mounted with their rulings at different angles so that the dispersed images from the HEG and MEG form a shallow X centered at the undispersed (zeroth-order) position: one leg of the X is from the HEG, and the other from the MEG. The HETG is designed for use with the ACIS spectroscopic array, ACIS-S, although other detectors may be used for particular applications. Figure 2.5 shows the raw and extracted HETG/ACIS-S spectrum of the well-known star, Capella. The lines cover a range of ionization states, indicating that Capella has plasma with a broad range of temperatures, from log $T = 6.3$–$7.2$ K (Canizares et al. 2000).

The Low Energy Transmission Grating Spectrometer (LETGS) comprises the LETG, a focal-plane imaging detector, and the HRMA. The *Chandra* HRC-S is the primary detector designed for use with the LETG. ACIS-S can also be used, though with lower quantum efficiency below ~0.6 keV and a smaller detectable wavelength range than with the HRC-S. The HETG used in combination with ACIS-S offers superior energy resolution and quantum efficiency > 0.78 keV.

The LETGS provides high-resolution spectroscopy ($\lambda/\Delta\lambda >$ 1000 Å) between 80 and 175 Å (0.07–0.15 keV) and moderate resolving power, $\lambda/\Delta\lambda \sim 20 \times \lambda$ at shorter wavelengths (3–50 Å, 0.25–4.13 keV). The nominal LETGS wavelength range accessible with the HRC-S is 1.2–175 Å (0.07–10 keV); useful ACIS-S coverage is

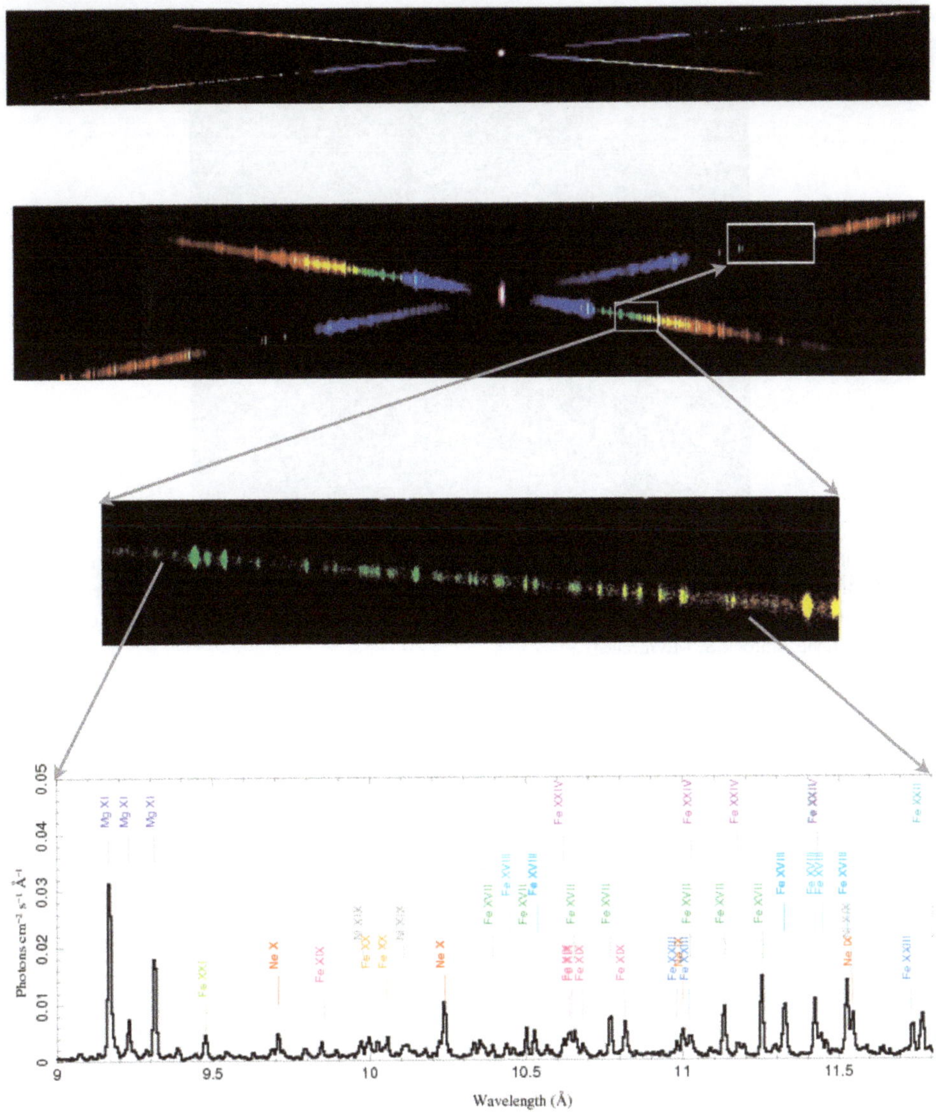


**Figure 2.5.** The raw (top), a zoomed region, and the extracted HETG spectrum of the star Capella with many of the spectral features identified. (Courtesy of NASA/CXC.)

1.2–60 Å (0.2–10.0 keV). Figure 2.6 shows the raw HRC-S/LETG spectrum of the black hole X-ray binary XTE J1118+480. The various structures are described in the figure caption. Figure 2.7 shows part of the extracted HRC-S/LETG spectrum of the active galaxy NGC 5548.

### 2.1.7 Anticipated Lifetime

*Chandra* was originally proposed for a five-year mission. It is currently in its 20th year of operations and continues to provide ground-breaking science results. Its

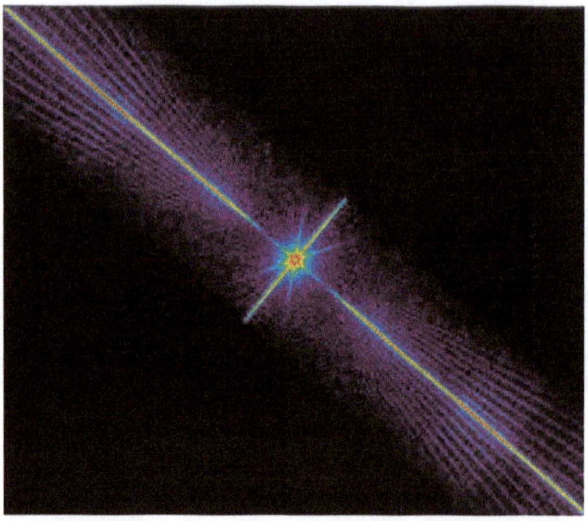

**Figure 2.6.** The raw LETG spectrum of the black hole X-ray binary, XTE J1118+480. The primary spectrum runs diagonally, down the middle of the rectangular array. The bright central image is the undispersed (zeroth order) spectrum of the target, with a star-shaped pattern due to diffraction by the coarse-support structure. The lines radiating from the dispersion axis are artifacts due to the grating fine-support structure. (Courtesy of NASA/CfA/J.McClintock & M.Garcia.)

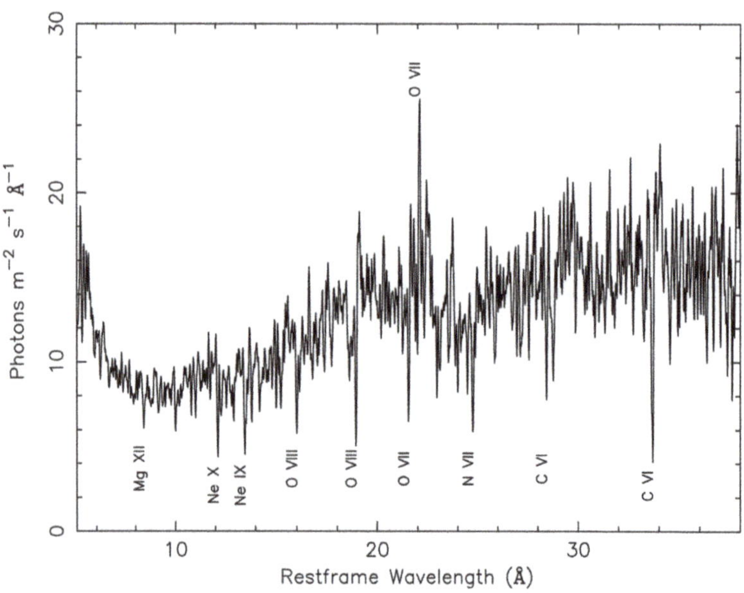


**Figure 2.7.** The extracted LETG spectrum of the active galaxy NGC 5548, with features labelled, showing details of the energy radiated by the hot gas flowing away from the Galaxy's central supermassive black hole. (Courtesy of NASA/SRON.)

subarcsecond spatial resolution remains unique among all operating or planned X-ray missions worldwide, giving *Chandra* the indispensable ability to resolve close-together celestial sources, particularly those in crowded fields, and sharp structures such as X-ray jets and the shock fronts in galaxies and clusters.

In 2015, a complete engineering review of all spacecraft subsystems concluded that there are no showstoppers to a 25+ year mission, i.e., continuing through 2025 and beyond. On the strength of this continuing success, NASA has recently negotiated and awarded SAO, in FY19, a 12 year contract for nine more years of *Chandra* operations, with a review every three years, potentially continuing operations until 2027. This is followed by a three-year closeout period to ensure that the *Chandra* data and software are uniformly archived and made available for the community ad infinitum.

While operations become more complex as the satellite ages, the observing efficiency continues at ~70%, the maximum possible given the ~17 hr *Chandra* spends in Earth's radiation zones during each ~3 day orbit.

## 2.2 *Chandra* Operations

The *Chandra* team (NASA's Marshall Space Flight Center (MSFC), Smithsonian Astrophysical Observatory (SAO), and major industry and instrument team subcontractors) has been in place since before launch and has worked seamlessly and effectively throughout the mission. A strong partnership among academia, government, and industry is essential to the success of a complex, state-of-the-art mission such as *Chandra*. The program organization has contributed materially to *Chandra*'s scientific and management success. MSFC manages the *Chandra* program for NASA's Science Mission Directorate. MSFC's *Chandra* Program Office oversees SAO, the prime contractor for the *Chandra* X-ray Center (CXC), while its Project Science team provides scientific oversight. The CXC is in contact with *Chandra* about three times per day via NASA's Deep Space Network (DSN). The DSN's three dishes, located around the globe, allow communication with *Chandra* throughout its orbit. MSFC arranges with NASA's Jet Propulsion Laboratory (JPL) to provide DSN services for the mission.

### 2.2.1 The *Chandra* X-Ray Center (CXC)

Both science and observatory operations are located at the CXC in Cambridge, MA, and managed for NASA by the Smithsonian Astrophysical Observatory (SAO). The CXC operates all aspects of the *Chandra* mission, including responsibilities for soliciting, holding a peer review of, and selecting observing proposals; supporting the science community across all aspects of their *Chandra*-related work; conducting mission planning and operations; processing, archiving, and disseminating science data; providing data analysis software (the *Chandra* Interactive Analysis of Observations, CIAO); and performing public communications. NASA's MSFC provides overall project management and science oversight. SAO subcontracts with several organizations to support portions of the mission, including Northrop Grumman Corporation, which provides the Flight Operations Team (FOT), systems engineering services, and support from the original spacecraft development team when needed. Subcontracts also include the institutions that host the Instrument Principal Investigator (IPI) teams that designed and built *Chandra*'s instruments. The IPI teams provide unique instrument expertise and work directly with the CXC on instrument operations and support.

*Chandra* was the first major NASA mission to have its operations center located outside a NASA center. The colocation of science and operations personnel has resulted in a seamless interface and a highly efficient observation-planning process, with excellent communication between science priorities and mission capabilities. In combination with a strong interface to the observing teams, this has resulted in a very small number of failed observations to date. The close working relationship also facilitates *Chandra*'s continued high observing efficiency despite the challenges of operating an aging spacecraft. Similarly, the close relationship built up over the years with the MSFC Project Office greatly contributes to the success of *Chandra* both in terms of operations and its excellent science output (Section 2.5).

### 2.2.2 Operations Control Center (OCC)

*Chandra* was the first, major NASA observatory to be operated outside of a NASA Center. In 1997 June, two years before launch, NASA awarded the SAO a contract extension to establish the Operations Control Center (OCC) as part of the CXC under the direction of NASA's MSFC. The CXC designed and built the OCC in Kendal Square, Cambridge, MA, a few miles from the CXC home base at the Center for Astrophysics | Harvard & Smithsonian. The CXC operated *Chandra* from there until 2019 May. In 2017 June, the OCC's landlord declined to renew the lease beyond 2019 September 30. Over the next two years, a new location was researched and acquired in Burlington, MA, and the new OCC was designed and constructed to incorporate the best features of the previous facility along with desired improvements based upon operational experience. For example, the new OCC brings all of the operational teams into one space to facilitate collaboration and situational awareness, but uses glass walls and physical separation to manage sound so individual team members can still effectively perform focused, technical work. Operations were formally switched over to the new OCC in Burlington following the successful completion of the Operations Readiness Review on 2019 May 30.

### 2.2.3 The *Chandra* Task Thread

The CXC operates all aspects of the *Chandra* mission (Section 2.2.1).

The CXC's activities are organized around a set of core tasks that form a coordinated sequence, or "thread" (Figure 2.8). This task thread constitutes the primary operations concept for the mission. The thread begins with the annual solicitation of proposals for *Chandra* observing time and research (*Chandra*-related archival and theoretical proposals) funding. Scientists worldwide submit 500–600 proposals each year. The CXC conducts a peer review to recommend the science program to the Director, who is the selection official. From the list of approved targets, the CXC's Science Mission Planning team generates a long-term schedule

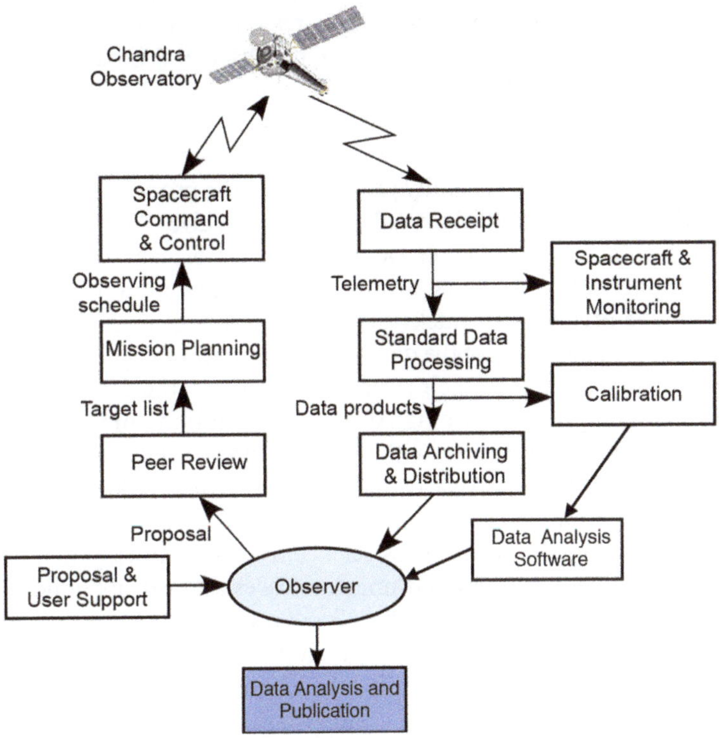


**Figure 2.8.** The CXC task thread, illustrating the path from soliciting proposals through observation to data delivery and community science support. (Courtesy of NASA/CXC.)

which fulfills the requirements of constrained targets throughout the annual cycle. In collaboration with the Flight Operations Team (FOT), weekly observing schedules are planned starting three weeks in advance. Once completed, detailed spacecraft command loads are generated, which on-console staff at the OCC transmit to *Chandra* via NASA's DSN.

The spacecraft's computers execute the command loads to carry out the observing plan and record the resulting data. OCC flight and ground operations staff conduct three daily contacts with *Chandra*, during which the online operations team checks the health of the spacecraft and instruments, performs any required real-time spacecraft procedures, and instructs the onboard recorder to transmit its stored data. The DSN receives and stores the resulting telemetry. The OCC's Ground Operations Team (GOT) retrieves and validates the telemetry and sends it to the FOT and IPI teams, typically within a half hour of contact, for health and safety analysis, and to the CXC Data System's (CXCDS) Operations (DSOps) team for data processing. The DSOps team runs pipeline software that processes the raw science data to yield a package of calibrated, verified data products suitable for archiving and analysis by scientific users. These data are generally available within a day of the observations.

The CXC's Calibration team uses ~5% of the available observing time to monitor and calibrate the Observatory. Calibration observations are included in the planning process outlined above as appropriate. The resulting calibrations are applied in standard data processing and are provided to observers via an online *Chandra* calibration database (CalDB).

The CXC's Archive Operations team delivers processed data products to the proposers, typically within ~20 hr of an observation, and enters them into the data archive. Observation data are generally proprietary to the proposing team for one year, after which they become publicly available. The archive (Section 2.3) makes all nonproprietary observation data, reprocessed versions of each observation, and all calibration data accessible to users worldwide by means of web-based search and retrieval tools. The archive also tracks the publication(s) of each data set, providing metrics for measuring the usage of the data, and the impact of *Chandra* on astrophysics.

To support *Chandra* users in their scientific endeavors, the CXC also provides an extensive suite of analysis software (CIAO), documentation, and help desk support.

Grants are issued to observers within about three weeks of a program's first observation. The SAO Grants Awards section administers ~800 grants at any given time. Funding for grantees at Federal facilities is administered by MSFC.

*Chandra* press and media staff review science results for press potential and work with PIs to develop regular press and image releases and other public products for further dissemination of the science. The *Chandra* public website[7] provides mission and science information to the public on multiple levels and in multiple ways. *Chandra* is also active on social media (e.g., @chandraxray on Twitter[8]).

### 2.2.4 The Observing Program

*Chandra* is a general observatory for which the majority (~90%, ~20 Ms per year) of the observing time is made available for science proposals submitted by the worldwide astronomical community. This practice was initiated by Dr. Riccardo Giacconi for the *Einstein X-ray Observatory*, and has since been adopted by all NASA Great Observatories and a number of other major observatories worldwide. About 1 Ms of time annually is used for calibration, in which regular observations of a few, carefully selected targets with well-understood characteristics, are observed and used to derive the required calibration products for all the instrument combinations and observing modes. A further ~1.2 Ms is assigned to the original Instrument Principal Investigator (IPI) teams, who built the *Chandra* instruments, to use for their science observations and to test out new and/or experimental observing modes.

To facilitate a wide variety of *Chandra* science, including short or long observations of individual sources, time-dependent and transient science, and large surveys of particular source types or of areas of the sky, the annual Call for Proposals[9] offers a variety of proposal categories. While the mix changes from cycle to cycle, the call generally includes General Observer (GO), Large (> 400 ks), Very

[7] http://chandra.si.edu/
[8] https://twitter.com/chandraxray
[9] http://cxc.cfa.harvard.edu/proposer/

Large (>1 Ms) or Visionary (1–6 Ms), Targets-of-Opportunity (ToO), and *Chandra*-related archive and theory proposals. The amount of time in each category is allocated in advance of the annual peer review based on balancing the level of demand with achieving a balanced program. GO programs generally include ~60% of the available (~20 Ms) observing time in a cycle, with the rest distributed amongst the other categories.

To address the growing need for timely, multiwavelength data, *Chandra* spear-headed, along with NASA's other Great Observatories, *Hubble* and *Spitzer*, joint observing programs whereby a single proposal submitted to the primary observ-atory's proposal call can request and be awarded time on other observatories (including e.g., *XMM-Newton*, *Hubble*, *Spitzer*, *NuSTAR*, *Swift*, NOAO, and NRAO), the data from which are required to address the science question(s) posed. *XMM-Newton*, *Hubble*, and NRAO similarly allocate *Chandra* time.

In addition to the annual proposal cycle, we have issued special calls for proposals on a number of occasions.[10] The most recent special call was for observing proposals with science results that would potentially support the science case for decadal mission studies for a *Chandra* Successor Mission[11] in Cycle 19. In 2018 October, we issued a different kind of call, for white papers[12] proposing lists or catalogs of *Chandra* Cool Targets (CCTs), seeking large lists of targets situated within 40° of the ecliptic plane and for which short ($\lesssim$ 30 ks) observations of a random subset would provide scientifically useful results. The CCT call and subsequent special scientific review generated a pool of targets at attitudes that allow most subsystems of *Chandra* to cool at some point during the year. CCT observations have no proprietary time. The *Chandra* website provides a list of the approved CCT programs[13] and of the CCT targets observed[14] to date.

To ensure that *Chandra* is able to maximize its observing efficiency, i.e., continue to maintain ~70% of wall-clock time, the amount of constrained time (i.e., observations that are required to be observed within a specific timeframe, whether absolute or relative to other observation(s)) approved by the peer review is limited.

Transient science is a major part of X-ray astronomy and the *Chandra* observing program. Scientists can request time to observe a particular target, or type of target, to be triggered in response to a specific source state or event, by submitting a ToO proposal to the annual cycle. Because such events can disrupt ongoing schedules and displace nontransient observations, the number of ToOs, particularly those requir-ing a fast turnaround, are limited among peer-review-approved observations. Approved ToO observations remain in the observation catalog until the specific trigger criteria are met, at which time the observation is triggered by the proposing science team. ToO targets remaining untriggered by the end of the annual cycle are discarded, often to be replaced by a similar, approved program in the next cycle.

[10] http://cxc.cfa.harvard.edu/cdo/special_calls.html
[11] http://cxc.cfa.harvard.edu/proposer/dec2017call
[12] http://cxc.cfa.harvard.edu/cdo/special_calls.html
[13] http://cxc.cfa.harvard.edu/target_lists/CCTS.html
[14] http://cxc.cfa.harvard.edu/cda/CCT.html

Director's Discretionary Time (DDT) is also available for the proposal of observations in response to unpredictable events, and new, unusual, and/or unique opportunities, throughout the annual cycle. The project reserves 1 Ms of DDT time annually for such proposals, which are assessed by the Director, Project Scientist, and Director's Office science staff. This allows *Chandra* the flexibility to respond to changing or unpredictable circumstances that may provide compelling science results at any time.

In support of *Chandra* observations approved as constrained whose science requires multiwavelength data, the CXC works with counterparts at multiple observatories worldwide to meet the proposal's scheduling requirements. This is the case whether or not the joint time was allocated as part of an approved *Chandra* proposal. In general, *Chandra* can plan simultaneous observations with many space-based observatories worldwide and with ground-based radio telescopes which often observe night and day. Observations coordinated/simultaneous with ground-based, night-time telescopes are scheduled as close together as is feasible.

### 2.2.5 Standard Data Processing (SDP)

*Chandra* observations are downloaded from the satellite, processed, archived, and delivered to the proposing team within a day of the observation. The DSOps team performs standard data processing (SDP) on all *Chandra* science data (Section 2.2.3). The processing runs in several stages or "levels," each of which is built on the results of the preceding level. A well-defined set of data products for each level is transferred to the archive.

Level 0 processing extracts time-tagged information from the telemetry stream into multiple parallel streams for each science instrument and spacecraft subsystem. Level 0.5 determines the as-run instrument configuration and on-sky science time to ensure that all received science data are included. Level 1 performs the main scientific data calibration, to map the X-ray event positions from the individual detector elements (i.e., the individual CCDs for ACIS and the individual micro-channel plates for HRC) onto a single, uniform "idealized" detector pixel grid on the focal plane and corrects for spatial distortions. Corrections are applied for the spacecraft dither pattern, a Lissajous figure on the sky, which minimizes the effect of bad pixels, and potentially improves the spatial resolution by enabling subpixel position determination. Other steps include flagging bad and noisy pixels for later removal, applying the gain calibration, and (for ACIS) correcting the CCD charge transfer inefficiency. The SDP determines the appropriate calibration product to apply based on the observation date and engineering data (e.g., focal-plane temperature). As a final step, the good time intervals, during which all science data collected are within valid limits on all spacecraft and instrument subsystems, are defined for each data set.

Level 2 is the final step of SDP in which the science data within good time intervals are extracted and any bad data filtered out. This is followed by an automated validation and verification (V&V) process to ensure that the observation was performed as requested by the observer and data processing did not encounter any errors. If automated V&V identifies an issue, then the data are sent for human review and resolution. In most cases, the level 2 (L2; or L1.5 for grating data) data files should be used for science data analysis.

All data products are recorded in the astronomical standard Flexible Image Transport System (FITS) format and comply with International Virtual Observatory Alliance (IVOA) standards that facilitate interoperability with data from other observatories and wavebands, and High Energy Astrophysics Science Archive Research Center (HEASARC) standards that promote compatibility with high-energy astrophysics data analysis packages, such as CXC's portable CIAO data analysis suite (Section 2.2.6). Science data processing is fully automated, with *Chandra* data being received, processed, archived, and made available to the observer typically within less than one day after the observation is completed.

The ChaSer[15] archive interface provides search and retrieval capabilities for all public data in the archive (Section 2.3.3). Proprietary data sets, i.e., those within a year of the completion of SDP, are only available to the original proposing teams. Further information about the *Chandra* standard data processing and the data products can be found on the CXC website.[16]

The level 3 (L3) products are determined for all significant *Chandra* sources derived by merging all public data sets from the archive for each region of sky to generate the *Chandra* Source Catalog (CSC).[17] CSC 2.0, including all data up to 2014, was released in 2019 and is described in detail in Section 2.4.

### 2.2.6 Data Analysis Software (CIAO)

The CXC provides a powerful, flexible, multidimensional set of data analysis software, CIAO.[18] CIAO has a wide range of functionality to allow users to analyze imaging and spectroscopic data, fit models in one or more dimensions, record results, and generate figures ready for publication. It has also proven useful for the analysis of data from other, non-X-ray missions, because of the mission independence that is the basis of the CIAO design.

CIAO has been developed with attention to interoperability with other X-ray packages, compliance with astronomical data standards, and with a level of data abstraction that permits flexibility. It is updated with a major release each year to incorporate the evolving understanding of the spacecraft and instruments and to adapt to changing trends in user data analysis.

The CIAO suite includes tools with specific *Chandra* telescope and instrument support. These include tools shared with the SDP pipeline (Section 2.2.5), which allow reprocessing of the data with the latest calibration products and/or non-default parameters settings; tools for generating spectral responses for individual sources, sensitivity (exposure) maps for full fields, and for handling data from the *Chandra*

[15] https://cda.harvard.edu/chaser/
[16] http://cxc.harvard.edu/ciao/dictionary/sdp.html
[17] http://cxc.cfa.harvard.edu/csc/
[18] http://cxc.harvard.edu/ciao/

gratings; and the MARX program for simulating source point spread functions (PSFs) including both mirror and detector properties.

CIAO also includes a set of advanced generic data manipulation tools (the Data Model (DM) tools) which operate both on ASCII text files and on FITS images and binary tables. It includes a set of tools for scientific analysis of event files and images (e.g., automated source detection, smoothing, and extraction and manipulation of spectra and light curves), which can be used on a variety of astronomical data, but have specialized support for X-ray data from any of the HEASARC-compliant missions. The Sherpa application is a general purpose, Python-based, model-parameter-fitting program, developed for spectral and spatial fitting of *Chandra* data, but with widely applicable functionality, including advanced Bayesian fitting methods. Python libraries and access methods are provided for multiple CIAO subsystems including spatial region filtering. For image-based analysis, the SAOImage DS9 program is developed alongside CIAO and integrated with it, including a GUI-based analysis menu supporting a variety of *Chandra* analysis tasks. DS9[19] can be used standalone, independently from CIAO, and is widely used by the astronomical community for visualization and analysis of non-*Chandra* data.

CIAO is available for several current platforms and can be downloaded directly from the CIAO website.[20] CIAO includes several levels of help including the details of individual tasks and scripts to perform commonly used analysis, step-by-step science threads to perform specific analysis tasks, general information about various aspects of the system as a whole (e.g., "Why" topics), and manuals. The CXC runs a help desk[21] to promptly address questions from individual users, and organizes frequent CIAO workshops[22] at SAO, and at meetings both in the USA and abroad, to help users learn how to work with CIAO.

## 2.3 Archives and Science

*Chandra*, by opening an unexplored region of the observational parameter space in the X-rays thanks to its still-unmatched instrumental specifications, has made possible major scientific advances time and again over its lifetime. However, the extent of the discovery parameter space accessible through any one facility is not limited by the usually unchangeable (at least for most space-borne missions) features of its detectors. As the number of single observations grow over the lifetime of a mission, new insights can be gained from the investigation of phenomena that have been repeatedly probed or observed in several different instances, as time coverage and statistics increase. Moreover, the combined analysis of multiple observations—through aggregation, like stacking or coadding—offers the chance to explore fainter regimes that single observations cannot probe. Modern data repositories, due to the impressive advancements in the availability and cost of storage, and web, and

[19] http://ds9.si.edu/site/Home.html
[20] http://cxc.cfa.harvard.edu/ciao/
[21] http://cxc.harvard.edu/helpdesk/
[22] http://cxc.cfa.harvard.edu/ciao/workshop/index.html

Virtual Observatory technologies are now key to facilitating "archival science," and, in doing so, maximizing the scientific output of their missions. The *Chandra* Data Archive (CDA) has striven to serve the mission during the first 20 years of its lifetime by quickly adapting to the continuous expansion of the archive to efficiently cater to newly emerging needs, both within the CXC and from the astronomical community. The CDA has also been central in the effort to standardize the structure of the data model used to represent *Chandra* data products with structure, headers, and keywords compliant with the FITS standard (Pence et al. 2010) and the NASA HEASARC recommendations. These measures guarantee cross-mission and cross-tool compatibility of *Chandra* data with data from other high-energy missions.

### 2.3.1 The *Chandra* Data Archive

The CDA is the ultimate repository of all data produced by the *Chandra* mission. It records, stores, distributes, preserves, and provides access to all public and proprietary *Chandra* observations, calibration data, and internal operational data throughout the mission. The CDA currently contains and provides access to ~15,000 distinct public scientific observations and ~2500 calibration observations. The ACIS instrument has been used for ~85% of the public observations, while the HRC accounts for the remaining ~15%—a total of 2060 ACIS or HRC observations inserted into the high-energy (HETG) or low-energy (LETG) gratings to obtain high spectral resolution X-ray spectra of their targets. ToO and DDT observations account for ~700 and ~600 of the ~17,500 *Chandra* scientific and calibration observations, respectively. The total volume of the CDA data holdings, including all levels of processed data from telemetry to science-ready products, exceeds 33 TB as of 2019 April: primary and secondary data products (L2) for the ~22,900 public observations (including engineering requests) account for ~14.5 TB, while the volume of the aggregated, value-added, stacked (L3) data products created for the CSC (Section 2.4 for details) has reached ~18.5 TB, outgrowing the size of the L2 data during 2018.

The CDA is heavily used worldwide. An average of ~14 TB data have been downloaded annually, from a total of 72 countries, since the start of the mission. *Chandra* has both revolutionized the scientific understanding of the high-energy universe, and promoted multiwavelength research at high sensitivity and spatial resolution. The CDA public repository includes ~2200 (~13% of all archival observations) obtained in coordination with one or more observatories. Targets include a very diverse range of astronomical sources, environments, and instrumental configurations. By virtue of its exquisite spatial resolution and sensitivity, the *Hubble Space Telescope* (*HST*) accounts for ~40% of all *Chandra*-coordinated observations, followed by other space-borne facilities: *XMM-Newton* (~11%) and *Spitzer* (~10%). The extensive networks of ground-based optical and radio telescopes operated, in the USA, by NOAO and NRAO, respectively, have been used in conjunction with *Chandra* for ~20% of all the coordinated observations. The number of observations coordinated with more than one other facility is ~150.

### 2.3.2 The *Chandra* Bibliography

One of the most important goals of the CDA is the collection and curation of bibliographic information of all scientific or technical publications using *Chandra* data. By connecting observations to papers, the archive has created a powerful tool (Winkelman & Rots 2016) to measure, quantitatively and objectively, (1) the impact of *Chandra* on science, and (2) the community of *Chandra* users who have enhanced the scientific value of the mission. The CDA has built and maintained an extensive record of cross-links between *Chandra* observations and the large body of scientific and technical literature that uses or cites these data. Taking advantage of the close collaboration with the NASA Astrophysical Data System (ADS),[23] CDA has collected, classified, and linked ~20,000 publications (as of the end of 2019 March) to *Chandra* data sets. Among these *Chandra*-related publications, ~7800 refereed papers are classified as "*Chandra* Science Papers" (CSPs), given the significant reliance of their results on *Chandra* data. The CSPs represent the depth, variety and significance of the scientific discoveries that *Chandra* has achieved throughout all areas of astrophysics. The CDA has identified and recorded ~95,000 literature–observation links in the CSPs, referring to ~13,400 observations. On average, each observation is referenced by 7.1 CSPs, and each CSP refers to 12.9 *Chandra* observations.

The vitality of the archival science supported by the CDA is demonstrated by the continued relevance of *Chandra* observations. After eight years in the archive, ~90% of the total exposure time of all observations has been published at least once (Figure 2.15), and ~60% of the time has been used in three or more publications (see Rots & Winkelman 2015, Winkelman et al. 2018, for details). The sustained rate of publication for most data demonstrates that the archive enables multiple scientific studies beyond those of the original proposers.

Chandra*'s Growing Community*

In addition to the scientific and technical publications, the CDA is a community builder, particularly for early-career astronomers. A total of 353 PhD and master's theses have used *Chandra* data, for an average of ~18 per year, with 113 of these theses granted by non-US institutions. Four PhD recipients were advised by *Chandra* PhD recipients, starting a second generation. The resulting growth of the *Chandra* community is confirmed as ~70% of thesis authors are also *Chandra* proposal PIs, and ~26% of them have served as a *Chandra* peer reviewer at least once.

### 2.3.3 Access to *Chandra* Data

The CDA provides a diverse range of interfaces for discovery, visualization, and access to *Chandra* public observations. These interfaces are designed, developed, and updated to address the needs of both X-ray astronomers and those interested in combining *Chandra* with multiwavelength data. The evolution of *Chandra*

[23] https://ui.adsabs.harvard.edu/

interfaces, enabled by technological progress, is also driven by the increasing complexity of the archival holdings, which motivate new types of searches. The main access points to the CDA are the following:

- ChaSeR[24] is the centerpiece, supporting searches that include target names and properties, observation identifiers, and instrumental parameters. It also allows filtering based on different categorizations of the archival publications, observational status, science category, instrument, type of observation (GO, GTO, DDT, ToO, etc.), and observation cycle. ChaSeR returns a list of metadata for all observations that satisfy the search criteria and provides direct access to the standard data products.
- The *Chandra* Footprint Service[25] is focused on spatial searches. It facilitates large-scale, coordinates-based queries that have become more scientifically interesting as the archive has grown over *Chandra*'s 20 year lifetime. Users can search for public observations within a specified distance of one or multiple positions in the sky, explore the footprints (sky area covered) of the observations, and decide to either directly retrieve the standard data packages or to display the metadata via the ChaSeR.
- The *Chandra* Bibliography Search[26] interface allows searches that combine a subset of observational parameters and literature-based criteria. It uses the CDA's bibliographic database to list and provide access to publications that report or use specific data sets.

The CDA has also devoted a significant amount of time to making *Chandra* data available through the IVOA[27] infrastructure, including a comprehensive set of protocols that make astronomical data searchable and discoverable, based on their metadata, across a wide range of VO-powered interfaces that are insensitive to the specific origin of the data. The CDA is fully compliant with the IVOA interoperability model, ensuring that all levels of *Chandra* data are discoverable and accessible through all VO-powered interfaces and portals. The CDA has also adopted VO-defined data standards for the visualization and analysis of complex astronomical data sets, that are opening new opportunities for the intuitive exploration of astronomical data. The *Chandra* Multi-Order Coverage maps (MOCs), for example, can be used to display the total footprint of all *Chandra* public observations and efficiently cross-match it with arbitrarily long lists of sky coordinates, to compare it with the footprints of other missions. The CDA has also produced the *Chandra* Hierarchical Progressive Survey[28] (HiPS), a new class of imaging product constructed to allow seamless, browser-friendly, hierarchical visualization of the *Chandra* sky across widely different angular scales. This has opened a new realm of exploration possibilities for astronomers interested in

[24] https://cda.harvard.edu/chaser/
[25] https://cxcfps.cfa.harvard.edu/cda/footprint/
[26] http://cxc.harvard.edu/cgi-gen/cda/bibliography
[27] http://www.ivoa.net/
[28] https://cdaftp.cfa.harvard.edu/cxc-hips/

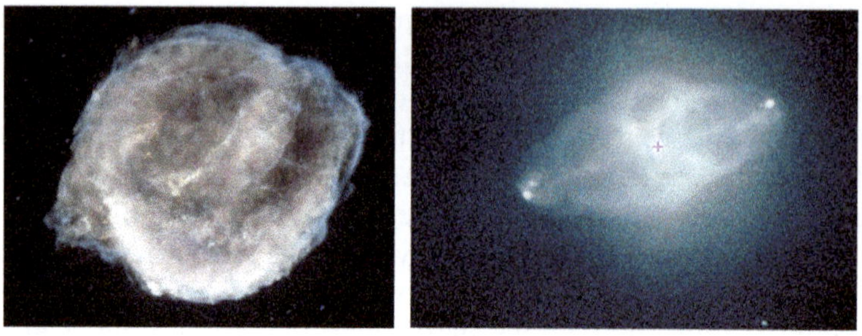

(a) KeplerSNR (b) CygnusAgalaxy

**Figure 2.9.** Two *Chandra*-observed sources displayed in the *Chandra* color HiPS (Section 2.3). (Courtesy of NASA/CXC.)

*Chandra* data (see Figure 2.9). The *Chandra* MOCs[29] have been available since 2017 May, while the HiPS became public in the summer of 2019.

### 2.3.4 The Legacy of the CDA

The lessons learned during the 20 years of CDA operations will inform and shape the design and implementation of data archives supporting future, similar missions, such as *Lynx* (Chapter 11). The archive design combined with forward-looking development of CDA operations has resulted in three distinct areas of progress.

- **Centrality.** Unlike most astronomical archives, which support only the final stages in the life cycle of the observations, the CDA underpins and supports the entire *Chandra* mission. This is based on the central role of the Observations Catalog (OCat), which supports multiple operational tasks (observation specification, mission planning, standard data processing, and archive) within the CXC, i.e., before, during, and after the observations are taken (Rots 2001; Rots et al. 2002). It also provides detailed status and parameter information on approved observations to the community at all stages. The resulting close working relations between CDA personnel and CXC staff working at all levels of operations has stimulated a high level of engagement and a deeper understanding of all aspects of *Chandra*.
- **Standardization**. The Data System (DS) within the CXC and, in particular, the CDA have heavily invested in the standardization of the data structure and their access and discoverability, by implementing interoperability standards and protocols at all levels. The adoption of a well-defined, general data model for the *Chandra* data, the use of FITS standards, and the intellectual and technical contributions to the nascent (at the beginning of the mission) VO, while adding significant cost compared to more easily workable but less general ad hoc solutions, have paid off abundantly in the long term. The community-wide adoption of VO standards has allowed the CDA to rapidly respond to the changing needs of the community it serves.

[29] http://cxc.cfa.harvard.edu/cda/cda_moc.html

- **Bibliography.** The emphasis and investment in a well-curated, comprehensive, and current bibliography of the *Chandra* data have set a new standard for tracking a mission's impact on both science and community. The comprehensive metrics defined to assess different facets of the science impact based on a mission's bibliography (Rots & Winkelman 2015; Winkelman et al. 2018) have not only measured *Chandra*'s impact, but also deeply influenced the literature in this field.

## 2.4 The *Chandra* Source Catalog

Over the past 20 years, the *Chandra X-Ray Observatory* has been observing the high-energy universe in the 0.5–8 keV X-ray energy band. The CDA now includes $>15,000$ imaging observations using two onboard cameras, ACIS and HRC (Section 2.3). The unique combination of subarcsecond spatial resolution for on-axis observations, the deep sensitivity resulting from low instrumental background, and the power of aggregation (stacking) between observations allows for the detection of many serendipitous sources in each observation, with sensitivity limits as low as 5 counts per source. Detecting, characterizing, and publicly releasing the properties of all these sources to the astronomical community is the primary motivation of the CSC, version 2.0 of which was released as part of the celebrations for the 20 years of *Chandra*.

### 2.4.1 A Catalog with Value to All Astronomers

Astrophysical investigations often require a large set of uniformly processed X-ray sources with observational properties derived using the same methods. For example, X-ray surveys efficiently detect large samples of AGNs, facilitating statistical studies that have resulted in the discovery of a long-sought, significant population of highly obscured AGNs (Masini et al. 2019). Similarly, comparing the measured X-ray luminosity function of intermediate- and high-redshift sources with model predictions facilitates studies of the growth over time of SMBHs (Griffin et al. 2019) and the assembly of galaxies. X-ray variability studies have revealed a broad range of physical phenomena related to the interaction between young stars and their surrounding protoplanetary disks (Sciortino et al. 2019; Winston et al. 2018). Despite the undeniable advantage of having uniformly processed X-ray sources, the typical researcher is often confronted with choosing between either carrying out a complex process of data retrieval, calibration, source detection, and properties derivation, or collating and compiling published data sets that have been heterogeneously processed by other colleagues. This process can be considerably more complicated for astronomers working in other wavelength regimes who are not familiar with the details of X-ray data reduction. The CSC provides an alternative that empowers all astronomers to perform high-quality research in all astronomical fields using science-ready data products that have been reliably calibrated, uniformly processed, and made accessible through the CDA.

### 2.4.2 Description of the Catalog

When it comes to maximizing the scientific output of *Chandra* data, every photon counts. Aggregation of individual observations significantly enhances the catalog's sensitivity. This section describes how aggregation is performed, outlines the general structure of the CSC, and provides a summary of the most important properties obtained for each source.

*The Power of Aggregation*

The first version, CSC 1.0 (Evans et al. 2010), performed source detection on individual *Chandra* observations taken over the first eight years of the mission, using the *wavdetect* algorithm (Freeman et al. 2002), which correlates a given data set with wavelets of different scales and then searches for significant correlations. This resulted in a list of >94,000 distinct X-ray sources distributed over a total sky area of >300 square degrees, with a sensitivity limit of $\sim 10^{-7}$ photons cm$^{-2}$ s$^{-1}$. Later versions of the CSC (v2.0 onwards) take advantage of data aggregation to significantly increase the catalog's sensitivity. For regions of the sky covered by two or more *Chandra* observations, all data sets are stacked prior to source detection (see Figure 2.10). *Chandra*'s PSF becomes significantly elongated away from the optical axis of the telescope (Allen et al. 2004). To avoid confusion between sources detected at very different off-axis angles, data sets are stacked together only if the targeted sky positions are within $\leqslant 1'$. When enough data sets are available, stacking can result in effective exposure times of > 1 Ms. Most stacks contain only a few observations, but in regions of great scientific interest (e.g., Sgr A*), they may contain >30 observations.

The overall sky coverage has increased in CSC 2.0, with approximately 600 deg$^2$ covered, including public observations through the end of 2014. The increased sensitivity and sky coverage result in >3× the number of sources, totaling >315,000 distinct, compact sources detected in ~7200 stacks, and a total exposure time >245.8

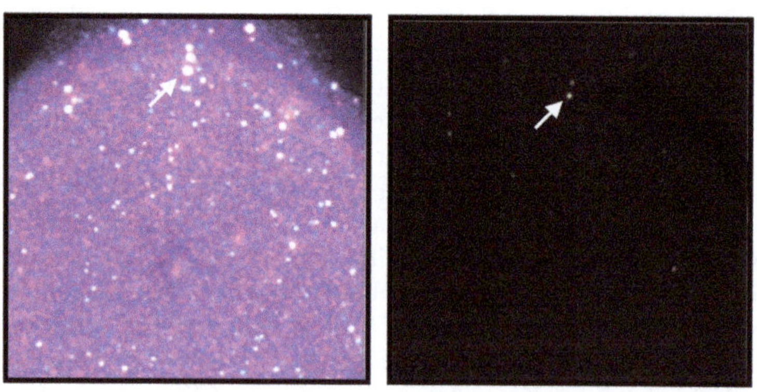

**Figure 2.10.** The effect of data aggregation in the *Chandra* Deep Field South (CDFS). The right panel shows a single image with an exposure time of 141 ks. With no aggregation, even the brightest sources in the CDFS were only marginally detected in individual fields in CSC 1.1. The left panel shows the stacked version of the same field, where 81 individual observations have been combined together for a total exposure time of 5.8 Ms. The number of CDFS sources detected increased from 555 to 978 within a 22′ radius. (Courtesy of NASA/CXC.)

Ms. In Figure 2.10 we show the dramatic effect of stacking on our ability to detect individual sources.

*The Structure of the Catalog*

Source detection is performed on the stacked data in energy bands broad (b, 0.5–7.0 keV), soft (s, 0.5–1.2 keV), medium (m, 1.2–2.0 keV), and high (h, 2.0–7.0 keV) for ACIS, and in the broad band (w, 0.1–10 keV) for HRC. If a detection is determined (in a probabilistic sense) to be a true source, the detected source position and the associated source region are used to compute the photometric, spectral, and variability properties in each band. A Bayesian approach estimates credible intervals (the Bayesian equivalent of confidence intervals) for all properties, by sampling the posterior probability density function using Markov Chain Monte Carlo (MCMC) methods. Source properties are computed for both the stacked detection and the individual observations. Finally, the master (most reliable) source properties are obtained by combining the estimates from different stacks in which the source is detected. In Figure 2.11, we show the hierarchical structure of the catalog.

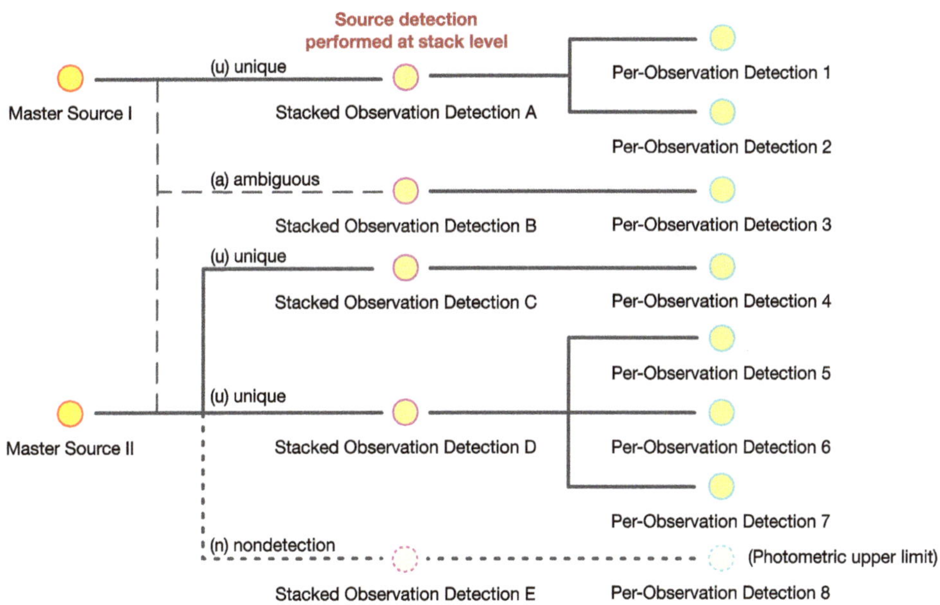


**Figure 2.11.** The source/detection hierarchy in CSC 2. The relationships between sources and their detection at the different levels are indicated. Source detection takes places in stacked observations. As illustrated, several per-observation detections can contribute to a single stacked observation detection. Similarly, several unique stacked source detections can contribute to a single master source. Also illustrated are a single, stacked source which is ambiguously defined, being associated with two different stacks having different off-axis angles and therefore different average PSF sizes, and the nondetection of a master source in one of the contributing stacks. CSC 2.0 contains tabulated properties and data products at observation, stack, and master levels. (Dataset identifier: ADS/Sa.CXO#CSC. Courtesy of NASA/CXC. Evans et al. 2010.)

**Table 2.1.** A Summary of Key Source Properties, Reported for Master, Stack, and Individual Detections

| Property Type | Properties |
|---|---|
| **Astrometry** | Source position, extent, detection significance, likelihood |
| **Photometry** | Band-specific aperture photon and energy fluxes, spectral model fluxes |
| **Spectral** | Hardness ratios, spectral fit parameters |
| **Variability** | Inter- and intra-observation variability probability, Bayesian blocks |

**Note.** These, and many more source properties, are available through the CSCview interface

*Catalog Properties*

Table 2.1 presents a summary of the catalog properties divided into four types: astrometry, photometry, spectroscopy, and variability. Each type is briefly described below.

***Astrometry and source shape***

Unlike sources in optical or infrared astronomy, the vast majority of X-ray sources are detected in the Poisson statistical regime, i.e., 5–10 photons often yield a significant detection. But since the errors are large for so few counts, determining astrometric source properties, such as position and source extent, is a challenging task. In CSC 2.0, source positions and extents are determined by fitting the local, band-specific PSFs to detections using state-of-the-art statistical algorithms, such as wavelet convolutions, maximum likelihood estimates, and MCMC. The average absolute astrometric uncertainty for CSC 2.0 sources is 0.29″.

***Aperture photometry***

A Bayesian formalism (Primini & Kashyap 2014) is used to estimate incoming X-ray flux densities in CSC 2.0, based on the detection of X-ray photons in the source and background regions. The resulting posterior PDFs are end-user products of the catalog. Photometry is estimated independently in each *Chandra* ACIS and HRC energy band.

***Spectral properties***

Spectral properties are characterized using two different approaches. First, hardness ratios are defined to compare the fluxes in two different bands, similar to the use of colors to define spectral slopes in optical and infrared data. Three hardness ratios are tabulated in the catalog, each using a pair of bands (*hm*, *hs*, and *ms*, where the hardness ratio, for e.g. the *hm* pair, is defined as $H_{hm} = (h - m)/(h + m)$). Hardness ratios provide a reliable, first-order approximation of the spectral shape of X-ray sources, and are a much more reliable diagnostic than spectral fits in the low count regime. In addition to the hardness ratios, CSC 2.0 sources with >150 detected counts are fitted with several canonical spectral models, e.g., an absorbed power law, an absorbed blackbody, and a bremsstrahlung model. In each case, the model parameters are fitted to adjust the observed X-ray spectrum using the Sherpa fitting tool (Freeman et al. 2001). In Figure 2.12, we show a fitted spectrum of a relatively bright CSC 2.0 source, together with the residuals to the fit.

***Variability***

Most X-ray sources are highly variable, and this variability should be characterized. Yet, we also require nominal, time-independent properties to describe each individual source. CSC 2.0 provides both time-independent properties and a rigorous characterization of time variability. The variability characterization is achieved by performing a number of statistical variability tests that return variability probabilities as well as the most significant light curve for each source. In Figure 2.13 we show the light curve of a CSC transient obtained from the Gregory–Loredo test (Gregory & Loredo 1992).

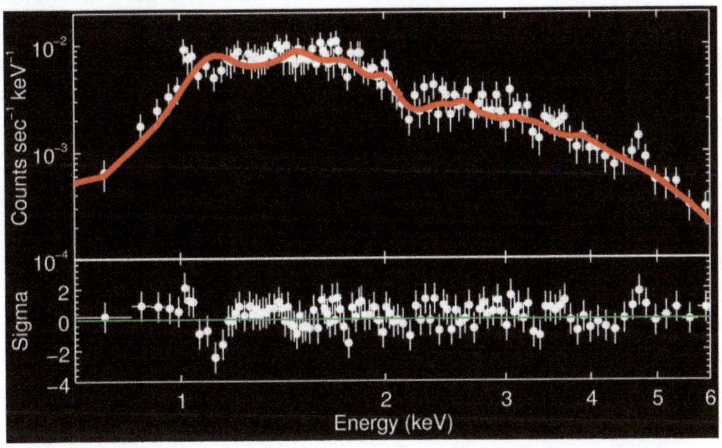


**Figure 2.12.** The spectrum of a CSC source showing the data in white, with $1\sigma$ error bars, and the best-fit, absorbed, power-law model in red. The residuals are shown in the lower plot, in units of $\sigma$. (Courtesy of NASA/CXC 2.0.)

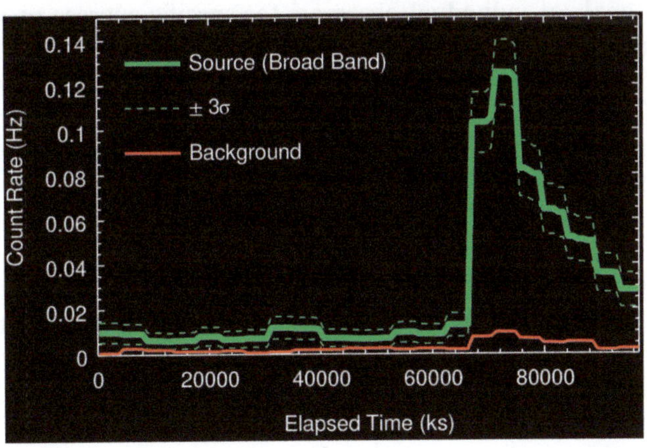


**Figure 2.13.** The light curves associated with a transient source obtained using the Gregory–Loredo test. Shown are the light curve of the source (green) with the $3\sigma$ confidence levels, and the background light curve (red). (Courtesy of NASA/CXC 2.0.)

In addition, the Bayesian blocks algorithm (Scargle et al. 2013) is used to group multiple observations of a source into time-ordered or flux-ordered “blocks” which share a consistent flux density, based on the flux distribution contained in the probability density functions (PDFs) of the individual observations. The time-independent, master source fluxes are then defined as the Bayesian block with the largest total exposure time (the “best block”).

### 2.4.3 Enabling Science with the *Chandra* Source Catalog

A thorough understanding of the most energetic astrophysical phenomena in the universe, from the birth of massive stars to the violent mergers of ancient galaxies, requires a multiwavelength approach. It is impossible to gather all of the relevant physics from data in a single band of the electromagnetic spectrum. For example, X-rays resulting from thermal bremsstrahlung or synchrotron emission in accreting AGNs are key to determining the fraction of bolometric emission associated with the accreting engine in luminous infrared galaxies. Also, X-ray emission from hot gas being swallowed by the most massive black holes in the universe (such as the recent submillimeter image of the horizon of the M87 SMBH; Event Horizon Collaboration et al. 2019)[30] is often among the first evidence for the presence of these massive engines (Perlman & Wilson 2005). The CSC therefore provides a powerful tool for scientific discovery. Here we describe some of the science that can be enhanced using the catalog.

*Multiwavelength Cross-correlations, Large Samples, and Population Studies*
Multiwavelength population studies of energetic phenomena in the Universe remain relatively rare. This is partly due to the lack of large, multiwavelength catalogs, particularly at high frequencies. Although X-ray and other space-borne observatories provide comprehensive archival repositories of their data sets (for *Chandra*, the CDA, Section 2.3), the methods for detecting and characterizing sources generally depend on the goals of individual observers for specific sources or samples. Thus, unless a comprehensive effort is made, a significant fraction of potentially detectable sources remains unexplored. The CSC provides a comprehensive list of all sources detected by *Chandra*, together with a thorough and uniform characterization of their properties. This enables a remarkable amount of new science. The common framework for the processing of all CSC sources, and the resulting uniformity in photometric, spectral, and time-domain properties of all ~315,000 compact sources, significantly improves the results of cross-correlations with catalogs in other wavelength regimes, facilitating well-defined populations of objects for statistical studies and joint analyses of large samples using the same tools. The resulting measurements of astrophysical properties are more robust against the heterogeneity of processing and analysis methods.

Resolving ambiguities in the identification and classification of astrophysical sources often requires cross-correlation between catalogs in different wavelength

[30] Also see Smithsonian Magazine.

regimes. For example, a reliable identification of AGNs among other types of X-ray sources benefits from combining X-ray fluxes with optical and/or infrared diagnostics (Section 8.6). The ratios of X-ray to optical emission, and/or thermal infrared emission from dust, can distinguish the AGN. In order to fully characterize astrophysical sources, it is therefore desirable to cross-correlate the CSC with surveys in infrared and optical wavelengths.

*Serendipitous Discovery with the CSC*

The CSC offers an unprecedented opportunity for serendipitous discovery. Of the ~315,000 compact sources included in CSC 2.0, two-thirds are new *Chandra* detections, and a significant fraction of the remaining objects has never been studied in detail. CSC sources are characterized by their X-ray fluxes, hardness ratios, variability, and spectral properties. They include known types of celestial objects ranging from young stars in star-forming regions to massive clusters of galaxies in the distant universe. The CSC also includes "weird" sources that could be either examples of a known class observed in rare or unknown stages of their evolution, or a previously unidentified X-ray source type. The recent explosion of methodologies based on machine learning and data mining, in combination with the rich data set defined by CSC 2.0, will facilitate the identification of sources and significantly expand the horizon for discovery and investigation in high-energy astrophysics.

## 2.5 *Chandra*'s Impact on Astronomy

A number of metrics have been developed over recent years to measure the level of impact telescopes and missions have on both science and the community. These metrics are now routinely used to define a mission's success and to make comparisons to other missions.

Metrics that track the level of *Chandra*'s impact include the number of papers published and citations to those papers, the demand for *Chandra* observing time, the number of people who use *Chandra* data, and the usage of *Chandra* data in the archive (see also Section 2.3). In addition, the public impact of press and image releases and the vigorous social media programs provide a measure of the impact on the broader community.

The number of refereed papers directly reporting *Chandra* observations per year since launch is shown in Figure 2.14. The total number of papers is 7759. Citations to these papers average 35 per paper after six years, increasing to 84 after 14 years. A full bibliography of *Chandra* refereed and non-refereed papers is available online.[31] CXC staff have introduced new metrics that measure the usage of the data (Rots et al. 2012), which also facilitate comparisons between observatories. For example, the median time from *Chandra* observation to publication is 2.4 yr. Tracking publications of individual data sets provides an estimate of the level of data usage, which is shown in Figure 2.15. For example, we can see that after 8.5 years, 90% of the data have been published, with 70% having two or more publications. The latter

[31] http://cxc.harvard.edu/pubs.html

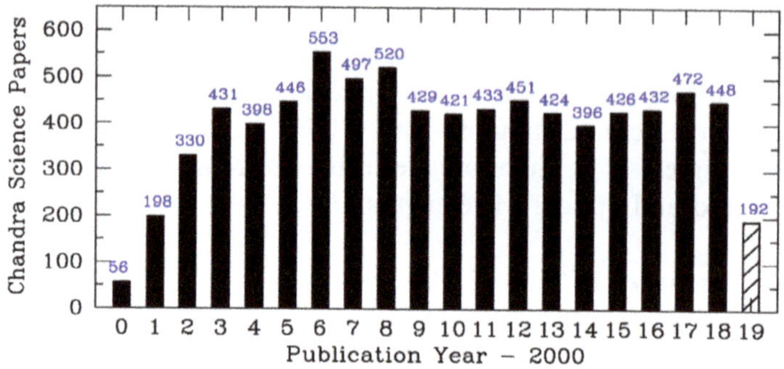


**Figure 2.14.** Number of refereed *Chandra* papers each year from the start of the mission (as of 2019 July 1). Note that the current year is incomplete and thus partially shaded. (Courtesy of NASA/CXC.)

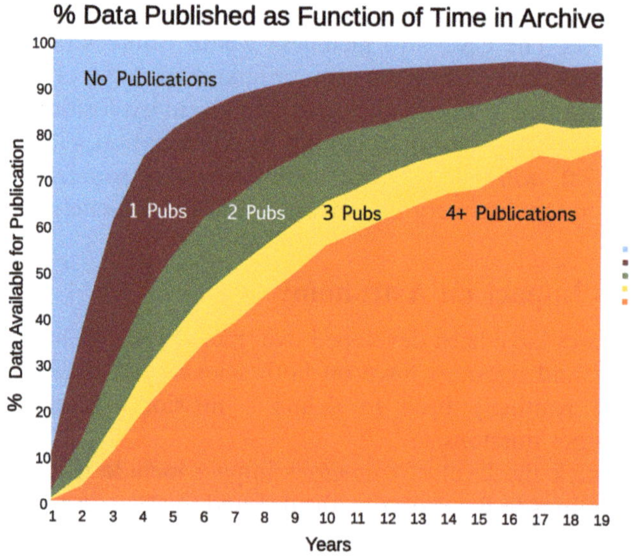


**Figure 2.15.** The percentage of available exposure time published in one, two, three, or four+ papers as a function of the time since the data were transferred into the archive. (Courtesy of NASA/CXC.)

number demonstrates the heavy use of the archive by the community in addition to the original proposers. The numbers are lower limits as they are based on linking papers to specific observation identification numbers, which are not always included in publications.

Demand for *Chandra* observing time and research support is consistently high. A total of 3818 of the 13,548 proposals submitted (Cycles 1–20) were accepted, ranging from 2 ks–3 Ms of observing time per proposal. The corresponding oversubscription, factors are 5.5 in observing time and 3.5 in proposals, has been high, even in Cycles 13 and 14, when the available observing time increased by ~25% due to the evolving orbit. Of the accepted proposals, 27% have foreign PIs from 40 different countries. As is standard for NASA-approved science projects, funding has been allocated to support successful, US-based proposing teams. To date, this has included 4638 teams (one or more per proposal) in 46 states. Statistics of the *Chandra* proposals for each cycle are reported in the annual newsletter.[32] We present only one top-level statistic here. Figure 2.16 shows that the overall success rate of *Chandra* proposal PIs has been indistinguishable for male and female PIs since Cycle 10. The fraction of PIs who are female has increased over the mission lifetime.

The worldwide community of *Chandra* Users currently includes ~4300 individual principal investigators (PIs) and co-investigators (co-Is) with an average over Cycles 9–19 of 178 new (distinct) investigators (42 new PIs) per year. This is a sizable fraction of the total astronomical community. These numbers do not include users of the *Chandra* archive unassociated with a proposal because the archive is freely accessible to all. The level of usage of the CDA is reported in Section 2.3.

*Chandra* also serves as a key training ground for young scientists. Our annual survey shows a count of 3911 worldwide (2001–2018): 952 undergraduates, 1618 graduate students, and 1341 postdoctoral fellows (all likely to be lower limits). We have recorded 375 (1999–2019, PhD theses[33]) that include *Chandra*-related research, with an average of $\sim 18\ \mathrm{yr}^{-1}$.

Another measure of the high impact of *Chandra* science is the broad reach of the press and public communications effort. To reflect the significant trend away from print, we now track IP (internet protocol) addresses accessing online news sites. Over the past three years, *Chandra* issued 51 press releases and 42 image releases, resulting in ~12,000 articles on the sites tracked. Based on the count of unique IP addresses,

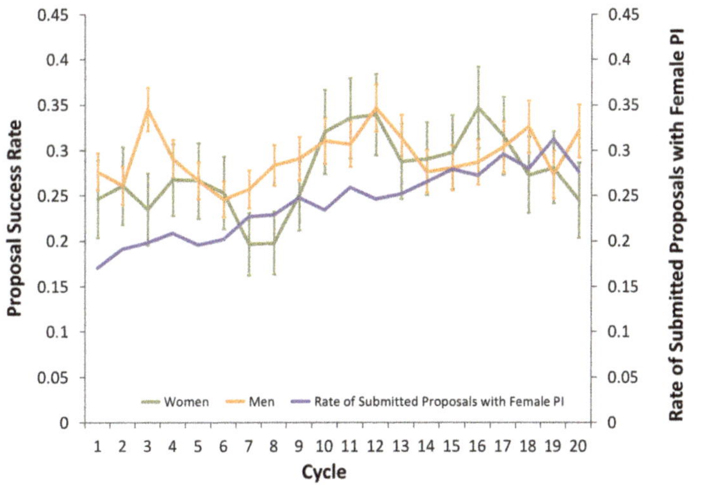


**Figure 2.16.** The success rate of *Chandra* proposals with female (green) versus male (orange) PIs over the mission lifetime, showing no statistically significant difference since Cycle 10 (2009). The blue line shows the fraction of female PIs, which has increased gradually during the mission. (Courtesy of NASA/CXC.)

[32] http://cxc.cfa.harvard.edu/newsletters/

[33] http://cxc.harvard.edu/cda/bibstats/biblists/phd_papers.html

the potential monthly readership averaged >500M. Our communication is also augmented with a vigorous web and social media program. Over the past three years, the average monthly web hits are 10.5M. Social media has acquired 285K followers on Facebook, 163K on Twitter, 248k on Instagram, and >3.5M on YouTube. Podcasts, our main download, have ~400K hits per month. Subscriptions number ~80K for *Chandra* images and ~60K for blogs.

## References

Allen, C., Jerius, D. H., & Gaetz, T. J. 2004, Proc. SPIE, 5165, 423–32

Canizares, C. R., Huenemoerder, D. P., Davis, D. S., et al. 2000, ApJ, 539, L41

Canizares, C. R., Davis, J. E., Dewey, D., et al. 2005, PASP, 117, 1144

Evans, I. N., Primini, F. A., Glotfelty, K. J., et al. 2010, ApJS, 189, 37

Event Horizon Telescope Collaboration; Akiyama, K., Alberdi, A., et al. 2019, ApJ, 875, L1

Freeman, P., Doe, S., & Siemiginowska, A. 2001, Proc. SPIE, 4477, 76–87

Freeman, P. E., Kashyap, V., Rosner, R., & Lamb, D. Q. 2002, ApJS, 138, 185

Gregory, P. C., & Loredo, T. J. 1992, ApJ, 398, 146

Griffin, A. J., Lacey, C. G., Gonzalez-Perez, V., et al. 2019, MNRAS, 487, 198

Hwang, U., Holt, S. S., & Petre, R. 2000, ApJ, 537, L119

Masini, A., Hickox, R., Comastri, A., & Civano, F. 2019, AAS/HEAD Meeting, 17, 106.56

Pence, W. D., Chiappetti, L., Page, C. G., Shaw, R. A., & Stobie, E. 2010, A&A, 524, A42

Perlman, E. S., & Wilson, A. S. 2005, ApJ, 627, 140

Primini, F. A., & Kashyap, V. L. 2014, ApJ, 796, 24

Rots, A. 2001, in ASP Conf. Ser. 225, Virtual Observatories of the Future, ed. R.-J. Brunner, S. G. Djorgovski, & A. S. Szalay (San Francisco, CA: ASP), 209–87

Rots, A., Winkelman, S., & Becker, G. 2012, in ASP Conf. Ser. 461, Astronomical Data Analysis Software and Systems XXI, ed. P. Ballester, D. Egret, & N. P. F. Lorente (San Francisco, CA: ASP), 771

Rots, A. H., & Winkelman, S. L. 2015, in ASP Conf. Ser. 492, Open Science at the Frontiers of Librarianship, ed. A. Holl, S. Lesteven, D. Dietrich, & A. Gasperini (San Francisco, CA: ASP), 71

Rots, A. H., Winkelman, S. L., Paltani, S., & DeLuca, E. E. 2002, Proc. SPIE, 4844, 172–9

Scargle, J. D., Norris, J. P., Jackson, B., & Chiang, J. 2013, ApJ, 764, 167

Schwartz, D. A., Marshall, H. L., Lovell, J. E. J., et al. 2000, ApJ, 540, 69

Sciortino, S., Flaccomio, E., Pillitteri, I., & Reale, F. 2019, AN, 340, 334

Winkelman, S., D'Abrusco, R., & Rots, A. 2018, Proc. SPIE, 10704, 1070418

Winkelman, S., & Rots, A. 2016, Proc. SPIE, 9910, 991006

Winston, E., Wolk, S. J., Gutermuth, R., & Bourke, T. L. 2018, AJ, 155, 241

# Chapter 3

# Mechanisms for the Production and Absorption of Cosmic X-Rays

**Wallace H Tucker**

This chapter summarizes the processes that can produce and absorb X-rays in a cosmic setting. After presenting some basic concepts of classical and quantum theory, specific processes are discussed. These include cyclotron and synchrotron radiation, electron scattering, blackbody radiation, bremsstrahlung, radiative and dielectronic recombination, and line emission following collisional excitation of ions by electrons.

## 3.1 Introduction

Almost all of our knowledge about the cosmos outside our solar system comes from the study of electromagnetic radiation from distant sources. In general, the details of the microscopic photon-producing processes are well understood, so a study of the radiation observed from cosmic sources allows us to deduce something about the nature of these sources.

This chapter considers the various physical processes that are important for the production and absorption of X-rays in a cosmic setting. The spirit of the discussion is to present simple, order-of-magnitude estimates to provide some insight into the basic physics involved and then state the exact results in a form useful for the remainder of the book. After summarizing some basic concepts, classical X-ray production processes are discussed, followed by quantum processes, and finally a discussion of X-ray absorption in a cosmic setting. Unless otherwise indicated, the units used are Gaussian (cgs).

Both thermal and nonthermal radiation processes play important roles in the production of cosmic X-rays in a wide variety of settings. By thermal radiation we mean radiation from particles with a Maxwellian distribution of momenta. This is in contrast to nonthermal radiation from particles, usually electrons, with very high

doi:10.1088/2514-3433/ab43dcch3

energies that are far from equilibrium and typically have a power-law distribution of momenta.

Thermal radiation from a plasma of temperature $T$ has a spectrum that peaks at $h\nu = \varepsilon \sim kT$, where $h$ is Planck's constant, $\nu$ is the frequency of the radiation, $\varepsilon$ is the energy of a photon, and $k$ is the Boltzmann constant (see Sections 3.4 and 3.6). Beyond the peak frequency, the radiation spectrum falls off exponentially, so production of X-rays with energies ~keV requires temperatures in the range

$$T \sim \varepsilon/k \sim 10\varepsilon_{\text{keV}}\ \text{MK}, \tag{3.1}$$

where $\varepsilon_{\text{keV}}$ is the photon energy in units of kiloelectronvolts. The conditions for the production of cosmic X-rays occur in a wide variety of cosmic settings.

Neutron star surfaces and optically thick accretion disks around neutron stars and black holes produce blackbody radiation with a temperature $T \sim$ 1 MK–10 MK. Optically thin hot plasmas with $T$ ranging from a few MK to a few 10 MK in stellar coronas and supernova remnants (SNR) produce a rich spectrum of X-ray lines from cosmically abundant elements such as oxygen, silicon, and iron. Clusters of galaxies, the largest gravitationally bound structures in the universe, are filled with 10 MK–200 MK plasma that is a strong source of thermal bremsstrahlung X-rays, and there is good evidence that a significant fraction of the baryons in the universe are in a diffuse cosmic web of hot plasma with temperatures in the range 0.1 MK–10 MK.

Cosmic nonthermal X-radiation is produced primarily by the synchrotron process (Figure 3.1), in which relativistic effects boost the frequency of the observed radiation from electrons by a factor $\gamma^2$, where

$$\gamma = \frac{1}{\sqrt{1 - (\text{v}/c)^2}} \tag{3.2}$$

is the Lorentz factor for a particle with velocity v and $c$ is the speed of light. The production of electron synchrotron X-rays with energies ~10 keV requires

$$B\gamma^2 \sim 6 \times 10^{11}, \tag{3.3}$$

where $B$ is the magnetic field strength in Gauss. For example, the observed X-radiation from supernova shock waves with $B \sim 10^{-4}$ G, requires electrons with $\gamma \sim 10^8$, corresponding to 40 TeV electrons.

Another important source of cosmic X-radiation occurs when low-energy photons undergo Compton scattering with high-energy electrons. This occurs in hot coronas above accretion disks around black holes, where lower-energy photons

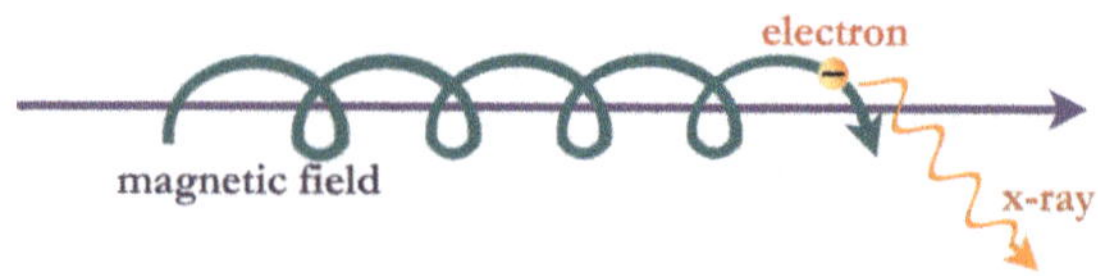


**Figure 3.1.** Synchrotron radiation. (Courtesy of NASA/CXC/SAO.)

from the accretion disk are Compton-scattered into the X-ray band by hot electrons in the corona, and on a much larger scale, when cosmic microwave background photons scatter off the relativistic electrons that produce synchrotron radio emission in giant radio sources.

When considering a specific radiation process, it is desirable to know whether or not classical physics provides an adequate description. In general, a classical treatment is valid when the de Broglie wavelength of the radiating particle is small compared to the typical dimension of the problem, i.e., when

$$\lambda_d \equiv \hbar/p \ll d, \tag{3.4}$$

where $p$ is the momentum of the radiating particle, $d$ is a typical dimension of the system, and $\hbar = h/2\pi$. When the de Broglie wavelength satisfies the above inequality, the particle can be considered point-like; otherwise, the finite extent of the particle's wave function must be considered, and a quantum-mechanical treatment must be used. The dimension $d$ in the above equation may refer to the size of the interaction, or in the case of scattering electromagnetic waves, the wavelength of the radiation. In the case of collisions between two or more different charged particles with masses $m_1$ and $m_2$, where $m_1 \ll m_2$, the momentum $p$ refers to the momentum of the least massive particle, because in general the radiative efficiency of a particle varies inversely as the square of the mass of that particle.

Another way of looking at the condition for the applicability of classical theory is in terms of the kinetic energy, $W$, of the radiating particle and the frequency $\nu$ of the emitted radiation. Because for nonrelativistic particles $W \sim p\mathrm{v}$ and $\nu \sim \mathrm{v}/d$, a classical treatment can be used when

$$h\nu \ll W. \tag{3.5}$$

This condition states that, in the classical limit, a particle cannot convert a significant fraction of its energy into one photon. The same condition remains valid in the relativistic regime where $\mathrm{v} \sim c$. In the context of cosmic X-ray sources, synchrotron and Compton scattering can usually be considered classically, whereas blackbody radiation and thermal radiation from hot plasmas require a quantum treatment.

## 3.2 Classical Radiation Processes

### 3.2.1 Electromagnetic Waves

The classical theory of radiation is based on Maxwell's theory of the electromagnetic field. For a given distribution of charge density $\rho$ and current density $\mathbf{J}$, the field is determined by Maxwell's equations:

$$\nabla \cdot \mathbf{E} = 4\pi\rho, \tag{3.6}$$

$$\nabla \cdot \mathbf{B} = 0, \tag{3.7}$$

$$\nabla \times \mathbf{E} = -(1/c)\partial\mathbf{B}/\partial t, \tag{3.8}$$

$$\nabla \times \mathbf{B} = (4\pi \mathbf{J}/c) + \partial \mathbf{E}/\partial t. \tag{3.9}$$

Here, vector quantities are denoted with a boldface, with **E** and **B** denoting the electric and magnetic field vectors. The motion of a charged particle is described by Newton's second law, together with the Lorentz force, $\mathbf{F}_{\mathrm{L}}$,

$$d\mathbf{p}/dt = \mathbf{F}_{\mathrm{L}} = e\mathbf{E} + e(\mathbf{v} \times \mathbf{B})/c. \tag{3.10}$$

Strictly speaking, the fields to be used in the Lorentz equation are the external fields as well as the fields produced by the charge itself. In general, the self-produced fields can be ignored, so only external fields are used.

The rate of change of the kinetic energy, $W$, of a charge in an electromagnetic field is

$$\frac{\mathrm{d}W}{\mathrm{d}t} = \mathbf{v} \cdot \frac{\mathrm{d}\mathbf{p}}{\mathrm{d}t} = e\mathbf{v} \cdot \mathbf{E}. \tag{3.11}$$

The magnetic field does not enter into the equation for the change in kinetic energy of a charged particle because the magnetic force acts perpendicular to the velocity of the particle, and so does no work on it.

The rate of increase of energy of all the particles in a unit volume is found by summing over all particles in that volume:

$$\frac{dU_{\mathrm{p}}}{dt} = \mathbf{J} \cdot \mathbf{E}, \tag{3.12}$$

where $U_{\mathrm{p}}$ is the kinetic energy density of the particles. In a given volume, $U_{\mathrm{p}}$ changes with time because of the changes in the energy of particles, and because of the flow of particles into and out of the volume. Thus,

$$\frac{dU_{\mathrm{p}}}{dt} = \frac{\partial U_{\mathrm{p}}}{\partial t} + \nabla \cdot \mathbf{Q} = \mathbf{J} \cdot \mathbf{E}, \tag{3.13}$$

where **Q** is the energy flux density vector. For a particle distribution function $f(\mathbf{r}, \mathbf{v}, t)$, **Q** is given by

$$\mathbf{Q} = \int Wf(\mathbf{r}, \mathbf{v}, t)dV. \tag{3.14}$$

Through the use of Maxwell's equations and some vector algebra, the equation for the rate of change of energy can be rewritten as

$$\frac{\partial(U_{\mathrm{p}} + U_{\mathrm{em}})}{\partial t} + (\nabla \cdot \mathbf{Q} + \mathbf{S}) = 0, \tag{3.15}$$

where

$$\mathbf{S} = c\frac{\mathbf{E} \times \mathbf{B}}{4\pi} \tag{3.16}$$

is the Poynting vector, and

$$U_{\rm em} = \frac{E^2 + B^2}{8\pi} \tag{3.17}$$

is the electromagnetic energy density.

### 3.2.2 Classical Dipole Radiation

For particle velocities $v \ll c$ and for distances large compared to the size of the source, the expression for the electric field of a point charge takes the form

$$\mathbf{E} = (e/R^2)\hat{\mathbf{n}} + (e/Rc^2)[\hat{\mathbf{n}} \times \hat{\mathbf{n}} \times \dot{\mathbf{v}}], \tag{3.18}$$

where $\hat{n} = \mathbf{R}/R$ is the unit vector directed from the position of the charge to the observer. The first term is the electrostatic field. It is independent of the velocity of the particle and varies at large distances as $1/R^2$. The second term is the radiation field. It depends on the acceleration $\dot{\mathbf{v}}$ and varies at large distances as $1/R$. In order of magnitude, the ratio of the two fields is

$$|\mathbf{E}_{\rm static}/\mathbf{E}_{\rm rad}| \sim (ct_{\rm c}/R)^2 \sim (\lambda/R)^2, \tag{3.19}$$

where $t_{\rm c}$ is the characteristic time for changes in the system, $R$ is the distance to the charge, and $\lambda$ is the characteristic wavelength of the radiation.

The radiation field dominates at distances, $R$, large compared to the wavelength of the radiation. The expressions for the electric and magnetic fields then take the form

$$\mathbf{E} = \hat{\mathbf{n}} \times (\hat{\mathbf{n}} \times \ddot{\mathbf{d}})/(Rc^2), \tag{3.20}$$

$$\mathbf{B} = \hat{\mathbf{n}} \times \mathbf{E}, \tag{3.21}$$

where

$$\mathbf{d}_{\rm e} = \Sigma(e_i\mathbf{r}_i) \tag{3.22}$$

is the electric dipole moment of the system. Note that the radiation is determined by the second derivative of the dipole moment, hence the name "dipole approximation."

The power radiated per unit solid angle $d\Omega = 2\pi \sin\theta \, d\theta$ for a single charge is given by

$$\frac{dP}{d\Omega} = P(\Omega) = R^2\mathbf{S} \cdot \hat{\mathbf{n}} = \frac{a^2e^2\sin^2\theta}{4\pi c^3}, \tag{3.23}$$

where $a$ is the acceleration of the charge. This is Larmor's formula for radiation from a nonrelativistic charge. Note that the angular distribution of the radiation is symmetric about the direction of the acceleration of the charge and independent of its velocity. Integrating over solid angle, the total power radiated is

$$P = \frac{2a^2e^2}{3c^3}. \tag{3.24}$$

### 3.2.3 Radiation from a Relativistic Charged Particle

In the frame of reference where a charged particle is at rest, the particle velocity $v \ll c$ and the nonrelativistic expression for the total power radiated applies. The total power radiated is Lorentz invariant, so transforming from the rest frame of the particle to the laboratory, or observer's frame, using the replacement

$$\left(\frac{dp_j}{dt}\right)\left(\frac{dp_j}{dt}\right) \rightarrow \left(\frac{dp_\mu}{d\tau}\right)\left(\frac{dp_\mu}{d\tau}\right), \tag{3.25}$$

where

$$d\tau = dt/\gamma \tag{3.26}$$

$$p_{1,2,3} = m\gamma v_{1,2,3} \tag{3.27}$$

and

$$p_4 = imc\gamma \tag{3.28}$$

yields, after some (well, quite a bit of) algebra,

$$P = \frac{2e^2}{3c^3}\gamma^6[\dot{v}^2 - [(\mathbf{v}/c) \times \dot{\mathbf{v}}]^2] = \frac{2e^2}{3c}\gamma^6[\dot{\boldsymbol{\beta}}^2 - (\boldsymbol{\beta} \times \dot{\boldsymbol{\beta}}^2)], \tag{3.29}$$

where

$$\boldsymbol{\beta} = \mathbf{v}/c. \tag{3.30}$$

The factor $\gamma^6$ in this equation shows the increased efficiency of relativistic particles for producing radiation. The force **F** on a particle is related to its acceleration by

$$\mathbf{F} = \frac{d\mathbf{p}}{dt} = mc\gamma[\gamma^2(\boldsymbol{\beta} \cdot \dot{\boldsymbol{\beta}})\boldsymbol{\beta} + \dot{\boldsymbol{\beta}}]. \tag{3.31}$$

When **F** is parallel to $\boldsymbol{\beta}$ and $\gamma \gg 1$, $\dot{\boldsymbol{\beta}} \rightarrow \mathbf{F}/mc\gamma^3$, so the power radiated is

$$P_{\parallel} \rightarrow \frac{2e^2}{3m^2c^3}F^2, \tag{3.32}$$

which is independent of the particle's energy. When **F** is perpendicular to $\boldsymbol{\beta}$,

$$P_{\perp} \rightarrow \frac{2e^2}{3m^2c^3}F^2\gamma^2. \tag{3.33}$$

Comparison of the expressions for $P_{\parallel}$ and $P_{\perp}$ shows that for comparable forces, perpendicular deflecting forces are more efficient at producing radiation.

For astrophysical applications, magnetic fields can provide the perpendicular deflecting force, so if particles with sufficiently high energy are present, a strong source of radiation, called synchrotron radiation because it was analyzed in detail in connection with synchrotron particle accelerators, can be produced. Using the expression for the Lorentz force (Equation (3.10)) and defining $\psi$ as the angle between the particle's motion and the magnetic field yields the power produced by synchrotron radiation from electrons of mass $m$:

$$P_{syn} = \frac{2e^4}{3\,m^2c^3}\beta^2\gamma^2B^2\sin^2\psi = \frac{(2r_0^2c)}{3}\beta^2\gamma^2B^2\sin^2\psi, \tag{3.34}$$

where $r_0 = e^2/mc^2$ is the classical electron radius.

## 3.3 Cyclotron and Synchrotron Radiation

Charged particles moving in a magnetic field experience acceleration as long as their motion is not parallel to the field, and emit electromagnetic waves. For historical reasons related to the development of particle accelerators, this radiation is called cyclotron radiation if the particles are nonrelativistic, and synchrotron or magneto-bremsstrahlung if the particles are relativistic. Synchrotron radiation is especially important in an astrophysical context, as it is responsible for the nonthermal radio emission observed from supernova remnants and radio galaxies. It is also observed in sporadic radio emission from the Sun and Jupiter. In addition, optical and X-ray synchrotron radiation is detected from many of these same sources. Cyclotron X-radiation plays an important role in regions of ultrastrong magnetic fields on neutron stars.

The equation of motion for an electron with momentum $\mathbf{p} = \gamma m_e\mathbf{v}$ is

$$d(\gamma m\boldsymbol{\beta})/dt = e\boldsymbol{\beta}\times\mathbf{B}. \tag{3.35}$$

The perpendicular component of the velocity obeys the equation

$$d\boldsymbol{\beta}_\perp/dt = \boldsymbol{\beta}_\perp\times(e\mathbf{B}/\gamma mc), \tag{3.36}$$

where

$$\beta_\perp = \beta\sin\psi \tag{3.37}$$

and $\psi$ is the pitch angle. The path of the electron is a circular helix of radius $R_\mathrm{L}$, called the Larmor radius, where

$$R_\mathrm{L} = m\gamma\beta_\perp c^2/eB. \tag{3.38}$$

The total power radiated is (See Section 3.2.3)

$$P_\mathrm{syn} = (2/3)r_0^2c\beta^2\gamma^2B^2\sin^2\psi = 1.6\times10^{-15}\beta^2\gamma^2B^2\sin^2\psi. \tag{3.39}$$

For an isotropic distribution of electron velocities,

$$P_\mathrm{syn} = (4/9)r_0^2c\beta^2\gamma^2B^2 = 1.6\times10^{-15}\beta^2\gamma^2B^2. \tag{3.40}$$

### 3.3.1 Cyclotron Radiation

When v $\ll c$, most of the power is radiated at a frequency

$$\nu_B = \frac{eB}{2\pi mc} = 3 \times 10^6 B_\perp \text{ Hz}, \tag{3.41}$$

or, in terms of photon energy

$$h\nu_B = 12 B_{\perp,12} \text{ keV}. \tag{3.42}$$

This is the 12–$B$–12 rule of thumb for cyclotron emission or absorption of X-rays: a magnetic field strength of $10^{12}$ G will produce or absorb 12 keV X-rays.

Electron cyclotron absorption lines have been observed in the spectra of ~30 neutron stars. These lines are formed in the atmosphere above hot spots on neutron stars and provide one of the best methods for determining the magnetic field strengths of neutron stars. Note that the observed energy of the cyclotron line must be increased by a factor $1 + z_{gr}$ to take into account the gravitational redshift, $z_{gr}$, which is ~0.3 for a neutron star. Typically, the inferred field strengths are ~ a few terragauss (Trümper et al. 1977; Mullen et al. 2017; Harding & Lai 2006; Schönherr et al. 2007; Schwarm et al. 2017a, 2017b; Maitra 2017). These magnetic field strengths are within about an order of magnitude of the critical magnetic field $B_{crit}$, at which point the electron spin–magnetic field interaction term $\mu \cdot B_{crit} \sim m_e c^2$, i.e.,

$$B_{crit} = m^2 c^3/(e\hbar) = 44 \times 10^{12} \text{ G}, \tag{3.43}$$

For $B \sim B_{crit}$, a relativistic treatment is needed.

In some sources, the cyclotron lines have been interpreted as proton cyclotron absorption lines, in which case the inferred magnetic field strength is a factor $m_p/m = 1836$ larger (Brightman et al. 2018).

### 3.3.2 Synchrotron Radiation

*Spectrum*

As discussed in Section 3.2.3, for relativistic motion, the power radiated due to acceleration perpendicular to the motion of a charged particle is a factor $\gamma^2$ larger than the power radiated for a comparable acceleration parallel to the motion. The radiation from an ultrarelativistic charge is, to a close approximation, the same as that emitted by a particle moving instantaneously along the arc of a circular path the radius of curvature of which is given by the Larmor radius.

If $t$ is the time at which radiation is emitted at the position of the particle and $t_0$ is the time at which the radiation arrives at the observer at a distance $R$, then $t_0 = t + R/c$. The observer sees a pulse of radiation that lasts for an interval

$$\delta t_0 = [1 + (dR/dt)/c]\delta t \approx [1 - \beta \cos\theta]\delta t \tag{3.44}$$

and detects the radiation only when the particle is on a segment of the arc with $\Delta\theta \sim 1/\gamma$, so $\delta t \sim R_L/(c\gamma)$, the quantity $[1 + (dR/dt)/c] \sim 1/\gamma^2$, and $\delta t_0 \sim R_L/c\gamma^3$.

The radiation will be observed with a frequency of order $\nu_{\rm syn}\sin\psi$, where $\psi$ is the pitch angle of the electron's motion with respect to the magnetic field, and

$$\nu_{\rm syn} = \nu_B\gamma^2 = 3\times 10^6 B\gamma^2 \sin\psi \ \ \text{Hz} \tag{3.45}$$

The synchrotron spectrum is composed of a series of discrete lines at integral multiples of $\nu_{\rm B}/\sin^2\psi$. However, for highly relativistic particles, most of the power is emitted at frequencies $\nu \gg \nu_{\rm B}$, so the spacing between the lines is sufficiently small that the spectrum may be regarded as continuous for frequencies of interest.

The derivation of the detailed spectrum involves expanding the electromagnetic field due to a moving charge into its frequency components using Fourier transforms, then computing the frequency components of the Poynting flux. For highly relativistic particles moving perpendicular to an external magnetic field, the emitted spectral power between $\nu$ and $\nu + d\nu$ is

$$P_{{\rm syn},\perp}(\nu)d\nu = \frac{2\sqrt{3}\,\pi e^2}{c}F_{\rm syn}(2\nu/3\nu_{\rm syn})d\nu, \tag{3.46}$$

where

$$F_{\rm syn}(x) = x\int_x^\infty K_{5/3}(\eta)d\eta \tag{3.47}$$

and $K_{5/3}$ is the modified Bessel function of order 5/3. $F_{\rm syn}(x)$ reaches its maximum value of 0.98 at $x = 0.29$. Well away from the maximum, $F_{\rm syn}(x)$ approaches the limits

$$F_{\rm syn}(x) = \frac{4\pi}{\sqrt{3}\,\Gamma(1/3)}(x/2)^{1/3} \ \text{for} \ x \ll 1 \tag{3.48}$$

and

$$F_{\rm syn}(x) = (\pi/2)^{1/2}x^{1/2}e^{-x} \ \text{for}\ x \gg 1, \tag{3.49}$$

where $\Gamma(1/3) = 2.68$ is the gamma function of argument 1/3. The synchrotron spectral power for pitch angles other than $\psi = \pi/2$ is, for $\psi \gg 1/\gamma$,

$$P_{\rm syn}(\nu) = \frac{2\sqrt{3}\,\pi e^2}{c}\sin\psi F_{\rm syn}[2\nu/(3\nu_{syn}\sin\psi)]. \tag{3.50}$$

*Synchrotron Radiation from a Power-law Distribution of Electrons*

The observed spectra of cosmic sources of synchrotron radiation are, in general, due to the superposition of radiation from an ensemble of electrons with a wide range of energies. The energy distribution of the radiating electrons can usually be described by a power law for $\gamma$ in a range $\gamma_1 < \gamma < \gamma_2$:

$$n_{\rm rel}(\gamma, \psi) = N_{\rm r}\gamma^{-q}g(\psi)/4\pi, \tag{3.51}$$

where $g(\psi)$ describes the angular distribution of the particles. For an isotropic distribution, $g(\psi) = 1$. The total spectral power emitted per unit volume $j_{syn}(\nu)$, assuming the radiation from the individual electrons is incoherent, is

$$j_{syn}(\nu) = \int\int n_{rel}(\gamma)g(\psi)P_{syn}(\nu, \psi)d\gamma d\Omega/4\pi. \quad (3.52)$$

When $\gamma_1$ and $\gamma_2$ are such that the end points of the integration over $\gamma$ do not contribute significantly, that is, when $\gamma_1^2 \ll (\nu/\nu_B) \ll \gamma_2^2$, then we can set $\gamma_1 = 0$ and $\gamma_2 = \infty$. For the case of local isotropy of the electron distribution, $g(\psi) = 1$, yielding, after integrating over $\gamma$,

$$j_{syn}(\nu)d\nu = 4\pi a(q)N_r r_0 eB(3\nu_B/2\nu)^{\alpha}d\nu. \quad (3.53)$$

where

$$\alpha = (q - 1)/2 \quad (3.54)$$

and $a_S(q)$ is a hairy expression involving gamma functions that depend on $q$. For the range of interest for most cosmic sources, $a_S(q)$ is a slowly varying function of $q$. Some representative values are $a_S(1.5) = 0.147$, $a_S(2) = 0.103$, $a_S(2.5) = 0.083$, $a_S(3) = 0.074$, and $a_S(5) = 0.087$. By a change of variables from $\nu$ to

$$x = 2\nu/3\nu_B, \quad (3.55)$$

the synchrotron emissivity can be written as

$$j_{syn}(x)dx = 24\pi a_S(q)N_r r_0^2 c(B^2/8\pi)x^{-\alpha}dx. \quad (3.56)$$

Synchrotron radiation from a power-law distribution of electrons accounts for the nonthermal radiation from supernova remnants over a broad range of frequencies, extending from the radio to the optical band. The production of X-rays of energy $\varepsilon_{keV}$ by synchrotron radiation requires

$$\gamma \sim \frac{3 \times 10^5 \varepsilon_{keV}^{1/2}}{(B \sin\psi)^{1/2}}, \quad (3.57)$$

corresponding to electron energies

$$W \sim \frac{150\varepsilon_{keV}^{1/2}}{(B \sin\psi)^{1/2}} \text{ GeV}. \quad (3.58)$$

The extremely high electron energies required to produce synchrotron X-rays has two effects: (1) radiative energy losses become important, and (2) the electrons involved in producing the X-rays may be near the upper limit of the electron energy distribution.

The radiative lifetime due to synchrotron radiation is

$$t_{\rm syn} = \frac{W}{P_{\rm syn}} \sim \frac{5 \times 10^8}{B^2\gamma} \text{ s.} \tag{3.59}$$

In terms of the photon energy $\varepsilon_{\rm keV}$,

$$t_{\rm syn} \sim \frac{1.7 \times 10^3}{B^{3/2}\varepsilon_{\rm keV}^{1/2}} \text{ s} = \frac{600}{B_{-4}^{3/2}\varepsilon_{\rm keV}^{1/2}} \text{ yr,} \tag{3.60}$$

where $B_{-4} = B/10^{-4}$.

For example, the electrons producing the ~100 keV X-rays in the ~$10^{-3}$ G magnetic field in the Crab Nebula have a radiative lifetime of ~2 years. This implies that the electrons producing the synchrotron radiation must be continuously accelerated on short timescales, a puzzle that was solved by the discovery of the Crab Nebula pulsar. Acceleration in supernova shock waves can also produce the energies required, e.g., Cas A, Tycho, Kepler, and SN 1006 (see the discussion in Chapter 5), but in this case the X-ray observations may be probing the limits of the acceleration process. The spectrum will then show an exponential decline above some critical frequency $\nu_c$,

$$\nu_c \sim \nu_B\gamma_m^2 = 3 \times 10^6\gamma_m^2 B \sin\psi \text{ Hz,} \tag{3.61}$$

corresponding to a critical photon energy

$$\varepsilon_{\rm keV,c} \sim 0.12(\gamma_{\rm m}/10^7)^2 B_{-4} \sin\psi = 3W_{\rm TeV,c}^2 B_{-4} \sin\psi, \tag{3.62}$$

where $W_{\rm TeV,c}$ is the critical electron energy in TeV, and $B_{-4} = 10^{-4}B$.

For more details on the theory of synchrotron radiation and its application to astrophysics, see the review articles by Blumenthal & Gould (1970) and Reynolds & Nowak (2003), and the books by Ghisellini (2013), Jackson (1975), Gould (2005), Ginzburg & Syrovatskii (1969), Landau & Lifshitz (1975), Rybicki & Lightman (1986), and Tucker (1975).

## 3.4 Brief Introduction to Quantum Radiation Processes

The classical approximation for radiation processes applies only when the frequency $\nu$ of the emitted radiation is much less than $W/h$. The classical formulas may be used for such astrophysically important applications as synchrotron radiation, electron scattering (to a certain extent), and low-frequency bremsstrahlung. However, many equally important processes such as line radiation from atoms and ions, high-frequency bremsstrahlung, and the photoelectric effect require a quantum-mechanical treatment. In this section, the basic concepts for quantum radiation processes are summarized.

### 3.4.1 Energy and Momentum of a Photon

The development of quantum mechanics was preceded by the quantum theory of radiation. Although many problems relating to the propagation of light could be understood within the framework of wave theory, a number of important phenomena relating to the emission and absorption of radiation remained unexplained. For instance, the energy spectrum of a blackbody derived on the basis of the wave theory was in contradiction with experiment. Quantum theory started in 1901 with Planck's hypothesis that radiation is emitted and absorbed in finite amounts called quanta or photons. With this assumption, the blackbody spectrum followed from thermodynamic arguments.

The energy $\varepsilon$ of a photon is proportional to the frequency $\nu$ of the oscillations of the radiation field:

$$\varepsilon = h\nu. \tag{3.63}$$

A few years later, Einstein showed the necessity of assigning to the photon a momentum $p = \varepsilon/c$. The direction of the momentum is given by the wave vector $k$, so

$$\mathbf{p} = \hbar\mathbf{k}. \tag{3.64}$$

The laws of conservation of energy and momentum in a collision between an electron, atom, or molecule with energy $W$ and momentum $\mathbf{p}$, and a photon with energy $h\nu$ and momentum $\hbar\mathbf{k}$ can be written as

$$h\nu_0 + W_0 = h\nu_1 + W_1, \tag{3.65}$$

$$\hbar\mathbf{k}_0 + \mathbf{p}_0 = \hbar\mathbf{k}_1 + \mathbf{p}_1. \tag{3.66}$$

The case $\nu_1 = 0$ refers to the absorption of a photon of energy $h\nu_0$; $\nu_0 = 0$ refers to the emission of a photon of energy $h\nu_1$. If both $\nu_0$ and $\nu_1$ are nonzero, the equations describe the scattering of radiation. Note that the quantum laws of conservation of energy and momentum are in conflict with both the wave and corpuscular concepts of radiation and cannot be interpreted within the framework of classical physics.

According to the wave theory, the energy $U_{em}$ of an electromagnetic wave is given by $U_{em} = (E^2 + B^2)/8\pi$, independent of the frequency $\nu$ of the wave. There is no general relation between the wave amplitude and the oscillation frequency, which would allow the energy of a photon to be related to the wave amplitude. The assumption that a photon is a particle located somewhere in space is also invalid. A photon, by definition, is associated with a monochromatic plane wave. Such a wave is a purely periodic process, infinite in both space and time, so the assumption that the photon is localized is in contradiction with the periodicity of the wave. Thus, the equations relating the energy and momentum of a photon to the frequency and wave vector imply that radiation must have both wave and corpuscular properties. These laws have been verified by numerous experiments. For example, in the photoelectric effect, the velocity of the photoelectron depends solely on the frequency $\nu$ of the light and not at all on the intensity of the incident light. These observations cannot be interpreted classically, but are easily understood in terms of the conservation laws for photons.

### 3.4.2 Blackbody Radiation

The elementary theory of radiation based on quantum ideas is due to Einstein. It is to some extent phenomenological, but his hypotheses are fully justified by modern quantum electrodynamics. Consider two states of a system, denoting one by the letter $m$ and the other by $n$. Let the energy of the first state be $W_m$ and that of the second be $W_n$. Assume $W_m > W_n$. A system can spontaneously jump from a higher state to a lower one, emitting in the process a photon with frequency

$$\nu = (W_m - W_n)/h. \tag{3.67}$$

The photon has a definite polarization and propagation vector $\mathbf{k}$ in a solid angle $d\Omega$. Summing over polarizations, the probability per second of a spontaneous transition $m \rightarrow n$ with emission of a photon of frequency $\nu = (W_m - W_n)/h$ into a solid angle $d\Omega$ can be defined as

$$dJ_{\mathrm{s}} = A_{mn} d\Omega/4\pi, \tag{3.68}$$

where $J_{\mathrm{s}}$ is the spontaneous emission probability and $A_{\mathrm{mn}}$ is the spontaneous emission coefficient.

Radiation incident on an atom can either be absorbed ($n \rightarrow m$) with a probability $dJ_{\mathrm{a}}$ or, if the atom is in an excited state, can induce a transition to a lower state ($m \rightarrow n$) with a probability $dJ_{\mathrm{i}}$. The probabilities for absorption and induced emission are proportional to the flux of incident photons:

$$dJ_{\mathrm{abs}} = B_{mn} I(\nu, \Omega)(d\Omega/4\pi), \tag{3.69}$$

$$dJ_{\mathrm{ind}} = B_{nm} I(\nu, \Omega)(d\Omega/4\pi), \tag{3.70}$$

where the coefficients have been summed over polarizations.

The emission and absorption probabilities depend on the nature of the system and can be calculated using quantum theory. However, general relations between the coefficients can be derived from thermodynamic arguments. Consider conditions in which there is equilibrium between the rate of absorptions and the rate of emissions. The relative number of atoms in states $n$ and $m$ is given by the Boltzmann formula:

$$N_n/N_m = (q_n/q_m)e^{(W_m - W_n)/kT}, \tag{3.71}$$

where $q_n$ and $q_m$ are the statistical weights of states $n$ and $m$. In equilibrium, the number of absorptions is equal to the number of emissions, so

$$N_m[A_{mn} + I(\nu, \Omega)B_{mn}] = N_n B_{nm} I(\nu, \Omega). \tag{3.72}$$

Using $W_m - W_n = h\nu$, the Boltzmann formula for $N_n/N_m$, and recognizing that as $T \rightarrow \infty$, $I(\nu, \Omega) \rightarrow \infty$, yield

$$I(\nu, \Omega) = \frac{A_{mn}/B_{mn}}{(e^{h\nu/kT} - 1)} \tag{3.73}$$

in the classical limit $h\nu \ll kT$, so

$$I \rightarrow (A_{mn}/B_{mn})\frac{kT}{h\nu}. \tag{3.74}$$

The relation between $A_{mn}$ and $B_{mn}$ can be determined from the classical limit. According to the equipartition theorem of classical statistical mechanics, the average energy associated with each independent vibration of the field in equilibrium is $kT$. The number of independent waves with frequency between $\nu$ and $\nu + d\nu$ that can fit into a sphere of radius $X$ is $(8\pi X^3/c^3)\nu^2 d\nu$, so the energy density of the radiation is

$$U(\nu) = (8\pi/c^3)kT\nu^2, \tag{3.75}$$

The intensity for an isotropic equilibrium radiation field is thus

$$I(\nu) = \frac{cU(\nu)}{4\pi} = 2kT\frac{\nu^2}{c^2}, \tag{3.76}$$

which gives

$$A_{mn}/B_{mn} = 2h\nu^3/c^2, \tag{3.77}$$

and the intensity of radiation in thermal equilibrium, also known as blackbody radiation, or Planck's law, is

$$I_{\rm BB}(\nu) = \frac{(2h\nu^3/c^2)}{e^{(h\nu/kT)} - 1}. \tag{3.78}$$

Blackbody radiation reaches a maximum at a photon energy given by

$$\varepsilon = h\nu_{\rm m} = 2.82\,kT, \tag{3.79}$$

i.e., at a photon energy slightly larger than the average energy ($= 3kT/2$) of the electrons. The corresponding peak frequency is

$$\nu_{\rm m} = 5.88 \times 10^{10}T \ \ {\rm Hz}, \tag{3.80}$$

so, e.g., a blackbody with a temperature $T = 10$ MK will produce radiation peaking at $\nu_{\rm m} = 5.88 \times 10^{17}$ Hz, or a photon energy $\varepsilon = 2.44$ keV. The intensity of blackbody radiation integrated over all frequencies is

$$I_{\rm BB} = \int_0^{\infty} I_{\rm BB}(\nu)d\nu = (\sigma/\pi)T^4, \tag{3.81}$$

where $\sigma$, the Stefan–Boltzmann constant, is given by

$$\sigma = 2\pi^5k^4/(15h^3c^2) = 5.67 \times 10^{-5}\ {\rm gs^{-3}\ K^{-4}}. \tag{3.82}$$

The energy density of the radiation is

$$U = 4\pi I/c = 4\sigma T^4/c. \tag{3.83}$$

The flux of radiation $dF/d\Omega$ across a unit area of surface at an angle $\theta$ to the normal of the surface is

$$dF/d\Omega = I\cos\theta. \tag{3.84}$$

The total flux from a blackbody is obtained by integrating over all solid angles in a hemisphere:

$$F_{\rm BB} = 2\pi\int_0^{\pi/2} I\cos\theta\sin\theta\, d\theta = \sigma T^4. \tag{3.85}$$

The luminosity of a blackbody of radius $R$ is

$$L_{\rm BB} = 4\pi R^2 F_{\rm BB} = 4\pi R^2\sigma T^4. \tag{3.86}$$

For example, the thermal radiation from a neutron star of radius $R$ = 10 km and temperature $T$ = 10 MK would be

$$L = 7.12\times 10^{36}{\rm erg\ s^{-1}}. \tag{3.87}$$

This is approximately the luminosity of the first extrasolar X-ray source discovered by Giacconi et al. (1962) and gave rise to the suggestion that this and similar sources were hot neutron stars (Shklovsky 1967). The sources did in fact turn out to be neutron stars, although the actual situation is more complicated than a uniformly radiating blackbody.

### 3.4.3 The Schrödinger Equation and Fermi's Golden Rule

In the classical limit, the position and momentum of an electron can be specified without any uncertainty at each point of space and at every moment. Once the external fields are specified, radiation processes can in principle be computed using Newton's second law and Maxwell's equations. In the quantum limit, we must take into account the dual wave–particle nature of particles and photons, as encapsulated by the concept of the de Broglie wavelength for particles. This can be done by the introduction of the wave function $\Psi$, which can be used to find the probability that a particle is at a given place with a given momentum at a given time, and replacing the energy equation for an electron with charge $-e$ moving in an electric field described by a potential $\phi$, e.g.,

$$W = (p^2/2m) - e\phi, \tag{3.88}$$

with an equation for the wave function.

The equation for the wave function is obtained by replacing the momentum and energy by differential operators that operate on the wave function:

$$\mathbf{p} \rightarrow -i\hbar\nabla \tag{3.89}$$

and

$$W \rightarrow i\hbar\partial/\partial t, \tag{3.90}$$

where $i = \sqrt{-1}$. The result is

$$-(\hbar/2m)\nabla^2\Psi - e\phi\Psi = \mathcal{H}\Psi = i\hbar\frac{\partial\Psi}{\partial t}, \tag{3.91}$$

where the Hamiltonian operator

$$\mathcal{H} = -(\hbar/2m)\nabla^2 - e\phi \tag{3.92}$$

has been introduced, by way of analogy with the Hamiltonian formulation of classical mechanics. This wave equation is called the Schrödinger equation for an electron in an electromagnetic field described by a scalar potential $\phi$. It is valid for a single nonrelativistic electron with no spin.

The quantization of the radiation field is more complicated but fortunately, most of the results needed to study cosmic X-ray production can be derived from a semiclassical treatment in which the behavior of the particles is treated quantum mechanically, but the wave–particle nature of photons is taken into account by means of the simple theory discussed in the previous two sections.

The wave function represents the state of a system in the sense that the probability $dw$ that a measurement of the various coordinates $x$, $y$, $z$ will give results lying in the range $x$ to $x + dx$, etc. at time $t$ is given by

$$dw(x, y, z, t) = |\Psi(x, y, z, t)|^2 dxdydz \tag{3.93}$$

and satisfies the normalization requirement

$$\int \Psi^*\Psi dxdydz = \int |\Psi|^2 dxdydz = 1, \tag{3.94}$$

where $\Psi^*$ is the complex conjugate of $\Psi$, and the integral is over all values of the coordinates.

States in which an isolated physical system has a definite energy are particularly important for our considerations. Although actual atomic systems and other systems are not quite isolated because they are coupled to the electromagnetic field, this coupling is weak for many problems, and the description of the system in terms of energy states is useful as a close approximation.

Functions representing energy states satisfy the equation

$$i\hbar\frac{\partial\Psi}{\partial t} = W\Psi. \tag{3.95}$$

The wave function $\Psi$ can be written as

$$\Psi(x, y, z, t) = u(x, y, z)e^{(i/\hbar)W_n t}, \tag{3.96}$$

where $u(x, y, z)$ is a function of the coordinates but not the time.

For an electron moving in an electric field, $u$ satisfies the equation

$$-(\hbar^2/2m)\nabla^2 u - e\phi u = Wu. \tag{3.97}$$

A general solution of this equation can be written in terms of energy eigenstates. The complete set of eigenfunctions $u(x, y, z)$ associated with the energies $W_k$ form a complete orthogonal set and as such satisfy the condition

$$\int u_k^* u_m dxdydz = \delta_{km}. \tag{3.98}$$

The wave function can be expressed in terms of the $u_k$ eigenfunctions:

$$\Psi(x, y, z, t) = \Sigma_k a_k u_k e^{-(i/\hbar)W_k t}, \tag{3.99}$$

where the coefficients $a_k$ are determined by the conditions of the problem. The Schrödinger equation can be solved exactly in only a few cases, so in general approximate methods, called perturbation theory, must be used.

For problems involving radiation, the Hamiltonian is time dependent, so time-dependent perturbation theory must be used. In this approach, the Hamiltonian is split into an unperturbed part and a smaller perturbing part:

$$\mathcal{H} = \mathcal{H}_0 + \mathcal{H}', \tag{3.100}$$

where the unperturbed part is usually

$$\mathcal{H}_0 = -(\hbar^2/2m)\nabla^2 - e\phi \tag{3.101}$$

and $\phi$ is the potential for some constant electric field, such that

$$\mathbf{E} = -\nabla\phi. \tag{3.102}$$

The perturbing part of the Hamiltonian is composed of terms containing first-order terms in the vector potential $\mathbf{A}$, where

$$\mathbf{B} = \nabla \times \mathbf{A} \tag{3.103}$$

and

$$\mathcal{H}' = -(ie\hbar/mc)(\mathbf{A} \cdot \nabla). \tag{3.104}$$

The effects of the electron spin are of a higher order.

The transition probabilities for a given process are obtained by expanding the wave function in terms of the eigenfunctions $u_k$ of the operator $\mathcal{H}'$:

$$\Psi(t) = \Sigma_k a_k u_k e^{-(iW_k t)/\hbar}. \tag{3.105}$$

Next, multiply the equation for $\Psi(t)$ by $\Psi^*\mathcal{H}_0$ and integrate over the coordinates to obtain

$$\int \Psi^*\mathcal{H}_0\Psi dxdydz = \Sigma_k |a_k(t)|^2 W_k. \tag{3.106}$$

The left-hand side of this equation is the average value of $\mathcal{H}_0$. The right-hand side is the sum of terms, each of which is a possible value of $\mathcal{H}_0$ multiplied by a term, which is the probability of that value of $\mathcal{H}_0$:

$$|a_f(t)|^2 = \text{probability that the system goes from state } i \text{ to } f \text{ in time } t. \tag{3.107}$$

The $a_k$ are determined from the wave equation by substituting

$$a_f(t) = a_f^0(t) + a_f'(t), \tag{3.108}$$

in the Schrödinger equation, assuming that the perturbation is zero for $t < 0$ and $t > \tau$, and keeping terms of first order:

$$a_f' = (2\pi/i\hbar)(\mathcal{H}'_{fi}(\omega_{fi})), \tag{3.109}$$

where

$$\omega_{f0} = (W_f - W_i)/\hbar \tag{3.110}$$

and

$$\mathcal{H}'_{fi} = \int u_f^* \mathcal{H}' u_i dxdydz. \tag{3.111}$$

The transition probability exhibits a resonance behavior. It is nonzero only for perturbations that contain frequencies equal to the eigenfrequencies of the system. For example, in the case of absorption of radiation, only those photons having an energy equal to $W_f - W_i$ will be absorbed.

For many problems, the characteristic time, $\tau$, for the interaction is much greater than the characteristic timescale for the system, which is $\sim\omega_{fi}^{-1}$. Then, the transition probability per unit time is

$$w = |a_f(t)|^2/\tau = (2\pi/\hbar)|\mathcal{H}'_{fi}|^2\delta(W_f - W_i). \tag{3.112}$$

This expression, often referred to as Fermi's "Golden Rule," or "Golden Rule Number Two," to use Fermi's expression, has many applications in quantum physics. The $\delta$-function guarantees energy conservation in the process. When the final states are very closely spaced, or continuously distributed, the transition probability per unit time becomes

$$w_{fi} = (2\pi/\hbar)|\mathcal{H}'_{fi}|^2(dn/dW), \tag{3.113}$$

where $dn/dW$ is the density of final states.

### 3.4.4 Absorption and Emission Probabilities

The probability of absorption of electromagnetic radiation is calculated using the perturbed Hamiltonian operator:

$$\mathcal{H}' = -(ie\hbar/2mc)(\mathbf{A}\cdot\nabla + \nabla\cdot\mathbf{A}) = -(ie\hbar/mc)(\mathbf{A}\cdot\nabla), \tag{3.114}$$

where the Coulomb gauge condition $\nabla\cdot\mathbf{A} = 0$ has been used.

Substituting this Hamiltonian operator into Fermi's Golden Rule expression for the transition, expanding the vector potential in terms of plane waves of frequency $\omega$

and amplitude $\mathbf{A}_0$, and expressing $|A_0(\omega)|^2 = (4c\tau/\omega^2)I(\omega)$, the transition probability per unit time for absorption is

$$w_{ab} = (4\pi^2 e^2/m^2 c)|\tilde{p}_{fi}|^2 \frac{I(\omega_{fi})}{(\hbar\omega_{fi})^2}, \tag{3.115}$$

where

$$\tilde{p}_{fi} = (e^{i\mathbf{k}\cdot\mathbf{r}}\hat{\mathbf{e}} \cdot \mathbf{p}) \tag{3.116}$$

and $\hat{\mathbf{e}}$ is a unit vector specifying the direction of polarization. The expression for $w_{ab}$ implies that the Einstein coefficient for absorption is, taking account of two directions of polarization, which gives a factor of 2,

$$B_{if} = \frac{8\pi^2 e^2 |\tilde{p}_{fi}|^2}{m^2 c\hbar^2 \omega_{fi}^2}. \tag{3.117}$$

The corresponding coefficient for spontaneous emission $A_{fi}$ is

$$A_{fi} = \frac{\hbar\omega_{fi}^3}{2\pi^2 c^2} B_{fi} = 4\alpha_f \omega_{fi} \frac{|\tilde{p}_{fi}|^2}{m^2 c^2}, \tag{3.118}$$

where

$$\alpha_f = \frac{e^2}{\hbar c} \tag{3.119}$$

is the fine structure constant.

### 3.4.5 Quantum-mechanical Dipole Approximation

The expression for the transition probabilities involve an integral which contains a factor $e^{i\mathbf{k}\cdot\mathbf{r}}$. The calculations are simplified for problems involving nonrelativistic velocities. In such cases,

$$\mathbf{k} \cdot \mathbf{r} \sim (\omega/c)(\hbar/p) \sim (W/pc) \sim (v/c) \ll 1, \tag{3.120}$$

so we can set $e^{i\mathbf{k}\cdot\mathbf{r}} = 1$, and $\tilde{\mathbf{p}}_{fi} = \mathbf{p}_{fi}$. Using the Schrödinger equation and integration by parts, $\mathbf{p}_{fi}$ can be written as

$$\begin{aligned} \mathbf{p}_{fi} &= \int u_f^* \mathbf{p} u_i dxdydz = \int u_f^*[\mathbf{r}\mathcal{H}_0 - \mathcal{H}_0\mathbf{r}]u_i dxdydz \\ &= (W_f - W_i) \int u_f^* \mathbf{r} u_i dxdydz. \end{aligned} \tag{3.121}$$

The transition probability for spontaneous emission into a solid angle $d\Omega$ is then

$$dw_s/d\Omega = (\alpha_f/2\pi c^2)|(\hat{\mathbf{e}} \cdot \mathbf{r})_{fi}|^2 \omega_{fi}^3 = (\alpha_f/2\pi c^2)|\mathbf{r}_{fi}|^2 \omega_{fi}^3 \sin^2\theta, \tag{3.122}$$

where $\theta$ is the angle between the direction of observation and the dipole moment vector $e\mathbf{r}$. Integrating over $d\Omega = 2\pi \sin\theta \, d\theta$ yields

$$w_{fi} = (4\alpha_f/3)|\mathbf{r}_{fi}|^2 \omega_{fi}^3. \tag{3.123}$$

The radiated power is

$$P = \hbar\omega w_{fi} = (4e^2/3c)|\mathbf{r}_{fi}|^2\omega_{fi}^4. \tag{3.124}$$

This formula is almost identical to the classical formula for radiation from an oscillating dipole described by $\mathbf{d} = e\mathbf{r} = \mathbf{d}_0 \sin \omega t$. The classical formula is recovered by replacing $|\mathbf{r}_{fi}|^2$ by the time average of the square of $\mathbf{r}$, $\overline{r^2} = r^2/2$, and using $\ddot{r} = ea = \omega^2 r$ Then,

$$dP = \frac{2a^2}{3c^3} \tag{3.125}$$

as in the classical case. For more on quantum radiation processes, see the books by Bethe & Jackiw (1997), Gould (2005), Rybicki & Lightman (1986), and Tucker (1975).

## 3.5 Scattering of Radiation by Free Electrons

When electromagnetic radiation is incident upon a charge, the charge is accelerated and emits radiation. This process is termed scattering. As the acceleration is inversely proportional to the mass of the charge and the power radiated is inversely proportional to the square of the mass, scattering by electrons (Figure 3.2), called Compton scattering, is much more important than scattering by nucleons.

### 3.5.1 Kinematics of Compton Scattering

Consider a reference frame in which the electron is initially at rest. The energy and momentum conservation equations are (Section 3.1)

$$h\nu_0 - h\nu_1 = mc^2(\gamma - 1) \tag{3.126}$$

and

$$\hbar\mathbf{k}_0 - \hbar\mathbf{k}_1 = \mathbf{p}_1 = m\gamma\mathbf{v}. \tag{3.127}$$

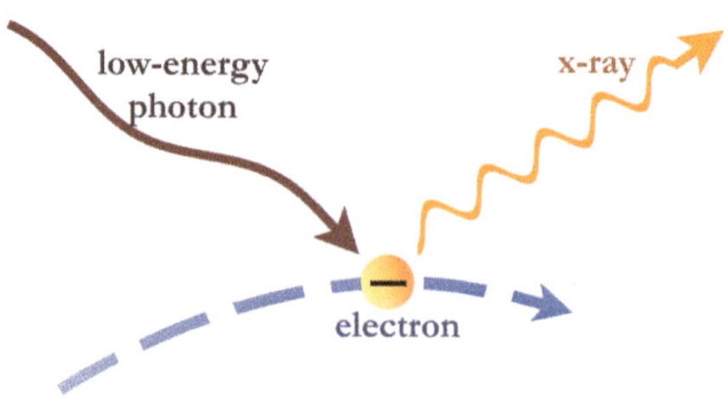


**Figure 3.2.** Production of high-energy photons by electron scattering. (Courtesy of NASA/CXC/SAO.)

Subtracting the squares of these two equations and using the relation $k_1 = 2\pi\nu_1/c$, etc., yields

$$h\nu_1 = \frac{h\nu_0}{[1 + (h\nu_0/mc^2)(1 - \cos\Theta)]}, \tag{3.128}$$

where $\Theta$ is the photon-scattering angle. In terms of photon wavelengths, the Compton-scattering relation can be written as

$$\lambda_1 = \lambda_0 + \lambda_c(1 - \cos\Theta), \tag{3.129}$$

where $\lambda_c = \hbar/mc$ is the Compton wavelength. The energy acquired by the scattered charge is

$$W = h\nu_0 - h\nu_1 = \frac{h\nu_0(h\nu_0/mc^2)(1 - \cos\Theta)}{[1 + (h\nu_0/mc^2)(1 - \cos\Theta)]}. \tag{3.130}$$

For photon energies $h\nu \ll mc^2$,

$$h\nu_0 - h\nu_1 = h\nu_0(h\nu_0/mc^2)(1 - \cos\Theta), \tag{3.131}$$

so, to first order in $(h\nu_0/mc^2)$, $\nu_1 = \nu_0$, i.e., the photon energy is unchanged in the scattering. Note that Planck's constant $h$ does not appear, indicating that the classical description is adequate in this limit. The scattering in this limit is called Thomson scattering (see Section 3.5.2).

For scattering from moving electrons, a transformation from the moving frame to the rest frame of the electrons, where the scattering conserves energy, and back to the moving frame yields the result, for incident and scattering angles $\psi_0$ and $\psi_1$ of the electrons,

$$\frac{\nu_1}{\nu_0} = \frac{(1 - \beta\cos\psi_0)}{[1 - \beta\cos\psi_1 + (h\nu/mc^2)(1 - \cos\Theta)]}, \tag{3.132}$$

which for $h\nu \ll mc^2$ becomes

$$\frac{\nu_1}{\nu_0} = \frac{1 - \beta\cos\psi_0}{1 - \beta\cos\psi_1}. \tag{3.133}$$

This equation shows that collisions of a photon with a moving electron can significantly boost the photon's energy. For example, for a head-on collision, corresponding to $\psi_0 = \pi$ and $\psi_1 = 0$,

$$\frac{\nu_1}{\nu_0} = \frac{1 + \beta}{1 - \beta} = \frac{(1 + \beta)^2}{(1 - \beta^2)} = \gamma^2(1 + \beta)^2. \tag{3.134}$$

Scattering in which the photon gains energy from the electron is often referred to as inverse Compton scattering.

### 3.5.2 Thomson Scattering

The limit of $h\nu_0 \ll mc^2$ is called the Thomson limit. The efficiency of Thomson scattering of radiation can be expressed in terms of the differential cross section, defined as the ratio of the power radiated per unit solid angle, divided by the incident energy flux. For a plane-polarized wave,

$$\frac{d\sigma}{d\Omega} = \frac{dP/d\Omega}{S} = \frac{r_0^2 c \sin \Theta^2 |E|^2/4\pi}{c|E|^2/4\pi} = r_0^2 \sin \Theta^2. \tag{3.135}$$

For unpolarized radiation, this expression must be averaged over all directions of the electric vector **E** in a plane perpendicular to the direction of propagation of the incident wave. The result is

$$\frac{d\sigma}{d\Omega} = \frac{r_0^2}{2}(1 + \cos^2\theta), \tag{3.136}$$

where $\theta$ is the scattering angle, i.e., the angle between the direction of the incident wave and the scattered wave. Integration over solid angle yields the Thomson cross section:

$$\sigma_{\rm T} = \frac{8\pi}{3} r_0^2. \tag{3.137}$$

### 3.5.3 Radiation Pressure and the Eddington Limit

When an electron scatters an electromagnetic wave, it absorbs momentum from the wave moving in a direction defined by the unit vector $\hat{\mathbf{n}}$:

$$d\mathbf{p}/dt = \mathbf{F}_{\rm rad} = \frac{\sigma_{\rm T} c |E|^2}{4\pi c}\hat{\mathbf{n}} = \sigma_{\rm T} U_{\rm rad}\hat{\mathbf{n}}, \tag{3.138}$$

where in the last equality we have introduced the quantity

$$U_{\rm rad} = \frac{|E|^2}{4\pi} \tag{3.139}$$

to denote the average energy density of the electromagnetic wave. The radiation pressure due to this force plays an important role in many astrophysical phenomena, ranging from the early universe, to the formation of stars, stellar structure, and the growth of black holes, where radiation pressure can limit the rate of accretion onto neutron stars or black holes.

In the accretion process, gravitational energy is released as a particle falls toward an object of mass $M$. This energy is transformed into radiant energy through interaction with the plasma around the object. The radiation pushes back on the infalling plasma, halting the accretion when the radiation force acting on the infalling matter becomes strong enough to offset the inward pull of gravity. Both the radiation force and the gravitational force are different for electrons and nuclei. This imbalance causes the charges to separate, setting up an electric field that opposes this separation. Very quickly, the system will come to an equilibrium in which the sum of the forces acting on the electrons and nuclei is the same. For a plasma composed of electrons and protons, the forces are, for purely radial flow,

$$F_e = -(GMm/r^2) + \sigma_T U_{rad} + eE_i \tag{3.140}$$

and

$$F_p = -(GMm_p/r^2) + \sigma_T U_{rad} - eE_i, \tag{3.141}$$

so, $F_e = F_p$ when $eE_i = -[(GMm_p/r^2] + \sigma_T U_{rad}$ and the net force on an electron or proton is

$$F_e = F_p = [(GMm_p/r^2) - \sigma_T U_{rad}]/2. \tag{3.142}$$

This equation shows that the net force on electrons will be zero for a critical radiation energy density,

$$U_{rad} = \frac{GMm_p}{\sigma_T r^2}. \tag{3.143}$$

Using the relation

$$U_{rad} = \frac{L}{4\pi r^2 c} \tag{3.144}$$

yields an expression for the critical luminosity called the Eddington luminosity, or the Eddington limit,

$$L_{Edd} = \frac{4\pi G m_p c M}{\sigma_T} = 1.3 \times 10^{38} M/M_\odot \quad \mathrm{erg\ s^{-1}}. \tag{3.145}$$

The Eddington limit is an extremely useful concept in many astrophysical contexts. It does, however, have its limits, based on the assumptions of isotropic and steady-state flow. See, for example, the discussion of ultraluminous X-ray sources in Chapters 6 and 7.

### 3.5.4 Compton Energy Exchange in a Hot Plasma

In addition to exerting pressure, collisions between photons and electrons can lead to significant energy exchange. The average energy transfer from photons to electrons per scattering for a cool plasma with $kT \ll h\nu$ is

$$(\Delta \bar{W}) = h\nu(h\nu/m_e c^2), \quad \text{for } kT \ll h\nu. \tag{3.146}$$

In the other limit, where $kT > h\nu$, the hot electrons can transfer energy to the photons. Averaging the expression for the photon energy gain from moving electrons over a Maxwellian distribution of electron velocities yields the average energy gain per scattering,

$$h\nu_1 - h\nu_0 = \frac{h\nu_0}{mc^2}(4kT - h\nu_0). \tag{3.147}$$

The amount of energy gained by the photon and the emergent spectrum of photons traversing a hot plasma of radius $R$ depends on the average number of scatterings $\mathcal{N}$ experienced by the photons and the energy gain per scattering:

$$\frac{d(h\nu)}{d\mathcal{N}} = \frac{h\nu_0}{mc^2}(4kT - h\nu_0). \tag{3.148}$$

For $kT \gg h\nu$, the energy of an escaping photon is

$$h\nu_f = h\nu_0 e^{(4kT/mc^2)\mathcal{N}}. \tag{3.149}$$

The value of $\mathcal{N}$ depends on the electron-scattering optical depth of the plasma:

$$\tau_T = n_e \sigma_T R = R/\lambda_T, \tag{3.150}$$

where

$$\lambda_T = \frac{1}{n_e \sigma_T} \tag{3.151}$$

is the mean free path for Thomson scattering. When $\tau_T \ll 1$, corresponding to $R \ll \lambda_T$, the plasma is optically thin to electron scattering, and only a small fraction of the photons are scattered,

$$\mathcal{N} = \tau_T \text{ for, } \tau_T \ll 1, \tag{3.152}$$

and the energy of a photon escaping the plasma after $\mathcal{N}$ scatterings is

$$h\nu_f = h\nu_0 e^{(4kT/mc^2)\mathcal{N}} \approx h\nu_0[1 + \tau_T(4kT/mc^2)]. \tag{3.153}$$

When $\tau_T \gg 1$, the plasma is optically thick, and photons diffuse out of the plasma, so

$$\mathcal{N} \approx \tau_T^2 \tag{3.154}$$

and

$$h\nu_f = h\nu_0 e^{(4kT/mc^2)\mathcal{N}} \approx h\nu_0 e^{(4kT/mc^2)\tau_T^2}. \tag{3.155}$$

The process of modifying a photon spectrum through Compton scattering is called Comptonization. It is customary to define the Comptonization parameter, $y_c$, to characterize the process:

$$y_c = \frac{4kT}{mc^2}\max(\tau_T, \tau_T^2). \tag{3.156}$$

The boost, or amplification, in a photon's energy by Comptonization in traversing a hot plasma is then

$$h\nu_f/h\nu_0 = e^{y_c}. \tag{3.157}$$

An interesting feature of Comptonization is that it can produce a power-law spectrum, even if the electrons doing the scattering have a thermal, or Maxwellian, distribution of energies. Consider the case when $\tau_T < 1$. Suppose we have a source radiating a spectrum $I_0(\nu)$ with a peak intensity $I_0$ at $\nu_0$ that is embedded in a cloud of hot electrons of density $N_e$, temperature $T$, and radius $R$, with $\tau_T < 1$. A fraction $1 - \tau_T$ of the photons will escape from the cloud without scattering, producing a spectrum of peak intensity $I_0 e^{-\tau_T} \approx I_0(1 - \tau_T)$. A fraction $\tau_T$ of the photons will scatter once, producing a spectrum of peak intensity $I_0\tau_T$ at a frequency $\nu_1 = \nu_0[1 + (4kT/mc^2)]$, a fraction $\tau_T^2$ with a peak intensity $I_0\tau_T^2$ at a frequency $\nu_2 = \nu_1[1 + (4kT/mc^2)] = \nu_0[1 + (4kT/mc^2)]^2$, etc. The slope of the spectrum $(I(\nu) \propto \nu^{-\alpha_c})$ on a log–log plot is approximately

$$\alpha_c = \frac{\ln(I_{n+1}/I_n)}{\ln(\nu_{n+1}/\nu_n)} = \frac{\ln(\tau_T^{n+1}/\tau_T^n)}{\ln[1 + (4kT/mc^2)]} = \frac{\ln \tau_T}{\ln[1 + (4kT/mc^2)]}. \quad (3.158)$$

In terms of $y_C$, the equation for the spectral index $\alpha_c$ becomes

$$\alpha_c = \frac{\ln \tau_T}{\ln[1 + (y_C/\tau_T)]} = \frac{\ln \tau_T}{\ln(y_C + \tau_T) - \ln \tau_T}. \quad (3.159)$$

The expressions for $\alpha_c$ illustrate how the spectrum of the radiation emerging from the source hardens as the temperature or optical depth increases, as expected, as the photons gain more energy from the hot plasma. For example, for a thin hot plasma with $kT \approx 125$ keV, $\alpha_c = 1.74$; for $y_c = \tau = 0.5$, $\alpha_c = 1$.

The Comptonization process for low optical depths, often referred to as unsaturated Comptonization, is considered a likely mechanism for producing the power-law components with an exponential decline at high energies that are characteristic of the X-ray spectra of accreting black holes. See Chapters 7 and 8 and the following references: Pozdnyakov et al. (1983), Shapiro et al. (1976), Sunyaev & Titarchuk (1980), Rybicki & Lightman (1986), and Titarchuk (1994).

When $\tau > 1$, the spectrum of photons emerging from a hot plasma must be computed from the Boltzmann equation. For nonrelativistic electrons, and small energy exchanges in collisions, the Boltzmann equation can be expressed as the Kompaneets equation (Cooper 1971; Weymann 1965):

$$\frac{\partial n_\gamma}{\partial t} = (n_e\sigma_T c)\frac{kT}{mc^2}\frac{1}{x^2}\frac{\partial}{\partial x}\left[x^4\left(n_\gamma + n_\gamma^2 + \frac{\partial n_\gamma}{\partial x}\right)\right] + Q(x) - \frac{n_\gamma}{t_{esc}}, \quad (3.160)$$

where $Q(x)$ describes the production of photons, $t_{esc}$ is time taken for photons to escape the source,

$$x = h\nu/kT, \quad (3.161)$$

and $n_\gamma$ is the photon phase-space density, defined such that the total number of photons in a volume $V$ at time $t$ is

$$N_\gamma(t) = \int\int n_\gamma dV d^3p. \quad (3.162)$$

$n_\gamma$ is related to specific intensity $I(\nu)$:

$$I(\nu) = \frac{2h\nu^3 n_\gamma}{c^3}. \tag{3.163}$$

For a steady state with $\partial n_\gamma / \partial t = 0$ and $Q = 0$, and in many cases of interest, $n_\gamma \ll 1$ so the $n_\gamma^2$ term can be ignored. Using the definition of $y_c$ yields a simplified equation:

$$\frac{1}{x^2}\frac{\partial}{\partial x}\left[x^4\left(n_\gamma + \frac{\partial n_\gamma}{\partial x}\right)\right] = \frac{n_\gamma}{y_c}. \tag{3.164}$$

When $x \ll 1$, corresponding to $h\nu \ll kT$, the $n_\gamma$ term can be ignored, so

$$\frac{1}{x^2}\frac{\partial}{\partial x}\left(x^4\frac{\partial n_\gamma}{\partial x}\right) = \frac{n_\gamma}{y_c}. \tag{3.165}$$

This equation has the solution

$$n_\gamma \propto x^q, \tag{3.166}$$

with

$$q = -\frac{3}{2} - \left(\frac{9}{4} + \frac{4}{y}\right)^{1/2}. \tag{3.167}$$

The Comptonized intensity spectrum is thus

$$I(\nu) \propto n_\gamma \nu^3 \propto \nu^{3+q} \propto \nu^{-\alpha_c}, \tag{3.168}$$

with

$$\alpha_c = -(3 + q) = -\frac{3}{2} + \left(\frac{9}{4} + \frac{4}{y}\right)^{1/2}. \tag{3.169}$$

Note that, as $y_c \to 1$, $\alpha_c \to 1$.

For photon energies such that $x \gg 1$, ($h\nu \gg kT$), the $\left(n_\gamma + \frac{\partial n_\gamma}{\partial x}\right)$ term dominates, so

$$n_\gamma \propto e^{-x} \tag{3.170}$$

and

$$I(\nu) \propto n_\gamma \nu^3 \propto \nu^3 e^{-h\nu/kT}. \tag{3.171}$$

### 3.5.5 Compton Scattering by Relativistic Electrons

For highly relativistic electrons, $\gamma \gg 1$. In this limit, for an isotropic distribution of incident photons, the average frequency of scattered photons is (see Section 3.5.1)

$$\langle \nu_1 \rangle = (4/3)\gamma^2 \nu_0, \tag{3.172}$$

so very high frequencies can be produced by scattering from relativistic electrons.

The rate of energy loss due to Compton scattering by highly relativistic electrons traversing a distribution of photons is just the average of the energy loss per collision times the number of collisions per second. This is equal to the scattered power $P_C$, which, for highly relativistic electrons, is a factor $\gamma^2$ greater than the energy per unit time removed from the ambient radiation field. For an isotropic distribution of photons, this is

$$dW_e/dt = P_C = (4/3)\sigma_T c \gamma^2 U_{rad}, \tag{3.173}$$

Substituting for $\sigma_T$ yields

$$P_C = (32\pi/9)\gamma^2 r_0^2 c U_{rad}. \tag{3.174}$$

This expression is identical to the expression (Equation (3.40)) for synchrotron losses with the replacement of the factor $B^2/8\pi$ by $U_{rad}$:

$$P_C = P_{syn}(U_{rad}/U_{mag}). \tag{3.175}$$

In general, both radiation and magnetic fields will be present, so relativistic electrons can produce radiation and lose energy through both Compton scattering and synchrotron radiation.

The time for a relativistic electron in a source of radius $R$ with luminosity $L$ to lose a significant fraction of its energy by Compton scattering is

$$t_C \approx \frac{mc^2\gamma}{P_C} = \frac{R}{c}\frac{3\pi}{\gamma C_c}, \tag{3.176}$$

where

$$C_c = \frac{\sigma_T L}{mc^3 R} = \frac{4\pi m_p}{m}\frac{L/L_{Edd}}{R/R_g} \tag{3.177}$$

is a dimensionless parameter called the "compactness" of the source. The compactness is basically a measure of the lifetime of electrons to cool by Compton scattering compared to the light-crossing time $R/c$.

Two settings where Compton scattering by relativistic electrons is particularly relevant to X-ray astronomy are the scattering of the cosmic microwave radiation by relativistic electrons in the extended lobes or jets of strong radio sources, and the Compton scattering of synchrotron radiation by the same electrons that produce the synchrotron radiation. The radiation in the latter instance is called synchrotron self-Compton radiation.

The total spectral power of the Compton-scattered radiation, assuming an isotropic distribution of electrons, is

$$j_C(\nu) = (1/4\pi)\int n_{\text{rel}}(\gamma)P_C(\nu)d\gamma, \tag{3.178}$$

where $P_C(\nu)$ involves an integral over the ambient radiation field $n_{\text{rad}}(\nu_0)$,

$$P_C(\nu) = 8\pi r_0^2 ch \int f(\nu/4\gamma^2\nu_0)n_{\text{rad}}(\nu_0)d\nu_0 \tag{3.179}$$

with

$$f(x) = x + 2x^2 \ln x + x^2 - 2x3 \tag{3.180}$$

and

$$n_{\text{rel}}(\gamma) = N_r\gamma^{-q}. \tag{3.181}$$

When the ambient radiation field $n_{\text{rad}}(\nu_0)$ is blackbody radiation and the end points of the $\gamma$ integration can be replaced by zero and infinity,

$$j_C(\nu) = \pi r_0^2 chb(q)N_r(kT/hc)^3(kT/h\nu)^{(q-1)/2}, \tag{3.182}$$

where $b(q)$ ranges between 5 and 10 as $n$ ranges from 2.0 to 3.0. The Compton-scattered radiation has the same power-law index as for synchrotron radiation.

For synchrotron self-Compton radiation, the ambient field is the synchrotron radiation field, which is usually a power law over a wide range of frequencies. For a uniform source of radius $R$,

$$j_C(\nu) = (N_r\sigma_T R)j_{\text{syn}}(\nu)G(\alpha), \tag{3.183}$$

where $G(\alpha) \approx 10$. Note that, although the shape of the spectrum is identical to that of synchrotron radiation, the frequency is shifted up by a factor $\sim\gamma^2$ (Blumenthal & Gould 1970):

$$\nu_C \sim \gamma^2\nu_{\text{syn}} \sim \nu_{\text{syn}}^2/\nu_B. \tag{3.184}$$

## 3.6 Bremsstrahlung

Bremmsstrahlung (German for "braking radiation" or "deceleration radiation") is produced by the deceleration of a charged particle when deflected by another charged particle, typically an electron by an atomic nucleus. The massive target particle is accelerated only slightly and radiates a negligible amount compared to the electron, so the process can be treated as the interaction of an electron with a fixed force field.

Bremsstrahlung (Figure 3.3) is an important process for the production of cosmic X-radiation from hot plasmas. In this setting, the energy of the emitted X-rays is comparable to the energy of the particles, so a quantum description is required.

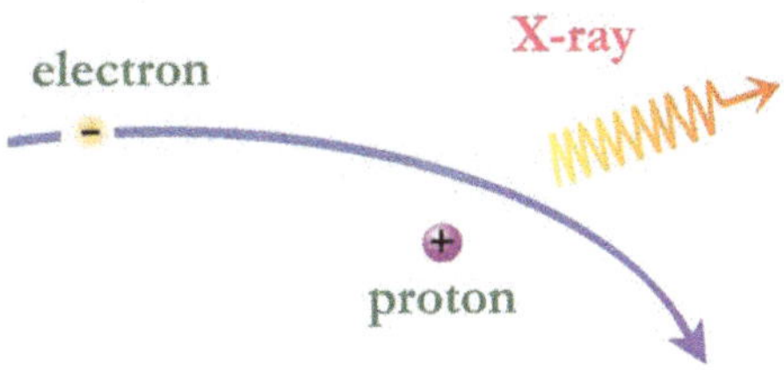


**Figure 3.3.** Bremsstrahlung. (Courtesy of NASA/CXC/SAO.)

However, a semiclassical description suffices to give insight into the process and is a good approximation.

When an electron with velocity v collides with a nucleus of charge $Ze$, most of the radiation occurs at the distance of closest approach, i.e., the impact parameter $b$, when the electron experiences an acceleration $a \sim Ze^2/mb^2$ over a time $\delta t \sim b/\mathrm{v}$. From Larmor's formula, the energy radiated in the collision is

$$\Delta W(b) \sim \frac{2e^2a^2}{3c^3}\frac{b}{\mathrm{v}}. \tag{3.185}$$

The power radiated by an electron flux $n_e v$ incident upon a medium with an ion density $N_Z$ is then

$$\frac{dP_\mathrm{B}}{dV} \sim 2\pi n_\mathrm{e} N_Z \mathrm{v} \int \Delta W(b) b db \sim (4\pi/3) n_\mathrm{e} N_Z Z^2 r_0^2 c (e^2/b_{\mathrm{min}}), \tag{3.186}$$

where $b_{\mathrm{min}}$ is the minimum impact parameter. Quantum effects come in through the requirement that the minimum impact parameter must be of the order of the spread in the wave packet of the electron as estimated by the uncertainty principle:

$$b_{\mathrm{min}} \sim \hbar / m_\mathrm{e} \mathrm{v} = \lambda_\mathrm{d}, \tag{3.187}$$

where $\lambda_\mathrm{d}$ is the de Broglie wavelength. Then,

$$\frac{dP_\mathrm{B}}{dV} \sim (4\pi/3) n_\mathrm{e} N_Z Z^2 \alpha_f m c^2 r_0^2 \mathrm{v} = (8\pi/3) n_\mathrm{e} N_Z Z^2 \alpha_f^3 \lambda_\mathrm{d}^2 W \mathrm{v}, \tag{3.188}$$

where $W$ is the initial energy of the electron.

The cross section is

$$\sigma \sim \frac{(dP/dV)}{n_\mathrm{e} N_Z W} = (8\pi/3) Z^2 \alpha_f^3 \lambda_d^2. \tag{3.189}$$

Note that the cross section is $\sim Z^2\alpha_f^3$ times the minimum impact cross section $\sim b_{\mathrm{min}}^2$, due to the probability of emitting a photon. An exact calculation yields

$$\frac{dP_\mathrm{B}}{dV} = \frac{16\pi}{3\sqrt{3}} n_\mathrm{e} N_Z Z^2 \alpha_f^3 \lambda_d^2 W \mathrm{v} g_\mathrm{B} \tag{3.190}$$

and

$$\sigma = \frac{16\pi}{3\sqrt{3}} Z^2 \alpha_f^3 g_B \lambda_d^2, \tag{3.191}$$

where $g_B \sim 1$ is the bremsstrahlung Gaunt factor (Gaunt 1930; Karzas & Latter 1961; Nozawa et al. 1998).

Bremsstrahlung is emitted in a flat spectrum up to frequencies $\sim v/b$, where it cuts off exponentially:

$$\frac{dP_B}{dVd\varepsilon} = \frac{16\pi}{3\sqrt{3}} Z^2 \alpha_f^3 \lambda_d^2 g_B v n_e N_Z. \tag{3.192}$$

For a Maxwellian distribution of electron velocities,

$$f(v)dv = 4\pi(m/2\pi kT)^{3/2} e^{-(mv^2/2kT)} v^2 dv; \tag{3.193}$$

the bremsstrahlung spectrum is given by

$$\frac{dP_B}{dVd\varepsilon} = \int \frac{dP_B}{dVd\varepsilon} f(v)dv. \tag{3.194}$$

Substituting for $dP_B/dVd\varepsilon$ and $f(v)$,

$$\frac{dP_B(\varepsilon, T)}{dVd\varepsilon} = \frac{32\pi}{3\sqrt{3}} n_e N_Z Z^2 \alpha_f^3 a_0^2 \left(\frac{I_H}{kT}\right)\left(\frac{2kT}{\pi m}\right)^{1/2} g_B(T, \varepsilon) e^{-\varepsilon/kT}, \tag{3.195}$$

where $a_0 = (\hbar/mc\alpha_f)$ is the Bohr radius, and $I_H = e^2/2a_0$ is the ionization potential of the hydrogen atom. The temperature-averaged Gaunt factor can be approximated by

$$\langle g_B(T, \varepsilon)\rangle \approx (kT/\varepsilon)^{0.4} \tag{3.196}$$

for $E \sim kT$.

The characteristic photon energy emitted by bremsstrahlung is

$$\varepsilon = h\nu \sim kT, \tag{3.197}$$

so temperatures $kT \sim$ keV are required to produce keV photons. Such temperatures are found in a wide range of cosmic conditions where bremsstrahlung can be the dominant emission mechanism: accretion disks around neutron stars and black holes, stellar coronas, supernova shock waves, and clusters of galaxies. Integrating over all photon energies yields the total bremsstrahlung power per unit volume:

$$\frac{dP_B}{dV} = \frac{32\pi}{3\sqrt{3}} n_e N_Z Z^2 \alpha_f^3 a_0^2 \left(\frac{I_H}{kT}\right)\left(\frac{2kT}{\pi m}\right)^{1/2} kT g_B(T). \tag{3.198}$$

Upon inserting numerical values for the atomic constants,

$$\frac{dP_B}{dV} = 1.4 \times 10^{-27} T^{1/2} n_e N_Z Z^2 g_B \text{ erg cm}^{-3}\text{ s}^{-1} \tag{3.199}$$

where $g_B(T) \approx 1.2$ (Nozawa et al. 1998).

In a plasma with a mixture of ions, the total bremsstrahlung emission involves the sum

$$S_B = \Sigma n_e N_Z Z^2. \tag{3.200}$$

For the cosmic abundances of the elements, $S_B \approx 1.4 n_e^2$ and

$$\frac{dP_B}{dV} = 2.4 \times 10^{-27} T^{1/2} n_e^2 = n_e^2 \Lambda_B \text{ erg cm}^{-2}\text{ s}^{-1}. \tag{3.201}$$

The cooling time of a plasma due to bremsstrahlung is

$$t_B \approx \frac{3n_e kT}{(dP_B/dV)} \approx \frac{2 \times 10^{11} T^{1/2}}{n_e} \text{ s}. \tag{3.202}$$

## 3.7 Radiative Recombination

In the radiative recombination process, a free electron makes a transition to a bound state of energy $-I_{(Z-1),n}$, with the emission of a photon of energy

$$\varepsilon = W_i + I_{(Z-1),n}. \tag{3.203}$$

The photon energy can assume any value greater than $I_{(Z-1),n}$, so the recombination spectrum is continuous, with edges or discontinuities at $\varepsilon = I_{(Z-1),n}$.

A gas that has been suddenly ionized, say by a flash of ionizing radiation, and is recombining will produce significant line emission as electrons that have recombined to excited states cascade down to the ground state. In addition, the recombination edges may appear like lines because of the low temperature. The computation of the entire recombination spectrum involves consideration of recombination to the individual levels $I_{(Z-1),n}$. In the limit where $W_f$ and $I_{(Z-1),n} \to 0$, the cross section for bremsstrahlung and recombination should be equal. This equality can be used to obtain an expression for the recombination cross section for recombination to a quantum level $n$ in hydrogenic ions,

$$\sigma_R(n) = (d\sigma_B/d\varepsilon)(d\varepsilon/dn) = (d\sigma_B/d\varepsilon) g_R \frac{2Z^2 I_H}{n^3}. \tag{3.204}$$

With the introduction of the recombination Gaunt factor $g_R \sim 1$, this expression is accurate even for captures to the ground state (Karzas & Latter 1961).

The total radiative recombination power spectrum is obtained by integrating over a Maxwellian distribution of electron velocities, summing over all ions, and levels $n$ for which $I_{Z,z,n} > \varepsilon$, where $z$ is the ionic charge. For nonhydrogenic ions, a reasonable approximation is to set $I_{Z,z-1,n} = Z^2 I_H / n^2$ and multiplying by a factor $\chi_n$, which represents the incompleted fraction of shell $n$. This yields

$$\frac{dP_{\mathrm{RR}}}{dVd\varepsilon} = 2.8 \times 10^{-6} n_{\mathrm{e}} N_{\mathrm{H}} T^{-3/2} e^{-\varepsilon/kT} X \text{ erg cm}^{-3}\,\mathrm{s}^{-1}, \tag{3.205}$$

where

$$X_{\mathrm{r}} = \Sigma_{Z,z,n} \frac{N_{Z,z}}{N_Z} \frac{N_Z}{N_{\mathrm{H}}} \chi_n g_{\mathrm{R}} \frac{Z^4}{n^3} e^{(Z^2 I_{\mathrm{H}}/n^2 kT)}. \tag{3.206}$$

For a gas with cosmic abundances, recombination with oxygen ions plays a large role in the sum, and $X_{\mathrm{r}}(T, E) \sim 30$. In general, for conditions under which the ionization state is determined by a balance of collisions and radiative recombination (collisional ionization equilibrium), radiative recombination dominates bremsstrahlung for $kT \ll 0.1$ keV and bremsstrahlung dominates for $kT \gg 1$ keV.

In a hot, optically thin thermal plasma that is out of ionization equilibrium, the strength of the radiative recombination features depends on the circumstances. In a plasma recently heated by a shock wave, there will be relatively fewer recombinations than in equilibrium, whereas a plasma rapidly cooling due to expansion will show strong radiative recombination features, since most of the ions are recombining.

## 3.8 X-Ray Line Emission

### 3.8.1 Timescales and Assumptions

The computation of X-ray line radiation from a hot plasma is complicated by the important role played by many different ions and atomic transitions, especially at temperatures below a few kiloelectronvolts. The various processes involved occur on different timescales, which allows for some simplification. In general, the following ordering of timescales holds:

$$\tau_{\mathrm{gs}} \sim \tau_{\mathrm{ms}} \gg \tau_{\mathrm{ex}} \gg \tau_{\mathrm{au}} \gg \tau_{\mathrm{ee}}, \tag{3.207}$$

where $\tau_{\mathrm{gs}}$ is the lifetime of the ground-state ion population (typically due to ionization or recombination rates), $\tau_{\mathrm{ms}}$ is the lifetime of metastable states; $\tau_{\mathrm{ex}}$ is the lifetime of ordinary excited states (those with energy lower than the ionization potential of the ion), $\tau_{\mathrm{au}}$ refers to auto-ionizing states (those with energy above the ionization potential, which can ionize spontaneously), and $\tau_{\mathrm{ee}}$ is the thermalization timescale for electron–electron collisions to create a Maxwellian energy distribution in the plasma. A common assumption for astrophysical plasmas is that the electrons have a Maxwellian energy distribution.

X-ray emission lines from low-density hot plasmas are produced primarily as a result of electron collisional excitation followed by radiative decay back to the ground state. The timescale for radiative decay is much less than the collisional excitation time, so the rate of line emission is governed by the collisional excitation rate.

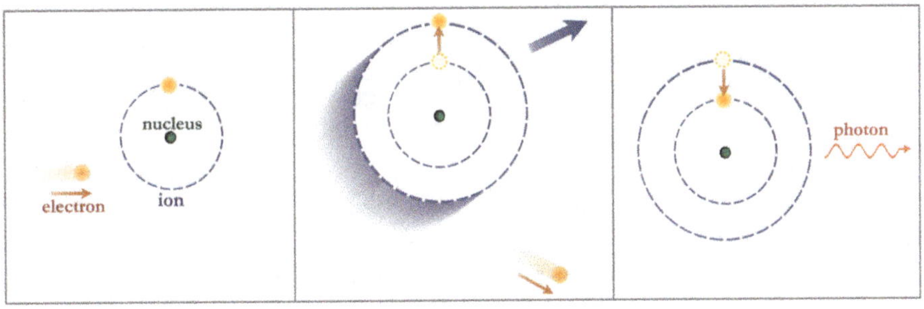


**Figure 3.4.** Collisional excitation followed by line emission. (Courtesy of NASA/CXC/S.Lee.)

### 3.8.2 Line Emission Following Collisional Excitation

A fairly accurate estimate of the cross section for collisional excitation (Figure 3.4) can be obtained by combining the classical treatment for energy transferred in a Coulomb collision between two electrons with the quantum-mechanical requirement that energy can be transferred only in discrete amounts. For a hydrogenic ion with a nuclear charge $Z$, the cross section for the collision with an impact parameter $b$, is

$$\sigma(b) \sim \pi b^2, \tag{3.208}$$

The momentum impulse is

$$\Delta p \sim (dp/dt)(\Delta t) \sim (Ze^2/b^2)(2b/\mathrm{v}) = 2Ze^2/b\mathrm{v}, \tag{3.209}$$

so the energy transferred is

$$\Delta W(b) \sim (\Delta p)^2/2\,m \sim 2Z^2e^4/(m\mathrm{v}^2b^2) \tag{3.210}$$

and

$$\sigma(\Delta W) \sim \pi b^2 \sim \frac{2\pi Z^2 e^4}{m\mathrm{v}^2\Delta W} = \frac{4\pi a_0^2 Z^2 I_{\mathrm{H}}^2}{W\Delta W} = 4\pi(a_0/Z)^2\frac{Z^4 I_{\mathrm{H}}^2}{W\Delta W}. \tag{3.211}$$

A quantum-mechanical calculation (nonrelativistic Born approximation) valid for conditions typical of X-ray-emitting plasma ($W \ll mc^2$ and $Z\alpha_f \ll 1$) yields

$$\sigma(\Delta W) = \frac{8\pi}{\sqrt{3}}\pi a_0^2\frac{Z^2 I_{\mathrm{H}}^2}{W\Delta W}\frac{f(Z, nn')\overline{g(Z, nn', W)}}{q_n}, \tag{3.212}$$

where $\Delta W$ is the excitation energy, $f(Z, nn')$ is the dipole oscillator strength for the transition, $\overline{g(Z, nn', W)}$ is the effective Gaunt factor, and $q_n = (2L + 1)(2S + 1)$ is the statistical weight of the initial state. For energies of interest for X-ray emission, $\overline{g(Z, nn', W)} \approx 0.2$.

Cross sections are often expressed in terms of the collision strength $\Omega(Z, nn', W)$, which is more generally useful because transitions that are forbidden in the dipole approximation can be important for collisional excitation:

$$\sigma(\Delta W) = \pi(a_0/Z)^2\left(\frac{\Omega(Z, nn', W)}{q_n}\right)\frac{Z^2 I_{\mathrm{H}}}{W}, \tag{3.213}$$

where

$$\Omega(Z, nn', W) = \frac{8\pi}{\sqrt{3}}\frac{I_H}{\Delta W}\overline{fg(W)}. \tag{3.214}$$

For $1s \rightarrow 2p$ transitions in hydrogenic ions at threshold $Z^2\Omega \approx 1.5$ and varying slowly with energy, $\Omega \propto W^y$ with $y \approx 0.6$ (Golden et al. 1981). For a Maxwellian distribution of electron velocities, the power emitted per unit volume due to collisional excitations of level $n'$ ions with charge $Z$ in the ground state $n$ is

$$\frac{dP_L(T)}{dV} = n_e N_Z \Delta W \int \sigma(v) v f(v) dv. \tag{3.215}$$

Using the definition for the collision strength yields

$$\frac{dP_L(T)}{dV} = 4\pi n_e N_Z a_0^2 Z^2 I_H \left(\frac{\Delta W}{kT}\right)\left(\frac{2kT}{\pi m}\right)^{1/2} \frac{\Omega(Z, nn')}{q_n} e^{-\Delta W/kT}. \tag{3.216}$$

Here, $\Omega(Z, nn')$ is the average of $\Omega$ over incident particle energies. For most applications, a good approximation is $\Omega(Z, nn') = \Omega(Z, nn', 1.5\Delta W)$ (Bely & van Regemorter 1970). Inserting values for the various atomic constants,

$$\frac{dP_L(T)}{dV} = 1.9 \times 10^{-16} n_e N_Z T^{-1/2} \left(\frac{\Delta W}{I_H}\right) \frac{\Omega(Z, nn')}{q_n} e^{-\Delta W/kT} \text{erg cm}^{-3}\,\text{s}^{-1}. \tag{3.217}$$

For nonhydrogenic ions, the hydrogenic expression is used, $N_Z$ is replaced by $N_{Z,z}$, the fraction of ions of nuclear charge $Z$ with ionic charge $z$, and the complexities of the exact atomic physics calculations are swept under the rug of the collision strength. For more discussion on the electron impact cross sections for hydrogenic, He-like, and complex ions, see the online databases, e.g., AtomDB (Foster et al. 2012), SPEX (Kaastra et al. 1996), and Chianti (Dere et al. 2019; Young et al. 2016). In some cases, the line is produced by a decay back to a state other than the ground state, in which case the excitation energy is not equal to the energy of the emission line, and a branching ratio giving the fraction of decays leading to the state of interest must be included in the expression for the radiated power.

A group of lines deserving special attention is the so-called He-like triplet (actually a quartet, but two of the lines are so close together they are usually lumped together as one line). The strongest line is the resonance line, an allowed $1s2p \rightarrow 1s^2$ transition. The forbidden $1s2s \rightarrow 1s^2$ line has a similar strength to the resonance line for the low-density conditions in supernova remnants and the intracluster medium, but it can be relatively enhanced in recombining plasmas or relatively diminished in higher density plasma like stellar coronal loops. In between both lines is the intercombination line. For high electron densities, e.g., $n_e \sim 10^9$ cm$^{-3}$ for O VII, the collisional excitation rate out of the excited state for the forbidden line is comparable to the radiative transition rate, so the relative strength of the forbidden line can provide a useful density diagnostic, especially for the study of stellar coronae, as discussed in Chapter 4.

### 3.8.3 Ionization Equilibrium and Spectrum of a Hot Plasma

The relative populations of the ions are not described by the thermal Boltzmann distribution but by a balance of the microprocesses of ionization and recombination. In many cases of interest, photoionization is unimportant, so the relative populations of the various stages of ionization are determined by balancing ionization by electron collisions with recombination processes. This is referred to as collisional ionization equilibrium. Because it applies to solar and stellar coronae, it is also sometimes called coronal ionization equilibrium. Collisional ionization equilibrium also applies to the hot intracluster medium.

The relevant recombination processes are radiative and dielectronic recombination. Dielectronic recombination occurs when an electron recombines with the ion in an excited state and excites a second electron while doing so. The ion is left in a highly excited state, which may then autoionize (thereby inverting the dielectronic recombination and converting the process into a simple scattering event), or it may radiatively decay when one of the excited electrons radiatively decays, creating a satellite line. A satellite line is so named because it will be at a slightly longer wavelength than the normal transition from an electron in that energy level. The reverse process, dielectronic ionization, can occur when a collision excites two electrons and one of the electrons is ejected as the other one returns to the ground state (Burgess & Summers 1969).

A similar process, called auto-ionization, occurs for Li-like and more complex atoms when a collision excites a K-shell electron to a higher level, leaving two electrons in an excited state with an energy greater than the ionization energy. The resulting atom is unstable, and a radiationless Auger transition can occur. The K-shell vacancy is filled by one of the outer shell electrons and another electron is ejected from an outer shell.

Balancing the rate for collisional ionization with the recombination rate,

$$c_{z-1}N_{z-1}n_{\rm e} = \alpha_z N_z n_{\rm e} \tag{3.218}$$

or

$$N_{z-1}/N_z = \alpha_z/c_{z-1}, \tag{3.219}$$

where $c_{z-1}$ is the collisional ionization rate, $\alpha_z$ denotes the sum of radiative and dielectronic recombination rates, and the subscript for the element $Z$ has been suppressed (Gaetz & Salpeter 1983). The collisional ionization rate is approximately

$$c_{z-1} \sim 4\pi a_0^2 \frac{kT}{m}\left(\frac{I_{\rm H}}{I_{z-1}}\right)^2 e^{-I_{z-1}/kT}. \tag{3.220}$$

The radiative recombination rate

$$\alpha_{\rm RR} \sim 5\alpha^3 \pi a_0^2 \left(\frac{kT}{m}\right)^{1/2}\left(\frac{I_z}{kT}\right), \tag{3.221}$$

and the dielectronic recombination rate is $\sim(10^9/Z^{-2}T)\alpha_{\rm RR}$.

For plasma temperatures and densities of interest for X-ray astronomy, the abundance of a given stage of ionization peaks, and line emission from that ionization stage peaks at temperatures in the range $kT \sim (0.3 - 0.8)I_Z$.

The calculation of the ionization balance and X-ray line emission from a hot collisionally dominated, optically thin plasma requires knowledge of the atomic transition rates and energies of the ions involved, as well as a computer code to calculate the interplay between the different rates. These codes amount to living documents, as they are continuously revised based on the availability of new atomic data made available by advances in both laboratory measurements and theoretical computations (Urdampilleta et al. 2017; Arnaud & Rothenflug 1985; Bryans et al. 2009; Mazzotta et al. 1998; Burgess & Seaton 1964; Kaastra et al. 2017, 2008; Landini et al. 1997; Sutherland & Dopita 1993; Landini & Fossi 1991;, Mewe et al. 1985).

Two widely used databases for the analysis of cosmic X-ray spectra are SPEX 2008 v3.0.5 (Kaastra et al. 1996) and AtomDB (Smith et al. 2001; Foster et al. 2012). These databases are updated on a regular basis to incorporate experimental and theoretical advances. Line emissivities from the latest AtomDB 3.0.9 release (2017 September 4) are now online.[1]

Line emission dominates the X-ray spectrum for temperatures $\lesssim$ 10 MK, even though the principal contributors are ions of elements (O, Ne, Mg, Si, S, and Fe) with abundances a factor of a thousand or more lower than the abundances of the H and He nuclei that dominate the bremsstrahlung and recombination emission. This is because of the much larger cross section for collisional excitation as compared with bremsstrahlung and radiative recombination.

The radiative cooling for a plasma with cosmic abundances is primarily due to line emission from iron ions in the $n = 3$ shell for 1 MK $< T <$ 3 MK and the $n = 2$ shell ions for 3 MK $< T <$ 10 MK. The radiative cooling time in this interval can be roughly approximated as

$$t_{\rm rad} \approx \frac{3kT}{n_{\rm e}\Lambda} \approx \frac{300T^{1.7}}{n_{\rm e}}\ {\rm s}. \tag{3.222}$$

This should be regarded as a lower limit, as nonequilibrium effects tend to reduce the radiative cooling rate. Comparison with the bremsstrahlung cooling time (Section 3.6) shows that for temperatures $\gtrsim$30 MK, bremsstrahlung dominates. However, even at high temperatures, line emission still plays an important role in determining the temperature and assessing departures from the cosmic abundances of the various elements.

Note that the application of collisional ionization equilibrium assumes that the age of the plasma, $t_{\rm age}$, is much greater than either the ionization, $t_{\rm ion}$, or recombination, $t_{\rm rec}$, timescales. There are cases, e.g., supernova remnants or black hole accretion disks and winds, where this assumption is not valid and the plasma can be in either an ionizing, or underionized state, $t_{\rm ion} \gg t_{\rm age}$, or a recombining

[1] http://www.atomdb.org/Webguide/webguide.php

$t_{rec} \gg t_{age}$ state. Because both the ionization and recombination timescales are proportional to $1/n_e$, the density-weighted timescale $n_e t_{age}$ is used to determine how close a plasma is to equilibrium. Calculations by Smith & Hughes (2010) show that $n_e t_{age} \sim 10^{11}$–$10^{12}$ for $T$ in the range $10^6$–$10^8$ K.

### 3.8.4 Line Broadening

Spectral lines are broadened because of three mechanisms: (1) the natural line width due to quantum-mechanical uncertainty in the energy $E$ of levels with finite lifetimes; (2) the reduction in the effective lifetime of a state due to collisions, sometimes called pressure broadening; and (3) Doppler or thermal broadening due to thermal or large-scale turbulent motion of the ions. The width due to quantum-mechanical uncertainty can be estimated from

$$\Delta\nu_{QM} \sim \frac{\Delta W}{h} \sim \frac{1}{2\pi\Delta t} \sim 10^9 Z^4 \text{ s}^{-1}, \tag{3.223}$$

where the last estimate uses the transition probability for hydrogenic ions. Collisional, or pressure, broadening occurs because the lifetime of the excited states is limited by the lifetime due to collisions. Roughly,

$$\Delta\nu_{coll} \sim \frac{1}{2\pi t_{coll}} \sim 10^{-11} n_e Z^2 T^{1/2} \text{ s}^{-1}, \tag{3.224}$$

which shows that, for temperatures characteristic of the emission of X-rays, extremely high densities are required for collisional broadening to be important. The motion of an emitting atom or ion relative to the observer will introduce a shift in frequency of the line due to the Doppler effect. For light emitted in the $x$-direction from a nonrelativistic ion, the shift is

$$\frac{\nu - \nu_0}{\nu} = \frac{v_x}{c}, \tag{3.225}$$

where $v_x$ is the velocity of the ion relative to the observer (positive velocity in the direction of the observer). For a Maxwellian distribution of ion velocities, the corresponding intensity distribution with frequency is

$$G(\nu) = \frac{e^{-(\nu-\nu_0)^2)/(\Delta\nu_D^2)}}{\sqrt{\pi}(\Delta\nu_D)}, \tag{3.226}$$

where

$$\frac{\Delta\nu_D}{\nu} = \left(\frac{2kT}{M_Z c^2}\right)^{1/2} = 4.3 \times 10^{-7} T^{1/2} A^{-1/2}, \tag{3.227}$$

and $M_Z$ is the mass of the ion Z, and $A_Z$ is its atomic weight. For conditions characteristic of cosmic X-ray sources, Doppler broadening is the dominant line-broadening mechanism. In addition to thermal motions, mass motions can cause Doppler broadening. Examples are the expansion of a supernova remnant, e.g.,

Tycho's supernova remnant, or turbulent motions within a source. In the latter case, the line broadening can estimated as a combination of thermal and turbulent Doppler broadening:

$$\frac{\Delta\nu_{\mathrm{D,turb}}}{\nu} = \left(\frac{2kT}{M_Z c^2} + \frac{\mathrm{v}_{\mathrm{turb}}^2}{c^2}\right)^{1/2}. \tag{3.228}$$

### 3.8.5 Charge Exchange

Another process for producing X-ray line emission is charge exchange (CX; Figure 3.5). In a typical charge-exchange process, a highly ionized ion with charge $Z$ will pick up an electron from a neutral atom or molecule. The ion will then be in an excited state and will decay with the emission of an X-ray characteristic of an excited state. For example,

$$O^{+8} + H \rightarrow O^{+7} + H^{+} \rightarrow O^{+7} + H^{+} + \gamma. \tag{3.229}$$

The cross section for CX is large (Wargelin et al. 2008),

$$\sigma_{\mathrm{CE}} \sim Z \times 10^{-15}\ \mathrm{cm}^2, \tag{3.230}$$

but the rate is usually relatively small because hydrogen and helium atoms, the most abundant targets for the process, are almost fully ionized under conditions appropriate for the emission of X-rays. However, charge exchange is important for solar wind interactions with planetary atmospheres and comets (Kuntz et al. 2015; Lisse et al. 2001; Mullen et al. 2017), and, as the sensitivity of X-ray telescopes has increased, has recently attracted interest in other contexts, such as supernova remnants, the Galactic Center, starburst galaxies (Zhang et al. 2014), and galaxy clusters (Gu et al. 2018).

## 3.9 Photoionization and X-Ray Absorption

### 3.9.1 Photoionization

X-rays emitted by sources outside the solar system can undergo absorption before they are detected by X-ray telescopes. Absorption occurs in the interstellar medium

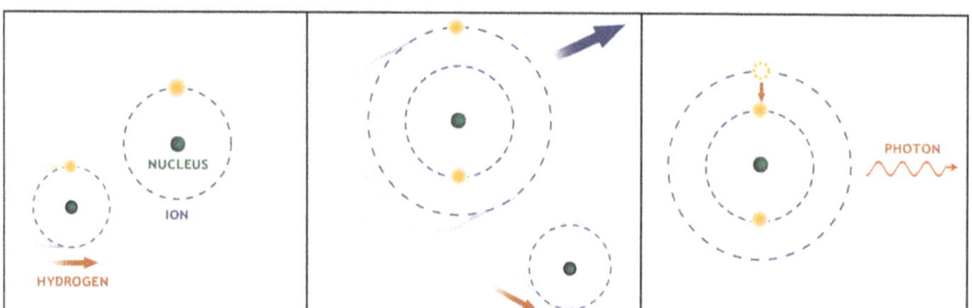


**Figure 3.5.** Photon emission following charge exchange. (Courtesy of NASA/CXC/S.Lee.)

and in many cases by matter surrounding the X-ray source. In both instances, the primary absorbing mechanism is photoionization (Figure 3.6). If a photon of energy $\varepsilon$ is incident on an ion with an ionization potential $I$, the ion can absorb the photon, and one of the electrons will be ejected with an energy

$$W = \varepsilon - I. \tag{3.231}$$

Because $W$ is a continuous variable, absorption is possible for a continuous range of photon energies. The cross section can be calculated exactly in various limits, depending on how far the energy of the incident electron is above the threshold for ionization. For hydrogenic ions, the cross section is

$$\sigma_{\rm PI} = \frac{64\pi}{3\sqrt{3}}\alpha_f(a_0/Z)^2(I/\varepsilon)^3 g_{\rm PI}(\varepsilon, n), \tag{3.232}$$

where $n$ is the principal quantum number and $g_{\rm PI}(\varepsilon, n)$ is the photoionization Gaunt factor, which has been tabulated by Karzas & Latter (1961). For more general ions, see Burgess & Seaton (1960). The optical depth $\tau_{\rm PI}$ due to photoionization is obtained from a sum of all the atoms and ions in the line of sight:

$$\tau_{\rm PI} = \Sigma_i \mathcal{N}_{\rm i}\sigma_{\rm i}(\varepsilon), \tag{3.233}$$

where $\mathcal{N}_{\rm i} = \int N_{\rm i}dx$ is the column density of ionic species $N_{\rm i}$ along the line of sight. Assuming no emission along the line of sight, the observed intensity $I_{\rm obs}(\varepsilon)$ is related to the emitted intensity $I_0(\varepsilon)$ by

$$I_{\rm obs}(\varepsilon) = I_0(\varepsilon)e^{-\tau_{\rm PI}}. \tag{3.234}$$

The photoionization cross section has two important characteristics: a threshold at $\varepsilon = I$ for each ion, and a strong decline, roughly $\propto \varepsilon^{-3}$, above threshold. This leads to the appearance of edges in the observed spectra. These absorption edges, as well as the magnitude and energy dependence of the observed absorption, are important diagnostic tools that can provide information about the interstellar medium and the environment of the source. *Chandra* and *XMM-Newton* spectrometer observations have led to major advances in the analysis of photoelectric absorption, with improved models for absorption edge structure, absorption due to resonance lines, absorption by dust, and depletion of elements in dust grains (see Section 6.4 and references cited therein).

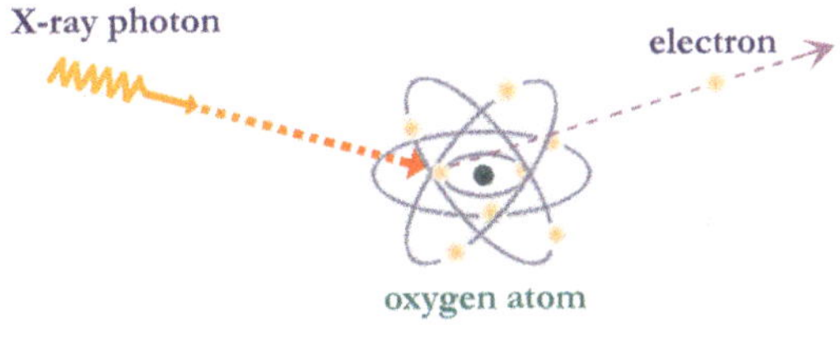


**Figure 3.6.** Absorption of an X-ray by photoionization. (Courtesy of NASA/CXC/S.Lee.)

### 3.9.2 Photoionization Equilibrium

In cases where a strong ionizing photon flux is present, collisional ionization equilibrium likely will not apply. Important examples in the context of X-ray astronomy are active galactic nuclei, X-ray binaries, and cataclysmic variables. In these situations, specific radiative transfer models may be needed to calculate the ionization, excitation, and heating of surrounding plasma by the ionizing photons. If the gas is optically thin, it is possible to scale the ionization conditions using the ionization parameter $\xi$:

$$\xi = \frac{L}{n_e r^2}, \tag{3.235}$$

where $L$ is the ionizing luminosity (Tarter et al. 1969).

For X-ray sources, $\xi$ is generally in the range $10^{-1} \leqslant \xi \leqslant 10^3$. The modeling code XSTAR developed and updated by Kallman and colleagues (Kallman & Bautista 2001) is commonly used to analyze photoionized emission and absorption spectra in the 0.1–10 keV range in active galaxies and X-ray binaries. Updates are available online.[2]

Cloudy, developed by Ferland and his collaborators, is more focused on modeling optical emission, but also has the capability to treat X-ray spectra (Ferland et al. 2017). Plasmas in photoionization equilibrium are characterized by a much lower temperature than for collisionally ionized plasmas for a given degree of ionization (Kallman & Bautista 2001). For example, in a collisionally ionized plasma, a temperature of ≈5 MK is required to ionize 90% of the O VIII ions, whereas a photoionized plasma exposed to an external ionizing radiation flux of the same spectrum will be heated to only ∼0.3 MK.

### 3.9.3 Absorption Lines

As with optical spectroscopy, absorption lines play an important role in the X-ray band and reveal information about absorbing material. Whether or not a given line appears in emission or absorption depends on the environment of the source. Generally speaking, if there is not a significant amount of cooler gas around a source, the lines will appear in emission. If, however, there is a substantial amount of cooler gas, as can appear in mass outflows from a source, absorption lines will be present. The intensity of the line is described by the equation of transfer,

$$I_L = I(0)e^{-\tau(0,\, s)} + \int_0^s (j_L + j_c)e^{-\tau(s',\, s)}ds'. \tag{3.236}$$

The optical depth is composed of line and continuum absorption coefficients. Continuum absorption is due to photoionization, so the optical depth is

$$\tau = \tau_{PI} + \tau_L, \tag{3.237}$$

[2] https://heasarc.gsfc.nasa.gov/xstar/xstar.html

where $\tau_L$ is the optical depth due to line absorption,

$$\tau_L = \int N\sigma_L ds, \tag{3.238}$$

and $\sigma_L$ is the line absorption cross section,

$$\sigma_L = \int \sigma_{L,\varepsilon}(\varepsilon)d\varepsilon. \tag{3.239}$$

$\sigma_{L,\varepsilon}(\varepsilon)$ can be expressed in terms of the oscillator strength for absorption, $f_{ij}$, that takes the ion from state $i$ to state $j$, and $\phi(\nu)$ is the line profile function that describes the variation of the absorption with frequency,

$$\sigma_L(\varepsilon) = \frac{\pi e^2}{mc} f_{ij}\,\phi(\varepsilon). \tag{3.240}$$

For most applications to X-ray astronomy, the two most important broadening mechanisms are natural line broadening, which follows from the uncertainty principle ($\Delta W \sim \hbar/\Delta t$), due to the finite lifetime of the excited state; and Doppler broadening, due to thermal motions in the absorbing medium. In the latter case,

$$\phi(\varepsilon) = \frac{1}{\sqrt{\pi}\,\Delta\varepsilon_D} e^{-[(\varepsilon-\varepsilon_0)/\Delta\varepsilon_D)]^2}, \tag{3.241}$$

where $\varepsilon_0$ is the line center, and

$$\Delta\varepsilon_D = \varepsilon_0 \left(\frac{2kT}{mc^2}\right)^{1/2}. \tag{3.242}$$

For hydrogenic and helium-like ions, the absorption oscillator strengths for the lowest level are $f(1s \rightarrow 2p) = 0.42$ and $f(1s1s \rightarrow 1s2p) = 0.74$. Absorption lines from highly ionized ions from iron and other elements have been detected in the spectra of winds flowing away from black holes, and the detection of absorption features from hydrogenic and helium-like ions of oxygen in the spectra of distant quasars is the most promising method for detecting the missing baryonic mass thought to be present in million-degree plasma in intergalactic space.

### 3.9.4 K-fluorescence Lines

The absorption of an energetic photon can lead to the ejection of an electron from the innermost, or K-shell of an atom or ion (Figure 3.7). The resulting ion is left in a highly excited state, and the ion may stabilize either by the emission of a photon or by the ejection of an electron through auto-ionization. The probability that an ion will emit a K-line photon is roughly given by

$$P(K) = \frac{Z^4}{Z^4 + 33^4}. \tag{3.243}$$

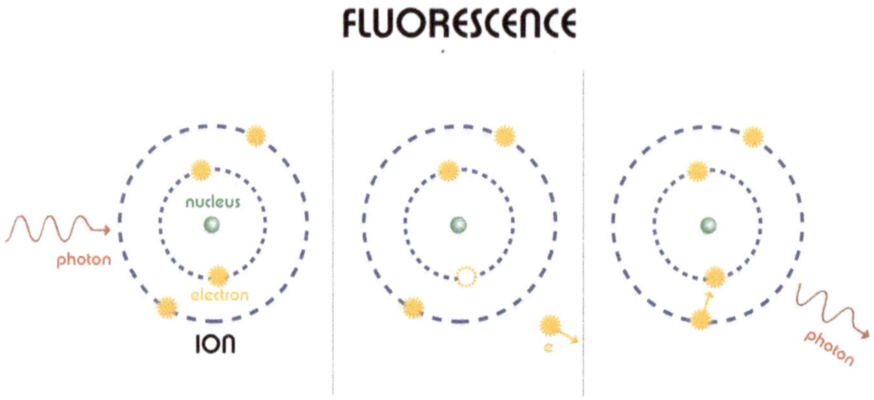


**Figure 3.7.** K-shell fluorescence. (Courtesy of NASA/CXC/M.Weiss.)

This shows that only high-$Z$ elements will be significant emitters of K-fluorescent lines. In an astrophysical context, the most significant K-fluorescent line comes from iron atoms and ions (Bambynek et al. 1972; Basko 1978). Observations of this line can be used to study Doppler and gravitational redshifts, which provide key information on the location and kinematics of the cold material within a few gravitational radii, or less, of the event horizon (see, e.g., the review by Reynolds & Nowak 2003 and other papers on this topic (Fabian & Miniutti 2005; Fabian et al. 2005; George & Fabian 1991; Krolik & Kallman 1987; Miller 2007).

## References


Arnaud, M., & Rothenflug, R. 1985, A&AS, 60, 425

Bambynek, W., Crasemann, B., Fink, R. W., et al. 1972, RvMP, 44, 716

Basko, M. M. 1978, ApJ, 223, 268

Bely, O., & van Regemorter, H. 1970, ARA&A, 8, 329

Bethe, H. A., & Jackiw, R. W. 1997, Intermediate Quantum Mechanics (Reading, MA: Addison-Wesley)

Blumenthal, G. R., & Gould, R. J. 1970, RvMP, 42, 237

Brightman, M., Harrison, F. A., Fürst, F., et al. 2018, NatAs, 2, 312

Bryans, P., Landi, E., & Savin, D. W. 2009, ApJ, 691, 1540

Burgess, A., & Seaton, M. J. 1960, MNRAS, 120, 121

Burgess, A., & Seaton, M. J. 1964, MNRAS, 127, 355

Burgess, A., & Summers, H. P. 1969, ApJ, 157, 1007

Cooper, G. 1971, PhRvD, 3, 2312

Dere, K. P., Del Zanna, G., Young, P. R., Landi, E., & Sutherland, R. S. 2019, ApJS, 241, 22

Fabian, A. C., & Miniutti, G. 2005, arXiv:astro-ph/0507409

Fabian, A. C., Miniutti, G., Iwasawa, K., & Ross, R. R. 2005, MNRAS, 361, 795

Ferland, G. J., Chatzikos, M., Guzmán, F., et al. 2017, RMxAA, 53, 385

Foster, A. R., Ji, L., Smith, R. K., & Brickhouse, N. S. 2012, ApJ, 756, 128

Gaetz, T. J., & Salpeter, E. E. 1983, ApJS, 52, 155

Gaunt, J. A. 1930, RSPTA, 229, 163

George, I. M., & Fabian, A. C. 1991, MNRAS, 249, 352

Ghisellini G. (ed) 2013, Lecture Notes in Physics, Vol. 873, Radiative Processes in High Energy Astrophysics (Berlin: Springer)
Giacconi, R., Gursky, H., Paolini, F. R., & Rossi, B. B. 1962, PhRvL, 9, 439
Ginzburg, V. L., & Syrovatskii, S. I. 1969, The Origin of Cosmic Rays (New York: Gordon and Breach)
Golden, L. B., Clark, R. E. H., Goett, S. J., & Sampson, D. H. 1981, ApJS, 45, 603
Gould, R. J. 2005, Electromagnetic Processes (Princeton, NJ: Princeton Univ. Press)
Gu, L., Mao, J., de Plaa, J., et al. 2018, A&A, 611, A26
Harding, A. K., & Lai, D. 2006, RPPh, 69, 2631
Jackson, J. D. 1975, Classical Electrodynamics (New York: Wiley)
Kaastra, J. S., Gu, L., Mao, J., et al. 2017, JInst, 12, C08008
Kaastra, J. S., Mewe, R., & Nieuwenhuijzen, H. 1996, in 11th Coll. UV and X-ray Spectroscopy of Astrophysical and Laboratory Plasmas, ed. K. Yamashita, & T. Watanabe (Tokyo: Universal Academy Press), 411–4
Kaastra, J. S., Paerels, F. B. S., Durret, F., Schindler, S., & Richter, P. 2008, SSRv, 134, 155
Kallman, T., & Bautista, M. 2001, ApJS, 133, 221
Karzas, W. J., & Latter, R. 1961, ApJS, 6, 167
Krolik, J. H., & Kallman, T. R. 1987, ApJ, 320, L5
Kuntz, K. D., Collado-Vega, Y. M., Collier, M. R., et al. 2015, ApJ, 808, 143
Landau, L. D., & Lifshitz, E. M. 1975, The Classical Theory of Fields (Oxford: Pergamon)
Landini, M., & Fossi, B. C. M. 1991, A&AS, 91, 183
Landini, M., Landi, E., & Fossi, B. M. 1997, in AIP Conf. Proc. 386, The XUV Spectral Code of Arcetri (Melville, NY: AIP), 421–2
Lisse, C. M., Christian, D. J., Dennerl, K., et al. 2001, Sci., 292, 1343
Maitra, C. 2017, JApA, 38, 50
Mazzotta, P., Mazzitelli, G., Colafrancesco, S., & Vittorio, N. 1998, A&AS, 133, 403
Mewe, R., Gronenschild, E. H. B. M., & van den Oord, G. H. J. 1985, A&AS, 62, 197
Miller, J. M. 2007, ARAA, 45, 441
Mullen, P. D., Cumbee, R. S., Lyons, D., et al. 2017, ApJ, 844, 7
Nozawa, S., Itoh, N., & Kohyama, Y. 1998, ApJ, 507, 530
Pozdnyakov, L. A., Sobol, I. M., & Syunyaev, R. A. 1983, ASPRv, 2, 189
Reynolds, C. S., & Nowak, M. A. 2003, PhR, 377, 389
Rybicki, G. B., & Lightman, A. P. 1986, Radiative Processes in Astrophysics (New York: Wiley)
Schönherr, G., Wilms, J., Kretschmar, P., et al. 2007, A&A, 472, 353
Schwarm, F.-W., Ballhausen, R., Falkner, S., et al. 2017a, A&A, 601, A99
Schwarm, F.-W., Schönherr, G., Falkner, S., et al. 2017b, A&A, 597, A3
Shapiro, S. L., Lightman, A. P., & Eardley, D. M. 1976, ApJ, 204, 187
Shklovsky, I. S. 1967, ApJ, 148, L1
Smith, R. K., Brickhouse, N. S., Liedahl, D. A., & Raymond, J. C. 2001, ApJ, 556, L91
Smith, R. K., & Hughes, J. P. 2010, ApJ, 718, 583
Sunyaev, R. A., & Titarchuk, L. G. 1980, A&A, 86, 121
Sutherland, R. S., & Dopita, M. A. 1993, ApJS, 88, 253
Tarter, C. B., Tucker, W. H., & Salpeter, E. E. 1969, ApJ, 156, 943
Titarchuk, L. 1994, ApJ, 434, 570

Trümper, J., Pietsch, W., Reppin, C., et al. 1977, in Annals of the New York Academy of Sciences, Eighth Texas Symp. on Relativistic Astrophysics, Vol. 302, ed. M. D. Papagiannis; New York: New York Academy of Sciences), 538
Tucker, W. 1975, Radiation Processes in Astrophysics (Cambridge, MA: MIT Press)
Urdampilleta, I., Kaastra, J. S., & Mehdipour, M. 2017, A&A, 601, A85
Wargelin, B. J., Beiersdorfer, P., & Brown, G. V. 2008, CaJPh, 86, 151
Weymann, R. 1965, PhFl, 8, 2112
Young, P. R., Dere, K. P., Landi, E., Del Zanna, G., & Mason, H. E. 2016, JPhB, 49, 074009
Zhang, S., Wang, Q. D., Ji, L., et al. 2014, ApJ, 794, 61

AAS | IOP Astronomy

The Chandra X-ray Observatory
Exploring the high energy universe
Belinda Wilkes and Wallace Tucker

# Chapter 4

# X-Rays from Stars and Planetary Systems

**Jeremy J Drake**

The phrase "stars and planetary systems" covers a vast array of astrophysical phenomena, from the relatively small scales of planetary science to the hundred-parsec scales of the most massive clusters of stars. This chapter attempts to summarize *Chandra's* major contributions to understanding energetic processes in these environments through their X-ray emission.

X-rays in the cosmos are primarily produced by matter heated to temperatures exceeding a million degrees, or else from matter accelerated to relativistic energies—see the preceding Chapter 3 for details on these mechanisms. It might then come as a surprise that many comparatively cool objects in the solar system are now known to shine in X-rays.

As we shall see, X-ray emission from stars themselves naturally divides them into the categories of "high" mass and "low" mass, the energetic radiation of which originates from fundamentally different astrophysical processes. The division occurs at spectral types where stars begin to develop substantial outer convection zones, which on the main sequence corresponds to type F and later, or a mass of about 1.5 $M_{\odot}$. The resulting convective flows maintain rotational shear, which sustains magnetic dynamo activity with energy eventually dissipated at the stellar surface in the form of chromospheric and coronal UV–X-ray emission and the driving of a magnetized wind. High-mass stars instead drive winds by radiation pressure, and their X-rays originate in plasma heated by shocks resulting from instabilities in this process, or by collisions within winds induced by magnetic channeling or with a wind-driving massive binary companion.

In between high- and low-mass stars lies an intriguing but quite narrow range of stellar masses that are X-ray dark, but which might sustain X-ray activity driven by natal differential rotation for a very brief period after the pre-main-sequence phase.

The neutron star or back hole remnants of high-mass stars are discussed in Chapters 5–7. Here, we briefly touch on the X-ray emission from the hot white dwarf

doi:10.1088/2514-3433/ab43dcch4  

remnants of lower mass stars and from white dwarfs in close, mass-transferring binary systems that form cataclysmic variables (CVs) and lead to nova explosions.

## 4.1 X-Rays from Solar System Bodies

X-rays from solar system bodies tend to be driven ultimately by the Sun, although there is emerging evidence that planetary systems can generate X-rays themselves (Dunn et al. 2016). The story of X-rays in the solar system forms an important segment in the historical narrative of X-ray astronomy itself. It was a proposed experiment to detect X-rays from the Moon that instead serendipitously made the first detection of a cosmic X-ray source—the low-mass X-ray binary Scorpius X-1 (Giacconi et al. 1962)—and ushered in the age of X-rays as a new window into the universe.

In the mid-1990s, only the Earth, the Moon, and Jupiter had been detected in X-rays, along with the Sun, which was ostensibly first photographed at X-ray wavelengths by a rocket-borne camera in 1949 (Burnight 1949)—we return to this type of X-ray source in Section 4.2 below. The Moon was studied using a proportional counter instrument from lunar orbit by the *Apollo 15* and *16* missions (Adler et al. 1973), while Jupiter was first observed by the *Einstein* satellite in 1979 (Metzger et al. 1983). *ROSAT* made the first detections of X-rays from comets (comet Levy/1990c) and Earth's geocorona, although those detections were not recognized until later (see Sections 4.1.2–4.1.4).

*Chandra* has played a major role in the X-ray exploration of the solar system. The tally of solar system objects detected in X-rays now includes asteroids, in addition to Mercury, Venus, Earth, Mars, Jupiter and its moons Io and Europa, the Io plasma torus, Saturn and its rings, and Pluto. *Chandra's* unique contribution to this field, as in the many other aspects of astrophysics discussed in this book, stems largely from its high spatial resolution capability combined with imaging spectroscopy. The reader is also referred to reviews of the general topic of X-rays from solar system objects presented by Bhardwaj et al. (2007), Dennerl (2014), and Branduardi-Raymont (2017).

### 4.1.1 X-Ray Emission Mechanisms in Solar System Bodies

Setting aside the Sun for now, the common thread for X-ray production by solar system objects is that they are essentially cold (<1000 K), and so the energy to generate X-rays has to come from elsewhere. X-ray emission from solar system bodies falls into the general categories of charge-exchange emission, elastic scattering of solar X-ray photons, X-ray fluorescence following inner-shell ionization by solar X-rays, and bremsstrahlung and collisionally excited line emission caused by the impact of energetic electrons and ions. The energy sources in these processes are electric potential energy, incident photon energy, and particle kinetic energy, respectively. Several solar system objects display X-rays produced by more than one of these mechanisms. X-ray emission processes in general are treated in Chapter 3, to which we again refer the reader for a more detailed introduction. We take these processes in turn below.

Charge-exchange (commonly abbreviated CX) X-ray emission occurs when a highly ionized, heavy ion encounters neutral material. Neutrals act as electron donors, and the electrons then combine with the ion in upper levels to leave the ion in an excited state. It is the decay from this excited state that leads to the emission of an X-ray photon. In the solar system, the Sun provides a copious supply of highly charged ions in the form of the solar wind. A schematic illustration of the process involving $O^{7+}$ ions interacting with neutral hydrogen atoms is shown in Figure 4.1. Solar wind CX emission (SWCX) was first identified as the origin of the X-rays detected from comet Hyatuke (Cravens 1997), and has since been observed from several other solar system objects, including a number of more recent comets and the exospheres of Venus, Earth, and Mars.

Coherent scattering of solar X-rays and fluorescence—essentially an inelastic scattering process—occurs to some extent on all solar system bodies illuminated by the Sun, although these might not be the dominant source of X-ray emission. In the soft X-ray spectral region that *Chandra* observes in, absorption cross sections are much larger than scattering cross sections, and the majority of the incident X-ray flux is consequently absorbed. Scattered X-rays have nevertheless been observed from Earth, the Moon, Jupiter, and Saturn. We will return to these objects below.

Fluorescence occurs when an atom is ionized in its inner shell, by either an X-ray photon or energetic electrons, protons, or ions. The "hole" created by the ionization event leaves the ion in an excited state that subsequently decays via the transition of an electron from a higher shell. If ionized in the $n = 1$ shell (the "K shell"), this transition can be either radiative, corresponding to the emission of a characteristic "K$\alpha$" photon, or nonradiative, corresponding to an Auger transition in which one or more Auger electrons are emitted. In the case of ionization of $n > 1$ inner shells (L, M, etc. shells), the decay channels get more complex, with transitions from higher subshells of the same shell also being possible—the so-called "Coster–Kronig" transitions. The probability of a radiative transition is very low for low-$Z$ elements, and increases strongly with increasing atomic number; the fluorescence yields for O "K$\alpha$" and Fe

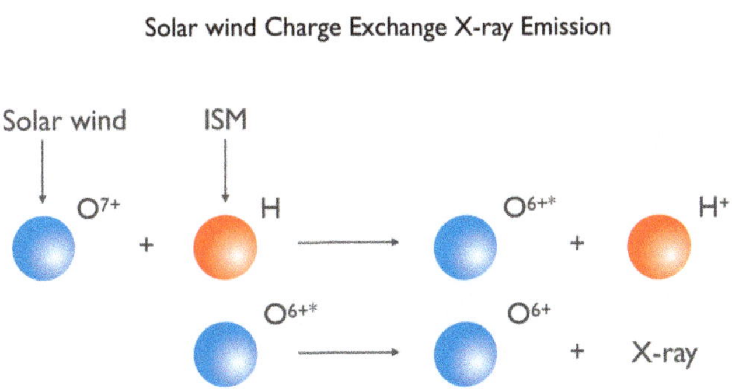


**Figure 4.1.** Illustration of the charge-exchange X-ray emission mechanism in which a highly charged ion interacts with a neutral and captures an electron into an excited state. In this case, an $O^{7+}$ solar wind ion interacts with a neutral H atom, and the decay of the resulting $O^{6+*}$ excited state produces a soft X-ray photon. The video shows an animation of the charge exchange X-ray emission mechanism. Additional materials available online at http://iopscience.iop.org/book/978-0-7503-2163-1.

"K$\alpha$" emission are $8.3 \times 10^{-3}$ and 0.34, for example. Consequently, the fluorescence process for the abundant solar system elements C, N, and O is very inefficient.

Collisional excitation by electrons and ions, leading to line emission or bremsstrahlung, can be driven by the precipitation of solar wind particles or when particles are accelerated by magnetospheric processes. Inner-shell ionization by particle impact produces characteristic X-ray line emission in the same way as fluorescence due to photoionization. We shall see that these processes are relevant for producing auroral X-ray emission.

### 4.1.2 Terrestrial Planets

*Venus*

The terrestrial planets subtend sufficiently large angular diameters—up to 66″ for Venus and 25″ for Mars—to be easily resolved by *Chandra*. While *Chandra* is able to observe as close as 45.5° to the Sun—closer than any other X-ray telescope past or present—elusive Mercury lives up to its name and, while subtending up to 13″, always resides too close to the Sun for *Chandra* to observe.

Venus generally lies much closer to the Sun than 45.5° and frustratingly out of reach of *Chandra*. However, its most extreme elongations extend to 48°, and *Chandra* was able to capitalize on these short windows of visibility by making the first X-ray observation of Venus in 2001 January (Dennerl et al. 2002; see Figure 4.2). The X-ray signal, obtained with the ACIS-I detector, was found to result from fluorescent scattering of solar X-rays in the Venusian thermosphere, as correctly predicted by Cravens & Maurellis (2001). Delightful confirmation of the fluorescence lines was provided by a high-resolution spectrum that was also obtained by the LETG and ACIS-S detector.

Later observations of Venus in 2006 March and 2007 October, when the Sun was close to minimum activity and X-ray flux, succeeded in detecting the expected much weaker SWCX emission from the interaction of the solar wind with the Venusian exosphere (Dennerl 2008).

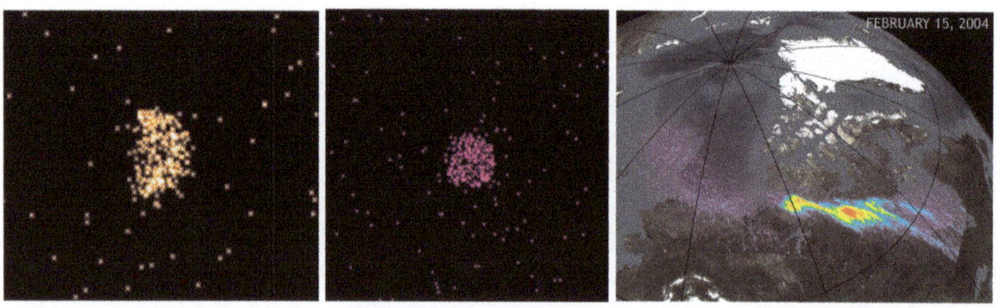


**Figure 4.2.** *Chandra* images of the terrestrial planets. From left to right: Venus, Mars, and a bright spot and X-ray-emitting arc in Earth's aurora superimposed over a visible light representation. From http://chandra.harvard.edu/photo/category/solarsystem.html. (Courtesy of NASA/MPE/K.Dennerl et al.; NASA/CXC/MPE/K.Dennerl et al.; and NASA/MSFC/CXC/A.Bhardwaj & R.Elsner, et al.; Earth model: NASA/GSFC/L. Perkins & G.Shirah)

*Mars*

Mars was detected in X-rays for the first time on 2001 July 4 in a *Chandra* ACIS-I observation, appearing "as an almost fully illuminated disk, with an indication of limb brightening at the sunward side, accompanied by some fading on the opposite side" (Dennerl 2002, p. 1119, Figure 4.2, center). The observation was timed when Mars was only 70 million kilometers from Earth and also near the point in its orbit when it is closest to the Sun. The ACIS spectrum of the disk of Mars was dominated by the O K$\alpha$ fluorescence line excited by solar X-rays absorbed at a height of approximately 120 km above the planetary surface. Dennerl (2002) also traced a faint X-ray-emitting halo out to three Mars radii, consistent with a thermal bremsstrahlung spectrum with a characteristic temperature of 0.2 keV. This spectral component was traced to SWCX, between highly charged heavy ions in the solar wind and exospheric hydrogen and oxygen around Mars. A global dust storm was intensifying while *Chandra* observed, but no sign of fluorescence lines from refractory elements such as Mg, Si, and Fe, which might be expected from atmospheric dust at very high altitude, was detected.

The detection of Mars by *Chandra* is testament to its remarkable sensitivity. The X-ray power emitted from the Martian atmosphere is very small indeed in comparison to the solar heating at Mars, amounting to only 4 MW. In a more terrestrial context, this corresponds to the X-ray power of about 10,000 diagnostic chest X-rays (Seibert 1997).

*Earth*

Earth differs from Venus and Mars in being a magnetized planet with a well-developed magnetosphere. Such a physical system displays a rich array of phenomena associated with the interaction of the solar wind and magnetosphere, the response of the upper atmosphere to ionization by solar extreme ultraviolet and X-ray irradiation, and the complex behavior of atmospheric and precipitated solar wind ions and electrons within this dynamic system.

An X-ray-emitting aurora on Earth has been known since balloon and rocket observations beginning in the late 1950s (e.g., Anderson 1958; Winckler et al. 1959) and observations by spacecraft since the 1970s. The X-ray aurora on Earth is generated by energetic electron bremsstrahlung (e.g., Berger & Seltzer 1972; Bhardwaj et al. 2007).

*Chandra* observed northern auroral regions of Earth using the HRC-I in 11 different 20 minute observations obtained between mid-December 2003 and mid-April 2004 (Bhardwaj et al. 2007). The data revealed a highly dynamic X-ray aurora, with multiple and variable intense arcs, and diffuse patches of X-rays at times visible and at times absent. In at least one of the observations an isolated blob of emission is observed near the expected cusp location. Lacking energy resolution, the HRC-I data could not probe directly the X-ray emission mechanism. However, one observation in 2004 January 24 during a bright arc seen by *Chandra* was accompanied, quite unplanned, by an overflight of the Sun-synchronous polar-orbiting Defense Meteorological Satellite Program satellite F13 that was able to obtain simultaneous energetic particle measurements. Bhardwaj et al. (2007) used those data to model the expected X-ray spectrum, finding that the observed soft X-ray signal was bremsstrahlung emission together with characteristic K-shell line emission of nitrogen and oxygen in the atmosphere produced by energetic electrons.

A further source of X-rays from Earth was only isolated in the mid-1990s: geocoronal SWCX emission resulting from interaction of the solar wind with neutrals in the geocorona. SWCX also occurs throughout the heliosphere as neutral gas from the interstellar medium flows by, contributing a significant fraction of the soft X-ray background (Cravens 2000). The contribution from the geocorona forms a variable background component in all X-ray observations from Earth's vicinity (Snowden et al. 1995; Cravens et al. 2001). Wargelin et al. (2004) utilized *Chandra* observations of the night side of the Moon to show that the very faint signal detected by *ROSAT* was not due to solar wind impact on the Moon, but to the geocoronal glow in the light of O vii K$\alpha$ and O viii Ly$\alpha$ resulting from SWCX. The SWCX model for geocoronal X-ray emission was further confirmed in detail by Wargelin et al. (2014) based on *Chandra* observations during times of strong solar wind gusts diagnosed by the *Advanced Composition Explorer* (*ACE*) spacecraft situated at the Sun–Earth L1 point about 0.01 au toward the Sun.

*The Moon*

The Moon was first observed in detail in X-rays using proportional counters on *Apollo 15* and *16* designed to detect fluorescent lines of abundant elements such as Mg, Al, and Si from reprocessed solar X-rays (Adler et al. 1973). The relative line strengths of these elements is dependent on their bulk chemical content on the lunar surface—a direct measurement not provided by mineral-dependent albedo at other wavelengths. While *Chandra* performed a pilot observation of the Moon using ACIS-I in 2001 and fluorescent lines of O, Mg, Al, and Si were detected by Wargelin et al. (2004), further lunar exploration that could have provided high-resolution geochemical lunar maps was unfortunately curtailed by the realization that the ACIS contamination layer might be polymerized by geocoronal H Ly$\alpha$ emission and potentially jeopardize any future attempts to remove the contaminant by thermal cycling ("bake out").

HRC-I observations performed in 2004 May, June, and July succeeded in detecting an albedo reversal with respect to images in visible light, in which optically dark maria are brighter in X-rays than optically bright highlands (Drake et al. 2004; Figure 4.3). It is not yet known whether this effect results from differences in weathering or chemical composition.

### 4.1.3 The Gas Giants

*Jupiter*

Like Earth, Jupiter has a comparatively strong magnetic field (7.8 G at the equator) and a well-developed magnetosphere. Coupled with the launching of gas from its moon Io that feeds the Io plasma torus it interacts with, in addition to interactions with the solar wind, Jupiter is arguably the most fascinating solar system object in X-rays. It was first detected in X-rays by the *Einstein* satellite (Metzger et al. 1983)

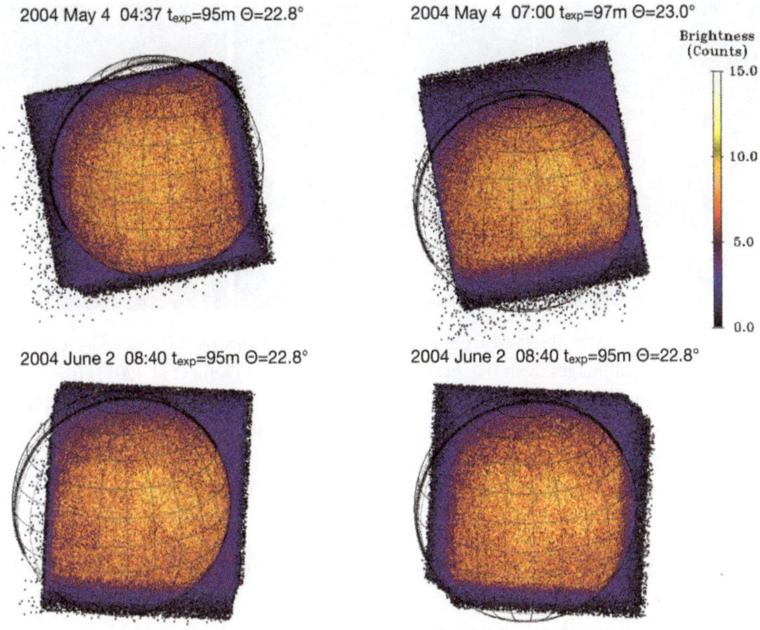


**Figure 4.3.** *Chandra* HRC-I images of the Moon obtained in 2004 May, June, and July illustrating X-ray albedo reversal in which maria appear brighter than highlands (from Drake et al. 2004).

and, at the time of writing, Jupiter had amassed a total of 44 separate *Chandra* observations over several different campaigns spanning the full mission, from 1999 to 2019.

X-rays from Jupiter arise from the full gamut of processes discussed in Section 4.1.1 above. The first *Chandra* observations found soft X-rays to be concentrated in a hot spot poleward of latitudes expected to be magnetically connected with the inner magnetosphere. This surprisingly placed their origin beyond 30 Jupiter radii from the planet, with subsequent observations suggesting the precipitating plasma originates beyond 60 Jupiter radii (>4 million km) from the planet (e.g., Kimura et al. 2016). The hot spot was also observed to be spectacularly pulsating with a period of about 45 minutes, a phenomenon that has also been seen in later observations (Dunn et al. 2016) and in energetic particle fluxes and Jovian radio emission (Cravens et al. 2003). Comparison of more recent observations from 2016 May–June with data obtained in 2007 March have also revealed a persistent southern X-ray hot spot that behaves independently of its northern counterpart and showed 11 minute periodic pulsations and uncorrelated changes in brightness (Dunn et al. 2017). Subsequent studies have shown that Jupiter's aurora exhibits these intriguing regular pulsations of a few tens of minutes in at least 30% of observations (Jackman et al. 2018). Snapshots of the northern and southern Jovian aurorae are illustrated in Figure 4.4. The origin of the pulsation behavior remains uncertain, though is thought to be due to either the "bounce" period for a magnetically trapped ion to repeat its north–south motion along a field line (e.g., Cravens et al. 2003), pulsed magnetic reconnection (Bunce et al. 2004; Dunn et al.

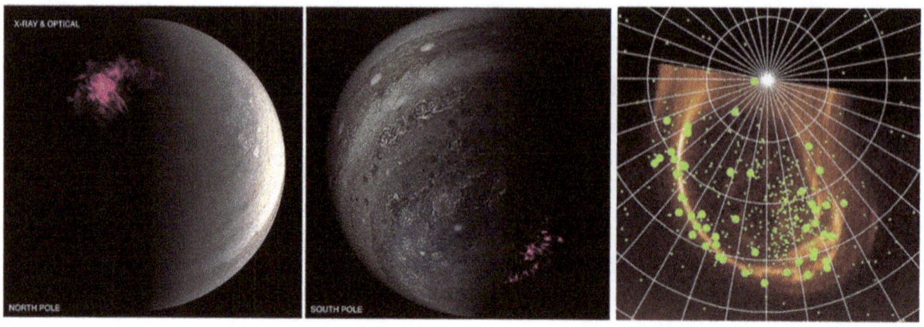


**Figure 4.4.** Left: X-ray emission observed from the north and south aurorae of Jupiter by *Chandra* (magenta) superimposed on visible light images from the *Juno* spacecraft. Based on Dunn et al. (2017) from https://chandra.harvard.edu/photo/2017/jupiter/. (South Pole image courtesy of NASA/JPL-Caltech/SwRI/MSSS/Gerald Eichstädt /Seán Doran; North Pole image courtesy of NASA/JPL-Caltech/SwRI/MSSS.) Right: polar projection of a *Hubble* STIS FUV image of Jupiter's northern aurora (orange) overplotted with the positions of X-ray photons detected by *Chandra* during simultaneous observations on 2003 February 24. Small and large green dots indicate photon energies $<2$ keV and $>2$ keV, respectively (reproduced from Branduardi-Raymont et al. 2008).

2016), or Kelvin–Helmholtz instabilities that cause Jupiter's magnetic field lines to resonate (Dunn et al. 2017).

Spectroscopic observations revealed the origin of the hot spot X-rays to be largely CX emission from energetic highly charged ions of oxygen, sulfur, and carbon (Elsner et al. 2005; Branduardi-Raymont et al. 2007; Kharchenko et al. 2008) that are co-spatial with the FUV emission seen by *Hubble* (see Figure 4.4, right panel). The question then is, what is the origin of the particles: the solar wind, or the Io plasma torus? Ion abundances can in principle be used to distinguish between them, the Io plasma having a much larger sulfur abundance relative to oxygen than the solar wind owing to its volcanic origin. Exploiting an elevated ram pressure from a CME with a *Chandra* target of opportunity observation, Dunn et al. (2016) found a factor of 8 enhancement in the Jovian X-ray aurora and periods of 26 minutes from S ions at lower latitudes and 12 minutes from C, O and S ions at the highest latitudes. Dunn et al. (2016) trace the former to precipitation of trapped magnetospheric plasma of Io origin and the latter to the solar wind and more open field lines.

Jupiter also exhibits steady, fainter, and more uniform non-auroral X-ray emission at low and mid-latitudes, although with some surface magnetic field strength dependence (Bhardwaj et al. 2006). The temporal variation of this emission follows variations in solar X-ray flux, tying the bulk of its origin to scattering of and fluorescence by solar X-rays, but with some additional component likely originating in ion precipitation from close-in radiation belts.

During *Chandra* studies aimed at Jupiter, detections of X-rays in the 0.25–2 keV band from the Galilean satellites Io and Europa, possibly Ganymede, and also the Io plasma torus, were made (Elsner et al. 2002). The latter appeared to be due to bremsstrahlung emission from nonthermal electrons in the few hundred to few thousand electronvolt range. In contrast, Elsner et al. (2002) found X-rays from the Galilean satellites, most likely due to bombardment by energetic H, O, and S ions from the Io plasma torus. This fluorescent signature could hold tremendous promise

**Figure 4.5.** Left: a *Chandra* X-ray image of the X-ray fluorescence and scattering from the surface of Saturn. Right: the X-ray shadow of Titan against the Crab Nebula observed during a transit on 2003 January 5 by Mori et al. (2004). Both images from http://chandra.harvard.edu/photo/category/solarsystem.html. (Courtesy of NASA/U. Hamburg/J.Ness et al; NASA/CXC/Penn State/K.Mori et al.)

for X-ray remote sensing of the composition of the Europa and perhaps Enceladus oceans with next-generation X-ray missions, or from X-ray spectrometers on future giant-planet missions.

*Saturn and Uranus*

Saturn was observed for 10 ks using the *Einstein Observatory* on 1979 December 17 but no X-ray emission was detected (Gilman et al. 1986). A marginal detection by the *ROSAT* PSPC was reported by Ness & Schmitt (2000), but the first unequivocal detection of Saturn in X-rays had to await a 70 ks observation on 2003 April 14–15 by *Chandra* and the ACIS-S detector (Ness et al. 2004, Figure 4.5). Unlike Jupiter, Saturn does not appear to drive strong X-ray-emitting aurorae (Branduardi-Raymont et al. 2013). Instead, its X-ray signal is similar at both low and high latitudes, including the polar caps.

In a *Chandra* observation obtained in 2004 January, Bhardwaj et al. (2005a) found the X-ray signal from Saturn to mirror the X-ray intensity from an M6-class solar flare originating in an active region associated with a sunspot that was clearly visible from both Saturn and Earth. Analyzing the full body of *Chandra*, *XMM-Newton*, and *ROSAT* X-ray data on Saturn available until the end of 2004, Bhardwaj et al. (2005a) noted that Saturn's X-ray signal is highly correlated with the solar 10.7 cm radio flux—a well-known proxy for coronal emission. This confirmed Saturn's X-ray emission as originating from fluorescence and back-scattering of solar X-rays (Bhardwaj et al. 2005b; Branduardi-Raymont et al. 2010).

The beguiling rings of Saturn might still hold some mysteries for future X-ray missions. First confirmed by Bhardwaj et al. (2005b), to shine in the light of O K$\alpha$ emission, for which fluorescent scattering of solar X-rays from oxygen atoms in the $H_2O$ icy ring material is the most obvious source mechanism, Branduardi-Raymont et al. (2010) found the ring O K$\alpha$ emission did not share the same dependence on the solar cycle as the disk emission, indicating a possible second excitation mechanism, such as lightning-induced electron beams.

*The X-Raying of Titan's Atmosphere*

One of the most spectacular *Chandra* observations of the solar system was made in 2003 when Saturn made a very rare transit of the Crab Nebula, a conjunction that we will have to wait until 2267 to see again. Mori et al. (2004) noted that, although a similar conjunction occurred in 1296 January, the Crab Nebula—the remnant of SN 1054—was probably too small to be occulted, rendering the 2003 event the first such transit.

Unfortunately, the transit of Saturn itself was during a passage of *Chandra* through the radiation belts and was missed. However, Mori et al. (2004) observed the occultation shadow of the largest of Saturn's moons, Titan, the only satellite in the solar system with a thick atmosphere. The shadow, illustrated in Figure 4.5, was clearly larger than the diameter of Titan's solid surface, indicating a thickness of the atmosphere of 880 ± 60 km—essentially consistent with or perhaps slightly larger than estimated from earlier *Voyager* observations at radio, IR, and UV wavelengths. The difference could partly be explained by Saturn being slightly closer to the Sun during the *Chandra* observation.

*No X-Ray Detection of Uranus Yet*

Uranus has so far proven to be elusive in X-rays. *Chandra* has observed it twice without detecting it: in 2002 August for 30 ks using ACIS-S and in 2017 November for 50 ks with the HRC-I during a CME impact. The observations aimed to see auroral emission, like that seen on Jupiter. Uranus' magnetic field is at 60° to the rotation axis and expected to produce a dynamic and constantly changing magnetosphere. Detection of any X-ray-emitting aurorae on Uranus must await longer exposures. While a fluorescent and scattering signal from processed solar X-rays is also expected, exposure times required for detection are also an order of magnitude longer than the existing observations.

### 4.1.4 Minor Planets and Comets

*The Remarkable X-Rays from Pluto*

The ice giants, Uranus and Neptune, likely have to await the arrival of a mission with higher sensitivity than *Chandra* or *XMM-Newton* (see, e.g., Snios et al. 2019). It is then quite remarkable that *Chandra* detected X-rays from Pluto (Figure 4.6).

Pluto was targeted in a campaign to support the *New Horizons* flyby in observations on 2014 February 24, and in three further visits from 2015 July 26 to August 03, for a total integration time of 174 ks. A total of eight X-ray photons were detected, all of which were in the 0.3–0.6 keV passband, amounting to a net signal of 6.8 counts after background subtraction (Lisse et al. 2017). The origin of this signal remains a mystery. Lisse et al. (2017) noted that the X-ray power from Pluto—approximately 200 MW—was comparable to that of other solar system X-ray emission sources such as auroral precipitation, solar X-ray scattering, and

**Figure 4.6.** An image of Pluto obtained by *New Horizons* in visible light, together with the X-ray image obtained by *Chandra* for observations obtained in 2014–2015. From https://chandra.harvard.edu/photo/2016/pluto/. (New Horizons image courtesy of NASA/JHUAPL; X-ray image courtesy of NASA/CXC/JHUAPL/R.McNutt et al.)

SWCX. However, Pluto appears to lack a significant magnetic field that could drive aurorae, and no auroral airglow has been detected. Moreover, the backscattered solar X-ray flux is expected to be two to three orders of magnitude below detection thresholds. While SWCX can produce the appropriate X-ray photon energies, the lower than expected neutral escape rate from Pluto found by *New Horizons* (Gladstone et al. 2016) leaves the observed solar wind flux in Pluto's vicinity a factor of about 40 too weak to account for the observed signal.

*Comets: The Rosetta Snowballs of Charge-exchange X-Ray Emission*

Comets were first detected in X-rays by *ROSAT*, beginning with C/1996 B2 (Hyakutake; Lisse et al. 1996) rapidly followed by several others found by Dennerl et al. (1997) in archival data. As noted in Section 4.1.1, these detections became of special general importance to X-ray astronomy because they established the CX process as an important source of X-ray emission (Cravens 1997). In turn, X-rays became an important diagnostic of cometary gas and dust production rates, as well as a means of probing the solar wind (Dennerl et al. 1997; Kharchenko & Dalgarno 2000; Snios et al. 2016). The SWCX signal originates with the interaction of solar wind heavy ions with outgassing cometary neutrals and probes the gas in the coma independently of the dust, which cannot easily be done at other wavelengths (Dennerl et al. 2012).

*Chandra* has played a vital role in cometary X-ray studies. While the SWCX mechanism proposed by Cravens (1997) remained the most likely explanation for cometary X-ray emission, the spectral resolution of the *ROSAT* PSPC was insufficient to resolve the expected line signatures from He-like and H-like C, N, and O.

It was not until *Chandra* observed C/1999 S4 (LINEAR) with ACIS-S that definitive spectroscopic evidence of these lines was obtained (Lisse et al. 2001; see Figure 4.7). The C/1999 S4 (LINEAR) spectrum also indicated the presence of a weak thermal bremsstrahlung component and subsequent *Chandra* observations of comets have confirmed that coherent scattering of solar X-rays by comet dust and ice particles contribute significantly to the signal at energies $>1$ keV (Snios et al. 2014, 2018).

Cometary SWCX emission is dependent on the elemental composition, ionization state, number density, and speed of the impacting solar wind, as well as the density of particles in the comet coma. Several studies have exploited this and used *Chandra* observations of comets to probe solar wind conditions. Christian et al. (2010) and Snios et al. (2016) found very different X-ray signals from comets at high and low solar latitudes. The former interact with the low-density fast solar wind that is deficient in highly ionized species and characterized by lower ion freeze-in temperatures. X-rays from high-latitude comets were commensurately softer and lacking in emission from species such as $O^{8+}$ and $Ne^{9+}$.

## 4.2 X-Rays from Low-mass Stars

Humans have been studying the hot, million-degree outer atmosphere, or "corona," of the Sun for hundreds if not thousands of years, albeit without realizing it until about 80 years ago. A recent spectacular example of this ground-based activity from the 2017 North American solar eclipse is shown in Figure 4.8. The white light corona we see with the naked eye, and rendered in spectacular detail in this composite exposure, is the light scatted by electrons in the million-degree plasma that makes up the outer corona and the solar wind. The impressive filamentary

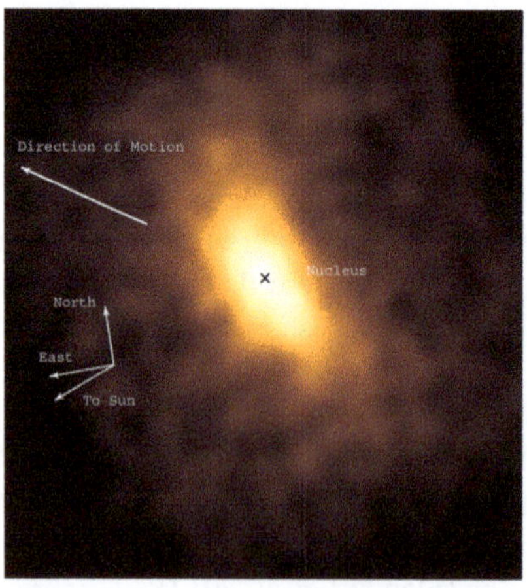


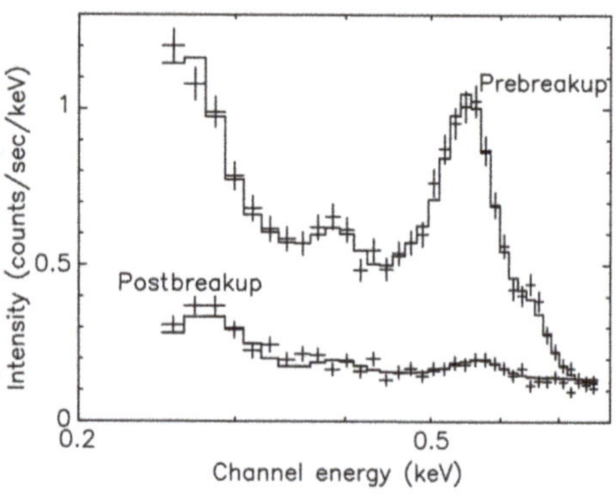


**Figure 4.7.** Left: image of the X-ray emission from comet C/1999 S4 (LINEAR) detected by *Chandra* in a 2.5 hr observation on 2000 July 14. Right: the ACIS-S spectrum of C/1999 S4 (LINEAR) fitted with a spectral model comprising six lines arising from CX of solar wind $C^{5+}$, $C^{6+}$, $N^{7+}$, $O^{7+}$, and $O^{8+}$ ions with neutrals in the comet coma together with a weak bremsstrahlung component. Both images from Lisse et al. (2001). (Courtesy of NASA/CXC/C.Lisse, S.Wolk, et al.)

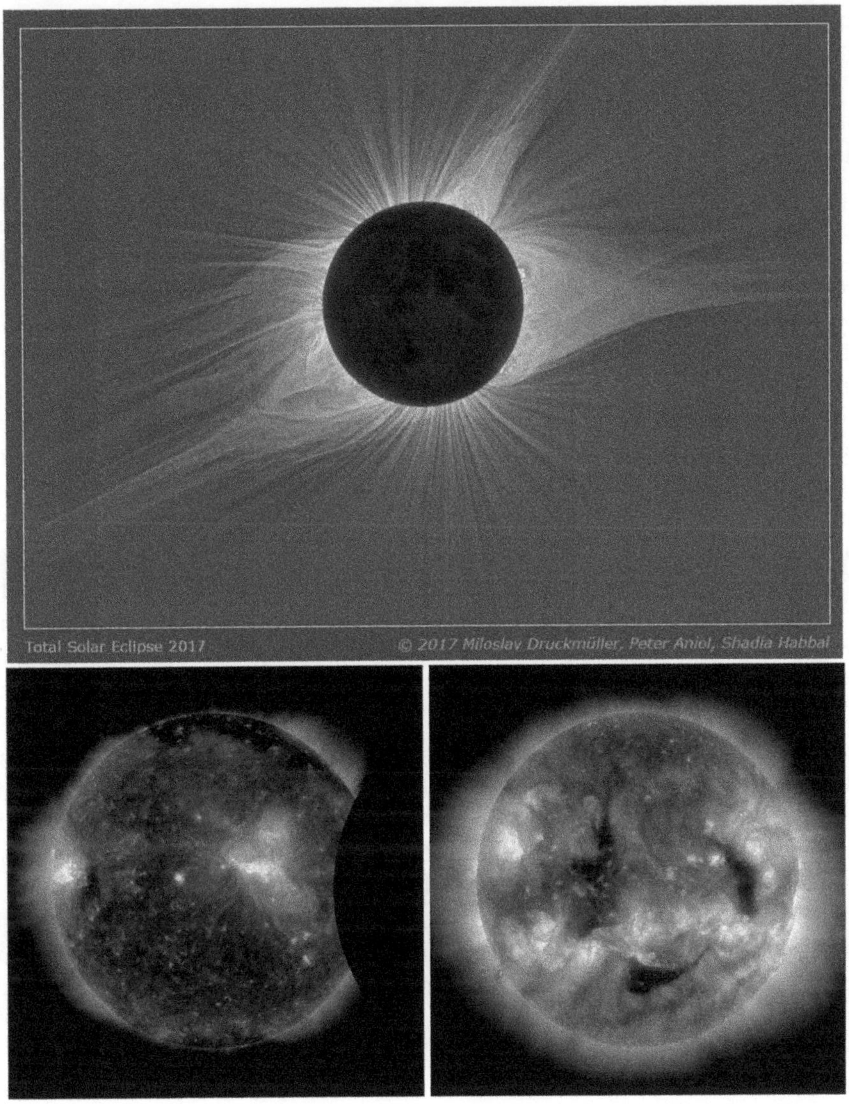


**Figure 4.8.** Top: white light composite image of the solar corona showing scattering of solar visible light by hot electrons obtained during the North American eclipse of 2017 August 21 (image credit: Nicolas Lefaudeux, https://hdr-astrophotography.com/high-resolution-2017-total-solar-eclipse/; Courtesy of Predictive Science Inc./Miloslav Druckmüller, Peter Aniol, Shadia Habbal/NASA Goddard, Joy Ng). Bottom left: the solar corona at the time of the eclipse, when the Sun was at a fairly low activity level, seen in the 211 Å filter of the SDO AIA. This filter is primarily sensitive to emission lines of Fe XIV formed at a temperature of approximately $2 \times 10^6$ K. The shadow of the Moon can be seen encroaching from the right. Because SDO is in an inclined geosynchronous orbit, it did not experience the totality witnessed on the ground. Bottom right: the AIA 211 Å image near solar maximum exactly four years earlier on 2013 August 21, exhibiting a rich array of active regions. (Courtesy of NASA/SDO and the AIA science team.)

structure betrays the beautiful complexity of the magnetic field to which the fully ionized plasma is tied and constrained.

While the white light corona has been appreciated for centuries, the beginning of the study of the hot outer atmospheres of stars can be traced to a breakthrough in the late 1930s and early 1940s when work by Edlén and Grotrian led to the identification of the "red" line at 6375 Å seen during eclipses with a forbidden line of Fe x and betraying a temperature of a million kelvins. Hunter (1942) commented that "the immediate reaction of most astronomers will no doubt be one of incredulity that such highly ionized matter as Edlén's proposals call for should exist in the outer envelope of a relatively cool star like the Sun." While decades of familiarity have diffused the incredulity, a large part of the research on stellar coronae since then has been devoted to trying to understand how such temperatures arise—the "coronal heating problem." While certain heating mechanisms are now known to operate in the solar corona, a definitive answer to the coronal heating problem has remained elusive.

The million-degree temperature of the solar corona was confirmed in suborbital experiments carried on V2 rockets captured from Germany following World War II. The first solar X-ray image is traditionally attributed to a photograph from a flight on 1948 August 5 by Burnight (1949), though the X-ray nature of the photographic signal was disputed by Friedman (1980), who argued that the rocket did not reach a high-enough altitude. The first actual detection of solar (and stellar!) X-rays was made a year later on 1949 September 29 with a V2 flight carrying photon-counting tubes (Friedman et al. 1951). Subsequent years have witnessed a myriad rocket and satellite instruments trained on the solar corona with increasing spectral and spatial resolution. Indeed, the impetus to improve spatial resolution imaging of the solar corona using grazing-incidence X-ray telescopes (Vaiana et al. 1973) played a significant role in the train of development of X-ray optics that led to the exquisite *Chandra* mirrors.

The history and development of the study of the solar corona is wonderfully summarized in the books by Golub & Pasachoff (2009) and Mariska (1992), and in numerous reviews (e.g., Vaiana & Rosner 1978) to which the reader is referred for more detailed accounts.

Also shown in Figure 4.8 is an image of the Sun obtained at the time of the eclipse by the *Solar Dynamics Observatory* (*SDO*) Atmospheric Imaging Assembly (AIA) in the 211 Å band. While this wavelength lies firmly in the extreme ultraviolet spectral region, the bandpass is designed to capture light from transitions of Fe xiv formed at a temperature of 2 million K—a temperature at which we are also accustomed to observing in X-rays. This can therefore be considered a good X-ray proxy image of the solar corona.

The X-ray corona comprises bright active regions characterized by plasma trapped within magnetic loops, dark coronal holes, small-scale bright points, and diffuse emission components. As we shall discuss below, stars like the Sun also exhibit rapid variability and flares, resulting from the sudden release of stored magnetic energy in the form of accelerated particles, heat, and ejections of mass.

The 2017 eclipse occurred when the Sun had declined considerably in activity from the last solar maximum in the summer of 2013 and was more than halfway toward minimum. The corona captured by AIA in Figure 4.8 is then rather weak, with very few active regions compared with conditions at solar maximum. Also shown is an image in the same 211 Å band near solar maximum exactly three years earlier, displaying a much richer array of bright active regions.

The conspicuous aspect of the images in Figure 4.8 is the magnetic nature of the corona. It might then be surprising to learn that it was not actually until the first X-ray survey of stars was made by the *Einstein Observatory* in the late 1970s that it was established that stellar coronae are magnetically driven. Despite the coincidence of bright X-ray-emitting regions with magnetically active regions on the solar disk, a common view up to the late 1970s was that the corona was heated acoustically (e.g., Biermann 1946; Schwarzschild 1948) and that the formation of coronae was distinct from magnetic field generation (e.g., Vaiana 1980). Vaiana et al. (1981) showed that the X-ray luminosities of stars of different spectral types were grossly inconsistent with acoustic heating models, while Pallavicini et al. (1981) established that X-ray luminosity was strongly correlated with stellar rotation known to drive magnetic dynamos. This picture of stellar coronae powered ultimately by a magnetic dynamo in the stellar interior had fully crystalized by 1980 (e.g., Rosner 1980).

A number of observatories following *Einstein* have added details to the picture. First and foremost is the *Roentgen Satellite*, which surveyed the sky at EUV and X-ray wavelengths and added many thousands of X-ray detections of stars (Voges et al. 1999). The *Extreme Ultraviolet Explorer* (*EUVE*) obtained EUV spectra and plasma and chemical abundance diagnostics of stellar coronae (Bowyer et al. 2000; Drake et al. 1996) that provided a precursor to the X-ray grating spectroscopy enabled by *Chandra* and *XMM-Newton*. The *Advanced Satellite for Cosmology and Astrophysics* (*ASCA*) and *BeppoSAX* made extensive X-ray observations of stars at low spectral resolution ("CCD resolution"), providing valuable temperature and abundance measurements. For reviews of these developments through to the *Chandra* and *XMM-Newton* era see, e.g., Drake (2001) and Güdel (2004).

By the year 1999 and the launch of *Chandra*, the scene for stellar coronae was largely set and awaited the execution of observations to push toward the next level of understanding.

### 4.2.1 Properties of Stellar Coronal Emission

Solar and stellar coronal spectra are characterized by emission lines from abundant ionized chemical elements superimposed on a continuum. The lines are formed by collisional excitation and subsequent decay; the continua are produced in recombination free–bound transitions. Solar and stellar outer atmospheres—the chromosphere, transition region, and corona—span a temperature range of $10^4$–$10^8$ K. The EUV spectral range is where the dominant emission of the transition region and cooler plasma of the corona resides (although lines of very highly charged ions can be found, such as those of Fe up to Fe XXIV; e.g., Drake 1999), while the X-ray range is the general regime of emission for plasma with temperatures $>10^6$ K.

The relatively low densities of the plasma in solar and stellar coronae (e.g., $n_e \sim 10^8$–$10^{10}$ in nonflaring plasma of the solar corona) render the emission optically thin and collision dominated. If the plasma is also in thermal equilibrium, then the form of the emergent spectrum for a plasma at a given temperature is in principle determined by only three parameters (albeit with the requirement of a substantial amount of data describing the collisional excitation and ionization processes involved): temperature, density, and chemical composition.

The lack of significant optical depth in a collision-dominated plasma means that any one volume element of plasma is radiatively decoupled from any other volume element; the plasma can then be thought of simply as a collection of quasi-isothermal plasmas of different temperatures, each occupying a different volume element. The emergent intensity of a given spectral line from one of these isothermal plasma elements simply depends on the volume-integrated product of the number density of the emitting ionic species, the number density of the line-exciting species (mostly electrons), and the appropriate excitation coefficient describing the efficiency of the line-excitation mechanism. Through the ionization state of the plasma and the relative abundance of the element in question, for a given transition, this product is proportional to the electron density squared, $n_e^2$.

More formally, for a transition $u \rightarrow l$, the intensity $I_{ul}$ is given by

$$I_{ul} = AK_{ul} \int_{\Delta T_{ul}} G_{ul}(T) n_e^2(T)\, dV(T), \tag{4.1}$$

where $A$ is the elemental abundance, $K_{ul}$ is a known constant which includes the wavelength of the transition and the stellar distance, and $G_{ul}(T)$ is the "contribution" function of the line containing all of the relevant atomic physics parameters (parent ion population and collisional excitation rates).

In the solar context, the integral in Equation (4.1) is often carried out over the plasma depth rather than the volume. The quantity $n_e^2(T)V(T)$—the product of the volume and electron density squared for plasma at temperature $T$—is more applicable to the stellar context when the stellar disk is not resolved and is usually referred to as the volume emission measure (VEM). Recasting this integral into one over the temperature, $T$, of the emitting plasma yields the expression for the total intensity of a spectral line:

$$I_{ul} = AK_{ul} \int_{\Delta T_{ul}} G_{ul}(T) \overline{n_e^2(T)} \frac{dV(T)}{d \log T}\, d \log T \ \mathrm{erg\ cm^{-2}\ s^{-1}}. \tag{4.2}$$

Here, the VEM has been transformed to its logarithmic differential form, the differential emission measure (DEM),

$$\mathrm{DEM}(T) = n_e^2(T) \frac{dV(T)}{d \log T}. \tag{4.3}$$

The term $G_{ul}(T)$ is often referred to as the line contribution function. Stellar coronae are low-density plasmas such that spontaneous decay rates usually greatly exceed collisional excitation rates. This "coronal approximation" enables the greatly simplifying assumption that essentially all ions are in their ground states and that collisional de-excitation is negligible. The contribution function can then be written

$$G_{ul}(T) = \frac{n_i(T)}{n_{\text{tot}}} C_{lu}(T) B_{ul}, \tag{4.4}$$

where $n_i$ is the number density of the ionization state $i$ of the element in question with a total number density $n_{\text{tot}}$, $C_{lu}$ is the collisional excitation rate coefficient for the transition $l \rightarrow u$, and $B_{ul}$ is the branching ratio for the $u \rightarrow l$ transition in relation to all other possible radiative de-excitation routes from level $u$. Under the coronal approximation, Equation (4.4) differs from a two-level atom model only in the branching ratio term, $B_{ul}$.

The rate coefficient $C_{lu}(T)$ represents the collisional excitation rate integrated over a Maxwellian velocity distribution for temperature $T$ and in modern assessments is usually the product of detailed quantum mechanical calculations and less commonly of laboratory experiments. Examples of the atomic data assessments and reviews have been presented by Dere et al. (1997), Boyle & Pindzola (2005), Kallman & Palmeri (2007), and Foster et al. (2010, 2012).

The dominant temperature dependence in Equation (4.4) arises through the ionization balance term $n_i/n_{\text{tot}}$, which is computed from the equilibrium between collisional ionization and radiative and dielectronic recombination (see, e.g., Jordan 1970; Arnaud & Rothenflug 1985; Bryans et al. 2009; Dere et al. 2009). For a given ion, $n_i/n_{\text{tot}}$ is strongly peaked at its characteristic temperature of formation. This is illustrated for the case of Fe ions in the temperature range of relevance for X-ray emission in Figure 4.9. Note in particular the somewhat broader population profiles of the ions with full outer shells, $Fe^{16+}$ and $Fe^{24+}$. The peak temperatures of formation of the ions of abundant elements are shown in Figure 4.10, from which the ions relevant for X-ray emission ($T \gtrsim 10^6$ K) can be seen. Thus, spectral lines of

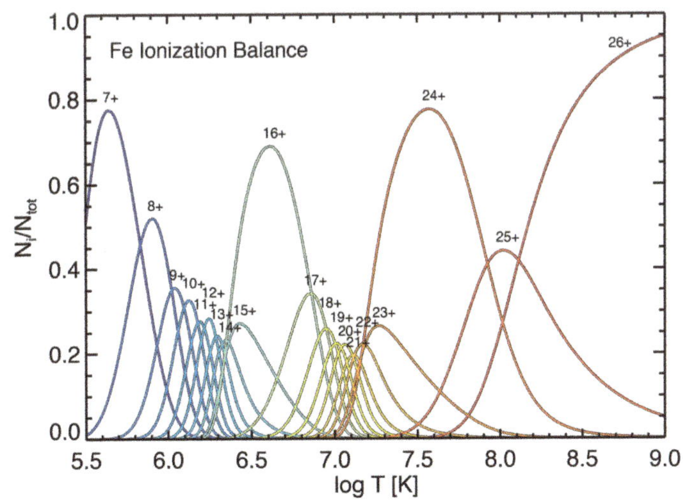


**Figure 4.9.** The ionization balance for Fe ions for the temperature range relevant for X-ray emission from the CHIANTI project (Dere et al. 2009; Landi et al. 2013). Note the broader temperature ranges over which the ions with filled outer electron shells dominate—in this case $Fe^{7+}$ (Ar-like isolectronic sequence), $Fe^{16+}$ (Ne like), and $Fe^{24+}$ (He like).

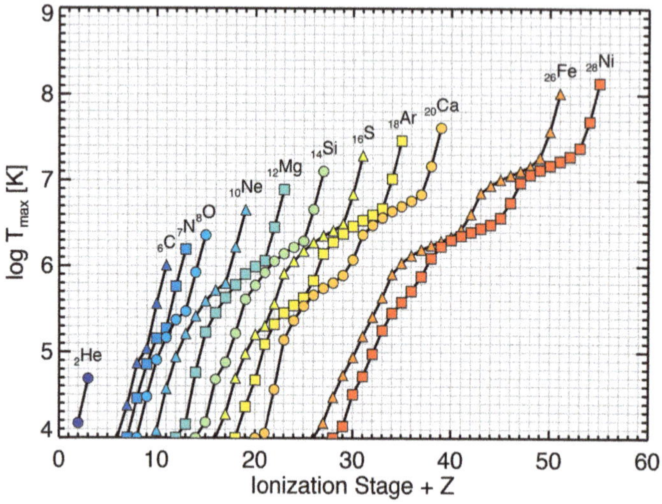


**Figure 4.10.** The temperature of the maximum ion population for abundant elements. The $x$-axis is the sum of the ionization stage and atomic number, such that $Fe^{16+}$ (Fe XVII), for example, lies at position 26 + 16 = 42. Species relevant for X-ray emission are those formed at temperatures of approximately $10^6$ K and higher.

a given ion provide a strong indication of the plasma temperature that produces them.

The radiation from a hot, optically thin plasma of temperature $T$ then comprises line emission from the ions present at that temperature, combined with bound–free and free–free continuum radiation; see Chapter 3 for further details. A useful quantity for understanding the astrophysical behavior of hot plasmas is the radiative cooling time, $\tau_{\rm rad}$,

$$\tau_{\rm rad} \simeq \frac{3kT}{n_{\rm e}\Lambda(T)}, \tag{4.5}$$

where $\Lambda(T)$ is the radiative loss function—essentially the total line + continuum emitted power as a function of temperature per unit emission measure (Chapter 3). For a plasma density of the order of $n_{\rm e} \sim 10^{10}$ cm$^{-3}$, the cooling time is of the order of 1000 s for a plasma with solar chemical composition. This is an upper limit to the true cooling timescale, which for stellar coronae also suffers conductive losses down to the chromosphere with a cooling timescale that can be approximated for a coronal loop of length $L$ as

$$\tau_{\rm cond} \simeq \frac{3n_{\rm e}kT}{\kappa T^{7/2}/L^2} = \frac{3n_{\rm e}kL^2}{\kappa T^{5/2}}, \tag{4.6}$$

where $\kappa$ is the plasma thermal conductivity.

Two example spectra obtained by the *Chandra* LETG+HRC-S for the very active K0 V star AB Dor and the inactive G2V $\alpha$ Cen A from an analysis by Wood et al. (2018) are illustrated in Figure 4.11. These spectra demonstrate the presence of hotter plasma in the corona of the more active star, as betrayed by the presence of

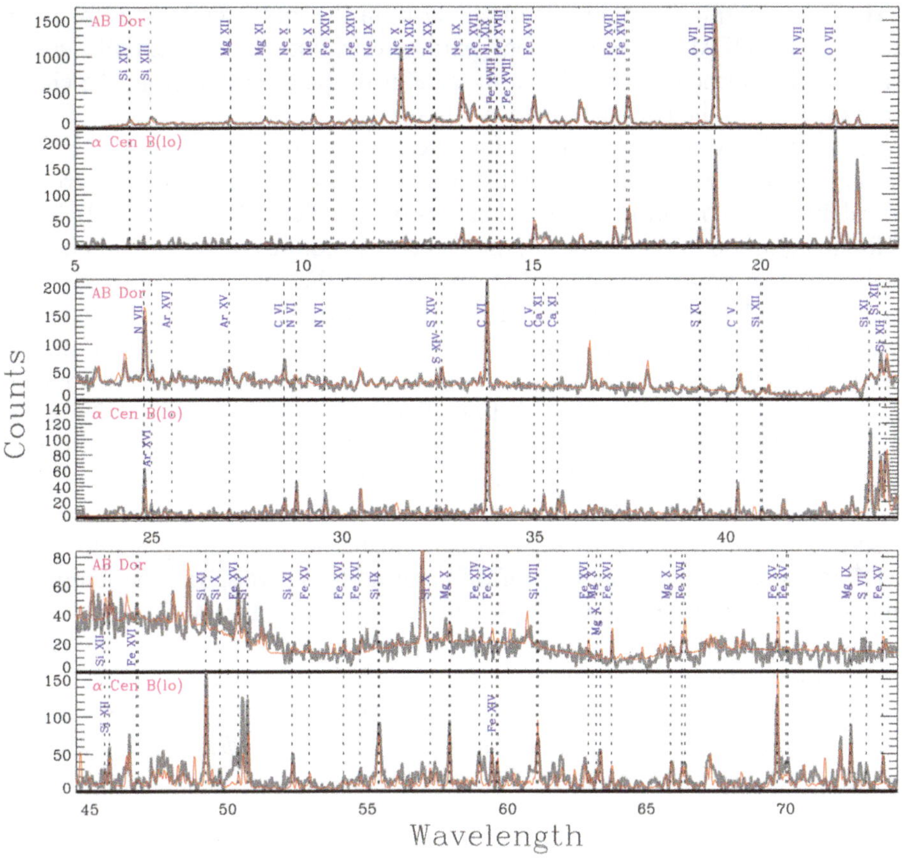


**Figure 4.11.** *Chandra* LETG+HRC-S spectra of the very active K0 V star AB Dor and the inactive G2V $\alpha$ Cen A. Each shows a series of lines from abundant ions superimposed on a continuum (which is difficult to see here) formed in the optically thin, collision-dominated plasma that makes up their coronae. Ions responsible for prominent lines are indicated. Note that the spectrum of the active AB Dor extends to shorter wavelengths and exhibits lines from plasma formed at much higher temperatures than the spectrum for the inactive $\alpha$ Cen A. The red traces are synthetic spectra computed from the DEMs derived from the spectra. Reproduced from Wood et al. (2018). 

species such as Fe xx, Mg xii, and Si xiv that are very weak or absent in the spectrum of $\alpha$ Cen A. Synthetic spectra generated using differential emission measure distributions derived from the observations show remarkably good agreement with the observed spectra, demonstrating the validity of the assumptions of emission from optically thin, collision-dominated plasmas.

*The Differential Emission Measure Distribution*

The sharp temperature range of line formation of a given ion is especially relevant for providing temperature-dependent diagnostics for the DEM in Equation (4.3). The DEM was first formulated by Pottasch (1963) and in different forms by several authors since. Craig & Brown (1976) provided the first rigorous definition of the

DEM as a weighting function, or source term, in the integral equation for the line intensity. The DEM is a convenient parameterization of a potentially complex plasma, and in principle, once its form is defined, the spectrum from the chromosphere through to the corona can be synthesized.

While the DEM provides great simplification to describing coronal emission, determining the form of the DEM from observations of individual or groups of spectral lines is a notorious integral inversion problem (Equation (4.2) being a Fredholm equation of the first kind) and generally requires some form of constraint to solve, such as enforcing artificial smoothness on the solution. Numerous studies have employed DEM analysis in solar and stellar research. For the former, see, e.g., Bruner & McWhirter (1988), Kashyap & Drake (1998), and Jordan (2000). Emission measure modeling of stars based on individual spectral lines was first applied to *EUVE* spectra (see, e.g., Drake et al. 1995; Sanz-Forcada et al. 2002, 2003; Bowyer et al. 2000). We shall return later in this section to DEM analysis based on *Chandra* observations. Subsequent studies based on high-resolution X-ray spectroscopy obtained by *Chandra* and *XMM-Newton* have helped to define the coronal emission measures for temperatures $>10^6$ K. A rough schematic of the approximate shape of the DEM, how it changes in stars of different activity levels, and the temperature range in which EUV emission is dominant is shown in Figure 4.12.

Many emission measure distribution analyses have been carried out based on the strengths of emission lines seen in *Chandra* grating spectra. One example is the work of Wood et al. (2018), who used the X-ray spectra of 19 main-sequence stars from early-F to mid-M spectral types observed using the *Chandra* LETG+HRC-S to investigate the emission measure distributions and chemical abundances in the corona. They further used the data to determine the general behavior in the DEM as

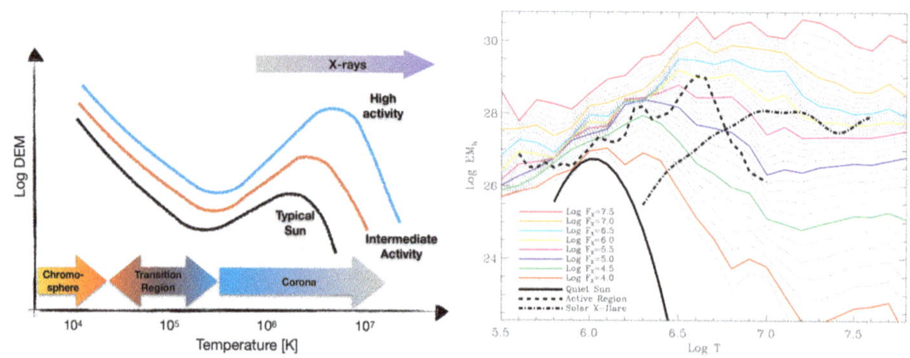


**Figure 4.12.** Schematic illustration of the rough shape of the emission measure distribution from the chromosphere to the corona. More magnetically active stars have larger emission measures stretching to higher temperatures. The form of the DEM is very uncertain in the transition region up until the corona and where temperatures reach in excess of $10^6$ K and are amenable to X-ray spectroscopy. Right: main-sequence star DEM distributions as a function of surface flux, $F_X$, based on the DEMs of 19 stars observed with the *Chandra* LETGS. These are compared with distributions for a solar active region (Warren et al. 2012), the quiet Sun (Kamio & Mariska 2012), and an average solar flare (Warren 2014). Reproduced from Wood et al. (2018). 

a function of activity level as described by surface X-ray flux using the Markov Chain Monte Carlo method of Kashyap & Drake (1998). The resulting DEMs are illustrated in Figure 4.12. It is important to note that the DEM at the ends of the temperature range covered by the available line diagnostics are generally very uncertain and depend on any applied smoothing and other assumptions made in the analysis.

*Density and Temperature Diagnostics of He-like Ions*

High-resolution X-ray spectra obtained by the *Chandra* diffraction gratings allow individual spectral lines to be resolved, opening up a wealth of plasma diagnostics afforded by the different behavior of some spectral lines to changes in plasma density and temperature. Perhaps the most powerful of these are density diagnostics—lines with excitation and resulting intensity depending, to an observable extent, on the plasma density in a different way from the simple $n_e^2$ dependence in Equation (4.2). While numerous lines in the X-ray range offer some degree of density sensitivity, they are often faint, difficult to observe, and blended with other lines, even at the resolution of *Chandra*'s gratings. We restrict our discussion here to the singlet and triplet lines of He-like ions that are by far the most commonly exploited density diagnostics in *Chandra* observations of collision-dominated, optically thin "coronal" plasmas; see, e.g., Foster et al. (2010, 2012) for further details of X-ray plasma diagnostics.

Due to their closed shell structure, He-like ions are abundant over wider temperature ranges than other ions in collision-dominated plasmas (see Figure 4.10 and 4.9). Owing to the large cosmic abundances of the elements C, N, and O, their He-like lines are often the most prominent features of soft X-ray spectra. He-like ions of Ne, Mg, Si, S, and Fe are also commonly observed, although *Chandra*'s gratings have insufficient resolving power to separate the different components for He-like Fe. A simplified energy-level diagram for He-like ions is illustrated in Figure 4.13.

The strongest transitions in He-like ions are between the $n = 2$ and $n = 1$ ground state. There are four lines of relevance: the resonance line, often referred to as $w$ or $r$ ($1s\,2p\;^1P_1 \rightarrow 1s^2\;^1S_0$), the intercombination lines, often called $x$ ($1s\,2p\;^3P_2 \rightarrow 1s^2\;^1S_0$) and $y$ ($1s\,2p\;^3P_1 \rightarrow 1s^2\;^1S_0$), or $i$ together, and the forbidden line $z$ or $f$ ($1s\,2s\;^3S_1 \rightarrow 1s^2\;^1S_0$). The blended $x$ and $y$ lines are often counted as one line, and the complex is commonly referred to as the He-like "triplet."

The diagnostic utility of the He-like triplet was first pointed out by Gabriel & Jordan (1969) in an analysis of this line complex for several elements in solar X-ray spectra. There are two ratios of combinations of the He-like triplet that are particularly useful: the density sensitive ratio, $R(n_e) = z/(x + y)$ (sometimes written $R(n_e) = f/i$), and the temperature sensitive ratio $G(T_e) = (z + (x + y))/w$ (sometimes $G(T_e) = (f + i)/r$).

The ratio $R(n_e) = f/i$ is illustrated for He-like O ($O^6$+ or O VII) in Figure 4.13. As density increases, the ratio is fairly constant at $R \approx 3.5$ until $n_e \sim 10^9\ \mathrm{cm}^{-3}$, after which it declines quite steeply until it reaches close to zero at $n_e \sim 10^{12}\ \mathrm{cm}^{-3}$. This behavior is a result of the small transition probability for the forbidden line $z$ and the

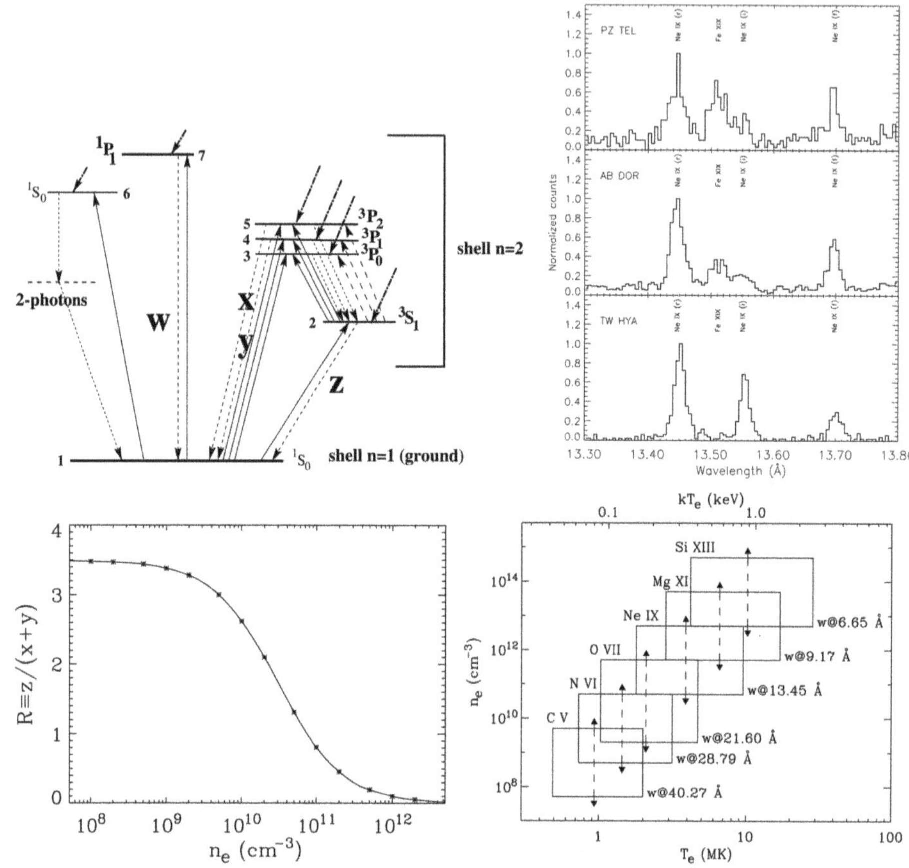


**Figure 4.13.** Top left a simplified energy-level diagram for He-like ions illustrating the resonance $w$ (or $r$), intercombination $x$, $y$ (or $i$), and forbidden $z$ (or $f$) line transitions, respectively. The various excitation and decay mechanisms are denoted by upward arrows for electron collisional excitation (solid arrows) and photoexcitation (dashed arrows). Radiative transitions are denoted by downward arrows. The thick dotted–dashed downward arrows denote the population of levels by radiative and dielectronic recombination and cascades. Top right: the Ne IX He-like complex observed in the classical T Tauri stars PZ Tel and TW Hya, together with the young active K1 dwarf AB Dor. TW Hya exhibits a high density of $n_e \approx 10^{13}\,\mathrm{cm}^{-3}$, thought to arise in an accretion shock (Kastner et al. 2002; see Section 4.2.7), whereas PZ Tel shows no signs of high densities and its spectrum resembles that of AB Dor. Reproduced from Argiroffi et al. (2004) © 2004. The American Astronomical Society. All rights reserved. Bottom: the density sensitivity of the ratio $R(n_e) = z/(x + y)$ for the He-like ion O VII (left), and the ranges of density and temperature sensitivities of the He-like isoelectronic sequence for abundant ions in cosmic plasmas (right). Figures are from Porquet & Dubau (2000), reproduced with permission © ESO; and Porquet et al. (2010), reprinted with permission from Springer, © Springer Science+Business Media B.V. 2010.

collisional excitation route from its upper $^3S_1$ level to the $^3P$ states, which then decay to the ground state. At densities of $n_e \sim 10^9\,\mathrm{cm}^{-3}$, this collisional excitation rate becomes comparable with the radiative decay rate from $^3S_1$ to ground, and at higher densities eventually dominates such that the forbidden line can be entirely quenched in favor of the $x$ and $y$ intercombination lines.

The ratio $G(T_e)$ decreases with increasing temperature (exciting electron energy) owing to an increasing excitation rate for the singlet $^1P_1$ state relative to the triplet $^3S_1$ and $^3P_{0,1,2}$ states.

The approximate ranges of density and temperature sensitivity of the He-like triplets of the abundant elements prominent in *Chandra* X-ray spectra are also illustrated in Figure 4.13. Higher $Z$ ions are only sensitive at higher temperatures and densities, and this limits their utility to some extent. For example, the Si XIII triplet is not sensitive to density below several $10^{12}$ cm$^{-3}$, which is much higher than ambient densities in stellar coronae (e.g., Testa et al. 2004; Ness et al. 2004) and probably only realized during the strongest flares.

The $x$ and $y$ upper levels can also be excited from $^3S_1$ radiatively (Gabriel & Jordan 1969), rendering the $R$ ratio also a potentially useful diagnostic of the intensity of the radiation field (e.g., Ness et al. 2001). In the O VII case, the transitions correspond to wavelengths of 1624, 1638, and 1640 Å; for lower Z He-like ions, the analogous transitions are at approximate wavelengths of 1900 and 2275 Å for N and C, respectively, and for higher Z ions at 1260, 1020, and 845 Å for Ne, Mg, and Si, respectively. Ness et al. (2002) showed that the radiation fields of late-type stars are negligible for O and higher $Z$ He-like triplets, but that for N and C, radiative excitation can be significant for early G-type and hotter stars.

While the above discussions of the DEM and of density diagnostics are by necessity brief, we are now in a position to understand the theoretical underpinnings of X-ray emission from low-mass stars. For a more complete treatment of coronal emission, see, e.g., Golub & Pasachoff (2009) and Mariska (1992).

### 4.2.2 The Rotation-powered Magnetic Dynamo

We noted in the introduction to this chapter that the magnetic nature of stellar coronae was essentially established by the *Einstein Observatory* and the realization that X-ray luminosity was highly correlated with stellar rotation (Vaiana et al. 1981; Pallavicini et al. 1981; Walter & Bowyer 1981). The groundwork for this was laid by chromospheric emission in Ca II H & K lines in the Sun showing line core emission fluxes varied linearly with surface magnetic field strength (Frazier 1970), and the realization that H & K fluxes declined linearly with rotation velocity (Kraft 1967) and with time $t$ approximately as $t^{1/2}$ (Skumanich 1972). The latter relation arises due to the gradual loss of angular momentum through the stellar analog of the solar wind (Kraft 1967; Weber & Davis 1967; Durney 1972; Mestel & Spruit 1987).

The remarkable short paper by Skumanich (1972), featuring only four Ca II data points, is a seminal work on the stellar rotation–activity relation and spawned a large volume of research on this topic.

#### *A Magnetic Activity Rossby Number*

The next major development was the inclusion of a "Rossby number" into flux–rotation relations. Ca II flux versus rotation period for late-type stars is subject to a systematic spectral-type-dependent scatter that Noyes et al. (1984, see also Noyes 1983) noticed was removed if the rotation period is replaced by the ratio of the convective turnover time to rotation period, $Ro = P_{rot}/\tau_c$, where this definition of the Rossby number is analogous to its more common fluid dynamics usage.

When the early *Einstein* X-ray–rotation results were bolstered with the advent of the *ROSAT* all-sky survey and the addition of many more X-ray luminosity data points for stars with a large range of ages and rotation periods, the scatter in the X-ray–rotation activity relation was also found to be greatly reduced when the ratio of stellar X-ray to total bolometric luminosity, $L_X/L_{bol}$, was plotted as a function of *Ro* (see, e.g., Pizzolato et al. 2003; Wright et al. 2011). This was borne out more dramatically when Wright et al. (2016, 2018) added *Chandra* observations of slowly rotating late-M dwarfs that are too faint to have been detected by *ROSAT*. The data are illustrated in Figure 4.14, plotted against both rotation period and Rossby number.

The association of chromospheric emission and coronal X-rays with an internal magnetic dynamo posits that some fraction of the magnetic energy created within the star by dynamo action and subject to buoyant rise is dissipated at the stellar surface and converted into particle acceleration and plasma heating. It must be emphasized that none of these processes are fully understood! In this context, Figure 4.14 is remarkable, showing that the extremely complex physical system of plasma and magnetic fields that make up a stellar corona, fed by a complex internal dynamo, behaves in bulk in a very simple way: at slower rotation rates, $L_X/L_{bol} \propto Ro^{\beta}$ up until a threshold at which point X-ray emission saturates, $L_X/L_{bol} \sim 10^{-3}$, close to $Ro = 0.13$. This saturation behavior was already apparent based on *Einstein* data (Vilhu 1984; Micela et al. 1985), though its origin has still not been firmly demonstrated. It is likely that it represents saturation of the dynamo itself, but other mechanisms such as centrifugal stripping of coronal plasma by rapid rotation, or saturation of the coronal surface filling factor, have also been suggested

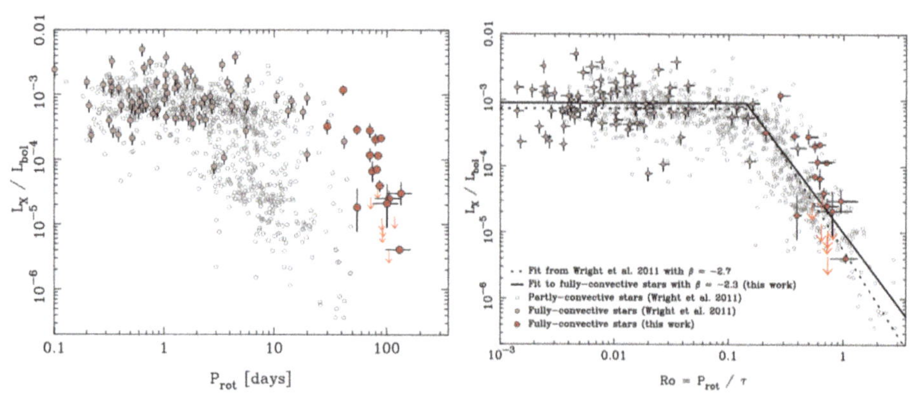


**Figure 4.14.** X-ray to bolometric luminosity ratio, $L_X/L_{bol}$, plotted against stellar rotation period, $P_{rot}$ (left) and Rossby number, $Ro = P_{rot}/\tau_c$ (right). Fully convective stars observed by Wright et al. (2016, 2018) are denoted with large red points, with other fully convective stars in the saturated regime shown as light red points. Partly convective stars are represented by empty gray circles. Error bars are shown for all fully convective stars. Upper $3\sigma$ limits are shown for the undetected fully convective stars observed as part of this work as red arrows. Reproduced from Wright et al. (2018), by permission of Oxford University Press on behalf of the Royal Astronomical Society.

to play a role (see, e.g., the discussion in Wright et al. 2011, and Blackman & Thomas 2015).

The implication from the correlation of stellar rotation with magnetic activity is that stellar dynamos are driven by rotation, or more properly, by differential rotation that results from convective transport in a rotating reference frame, providing an elegant qualitative confirmation of the elementary "$\alpha\Omega$" dynamo theory proposed by Parker (1955) and demonstrated in the solar context by, e.g., Babcock (1961). An $\alpha\Omega$ dynamo comprises differential rotation, which stretches an initially poloidal field to produce a toroidal field (the $\Omega$ effect), and cyclonic convection, which stretches the field as it rises due to magnetic buoyancy (the $\alpha$ effect), regenerating the poloidal field. The topic of stellar dynamos now comprises an immense literature, and its full discussion requires an entire book of its own; the reader is instead referred to reviews by Ossendrijver (2003) and Charbonneau (2014).

The Rossby number approach of Noyes et al. (1984) leads to a dynamo efficiency (or "dynamo number") proportional to $Ro^{-2}$, such that on the unsaturated part of the $L_X/L_{bol}$–$Ro$ relation, $L_X$ should depend on $Ro$ as $L_X \propto Ro^{-2}$. Montesinos et al. (2001) further refined the expression for the dynamo number, and Wright et al. (2011) showed that this can be approximated by

$$N_D \propto \frac{1}{Ro^2}\frac{\Delta\Omega}{\Omega}, \tag{4.7}$$

where $\Omega$ is the angular rotation rate $2\pi/P_{rot}$ and $\Delta\Omega$ is an effective mean differential rotation. Based on the stellar data, Wright et al. (2018) found $L_X \propto Ro^{\beta}$ with $\beta = 2.3^{+0.4}_{-0.6}$. This implies that, to within the accuracy of the measurement, the differential rotation rate relevant to stellar dynamos is proportional to the rotation rate, $\Delta\Omega/\Omega \propto \Omega$, thus recovering $L_X \propto Ro^{-2}$.

It is presently thought that the site of the differential rotation responsible for the $\Omega$ effect in $\alpha\Omega$ dynamos is the "tachocline"—a thin shear layer at the base of the convection zone at the radiative–convective boundary found through helioseismology (Brown et al. 1989; Goode et al. 1991; see, however, the critique of Spruit 2011). This poses an interesting issue for main-sequence stars with masses $M \lesssim 0.35\ M_{\odot}$, corresponding to spectral types later than M3.5 V, with fully convective internal structures: because they possess no tachocline, their dynamo behavior might naively be expected to be different. The internal structure of stars with and without a tachocline are illustrated schematically in Figure 4.15. The slowly rotating late-M dwarfs observed by *Chandra* show the same dependence of $L_X$ on Rossby number as higher mass stars, implying that a tachocline is not a necessary ingredient in solar-like dynamos.

*Supersaturation*

The X-ray luminsosity data for stars in Figure 4.14 show one further interesting feature: at very fast rotation rates and small values of $Ro$, there is a hint of a decline in $L_X/L_{bol}$ with increasing rotation in G and K dwarfs. Dubbed "supersaturation," this phenomenon first came to light in *ROSAT* stellar surveys (e.g., Randich et al. 1996). Wright et al. (2011) examined two possible mechanisms for supersaturation,

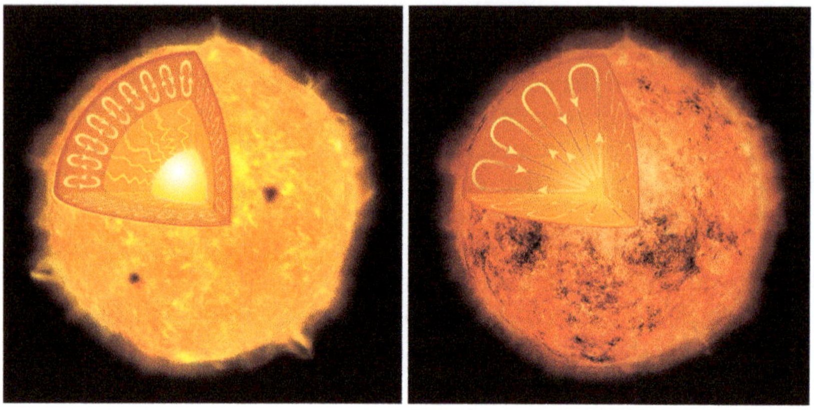

**Figure 4.15.** Illustration of the internal structure of a Sun-like star, with an outer convection zone and a radiative, convectively stable, core (left), and a low-mass, late-M dwarf with an interior that is unstable to convection through to the center. Conventional dynamo theory seats the dominant differential rotation amplification Ω of the magnetic field at the base of the convection zone. From http://chandra.harvard.edu/photo/category/stars.html. Courtesy of NASA/CXC/M.Weiss.

centrifugal stripping of coronal loops and poleward migration of magnetic flux due to rotation-induced polar updrafts, but found the data insufficient to distinguish between them.

The answer might have been provided by Argiroffi et al. (2016), who used *Chandra* to investigate the rotation–activity relation in the young ~13 Myr old cluster h Persei. Stars in h Persei have ended their T Tauri accretion phase (see Sections 4.2.7 and 4.2.7) and those of approximately solar mass have developed a radiative core. The expectation is that their dynamos should operate with the same $\alpha\Omega$ mechanism as the solar dynamo, as discussed in Section 4.2.2. Some of them have already experienced rotational braking such that the cluster samples from the most rapid rotators through to the slower rotators expected to be in the unsaturated regime.

Argiroffi et al. (2016) found that h Per members in the mass range $1.0\,M_{\odot} < M < 1.4\,M_{\odot}$ indeed sample all three regimes of the rotation–activity space: unsaturated, saturated and supersaturated. Supersaturation in $L_X/L_{\rm bol}$ was better described by the rotation period, $P_{\rm rot}$, than by $Ro$, and the distribution in $L_X/L_{\rm bol}$ at the fastest rotation periods was compatible with the centrifugal stripping of coronal loops. Compact loops with sizes significantly less than the stellar radius should be stable to centrifugal stripping; Argiroffi et al. (2016) concluded that a significant fraction of the X-ray luminosity in active stars originates in structures as large as two stellar radii above the stellar surface.

*X-Ray Magnetic Cycles*

A salient aspect of solar magnetic activity is the 22 year cycle in which the amplitude of activity indices is strongly modulated with an 11 year period, and the magnetic field polarity undergoes a full reversal and return cycle. The X-ray luminosity of the

Sun varies over the cycle from sunspot minimum to maximum by a factor of 5–10, depending on the bandpass of the measurement (e.g., Ayres 2014).

Stellar magnetic cycles have been searched for since 1966 and the beginning of the "HK Project" at Mount Wilson Observatory to monitor the chromospheric activity diagnostic Ca II H & K line cores in stars (Wilson 1978). Cycles can in principle provide key information on the dynamo process at work in the stellar interior. In most cool stars (F–M), they are thought to arise from the interplay of large-scale shear arising from differential rotation, small-scale convective helicity, and meridional circulation in an $\alpha\Omega$ dynamo (Charbonneau 2014).

Clear magnetic cycles from H and K lines turn out to be less common than flat or chaotic activity trends, with the occurrence rate tending to increase with stellar age and rotation period. The first detection of an X-ray cycle was made based on *XMM-Newton* monitoring of the G2V star HD 81809, which also has a clear Ca II cycle (Wilson 1978). The stars with detected X-ray cycles at the time of writing are illustrated in Figure 4.16 from Wargelin et al. (2017). *Chandra* has contributed the key data of the $\alpha$ Cen system to this collection (Ayres 2014; Wargelin et al. 2017).

The X-ray amplitude of cycles tends to decrease with increasing stellar activity such that they become challenging to detect; the cycle for Proxima Cen, for example, required careful filtering out of flaring in order to see the underlying cyclic trend. Wargelin et al. (2017) compared X-ray amplitude to various stellar parameters, such as mass and rotation period, and found that the best correlation was with Rossby number, *Ro*. The best-fit power law to the data yields $L_X^{max}/L_X^{min} = 2Ro^{1.4}$.

That X-ray cycle amplitude decreases with increasing magnetic activity level meshes with the observation that the most active stars seem to have essentially constant levels of quiescent (nonflaring) X-ray emission. The longest sequence of

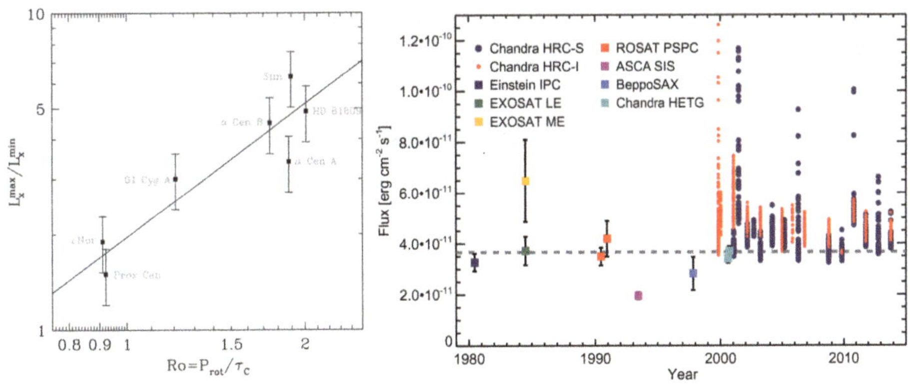


**Figure 4.16.** Left: X-ray cycle amplitude versus Rossby number for stars with X-ray cycle detections. The fitted power law corresponds to $L_X^{max}/L_X^{min} = 2Ro^{1.4}$. Cycle amplitude uncertainties have not been properly determined for these data and ±20% error bars are shown for illustrative purposes. Reproduced from Wargelin et al. (2017), by permission of Oxford University Press on behalf of the Royal Astronomical Society. Right: the X-ray flux measured for the RS CVn-type active binary AR Lac, showing a remarkably constant level of coronal activity over a period of 33 years. The sharp peaks in the *Chandra* data correspond to flares on top of this steady quiescent base-level emission. Reproduced from Drake et al. (2014a). 

X-ray observations of a star other than the Sun is of the very active RS CVn-like binary system AR Lac, which is used for short *Chandra* calibration and monitoring observations. AR Lac comprises a G2 IV primary star and a K0 IV secondary in a close orbit with a period of 1.98 days, and is well inside the saturated activity regime.

Drake et al. (2014a) combined *Chandra* data with older *ASCA*, *Einstein*, *EXOSAT*, *ROSAT*, and *BeppoSAX* observations (see Figure 4.16) and found the level of quiescent, nonflaring coronal emission at X-ray wavelengths to have remained remarkably constant over 33 years, with no sign of variation due to magnetic cycles. Variations in base-level X-ray emission seen by *Chandra* over 13 years were only ~10%, while variations back to pioneering *Einstein* observations in 1980 amounted to a maximum of 45% and more typically about 15%.

*X-Rays in Time*

Magnetic activity turns out to be the cause of its own demise. Figure 4.14 can be thought of as a roadmap of the X-ray activity of a star through time, beginning somewhere toward the left at the zero-age main sequence and slowly evolving in Rossby number to the right as angular momentum is leached from the star by its wind that is itself powered by magnetic dissipation. Stars of different masses evolve in Rossby number at different rates, with F stars evolving much more quickly than M stars. This can be seen from the Rossby number definition itself: for a given rotation period, *Ro* is proportional to the inverse of the convective turnover time, which is a few days for stars of spectral-type F but up to 1000 days for late-M dwarfs (e.g., Wright et al. 2011). Higher mass stars of a given rotation period then lie farther to the right in Figure 4.14 than lower mass stars rotating at the same rate. A solar-mass star, with $\tau_c \sim 10$ days, desaturates rapidly in a time of about 200 Myr. The lowest mass stars instead do not leave the saturated regime until their rotation periods reach of the order of 100 days, which can correspond to timescales of several gigayears (e.g., Newton et al. 2016).

### 4.2.3 Inference of Coronal Structure from Density Diagnostics

Plasma density diagnostics such as those described in Section 4.2.1 have now been applied to a wide range of stars based on high-resolution *Chandra* grating spectra (e.g., Ness et al. 2004; Testa et al. 2004). Their diagnostic power lies partly in helping understand the geometry and morphology of stellar coronae. As we have seen in Section 4.2.2, the Sun is a comparatively inactive star, and the most active stars can have X-ray luminosities 10,000 times brighter than the average solar X-ray output. Since their first detection, an important question has been how are these coronae structured? Are their coronal volumes 10,000 times larger, such that coronae are significantly extended relative to the stellar radius? Or are they more dense and therefore compact? By understanding the density, $n_e$, the emitting volume drops readily out of the volume emission measure, $n_e^2(T)V(T)$.

Ness et al. (2004) derived densities for a sample of 42 stars observed in 22 *XMM-Newton* RGS spectra, and in 16 *Chandra* LETGS and 26 HETGS spectra scattered between $\log n_e \approx 9.5$–11 based on the O VII triplet and between $\log n_e \approx 10.5$–12 from

Ne IX. Testa et al. (2004) analyzed O VII, Mg XI, and Si XIII X-ray spectra of a sample of 22 active stars observed with the *Chandra* HETG. Mg XI lines indicated the presence of high plasma densities up to a few times $10^{12}$ cm$^3$ for most of the more active sources with X-ray luminosity $L_X > 10^{30}$ ergs s$^{-1}$, where stars with higher $L_X$ and $L_X/L_{bol}$ have higher densities at high temperatures. These densities indicate remarkably compact coronal structures. O VII lines instead yielded much lower densities of a few $10^{10}$ cm$^{-3}$, showing that cooler and hotter plasmas occupy physically different structures.

Based on understanding the coronal emitting volume, coronal "filling factors"—the fraction of the stellar surface covered by X-ray emission—can be derived by assuming a scale height for the coronal plasma. Testa et al. (2004) adopted a scale height based on the lengths of quasi-static uniformly heated coronal loops (Rosner et al. 1978) and found filling factors ranging from $f_{MgXI} \approx 10^{-4}$ to $10^{-1}$ for Mg IX and the hot ($\sim 10^7$ K) plasma, and $f_{OVII}$ from a few $10^{-3}$ up to 1 for O VII and the cooler ($2 \times 10^6$ K) plasma. These results are shown as a function of X-ray surface flux in Figure 4.17. Remarkably, $f_{OVII}$ approaches unity at the same stellar surface X-ray flux level as that which characterizes solar active regions ($F_X \sim 10^7$ erg cm$^2$ s$^{-1}$), suggesting that at this activity level, these stars become completely covered by active regions. At the same surface flux level, $f_{MgXI}$ increases more sharply with increasing surface flux. Testa et al. (2004) concluded that hot, dense $10^7$ K plasma in active coronae arises from flaring activity and that this flaring activity increases markedly once the stellar surface becomes covered with active regions, which then increase surface magnetic interactions.

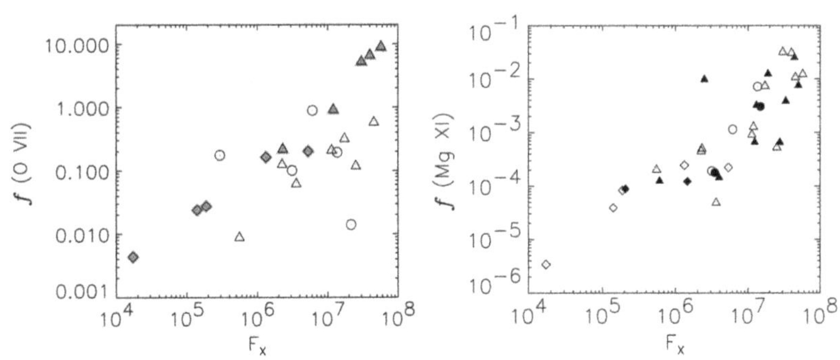


**Figure 4.17.** Surface filling factors derived from the He-like density diagnostics O VII (left) and Mg XI (right) versus stellar X-ray surface flux in a sample of active stars observed by the *Chandra* HETG and analyzed by Testa et al. (2004). Circles denote dwarfs, diamonds denote giants, and triangles refer to binaries. For the O VII data, gray symbols refer to stars with O VII lines that were not measurable and a density $n_e = 2 \times 10^{10}$ cm$^{-3}$ was assumed instead. Filling factors greater than 1 are unphysical and are likely a result of the true density being higher than this for those stars. In the case of the Mg XI data, filled symbols are results from the HEG while empty symbols denote MEG results. Remarkably, the filling factor for hot $10^7$ K plasma indicated by the Mg XI lines increases sharply once the lower temperature plasma reaches a filling factor close to unity. Reproduced from Testa et al. (2004) © 2004. The American Astronomical Society. All rights reserved.

### 4.2.4 Magnetic Reconnection Flares

Stellar coronae are almost all observed to undergo flaring in X-rays, much like the Sun. The flares are caused by the impulsive release of magnetic energy that is gradually built up by convective and other surface motions on the stellar surface within which the magnetic field is anchored. Magnetic reconnection flares then represent a resetting of the corona to a lower magnetic potential energy state. The theoretical challenge of flare studies is to understand the processes involved and how the energy is partitioned between the different loss mechanisms—optical through to X-ray radiation, energetic particles, MHD waves, mass motions, and associated CMEs.

Flares on the Sun are observed over a very wide range of X-ray energies, ranging from less than $10^{27}$ erg in the *Geostationary Operational Environmental Satellite* (*GOES*) 1–8 Å band (~1.5–12 keV) to $10^{31}$ erg, and likely more, for historical solar events such as the 1859 Carrington event (e.g., Moschou et al. 2019). Shorter events appear to decay on radiative cooling timescales (Equation (4.5)) of tens of minutes to an hour or so, whereas the largest flares on the Sun—the so-called "two-ribbon flares"—can last significantly longer than this. These events occur in more complex loop or arcade systems and imply that continued heating after the flare onset is applied. In the canonical flare picture, this is by continuous reconnection of nested, initially open, magnetic fields at successively greater heights. The literature on solar flares is vast; the reader is referred to Benz (2008) and Shibata & Magara (2011) for thorough reviews.

There are many examples of flares in *Chandra* observations of stars, and we touch on this topic again in Section 4.2.7 below. Here we cite two examples.

Figure 4.18 illustrates the *Chandra* X-ray light curve showing flares on the active M dwarf "flare star" EV Lac based on a 96 ks observation with the HETGS obtained by Huenemoerder et al. (2010) in 2009 March. EV Lac is a nearby (5 pc) dM3.5e single star and among the most X-ray of its type, with a mean $L_X \sim 4 \times 10^{28}$ erg s$^{-1}$. The term "flare star" originates from the early to mid-19th century when nearby red dwarfs, such as AT Mic and UV Ceti, were first noticed to undergo unpredictable and dramatic increases in brightness. These variations in the optical are the stellar analogs of the white light component of flares—essentially the photospheric and chromospheric response to fluxes of energetic electrons and protons accelerated in the magnetic reconnection process (e.g., Benz 2008).

The X-ray brightness variations in Figure 4.18 show structure on different scales. The beginning of the observation is characterized by the decay of a large flare with an X-ray peak that likely occurred prior to the observation start. The decay timescale of this event is of the order of 10 ks. In contrast to this, several smaller events are seen, all with much shorter decay timescales of 1 ks or less. Surprisingly, Huenemoerder et al. (2010) found the shorter flares to be hotter than large, long events. To place these EV Lac flares in the solar context, their peak flare power is about $3 \times 10^{29}$ erg s$^{-1}$, and their total energies are ~$10^{32}$–$10^{34}$ erg—10 to 1000 times more energetic than the largest solar flares ever recorded.

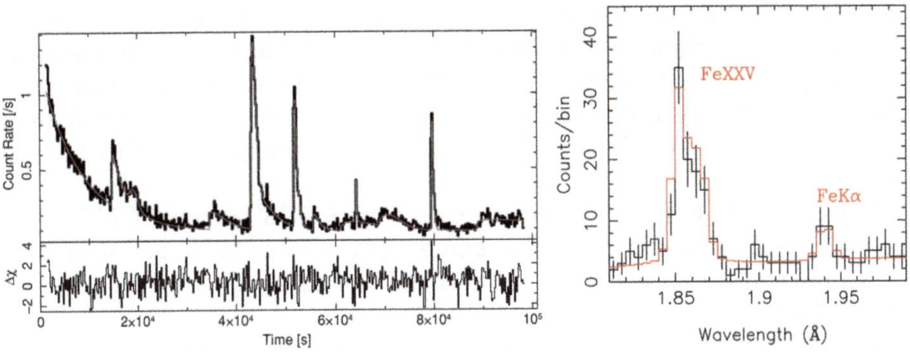


**Figure 4.18.** Left: HETG 1–25 Å events light curve of EV Lac from a 2009 March observation by Huenemoerder et al. (2010) © 2010. The American Astronomical Society. All rights reserved. Flares are frequent with an approximate rate of 0.4 $hr^{-1}$ for easily discernible events. An empirical function modeling the flares is superimposed together with residuals (lower panel). Right: a portion of the HEG spectrum of HR 9024 on which a large flare was observed during the *Chandra* observation. In addition to the prominent He-like resonance line of Fe XXV, a cool fluorescence line resulting from reprocessing of the X-rays by the stellar photosphere was seen that provided geometrical constraints on the flare size, limiting the height of the source above the photosphere to $0.3R_\star$ or less. Reproduced from Testa et al. (2008) © 2008. The American Astronomical Society. All rights reserved.

An even more energetic flare was observed by the *Chandra* HETG on the single intermediate-mass active K1 giant HR 9024 by Testa et al. (2007), peaking at $3.5 \times 10^{32}$ erg $s^{-1}$. The flare lasted about 50 ks, and the integrated X-ray energy was approximately $6 \times 10^{36}$ erg—five orders of magnitude more energetic than the largest solar events. Testa et al. (2008) detected the fluorescent line of Fe produced by inner-shell ionization of the iron in the chromosphere. The strength of this line depends on the height of the ionizing source, and Testa et al. (2008) were able to use the line to constrain the height of the flaring loop, or loops, to within 0.3 stellar radii. Testa et al. (2007) noted that the radius of HR 9024 is about 14 $R_\odot$, giving the height of the flaring loops of up to 4 solar radii. This was broadly consistent with the hydrodynamic flare model investigated by Testa et al. (2007).

Argiroffi et al. (2019) subsequently detected Doppler shifts in S XVI, Si XIV, and Mg XII lines that betrayed upward and downward plasma motions with velocities of 100–400 km $s^{-1}$ within the flaring loop, again in broad agreement with a hydrodynamic model. Perhaps even more fascinating was a later blueshift seen in O VIII, which reveals a line-of-sight upward motion with velocity 90 km $s^{-1}$ that Argiroffi et al. (2019) ascribed to a CME, representing the first direct X-ray detection of a CME event on a star other than the Sun. The estimated CME mass was $10^{21}$ g, and the inferred kinetic energy was of the order of $5 \times 10^{34}$ erg. Again, comparison with the solar case is instructive: the largest observed solar CMEs have masses of $10^{17}$ g and kinetic energies of the order of $10^{32}$–$10^{33}$ erg. The HR 9024 CME candidate was then more massive than the largest solar CMEs by four orders of magnitude, but only 50 times more energetic.

At face value, the Argiroffi et al. (2019) result suggests that the partitioning of energy is different for CMEs on the Sun and on very active stars, and in particular that CME kinetic energy is much lower than might be expected by extrapolating solar flare–CME relations. While it must be cautioned that solar CME parameters show a large scatter, the relatively low HR 9024 CME candidate kinetic energy compared with its larger inferred mass is consistent with the general picture from a compilation of stellar CME candidates on active stars analyzed by Moschou et al. (2019).

The HR 9024 flare is at present the only example of a potential CME inferred from Doppler shifts in *Chandra* observations; the challenge for future missions will be to garner a larger sample of such events and build a more secure foundation for theoretical models.

The vigor of flare activity is strongly related to the underlying magnetic activity level of the star itself: more active stars experience more energetic and more frequent flaring owing to the larger reservoir of stored magnetic energy in their coronae. Red dwarf "flare stars" are conspicuous for two reasons: their timescales for activity decline are much longer than those for higher mass stars (Section 4.2.2), and their photospheres are comparatively red and faint, such that the contrast between white light produced during flares and photospheric emission is much stronger and more readily visible.

### 4.2.5 Stellar Coronal Chemical Compositions

Following the launch of the *ASCA* and *EUVE* satellites in the early 1990s, a picture began to emerge of the abundances of elements in stellar coronae differing from those in the underlying photosphere (e.g., Drake 1996, 2002). There is a considerable history of abundance anomalies occurring in the solar corona (see, e.g., the review by Laming 2015), so in some respects the results were not a surprise. In the Sun, the abundance anomaly is referred to as the "First Ionization Potential (FIP) Effect": elements with low FIP (FIP $\leqslant$ 10 eV; e.g., Si, Mg, Fe) are enhanced by factors of 2–4 relative to elements with high first ionization potentials (FIP $\geqslant$ 10 eV; e.g., N, Ne, Ar).

The FIP effect was indeed seen in a small handful of low-activity stars, such as $\alpha$ Cen AB (Drake et al. 1997). But more commonly, because they are brighter in the EUV and X-rays, active stars were selectively observed more and their abundances appeared to show the reverse of this! Fe in particular appeared depleted in the coronae of active stars.

High-resolution spectra obtained by the *Chandra* and *XMM-Newton* gratings that enabled individual lines for different elements to be easily measured provided more definitive observations of coronal abundance anomalies, and one of the first results was of a what is now termed an inverse FIP (iFIP) effect in the corona of the active RS CVn-type binary HR 1099 (Brinkman et al. 2001; Drake et al. 2001). Elements with low FIP (FIP $\leqslant$ 10 eV; e.g., Si, Mg, Fe) were observed to be depleted relative to elements with high FIPs (FIP $\geqslant$ 10 eV; e.g., N, Ne, Ar). This pattern has also now been extensively observed in the very active T Tauri stars (Maggio et al. 2007; Flaccomio et al. 2019).

The Ly$\alpha$ resonance line of Ne X was particularly conspicuous in the *Chandra* grating spectra, and Drake & Testa (2005) showed that stars over a large range of

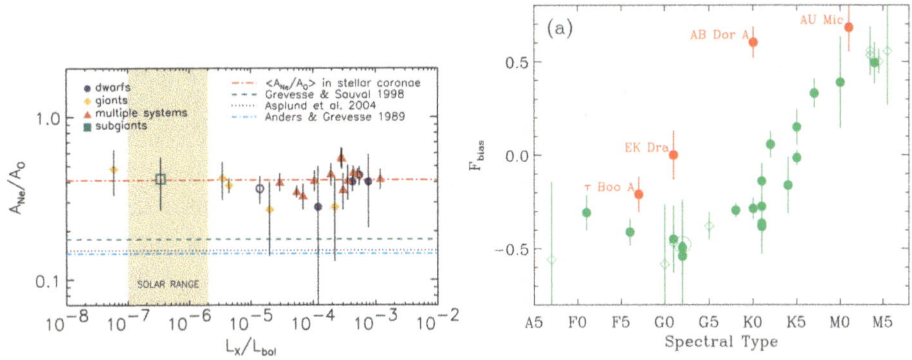


**Figure 4.19.** Left: Ne/O abundance ratios in coronae with different activity levels from the analysis of Drake & Testa (2005), demonstrating a "saturation" in the Ne/O ratio. Some more recent measurements, including of the Sun, indicate that Ne/O declines toward the lowest activity levels. Right: the "FIP bias," the ratio of an average of high FIP element abundances relative to Fe, plotted as a function of spectral type, illustrating a spectral type and activity dependence of the FIP effect. Active stars are shown as red points and also define a dependence of FIP bias on activity level. Reproduced from Wood et al. (2018) © 2018. The American Astronomical Society. All rights reserved.

activity levels appeared to have Ne/O abundance ratios about twice that typically seen in the solar corona (Figure 4.19), although the solar Ne/O ratio is also observed to vary from region to region. This picture has evolved, and it now appears that other low-activity stars also exhibit more solar-like Ne/O ratios.

As more stars were observed, an additional dependence of the FIP and iFIP effects on stellar spectral type emerged (see Wood et al. 2018, and references therein). The "FIP bias"—an average of high FIP element abundances relative to the low FIP element Fe compared with a solar photospheric mixture—based on an analysis of a sample *Chandra* LETG+HRC-S spectra by Wood et al. (2018) is illustrated in Figure 4.19. The FIP bias is negative for the solar FIP effect in which low FIP elements are relatively enhanced, and positive for the iFIP effect. A strong trend of increasing FIP bias with later spectral type is observed, together with a strong increase in FIP bias with activity level, which is reminiscent of the early stellar results.

The mechanism underlying the chemical fractionation responsible for FIP-based effects has not yet been identified with certainty. The characteristic FIP at which the abundances change corresponds to the temperatures in the chromosphere (~10,000 K). The most promising explanation is based on the ponderomotive forces experienced by ions in the chromosphere as a result of Alfvén waves initiated both from the photoshere through convection and turbulence and above in the corona through magnetic reconnection (Laming 2015). The variation in FIP effect with spectral type might then be due to the change in Alfvén wave spectrum and intensity as the characteristics of the convection zone change with effective temperature. Because Alfvén waves are thought to be important sources of coronal heating and for driving the solar wind, the possibility of using coronal abundance anomalies as diagnostics of Alfvén waves is potentially very important.

### 4.2.6 The End of the Main Sequence and Beyond

We noted in Section 4.2.2 that stars with masses below that at which they become fully convective appear to behave the same as a function of Rossby number to partially convective, higher mass stars. An important questions is whether or not this behavior persists all the way into the substellar brown dwarf (BD) regime. Magnetic activity of BDs is not only of fundamental astrophysical interest, but is also important for understanding the possible influence of magnetic star spots in the interpretation of surface features on BD generally interpreted as clouds.

BDs never achieve sufficient core temperatures and densities to fuse hydrogen and subsequently cool down and become fainter and fainter as they age. From a H$\alpha$ survey of mid-M to L field dwarfs, Mohanty & Basri (2003) found a drastic drop in activity and a sharp break in the rotation–activity relation. The H$\alpha$ emission levels were found to be much lower than in earlier types, and often undetectable, even in very rapidly rotating objects. They argued that chromospheric emission may shut down below a critical temperature because the atmosphere becomes too neutral to provide sufficient coupling between the gas and magnetic field.

Very young BDs have been quite regularly detected in sensitive X-ray surveys of young star-forming regions (e.g., Preibisch et al. 2005a). In these cases, the objects are still in the contracting phase and are of higher luminosity and earlier spectral type—the stellar/substellar boundary being at a spectral type of about M6 at an age of 1 Myr—than mature, fully collapsed objects. Their X-ray emission might also be in part due to accretion. Preibisch et al. (2005a) found the BDs in the *Chandra* survey of the Orion Nebula Cluster (see Section 4.2.7 below) to have similar X-ray properties to field M dwarfs of the same spectral type, suggesting that the effective temperature, rather than mass, is the most important parameter for dynamo action.

*Chandra* made the first X-ray detection of an isolated, fully contracted mature BD when a flare was detected on LP 944-20 in a 44 ks observation on 1999 December 15 (Rutledge et al. 2000). LP 944-20 is a ~500 Myr old, rapidly rotating ($v \sin i = 28$ km s$^{-1}$, $p \sim 4.4\, h \sin i$) BD at a distance of 5 pc with an effective temperature of approximately 2500 K and a bolometric luminosity of $\sim 6 \times 10^{29}$ erg s$^{-1}$. A total of 15 counts were detected from the source during a period of 1–2 hr, with an expectation of 0.14 background counts. The estimated temperature of the signal was 0.26 keV ($3 \times 10^6$ K), which is rather cool compared with the flares typically observed on stars by *Chandra* (Section 4.2.4). The source was not detected prior to the flare, setting a $3\sigma$ upper limit to the relative X-ray and bolometric luminosities of $L_X/L_{bol} < 2 \times 10^{-6}$ erg s$^{-1}$, to be compared with $L_X/L_{bol} \approx 2 \times 10^{-4}$ erg s$^{-1}$ at the flare peak.

The detected flare is important as it requires that a relatively strong persistent magnetic field be present on LP 944-20 and that at least occasionally, this field must be perturbed into a configuration in which reconnection and flare dissipation of stored magnetic energy occurs. However, despite the fast rotation of LP 944-20, its extremely low quiescent $L_X/L_{bol}$ ratio indicates that it is severely out of line with the saturation value of $L_X/L_{bol} \sim 10^{-3}$ seen in fully convective M dwarfs (Figure 4.14). While LP 944-20 is obviously very faint in X-rays, the upper limit to its fractional

X-ray luminosity is still consistent with that of the Sun, which is about $L_X/L_{bol} \sim 10^{-6}$ erg s$^{-1}$ at solar maximum.

Further surveys of field BDs with *Chandra* have fleshed out the details of this $L_X/L_{bol}$ difference and highlighted a precipitous decline in X-ray emission at a spectral type of about M9, which is illustrated in Figure 4.20 (see, e.g., Stelzer et al. 2006; Berger et al. 2010). Cook et al. (2014) found an anticorrelation between rotation and X-ray activity reminiscent of supersaturation (Section 4.2.2), and that the scatter $L_X/L_{bol}$ X-ray activity at a given rotation rate is three times larger than for earlier-type stars.

While X-ray emission appears to be comparatively weak in BDs, radio emission persists and the luminosity remains relatively unchanged from spectral types M0 to L4, reflecting a substantial increase in $L_{rad}/L_{bol}$ with later spectral type, corresponding to radio overluminosities of up to a factor of 100 (Figure 4.20; see, e.g., Williams et al. 2014). The persistence of radio emission is a testament to the continued presence of magnetic fields and particle acceleration.

Williams et al. (2014) and Cook et al. (2014) suggested that magnetic field topology could be key and that the large scatter in X-ray fluxes reflects the presence of two dynamo modes that produce distinct magnetic topologies. While more detections of BDs in X-rays would help understand the relative partition of radio and X-ray magnetic dissipation, substantial progress likely must await a next-generation mission with greater sensitivity.

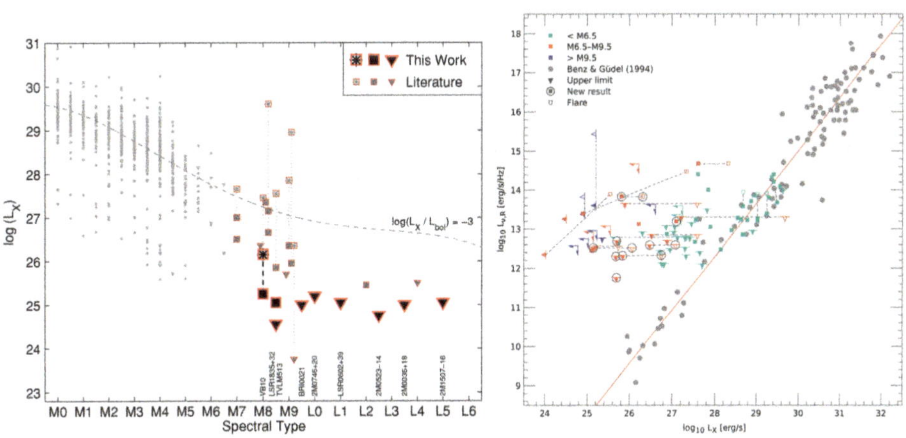


**Figure 4.20.** Left: X-ray luminosity of M and L dwarfs based largely on *Chandra* surveys as a function of spectral type. Literature values for earlier M dwarfs are shown as gray dots. Quiescent emission of ultracool dwarfs (spectral type >M7) is denoted by squares, upper limits by inverted triangles, and flares by asterisks. Dotted lines connect objects for which both flares and quiescent emission were observed. Reproduced from Berger et al. (2010) © 2010. The American Astronomical Society. All rights reserved. The dashed line marks the canonical saturation ratio for F–M stars of log $L_X/L_{bol} = -3$. Right: relationship between X-ray, $L_X$, and radio, $L_{\nu,R}$, luminosities from Benz & Guedel (1994; gray) overlaid with data for ultracool dwarfs from the survey of Williams et al. (2014). Upper limits are denoted by triangles pointing down, to the left, or both down and left. Gray points with $L_{\nu,R} < 12$ are from solar flares. Reproduced from Williams et al. (2014) © 2014. The American Astronomical Society. All rights reserved.

### 4.2.7 Young Stars, Protostars, Disks, and Jets

Young stars in the pre-main-sequence phase—the so-called T Tauri stars named after the prototypical example in Taurus—are observed to rotate rapidly, with typical periods of one to several days (e.g., Herbst & Mundt 2005), and they are also vigorously convecting with typical convective turnover times in the 100–200 day range (Preibisch et al. 2005b). On this basis, with Rossby numbers $Ro \lesssim 0.1$, they would be expected to be vigorous X-ray emitters in the saturated regime. This is just what was found when the *Einstein Observatory* was pointed toward nearby star-forming regions, and many T Tauri stars were detected with relative X-ray luminosities of $L_X/L_{bol} \sim 10^{-3}$ (e.g., Ku & Chanan 1979; Feigelson & Decampli 1981; Walter & Kuhi 1981). Some years in advance of the detailed picture of coronal X-ray emission we are privileged to have today, Walter & Kuhi (1981) first made the connection between solar-like coronal emission and T Tauri X-rays.

The early work using *Einstein* was a watershed moment: because the X-ray luminosity of a solar-mass star declines rapidly over time by orders of magnitude—within a few hundred megayears—X-rays provide an extremely efficient method for distinguishing between young stars and the myriad field stars along the lines of sight to star-forming regions. Moreover, X-rays from young stars can penetrate through all but the most opaque lines of sight in the Galaxy and deep into molecular clouds, the obscuration from which stymies observations at other wavelengths.

The power of X-ray observations for studying young stars was cogently demonstrated when Walter (1986) used X-ray-selected objects in the Taurus, Ophiuchus, and Corona Australis star-forming regions to first classify the "naked" T Tauri stars. The main defining features of T Tauri stars had been the presence of infrared excesses from a circumstellar accretion disk (also known as a protoplanetary disk, or "proplyd" for short) and emission lines from the accretion process. Many of the sources being detected in star-forming regions did not exhibit these characteristics and were initially thought to be "post-T Tauri" stars, on their way to the main sequence. Walter (1986) showed that these stars are still in the T Tauri evolutionary phase and are simply bereft of circumstellar material. The "naked" monicker eventually passed out of common usage and the terms "classical" and "weak-lined" are now used to refer to T Tauri stars with or without substantial circumstellar material and accretion.

*ROSAT* added substantially to the study of T Tauri stars and X-rays from star-forming regions (see, e.g., Krautter et al. 1994; Neuhäuser 1997), and by the end of the 1990s, the time was ripe for *Chandra* to bring its full arsenal of high angular resolution and X-ray imaging spectroscopy to bear on crowded regions of star formation. *Chandra* results on star-forming regions are now so extensive that here we must limit ourselves to some "greatest hits."

*The Detection of X-Rays from Protostars*

One key question in the magnetic activity of young stellar objects is the evolutionary phase at which X-ray emission sets in. Ionization of infalling gas by X-rays can tie it to magnetic field lines and potentially slow the speed of gravitational collapse. More locally to the protostar itself, entrainment of the gas to field lines could be important for the launching of outflows and jets. The lesson from main-sequence stars in Section 4.2.2 is that rotation and convection are the key ingredients for dynamo activity and associated X-ray emission, both of which should be present in protostars.

In the evolutionary framework defined by the shape of the infrared spectrum (e.g., Lada 1987) running from Class 0, which are highly obscured and only appear in the far-infrared, to Class III—the weak-lined T Tauri stars with blackbody-like infrared spectra—Class 0 protostars have never been definitively detected, although it is often difficult to distinguish between classes 1 and 0 without far-infrared data.

Class I protostars, despite often being still deeply embedded in natal clouds and circumstellar material, are instead commonly picked up as heavily absorbed, and therefore spectrally quite hard, X-ray objects (e.g., Imanishi et al. 2001; Winston et al. 2007; Romine et al. 2016). These objects, with typical ages of $1–5 \times 10^5$ yr, are still deriving a significant fraction of their luminosity from accretion. Upper limits to X-ray fluxes for Class 0 objects, with lifetimes an extremely short $10^4$ yr or so, are similar to those for Class I objects, and it is still possible that X-ray do turn on during the Class 0 phase.

Class 0 and I protostars are also observed to drive bipolar outflows with velocities up to several hundred km $s^{-1}$ that interact and shock-heat gas in the ambient interstellar cloud, producing what are known as Herbig–Haro objects (e.g., Reipurth & Bally 2001). The preshock velocities are often sufficiently high to heat gas to million-degree X-ray-emitting temperatures (e.g., Raga et al. 2002), and *Chandra* has detected these shocks in several Herbig–Haro flows (e.g., Pravdo et al. 2001; Favata et al. 2002; Pravdo et al. 2004), as well as from the base of the flow in other cases (Güdel et al. 2007). The X-ray detections represent only a few percent of the Herbig–Haro objects observed, likely because the emission is too soft to penetrate the surrounding gas.

X-ray-emitting outflows are also observed from more evolved objects. The classical T Tauri star DG Tau shows X-ray emission from a jet complex extending several hundred astronomical units in either direction, in addition to an inner X-ray emission region located only a few 10s of astronomical units from the star itself (Güdel et al. 2008; Schneider & Schmitt 2008; Figure 4.21).

*X-Ray Properties of T Tauri Stars*

The field of view of *Chandra* is well matched to typical angular sizes of nearby regions of star formation, and many have been observed during the course of the mission. Arguably the most important for understanding the X-ray properties of T Tauri stars has been the Orion Nebula Cluster (ONC; sometimes also known as the Trapezium Cluster or Ori Id OB Association). At a distance of approximately 450 pc, the ONC is the most well-studied star-forming region in the sky and has been extensively cataloged at all wavelengths. It contains a dense and rich population of approximately 2000 pre-main-sequence stars within a spherical volume 2 pc across, 80% of which are younger than 1 Myr (Hillenbrand 1997). Its combined unresolved X-ray emission was originally detected by the *Uhuru* satellite (Giacconi et al. 1972).

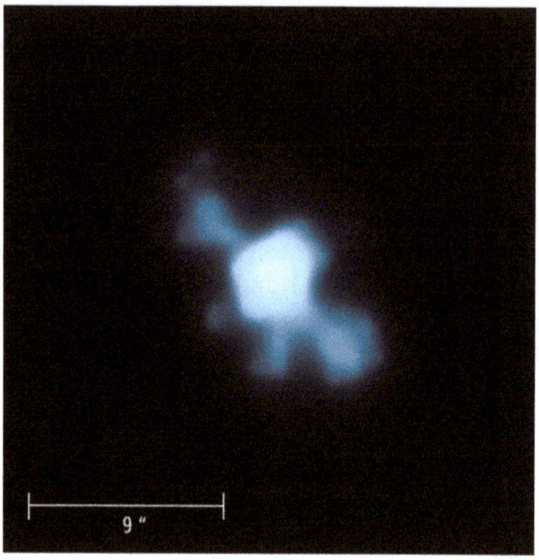


**Figure 4.21.** The *Chandra* X-ray image of the spectacular X-ray-emitting jet of the classical T Tauri star DG Tau from the study by Güdel et al. (2008). At the 140 pc distance of DG Tau, the 9″ scale bar shown corresponds to a size of 1260 AU. Image from http://chandra.harvard.edu/photo/2008/dgtau/. Courtesy of NASA/CXC/ETH Zuerich/M.Guedel et al.

*Chandra* has observed the ONC on several different occasions, using both HRC-I (Flaccomio et al. 2003) and ACIS-I. Flaccomio et al. (2003) detected 742 X-ray sources in a 63 ks HRC-I pointing and established that X-ray luminosities of low-mass stars and BDs ($M \leqslant 3\ M_{\odot}$) increase with increasing mass and decreasing stellar age. Stars of intermediate mass 2–4 $M_{\odot}$ were conspicuously fainter, which Flaccomio et al. (2003) attributed to their being essentially fully radiative and unable to sustain a magnetic dynamo (see Section 4.5). Similar results were obtained in 83 ks of data ACIS-I data obtained by Feigelson et al. (2002) in 1999 October and 2000 April, with the addition of photon energy information afforded by the ACIS-I detector. T Tauri plasma temperatures were found to be often very high even outside of obvious flares, sometimes reaching $10^8$ K and beyond.

The rich data reaped by the initial *Chandra* observations provided impetus for a much deeper 1 Ms exposure with ACIS-I in what was dubbed the *Chandra* Orion Ultradeep Project (COUP; Figure 4.22; Getman et al. 2005; Feigelson et al. 2005). COUP garnered 1616 X-ray sources in a region of the sky the size of the ACIS-I detector (16′ × 16′), of which 1408 were identified with cluster members, providing a large coeval sample of pre-main-sequence stars down to below the stellar limiting mass.

While the completeness of the COUP survey was still mass dependent—higher mass, X-ray brighter stars were preferentially detected compared with lower mass, fainter stars—it provided the most unbiased census of T Tauri stars yet obtained. Preibisch et al. (2005b) studied nearly 600 of the COUP X-ray sources that were reliably identified with well-characterized T Tauri stars. The detection limit was $L_X \geqslant 10^{27.3}$ erg s$^{-1}$ for the least absorbed sources, leading to a detection

**Figure 4.22.** The Orion Nebula Cluster seen in a composite *Hubble* optical (red–purple) and *Chandra* X-ray image (left), and just in X-rays observed by *Chandra* (right). The interactive figure shows separately the optical emission (purple), the X-ray emission (color-coded by energy band: 0.3–1.0 keV, red; 1.0–3.0 keV, green; 3.0–8.0 keV, blue), and the composite image. A total of 1616 X-ray sources were detected, of which 1408 were identified with cluster members (Feigelson et al. 2005). Images from http://chandra.harvard.edu/photo/2007/Orion/. X-ray image courtesy of NASA/CXC/Penn State/E.Feigelson & K.Getman et al.; Optical courtesy of NASA/ESA/STScI/M.Robberto et al. Additional materials available online at http://iopscience.iop.org/book/978-0-7503-2163-1.

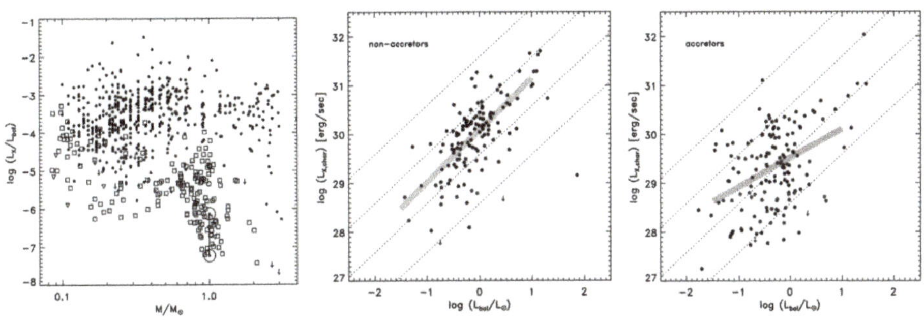


**Figure 4.23.** The X-ray luminosities for T Tauri stars in Orion from the COUP survey reproduced from Preibisch et al. (2005b) © 2005. The American Astronomical Society. All rights reserved. The left panel illustrates the fractional luminosity, log $L_X/L_{bol}$, as a function of stellar mass for COUP stars (solid dots, arrows for upper limits) and for a sample of nearby field stars observed by *ROSAT* (open squares, triangles for upper limits). The right two panels show what Preibisch et al. (2005b) termed the "characteristic" X-ray luminosity, computed to be that remaining after obvious flares are removed, as a function of stellar luminosity for weak-lined T Tauri stars ("non-accretors") and classical T Tauri stars ("accretors"). The range of $L_X/L_{bol}$ exhibited by the Sun through its cycle (Judge et al. 2003) is denoted by the joint hollow circles.

completeness of 97% of cluster stars with spectral types in the range F–M. The immediate deduction here is that all T Tauri stars are X-ray bright and very active, and there is no hidden faint population with suppressed magnetic activity.

The fractional X-ray luminosity, $L_X/L_{bol}$, for the Preibisch et al. (2005b) sample is compared as a function of stellar mass with field stars detected by *ROSAT* (Schmitt & Liefke 2004) in Figure 4.23. There are several important results from these data. From a solar perspective, the most conspicuous feature is the absolute value of the luminosities in comparison to the Sun: T Tauri stars are three to four orders of magnitude brighter than the Sun in X-rays. There is no trend in T Tauri log $L_X/L_{bol}$ with stellar rotation velocity, as expected because all T Tauri stars are in either the saturated or supersaturated regime. The fractional X-ray luminosities are relatively flat as a function of stellar mass, with a slight rising trend of $\log L_X/L_{bol} = -3.65(\pm 0.05) + 0.40(\pm 0.10)\log(M/M_\odot)$ but with a very large scatter over the range $10^{-5} \leqslant L_X/L_{bol} \leqslant 10^{-2}$, while in the saturated regime, these X-ray luminosities are slightly below those for saturated main-sequence stars in Figure 4.14. The best estimate for how the X-ray luminosity itself varies with stellar mass is $\log L_X = 27.58(\pm 0.07) + 1.25(\pm 0.15)\log(M/M_\odot)$.

The origin of the large X-ray luminosity scatter, and whether T Tauri stars change their X-ray luminosities by factors of 100 or more over timescales longer than the observation length (approximately 13 days total elapsed time in the case of COUP), remains a subject of debate. Some fraction can be attributed to stochastic variability, and Flaccomio et al. (2012) established that in all subsamples, the variability amplitudes increase with increasing timescale at least up to the elapsed time over which the observations were taken. However, comparison of the COUP X-ray luminosities to those in the earlier shorter pilot study by Feigelson et al. (2002) revealed only a factor of 2 or so difference over the three years between the respective observations.

Part of the scatter can be attributed to a difference between classical (accreting) and weak-lined (non-accreting) T Tauri stars. Figure 4.23 also shows the X-ray luminosities of the Preibisch et al. (2005b) sample as a function of stellar luminosity, divided into accreting and non-accreting stars. The former are typically a factor of 2–3 brighter than the latter, confirming results based on earlier studies of smaller samples of T Tauri stars in Taurus observed with *ROSAT* and *Einstein* (Neuhaeuser et al. 1995; Damiani & Micela 1995).

Also clear from Figure 4.23 is the much larger scatter in the X-ray luminosities of the classical T Tauri stars. In fact, Preibisch et al. (2005b) and Flaccomio et al. (2012) argue that the scatter in the weak-lined T Tauri star X-ray luminosities is consistent with the variability observed over short and long timescales combined with the uncertainties in derived parameters, such as mass and luminosity. However, this does not explain the classical T Tauri stars. Flaccomio et al. (2012) found that classical T Tauri stars also showed more variability, and the most promising explanation to date for the luminosity scatter is that circumstellar material is responsible for obscuration of some fraction of their coronae, leading to both a lower $L_X$ on average, and to a larger secular change in $L_X$.

*Giant Flares and the Ionization of Protoplanetary Disks*

The high, saturated activity level of T Tauri stars is accompanied by vigorous flaring as expected from analogous behavior of their active main-sequence counterparts. Figure 4.24 illustrates the *Chandra* X-ray light curves of two T Tauri stars observed to flare during the COUP survey from the study of Favata et al. (2005). The peak count rate in the largest flare (left lower panel; source COUP 891) of 0.75 count $s^{-1}$

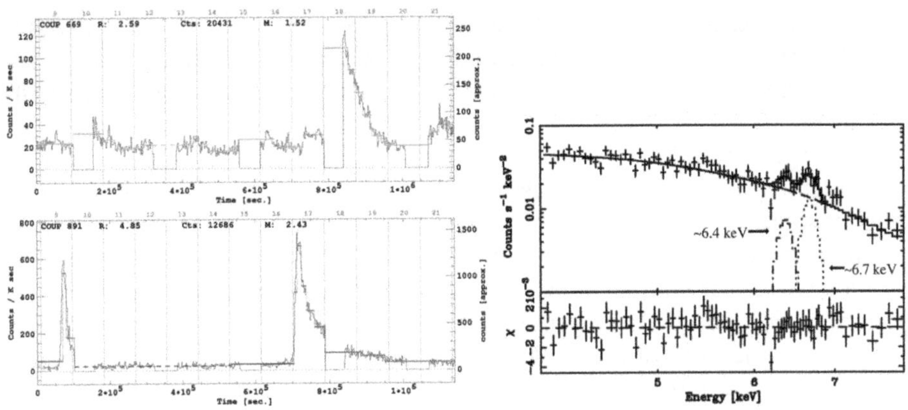


**Figure 4.24.** Left: the X-ray light curves of two bright flares from ONC stars analyzed by Favata et al. (2005) © 2005. The American Astronomical Society. All rights reserved. In the upper light curve, the structure reveals a low-level variability that can be attributed to a superposition of continual flaring. This corresponds to an X-ray flux of about $2 \times 10^{32}$ erg, or 5% of the solar X-ray luminosity. Right: the *Chandra* ACIS X-ray spectrum of the Class I protostar YLW 16A in the $\rho$ Ophiuchi cloud exhibiting a strong Fe K$\alpha$ fluorescence line originating from flare ionization of the protoplanetary disk. Reproduced from Imanishi et al. (2001) © 2001. The American Astronomical Society. All rights reserved.

corresponds to an X-ray flux of about $2 \times 10^{32}$ erg. This is 5% of the solar bolometric luminosity! In the context of large solar flares, denoted as "X-class" and having a peak flux $\geqslant 10^{-4}$ W m$^{-2}$ in the 1–8 Å band, the COUP 891 flare amounts to $10^2$ W m$^{-2}$, or a million times more energetic.

Favata et al. (2005) found a broad range of decay times for the brightest flares, ranging from 10 to 400 ks. Peak flare temperatures were often extremely high, with half having $T > 10^8$ K. Analyzing the flares in terms of simple quasi-static loop models, Favata et al. (2005) found significant sustained heating was required for the majority of the flares, and that flaring loops were extremely large, with semi-lengths up to $10^{12}$ cm—several times the stellar radius. Such loops would not be stable due to centrifugal forces, and it was speculated that loops could be anchored between the star and its disk.

The flare frequency as a function of energy for the Orion T Tauri flares was investigated by Caramazza et al. (2007). The frequencies of flares on the Sun and stars are typically distributed as a function of energy in a power law:

$$\frac{dN}{dE} = k\,E^{-\alpha} \quad \text{where} \quad \alpha > 0 \tag{4.8}$$

where $N$ is the number of flares with energies between $E$ and $E + dE$, emitted in a given time interval. If the index of the power law, $\alpha$, is larger than 2, then a minimum flare energy must be introduced in order to keep the total emitted energy finite and, depending on this cutoff, even very high levels of apparently quiescent coronal X-ray emission can be obtained from the integrated effects of many small flares.

Caramazza et al. (2007) found a power-law index of $\alpha = 2\pm$ for the Orion T Tauri flares, consistent with the conclusion that the T Tauri X-ray light curves are entirely composed of overlapping flares. While the more intense flares can be individually detected, weak flares merge together and form a pseudo-quiescent emission level. Low-mass and solar-mass stars of similar X-ray luminosity have very similar flare frequencies.

The energetic radiation of T Tauri stars is important for understanding their protoplanetary disks, within which planets form. It had been thought that the dominant source of ionization of protoplanetary disks would be by cosmic rays, but an analysis by Glassgold et al. (1997) found that stellar X-rays are the dominant ionization source throughout much of the disk volume. The disks are predominantly heated by optical and IR light from the stellar photosphere, which give them a flared geometrical structure with a scale height that increases with increasing radial distance (e.g., Armitage 2011). This geometry favors the penetration of X-rays, ionization, and coupling of the gas to magnetic fields, and ultimately likely contributes to disk dispersal through photoevaporation (e.g., Ercolano et al. 2008). X-ray ionization influences chemistry and drives ion–molecular reactions in the outer disk layers (e.g., Semenov et al. 2004; Walsh et al. 2012). Gas–magnetic field coupling is important for the induction of magnetohydrodynamic turbulence via the magnetorotational instability (Balbus 2003) and allows angular momentum transport that can drive accretion. More recent studies have favored the action of magnetized disk winds in driving angular momentum loss and accretion, and contributing to disk dispersal (Bai 2016). Regardless of the exact mechanisms and their relative importance, X-ray ionization appears to be crucial.

That flare X-rays are indeed absorbed by the surrounding disk is illustrated graphically in the *Chandra* ACIS X-ray spectrum of the Class I protostar YLW 16A in the $\rho$ Ophiuchi cloud (Imanishi et al. 2001). A prominent fluorescence line is present at 6.4 keV originating from inner-shell ionization of cold, neutral, or near-neutral Fe atoms in the disk by flare X-rays (see also Section 4.1.1).

*X-Ray Evidence for Accretion from Protoplanetary Disks*

The protoplanetary gas accretion disk of T Tauri stars is expected to be truncated by the stellar magnetic field at a distance of several radii from the stellar surface near the corotation radius (Uchida & Shibata 1984). From there, accretion is thought to proceed down open magnetic field lines and form an accretion shock at the stellar surface. This accretion is ultimately the source of the elevated UV emission of T Tauri stars for their spectral type and the "veiling" and filling in of absorption lines (e.g., Calvet & Gullbring 1998).

The free-fall velocity, $v_{\rm ff}$, from several stellar radii can amount to 150–400 km s$^{-1}$ or so, and because the Mach number is high, the gas shock temperature, $T_{\rm s}$, can be estimated under strong shock conditions and is given for a star of mass $M$ and radius $R$ by (e.g., Calvet & Gullbring 1998)

$$T_{\rm s} = \frac{3\mu m_{\rm H}}{16k} v_{\rm ff}^2 = 8.6 \times 10^5\,{\rm K}\left(\frac{M}{0.5\,M_\odot}\frac{2\,R_\odot}{R}\right). \qquad (4.9)$$

For a typical T Tauri star, the term in the brackets on the right-hand side of Equation (4.9) is approximately unity, and the immediate postshock temperature is about a million degrees kelvins.

For a strong shock, the postshock density is four times the preshock density, and for a mass accretion rate $\dot{M}$ in solar masses per year through an accretion column covering a fraction $f$ of the stellar surface, it is given by

$$n_{\rm H} \approx 2 \times 10^{13} \left(\frac{0.01}{f}\right)\left(\frac{\dot{M}}{10^{-8}}\right)\left(\frac{0.5\,M_\odot}{M}\right)^{1/2}\left(\frac{2\,R_\odot}{R}\right)^{3/2} \mathrm{cm}^{-3}. \tag{4.10}$$

Typical accretion spot filling factors, $f$, estimated from UV observations are a few percent, while accretion rates tend to fall in the range $10^{-8}$–$10^{-10} M_\odot\,\mathrm{yr}^{-1}$. With the rest of the terms in Equation (4.10) being approximately unity, the postshock density is expected to be of the order of $10^{11}$–$10^{13}$ cm$^{-3}$—considerably higher than typical ambient plasma densities in stellar coronae (Section 4.2.1).

The first and, as it turns out, best example of accretion shock X-ray emission was found in the closest known classical T Tauri star TW Hya ($d = 60$ pc) by Kastner et al. (2002). The He-like Ne IX lines were illustrated earlier in Figure 4.13; the intercombination lines $x$ and $y$ are clearly enhanced relative to the coronal case (AB Dor in the case of Figure 4.13), and the $w/(x + y)$ ratio indicates a density of $n_e \approx 10^{13}$ cm$^{-3}$. Pushing *Chandra* HETG spectral resolution to its limits by means of comparative spectroscopy with non-accreting coronal X-ray sources, Argiroffi et al. (2017) measured the line-of-sight redshift in the postshock accreting gas of $38.3 \pm 5.1$ km s$^{-1}$, indicating the accretion occurs at low latitudes. TW Hya possesses a well-resolved face-on dusty disk, and the Doppler shift indicates that accretion streams are close to the disk plane.

High-density accretion signatures have been seen in several other classical T Tauri stars, but they are not ubiqitous among the class. The most likely reason for this is the shocks being formed too far into the photosphere and getting obscured from view. The shock height depends on the accretion rate through both the ram pressure of the stream and the cooling time of the shocked plasma: higher accretion rates have larger ram pressures that penetrate farther into the photosphere, and smaller scale heights due to higher densities and more rapid cooling (Drake 2005). As not all X-rays from accretion necessarily escape, determining accretion rates from X-rays is hazardous. Bonito et al. (2014) examined the observability of accretion shocks based on the results of detailed hydrodynamic simulations and found lower fluxes than expected because of the complex absorption by the optically thick chromosphere within which the shocks can be embedded, and of the unperturbed accretion stream.

The TW Hya accretion shock Ne/O abundance ratio is anomalously high (Drake et al. 2005), perhaps due to the accretion of grain-depleted gas in a mature old protoplanetary disk. However, other accretion shocks do not necessarily show the same pattern (Argiroffi et al. 2007), and the origin of the anomaly remains unclear. More extensive *Chandra* observations of TW Hya using the HETG have also revealed gas emitting in O VII that appears to be ambient corona plasma heated indirectly by the accretion process (Brickhouse et al. 2010), although this might also be explained by complex accretion shocks (Bonito et al. 2014). An excess in O VII emission had indeed been noticed based on earlier *XMM-Newton* observations of T Tauri stars in the Taurus Molecular Cloud that likely arises from the same processes.

*Protoplanetary Disk Destruction in High-mass Star-forming Regions*
The ability of *Chandra* to see all young stars in a star-forming region in a reasonably unbiased fashion enables an assay of the fraction of stars that have retained protoplanetary disks. The disks are conspicuous by their infrared-excess emission on top of the stellar photospheric radiation field and can be readily detected in infrared surveys by *Spitzer* and to some extent by ground-based near-infrared *K*-band excesses (e.g., Guarcello et al. 2013).

Protoplanetary disks are gradually dissipated by accretion, disk winds, and photoevaporation driven by host star UV–X-ray irradiation, by dynamical interactions in very dense star-forming regions, and if intense enough, by photoevaporation by the external radiation field. Typical disk lifetimes are a few megayears, with essentially all disks having dissipated by 10 Myr (e.g., Ribas et al. 2014). Modeling of dynamical interaction effects indicate they only become important in the densest stellar environments exceeding $\sim 10^3$ $pc^{-3}$ (e.g., Rosotti et al. 2014).

The relative importance of radiative processes leading to disk dispersal has been more difficult to determine. This poses problems for understanding the types and configurations of planetary systems that remain after the disk is gone, because they likely have a large influence on both the planet formation process itself, which depends on the reservoir of building material, and on planet migration, due to planet/planetesimal–gas interaction (e.g., Ribas et al. 2015). A key part of the problem is whether or not the strong UV and EUV radiation fields of high-mass stars are significant sources of disk photoevaporation and loss.

Evidence of photoevaporation in regions of clusters close to high-mass stars based on *Spitzer* and combined *Spitzer* and *Chandra* observations was uncovered by, e.g., Balog et al. (2007; NGC 2244) and Guarcello et al. (2010; NGC 6611). However, the most definitive evidence to date for disk photoevaporation was uncovered by Guarcello et al. (2016) based on a multiwavelength analysis as part of the *Chandra* Cygnus OB2 Legacy Survey (Drake et al. 2019). At a distance of 1.4 kpc and a stellar content of about 17,000 $M_{\odot}$, including at least 52 O-type stars and three Wolf–Rayet stars (Wright et al. 2015), Cygnus OB2 is the nearest truly massive star-forming region to Earth. Despite being so massive, it is of quite low stellar density and consequently is dynamically relatively unmixed and dynamical interactions are negligible. Guarcello et al. (2016) found the protoplanetary disk fraction to decline from 40% to $\leqslant$20% as a function of local FUV and EUV fluxes (Figure 4.25). The FUV radiation dominates disk dissipation timescales within a parsec of the O stars. In the rest of the association, EUV photons likely induce significant disk mass loss across the entire association, modulated only at increasing distances from the massive stars due to absorption by the intervening intracluster material.

At face value, the Guarcello et al. (2016) study indicates that massive star-forming regions could be hostile to habitable planet formation. However, if

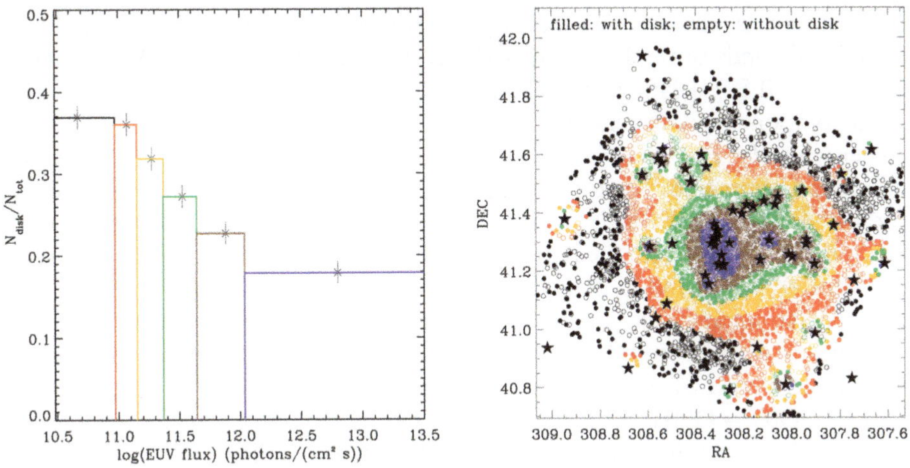


**Figure 4.25.** The protoplanetary disk fraction as a function of the local EUV radiation field in the massive Cygnus OB2 association from a *Chandra*-based multiwavelength survey. The spatial distribution of the stars in the *Chandra* survey field is shown in the right panel. Stars are coloured according to the EUV flux histogram bin in which they fall in the left panel. The star symbols mark the positions of the O stars. A sharp decline in disk fraction is seen with increasing EUV flux from high-mass stars. Reproduced from Guarcello et al. (2016).

terrestrial planet formation timescales are shorter than photoevaporation timescales, the disk gas loss might only affect giant-planet formation (e.g., Ribas et al. 2015).

*Large-scale and Small-scale Diffuse X-Ray Emission in Star-forming Regions*
The sharp *Chandra* PSF not only provides high sensitivity for detecting point sources, but also enables accurate removal of point sources in survey data. This capability is crucial for understanding the origin of residual diffuse or unresolved X-ray emission.

Star-forming regions hosting high-mass stars present tantalizing targets for probing diffuse X-ray emission. From a theoretical perspective, high-mass stars with supersonic winds and high mass-loss rates are expected to blow stellar wind bubbles in the interstellar medium. A reverse shock from interaction with the ambient ISM should heat the gas to temperatures of $10^6$ K and higher (e.g., Freyer et al. 2006). Such emission is in principle detectable with *Chandra*, provided the surface brightness is sufficiently high.

Diffuse and mostly thermal X-ray emission has indeed now been detected from several star-forming regions (see, e.g., the summary by Townsley et al. 2011a), although it is not always clear that such emission originates from shocks induced by massive stellar winds. Wolk et al. (2002), for example, find evidence for nonthermal synchroton diffuse emission in the embedded region RCW 38 possibly associated with a supernova remnant, and similar nonthermal emission is present in NGC 3576 (Townsley et al. 2011a). Both the thermal and nonthermal emission are seen to fill voids between bright dust emission revealed by *Spitzer*, as might be expected by a flow seeking the path of least resistance.

Albacete Colombo et al. (2018) detected large-scale thermal X-ray diffuse emission in the Cygnus OB2 association that follows the spatial distribution of massive stars and similarly fills a cavity that appears to have been excavated by the accumulated winds of the 169 cataloged OB stars of the region. The X-ray emission in the broad 0.5–7.0 keV energy band was reasonably well matched by thermal plasma components with temperatures of 0.11, 0.40 and 1.2 keV and is illustrated together with *Herschel* 500 $\mu$m (corresponding to a temperature $T \approx 10$ K) cold gas emission in Figure 4.26. The total luminosity of the diffuse emission was found to be $4.2 \times 10^{34}$ erg s$^{-1}$, which is $10^4$ times less than the estimated total OB star wind kinetic energy injected into the region (Drake et al. 2019). Remarkably, this fraction is the same as that found in the hydrodynamic modeling study of the interaction of massive stellar winds with the ambient medium by Freyer et al. (2006).

The biggest surprise from Cygnus OB2, however, was the detection of diffuse X-ray halos around some evolved massive stars. Figure 4.26 also shows an example around the massive O6V + O9III star Schulte 3 that appears to be a direct consequence of the collision and interaction of a fast, dense wind and surrounding gas.

Townsley et al. (2011a) also noted evidence for CX emission in all of the star-forming regions they studied that exhibited diffuse X-ray emission which they interpreted to be a result of interaction of the hot plasma with gas at the interfaces of cold neutral pillars, ridges, and clumps. Similar discrepancies were also found in Cyg OB2 by Albacete Colombo et al. (2018), reinforcing the conclusion that CX emission is a ubiquitous feature of massive star-forming regions.

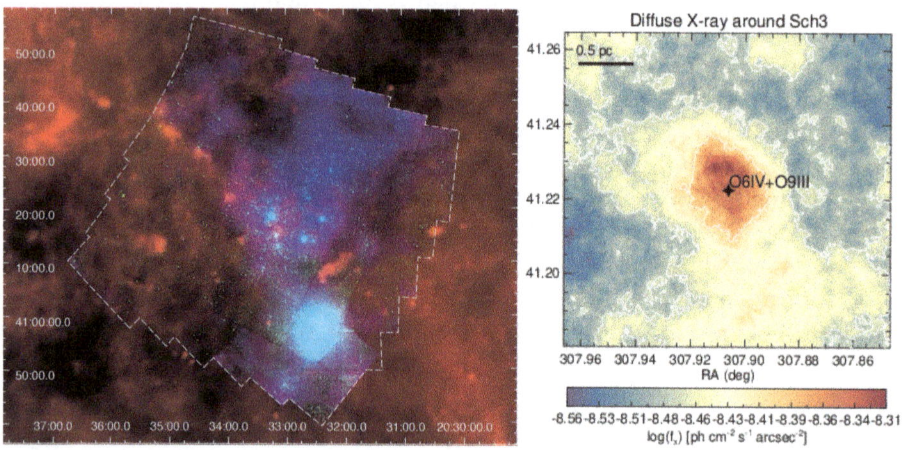


**Figure 4.26.** The neighborhood of Cygnus OB2. The ACIS-I mosaic of the Cygnus OB2 survey (Drake et al. 2019) is outlined in white. The false RGB color image is composed of the *Herschel* 500 $\mu$m ($T \approx 10$ K) cold gas emission in red, the mosaic of *Chandra* observations with the point sources in the 0.5–7.0 keV band in green, and the diffuse X-ray emission in the 0.5–2.5 keV energy range in blue. The diffuse X-ray emission fills a cavity in the infrared *Herschel* data. Right: observed X-ray flux photon (ph cm$^{-2}$ s$^{-1}$ arcsec$^{-2}$) diffuse map computed with an adaptive smoothing algorithm to achieve a signal-to-noise ratio of 16. The black filled star symbol indicates the position of the massive O6V + O9III binary star Schulte 3 that possesses a bright X-ray-emitting halo. Reproduced from Drake et al. (2019)  The interactive figure shows separately the cold gas emission (500μm, red), the X-ray point sources (0.5–7.0 keV, green), the diffuse X-ray emission (0.5–2.5 keV, blue), and the composite image. Additional materials available online at http://iopscience.iop.org/book/978-0-7503-2163-1.

## 4.3 X-Ray Studies of Exoplanet Systems

The explosion in exoplanet discoveries and science over the last two decades have sparked interest in the relevance of X-rays for exoplanetary systems. There are at least four X-ray aspects to exoplanets: the effect of stellar X-rays on planets, the effects of planets on their host stars and possibly on their X-ray emission, the use of stellar X-rays to probe exoplanets, and the possibility of intrinsic exoplanetary X-ray emission analogous to the solar system processes discussed in Section 4.1. The latter remain well beyond the capabilities of *Chandra* and must await future missions to explore.

### 4.3.1 X-Ray Induced Atmospheric Loss

One important, and potentially critical, effect of stellar coronal emission on exoplanets is photoevaporation of the atmosphere. The rate at which gas is lost from an exoplanet's atmosphere is critical for the survivability of surface water. Atmospheric mass loss can be driven by both thermal and nonthermal processes, all of which depend upon the radiation and winds of their host stars. The dominant thermal process is hydrodynamical outflow energized by extreme ultraviolet (EUV; 100–912 Å) and X-radiation (0.1–100 Å). This short-wavelength radiation is absorbed in the very top layers of the atmosphere, in the thermosphere, and the energy input can be sufficient to levitate gas against the exoplanetary gravitational potential (e.g., Owen & Jackson 2012).

A simple "energy limited" atmospheric mass-loss rate for an X-ray flux $F_{\mathrm{x}}$ received by a planet can be written as

$$\frac{dM}{dt} = \frac{4\pi R_{\mathrm{pl}}^3 F_{\mathrm{X}}}{G\mu M_{\mathrm{pl}}}, \tag{4.11}$$

where $\mu$ is the mean atmospheric particle mass, $M_{\mathrm{pl}}$ and $R_{\mathrm{pl}}$ are the planetary radius and mass, respectively, and $G$ is the gravitational constant. Equation (4.11) assumes that the bulk of the energy from the X-ray (and in principle, also EUV) heating goes into the escape of the gas, and that the radius at which X-rays are absorbed is not significantly different from the planetary radius. Using Equation (4.11), Penz et al. (2008) showed that a significant fraction of the mass of a close-in gas giant can be lost to X-ray photoevaporation over several gigayears.

Empirical evidence that X-ray photoevaporation is important was presented in a statistical study of exoplanet mass for stars observed by *Chandra* and *XMM-Newton* by Sanz-Forcada et al. (2010). Tracing the accumulated X-ray dose over the lifetime of each system, they found a distribution of planetary mass with X-ray dose (Figure 4.27) consistent with a scenario in which the bulk of the mass of the lightest systems has been eroded away, with the most massive planets tending to have suffered the smallest X-ray doses.

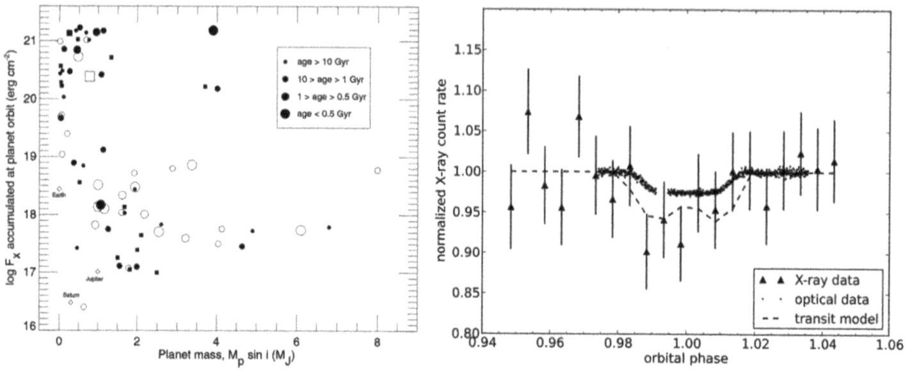


**Figure 4.27.** Left: distribution of planetary masses ($M_{\rm pl}$ sin $i$) with accumulated X-ray dose from an age of 20 Myr to present. Filled symbols (squares for subgiants, circles for dwarfs) represent *Chandra* and *XMM-Newton* data, arrows denote upper limits, and open symbols are *ROSAT* data. From Sanz-Forcada et al. (2010), reproduced with permission © ESO. Right: combined data for six *Chandra* and one *XMM-Newton* X-ray transits of the HD 189733 hot Jupiter system in comparison with the optical transit. Dashed lines show the best fit to a limb-brightened transit model. Reproduced from Poppenhaeger et al. (2013) © 2013. The American Astronomical Society. All rights reserved.

### 4.3.2 Star–Planet Interaction?

Close-in planets can interact with their host stars either gravitationally or via their respective magnetic fields, with the latter expected to dominate any enhancement in the stellar X-ray signal. Some weak evidence for such an enhancement based on a survey of exoplanet host X-ray fluxes was found by Kashyap et al. (2008), although subsequent studies with larger samples have found no positive statistical evidence that the effect is really attributable to planets (e.g., Poppenhaeger & Schmitt 2011; Miller et al. 2015). Firm evidence for significant star–planet interaction is still lacking. *XMM-Newton* observations of the hot Jupiter host HD 189733 that revealed three flares within a fairly narrow range of planetary orbital phase prompted Pillitteri et al. (2014) to suggest planetary magnetic interaction with stellar surface active regions as a possible mechanism. Maggio et al. (2015) detected evidence of X-ray enhancement at periastron in the eccentric HD 17156 system hosting a Jupiter-mass planet. In this case, Maggio et al. (2015) suggested that accretion of material evaporated from the planet onto the star might be the cause.

### 4.3.3 X-Rays as Probes of Exoplanet Atmospheres

X-rays provide potentially powerful diagnostics of planetary upper atmospheres through X-ray transit observations. X-ray absorption through an atmosphere measures the absorbing column depth along the line of sight. Because the absorption cross-section as a function of energy in the X-ray range depends on chemical composition, gas bulk chemical composition can also be determined. This type of measurement is unique to the X-ray range. Because even the largest exoplanets—the so-called "hot Jupiters"—have cross-sectional areas of only a few percent of that of their stellar hosts and their transits across stellar disks typical last only a few hours, X-ray transit measurements are very challenging.

The transit of the hot Jupiter HD 189733b in front of its K-type host star was detected through X-ray absorption by oxygen in *Chandra* ACIS-S observations by Poppenhaeger et al. (2013). The transit in soft X-rays, illustrated in Figure 4.27, was significantly deeper than observed in the optical—removing 6%–8% or so of the stellar intensity versus a broadband optical transit depth of 2.41%. The deep transit was inferred to be caused by a thin outer planetary atmosphere that is transparent at optical wavelengths but sufficiently dense to be opaque in X-rays. The X-ray radius was also larger than observed at far-UV wavelengths, most likely due to the outer atmosphere being largely ionized, rendering the gas transparent in the UV but not in X-rays.

The exciting promise of using X-rays as a new probe of exoplanet atmospheres has been stymied by the build-up of a contamination layer on the ACIS instrument with a consequent loss of low-energy sensitivity that has prevented *Chandra* from pursuing further X-ray transit observations.

## 4.4 X-Rays from High-mass Stars

X-rays from high-mass stars were discovered serendipitously with the *Einstein* satellite shortly after launch during calibration observations using the powerful high-mass X-ray binary "microquasar" Cyg X-3 (Harnden et al. 1979). *Einstein* had seen some of the stars in the nearest region of truly massive star formation, the Cygnus OB2 association. The early *Einstein* and later *ROSAT* observations established that the X-ray luminosity of O and early B-type stars scales with their bolometric luminosity approximately following the relation $L_X/L_{bol} \sim 10^{-7}$ (Long & White 1980; Berghoefer et al. 1997).

O-type to early B-type stars have masses of approximately 10 $M_\odot$ or more. They evolve rapidly, eventually becoming Wolf–Rayet stars and have total lifetimes of typically only 4–10 Myr. OB stars emit copious ultraviolet radiation, dominating the ionization of the interstellar medium in their vicinity. Their intense radiation fields also drive powerful and massive stellar winds through line opacity, with terminal velocities exceeding 1000 km $s^{-1}$. Through their radiative and mechanical energy injection into the ISM during their lifetimes, together with their ultimate demise in colossal supernova explosions that enrich their surroundings in metals as well as in kinetic energy, they play a pivotal role in the evolution of their host galaxies.

The ubiquity of rapidly expanding stellar winds from OB stars were one of the most unexpected and important discoveries of the early NASA space program (e.g., Snow & Morton 1976). The soft X-ray emission subsequently discovered by *Einstein* was originally understood in terms of shock instabilities in line-driven winds (Lucy & White 1980); however, understanding of the processes that lead to the shocks and X-ray emission are still incomplete. The instabilities arise because of "shadowing" in the continuum, due to absorption in the photosphere such that a Doppler shift in the wind material out of the line results in an increase in driving force—the so-called "line de-shadowing instability" (see, e.g., the review by Puls et al. 2008). In cases where emission-line driving is important, Doppler shifting out of the rest frame of the emitting source leads to a decrease in driving force. In both cases, the variation of acceleration and velocity among streams and parcels of wind material lead to collisions and hydrodynamic shocks, and a highly inhomogeneous clumpy wind (e.g., Feldmeier et al. 1997).

The X-ray luminosity of OB stars declines rapidly for stars later than B type: a *ROSAT* OB star survey found detection rates of close to 100% for O stars but less than 10% for stars B stars later than B3 (Berghoefer et al. 1997). The weaker radiative driving power of later spectral types produces winds that are simply too tenuous to generate copious X-ray-emitting shocked gas and X-ray emission is predicted to drop sharply (see Section 4.4.1 below).

As we shall see below, this picture has expanded to encompass the effects of stellar magnetic fields and the collisions of winds in high-mass binaries, as well as the notorious "weak wind" problem uncovered for massive stars in which UV line diagnostics show clear wind signatures but derived mass-loss rates are lower by more than an order of magnitude compared with similar stars and with theoretical expectations (e.g., Puls et al. 1996).

### 4.4.1 Universality of the $L_X$–$L_{bol}$ Relation

Massive stellar clusters containing numerous coeval objects at a known distance are valuable targets for probing the high-energy properties OB star winds. The *Chandra* ACIS-I survey of the massive star formation region in the Carina OB1 association, dubbed the *Chandra* Carina Complex Project (CCCP; Townsley et al. 2011b), garnered the largest sample of OB stars in a single survey to date, detecting 129 O and B stars (Nazé et al. 2011). Of these, 78 had sufficient counts for spectral characterization and all of those were of spectral type B3 or earlier. Based on the 60 of these that are O stars, Nazé et al. (2011) found a relation $\log L_X/L_{bol} = -7.26 \pm 0.21$, in good agreement with the rough canonical $\sim -7$ hewn out from early *Einstein* data. The large number of stars enabled the study of subsets of them. They derived similar $L_X/L_{bol}$ ratios for both bright and fainter objects, and for stars of different luminosity classes or spectral types.

The universality of the $L_X/L_{bol}$ relation for O stars is further highlighted in the study of a large number of OB stars in the Cygnus OB2 association. Rauw et al. (2015) obtained $\log L_X/L_{bol} = -7.21 \pm 0.24$ from 40 O stars bright enough for spectral characterization—essentially identical to the Carina OB1 result (Figure 4.28).

Despite having essentially been established since the early 1980s, the physical origin of the $L_X$–$L_{bol}$ relation has proven difficult to reproduce from a theoretical perspective. Owocki et al. (2013) note that a wind subject to the line de-shadowing instability can be viewed as causing a small fraction of the wind material (say $10^{-3}$) to pass through an X-ray-emitting shock. This rough picture has been verified in detail based on an analysis of line profiles in high-resolution *Chandra* HETG X-ray spectra by Cohen et al. (2014b), who found that the average wind mass element, as it advects through the wind, passes through roughly one shock that heats it to at least $10^6$ K. In this case, if the shocks cool purely radiatively, X-ray luminosity is limited by the kinetic energy flux and should scale approximately with the mass-loss rate in

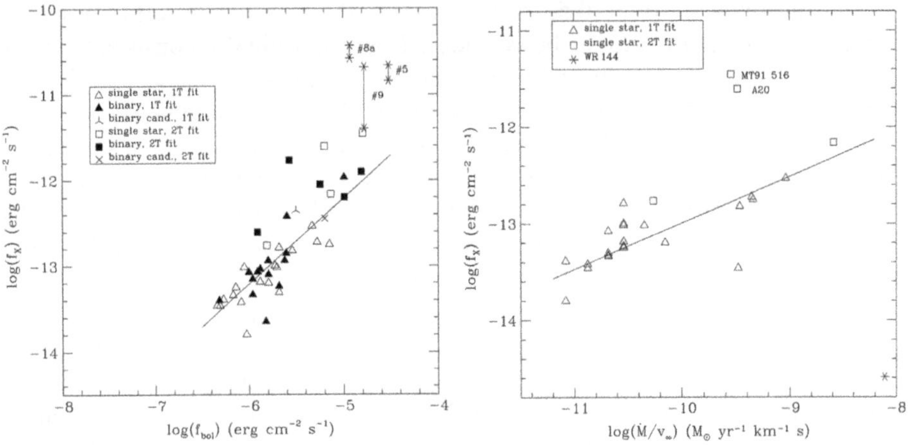


**Figure 4.28.** Left: the relation between X-ray and bolometric luminosities for bright O stars with more than 30 counts in the *Chandra* Cygnus OB2 survey analyzed by Rauw et al. (2015). X-ray luminosities are for the 0.5–10 keV band and have been corrected for interstellar absorption. The straight line corresponds to the unweighted best-fit scaling relation $\log L_X/L_{\rm bol} = -7.21 \pm 0.24$. The variation of X-ray fluxes seen in *XMM-Newton* observations for the multiple systems Cyg OB2 #5, #8a, and #9 that are likely colliding wind sources (see Section 4.4.2) are indicated by asterisks. Right: the X-ray fluxes of presumably single stars in the Cygnus OB2 survey corrected for interstellar absorption as a function of the logarithm of the ratio of the wind mass-loss rate and terminal velocity, $\log(\dot{M}/v_\infty)$. Both figures reproduced from Rauw et al. (2015) 

the wind, $L_X \sim \dot{M}$. The problem with this scaling is that the standard model for radiatively driven stellar winds predicts, to first order, a mass-loss scaling with luminosity as $\dot{M} \sim L_{\rm bol}^{1.7}$, implying a much steeper $L_X \sim L_{\rm bol}^{1.7}$ than is observed. The situation is even worse if shocks are cooled adiabatically, with an $\dot{M}^2$ dependence expected, and therefore $L_X \sim L_{\rm bol}^{3.4}$.

Owocki et al. (2013) proposed a solution to the conundrum through mixing of cooler unshocked gas with the shocked gas through the thin-shell instability. Kee et al. (2014) showed that the deformation of the shock compression by the instability leads to extended shear layers and fingers of cooled, denser material. Strong X-ray emission is limited to the tips and troughs of these fingers, and reduces the X-ray emission well below analytical expectations from radiative shocks without consideration of the instability. Assuming the strength of this mixing-induced quenching of X-ray emission scales with the shock cooling length raised to a power $m$, termed the "mixing exponent" by Owocki et al. (2013), they show that if $m$ lies in the range 0.2–0.4, this leads to an X-ray luminosity scaling with mass loss as $L_X \sim \dot{M}^{0.6}$, thereby recovering $L_X \sim L_{\rm bol}$.

According to the Owocki et al. (2013), the X-ray luminosity should also scale with $(\dot{M}/v_\infty)^{1-m})$ for O stars. Figure 4.28 illustrates the relation between the observed X-ray flux and $\log(\dot{M}/v_\infty)$ for which Rauw et al. (2015) obtained

$$\log f_{\rm x} = (0.48 \pm 0.10)\log\frac{\dot{M}}{v_\infty} - 8.19 \pm 0.97, \tag{4.12}$$

which corresponds to $m = 0.52 \pm 0.1$. Given the various uncertainties in the determination of $f_x$, this might be considered very marginally consistent with the upper limit to the range proposed by Owocki et al. (2013).

While promising for explaining the observed X-ray to bolometric luminosity scaling for O-star winds, the Owocki et al. (2013) scenario does involve some ad hoc assumptions that can only be properly verified with detailed three-dimensional modeling. Importantly, their model predicts different behaviors of the scaling of $L_X$ and $L_{bol}$ toward both low and high mass-loss limits of O-star winds. At the former, corresponding to low luminosities and later (B) spectral types, shocks should become adiabatic and the X-ray luminosity should drop steeply with $\dot{M}^2$. Such a drop is indeed observed for stars of spectral type later than B3 (Cohen et al. 1997).

At the latter high-luminosity end, winds should become optically thick and the X-ray luminosity should saturate and for very high mass-loss-rate winds that should be optically thick, even decrease with increasing luminosity and mass-loss rate approximately as $L_X \sim 1/\dot{M}$. No such drop is apparent in the *Chandra* Cygnus OB2 sample, although Rauw et al. (2015) suggested that this is likely because the transition occurs at still higher luminosities than reached by those stars. Owocki et al. (2013) suggested that the analysis of *Chandra* HETG X-ray data for the O2If star HD 93129A—one of the earliest and most massive stars known in the Galaxy—by Cohen et al. (2011) provides a potential example approaching the high mass-loss rate limit. While the X-ray luminosity was still well approximated by the $L_X \sim 10^{-7} L_{bol}$ relation, absorption by bound–free opacity in the dense wind makes the spectrum harder. Based on a catalog of O stars from *XMM-Newton* data, Nebot Gómez-Morán & Oskinova (2018) found instead that the $L_X$–$L_{bol}$ relation tends to break down for O supergiants, with a large scatter in X-ray luminosities over three orders of magnitude for stars of rather similar bolometric luminosity. Whether or not this is a symptom of the prediction of the Owocki et al. (2013) model is as yet uncertain.

*Insights from High-resolution X-Ray Spectroscopy*

Massive stars present compelling cases for high-resolution spectroscopy with the *Chandra* diffraction gratings. With wind outflow velocities in excess of 1000 km s$^{-1}$, line profiles are expected to show significant Doppler broadening. The intercombination of the forbidden line ratios of He-like ions, $z/(x + y)$, is sensitive to both plasma density and the intensity of the local radiation field (Section 4.2.1) and are expected to provide powerful diagnostics of the location of the shocked plasma X-ray source. It will come as no surprise that there are a myriad observations at high resolution with both *Chandra* and *XMM-Newton*; here we provide just some brief examples. Extensive discussion of further results can be found in the review by Güdel & Nazé (2009).

An early indication of the diagnostic power of high-resolution X-ray spectra was provided by the HETG spectrum of $\theta^1$ Ori C, the most massive of the Orion "Trapezium" stars and a close binary comprising O6Vp and B0V components. Schulz et al. (2000) found temperatures ranging from 0.5–6 $\times$ 10$^7$ K from a rich emission-line spectrum. Lines were significantly broadened, with velocities ranging from 400 to 2000 km s$^{-1}$, while the He triplets indicated electron densities

$n_e > 10^{12}$ cm$^{-3}$. While some of these features confirmed the wind shock model for X-ray emission, the surprisingly high temperatures and densities were not expected. These features indeed later helped confirm that $\theta^1$ Ori C is a case of a magnetically confined wind shock (see Section 4.4.3 below).

Similar broad lines were obtained for the O 9.7Ib star $\zeta$ Ori by Waldron & Cassinelli (2001), and for the O4 f star $\zeta$ Pup by Cassinelli et al. (2001), in agreement with shocked wind models. $\zeta$ Pup showed the classic line-profile shapes expected from a shocked wind model: a broadened profile asymmetric and blueshifted significantly due to absorption of the receding redshifted part of the wind hidden behind the blueshifted material. This "classic" case turned out to be a rarity, with the great majority of high-resolution O-star spectra showing more symmetrical profiles with little or no blueshift (e.g., Waldron & Cassinelli 2007)!

One of the best-quality spectra obtained for a massive star is the *Chandra* HETG spectrum of the triple system $\delta$ Ori A, which is dominated by $\delta$ Ori Aa1 in X-rays (Figure 4.29). Corcoran et al. (2015a) also found largely symmetrical line profiles. He-like triplets were affected by UV excitation, indicating formation within a few stellar radii and increasing in distance with decreasing temperature of formation. Lines were broadened by 550–1000 km s$^{-1}$, which is 0.3–0.5 times the wind terminal velocity, indicating formation within the acceleration zone and again in broad agreement with the radiatively driven wind shock scenario.

Two possible solutions to the lack of expected line asymmetries are wind clumping (see Figure 4.29) and lower mass-loss rates, with both having the effect that winds are effectively less opaque and the receding redshifted wind visible. The answer appears to be a combination of both. Cohen et al. (2014a) revisited the HETG spectra of single O stars and applied a spherically symmetric wind model to derive mass-loss rates from the observed spectral line profiles in a way that is not sensitive to optically thin clumping. Two such profiles are illustrated in Figure 4.29. They derived mass-loss rates are lower by an average factor of 3 than the theoretical models of Vink et al. (2000) and consistent with clumping-dependent H$\alpha$ measurements if clumping factors of $f_{\rm cl} \sim 20$, where $f_{\rm cl} = \langle\rho^2\rangle/\langle\rho\rangle^2$ for the gas density $\rho$, are assumed. The same profile-fitting technique found an onset radius for X-ray emission in the wind of typically at $r \approx 1.5R_\star$.

### 4.4.2 Colliding Winds

High-mass stars tend to have a high multiplicity and most are expected to reside in binary systems. Supersonic winds from two stars in relative proximity are expected to interact with each other leading to high Mach number shocks, giving rise to a variety of observational signatures from radio to $\gamma$-rays. X-rays from "colliding wind binaries" were in fact predicted by Prilutskii & Usov (1976) and Cherepashchuk (1976) two years before the detection of X-rays from massive stars (see, e.g., the review by Rauw & Nazé 2016).

Early observations by *Einstein* and *ROSAT* came as a surprise, finding the X-ray luminosities of known binary systems, both of Wolf–Rayet type and O-type stars, to be significantly lower than theoretical colliding wind calculations had suggested, and

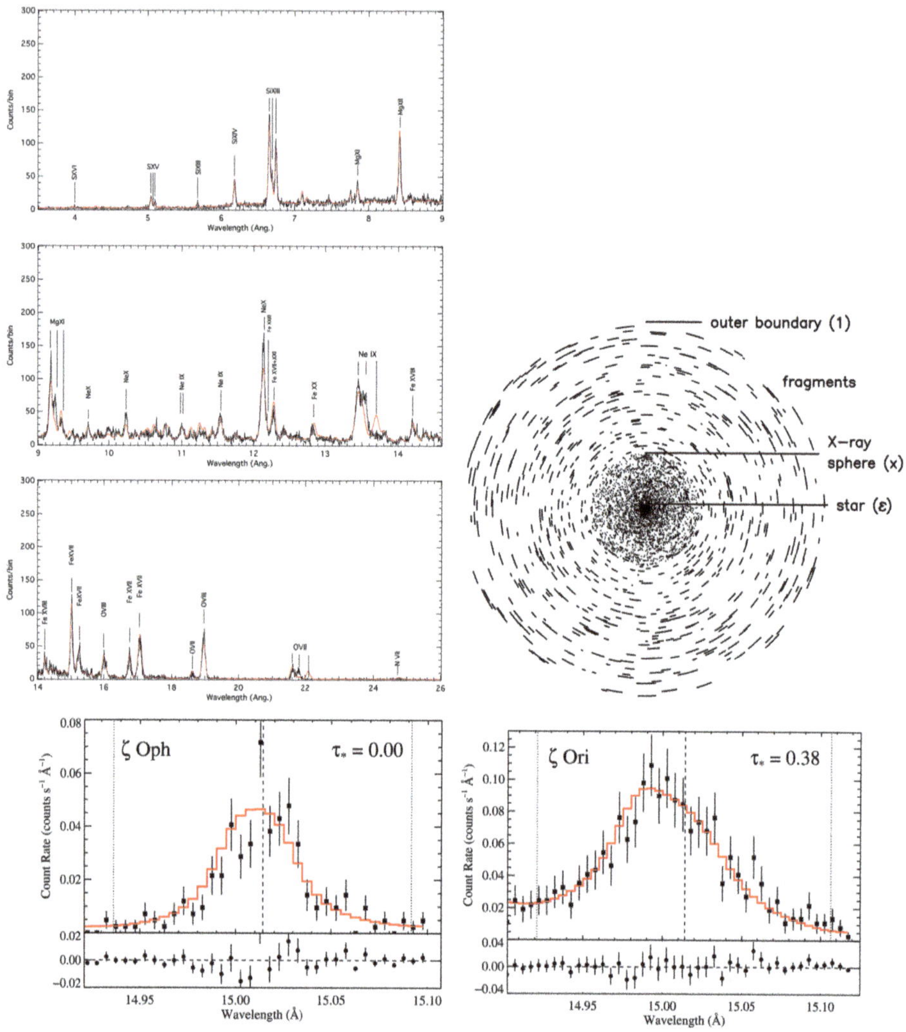


**Figure 4.29.** Top left: *Chandra*-combined MEG+HEG spectrum of the binary $\delta$ Ori Aa (black) from a 500 ks exposure covering nearly an entire orbit, demonstrating a line-dominated thermal spectrum with significant line broadening, due to wind outflow velocities in the range 550–1000 km $s^{-1}$. Model spectra shown in red assume low density and do not include the effects of UV photoexcitation. Consequently, the model overestimates the strength of the forbidden lines and underestimates the strengths of the intercombination lines in He-like ions. Reproduced from Corcoran et al. (2015a) © 2015. The American Astronomical Society. All rights reserved. Top right: a sketch of the porous clumped wind model from Feldmeier et al. (2003), reproduced with permission © ESO, suggested to help explain the observed line profiles for O-star winds. The "X-ray sphere" corresponds to the region within which X-rays originate. Bottom: spectral line profiles for the Fe XVII 15.02 Å resonance line in $\zeta$ Oph and $\zeta$ Ori. Line center is denoted by the vertical dashed line, and red histograms indicate the best-fit spherical shocked wind model. The line profile for $\zeta$ Oph appears symmetrical, due to a weaker wind compared with the $\zeta$ Ori profile that shows wind self-absorption of the redshifted, receding emission. Reproduced from Cohen et al. (2014b), by permission of Oxford University Press on behalf of the Royal Astronomical Society.

in many cases no larger than expected from single stars (Pollock 1987; Chlebowski & Garmany 1991). However, some of the more luminous binaries did turn out to be significantly brighter in X-rays and this overluminosity is now understood in terms of colliding winds.

The collision of two winds broadly results in two high Mach number shocks separated by an interface region with shape depending on the relative ram pressures of the two winds. The general picture of the interaction has been investigated for different parameters using 3D hydrodynamical simulations (e.g., Pittard & Parkin 2010). The shocks raise the gas involved to significantly higher temperatures than the line de-shadowing instability. For winds in which X-ray optical depth effects are important, the observed X-ray emission can also be phase dependent as the line of sight to the shock region changes. In eccentric binaries, the emission is also expected to vary with orbital phase as the conditions in the collision shock regions vary.

Stevens et al. (1992) introduced a cooling parameter, $\chi = t_{\rm cool}/t_{\rm esc}$, for the ratio of the cooling time to the escape time for gas to flow from the system. In systems where $\chi > 1$, the gas does not have time to cool before it escapes and so the wind collision region behaves like an adiabatic shock and is large and extended. Instead, in denser interaction regions, when radiative cooling is important and $\chi < 1$, shocks are radiative and confined to relatively thin shells. In the case of adiabatic shocks, Stevens et al. (1992) showed that the X-ray luminosity should scale with the inverse of the stellar separation, $L_{\rm X} \sim 1/D_{\rm sep}$.

While progress with theory and modeling of colliding wind binaries has been strong, the observations have tended to be difficult to understand within the theoretical framework. Again, the large samples of stars collected in *Chandra* surveys have been pivotal in understanding the importance of colliding winds in explaining the X-ray emission of high-mass stars.

Nazé et al. (2011) found that high-mass binaries in the Carina OB1 survey that have colliding winds and therefore might be expected to appear X-ray bright were only marginally more X-ray luminous and harder than single stars to an extent that they declared them not statistically significant. Similarly, Rauw et al. (2015) also found no conspicuously large excess in X-ray luminosity in binaries as compared to single stars in the Cygnus OB2 association. However, they did find evidence for modest X-ray overluminosity as a function of wind kinetic power for the more massive known binaries. Examples are Cyg OB2 #5, #8a, and #9, shown in Figure 4.28.

Nazé et al. (2011) noted that winds tend to collide at comparatively low speed in close binaries whereas the collision in wide systems is adiabatic and therefore not so X-ray bright. This leaves a fairly small region of parameter space for X-ray-bright wind–wind collisions. They also note that in cases where the wind momenta of the components are very different, modeling has shown that the stronger wind penetrates directly onto the companion (Pittard & Parkin 2010).

*The Spectacular Case of* η *Carinae*

$\eta$ Carinae is one of the most remarkable stars in the Galaxy. It is the nearest example of a luminous blue variable—a supermassive, superluminous, unstable star, and likely the most luminous and massive object within the nearest 3 kpc or so. Mass loss from $\eta$ Carinae shapes its circumstellar environment. While its current mass loss is in a slow ~500 km $s^{-1}$, dense $\dot{M} \sim 10^{-3} M_{\odot}$ $yr^{-1}$ stellar wind, it is also prone to eruptions during which large amounts of mass are expelled by processes that are not yet entirely understood (Davidson & Humphreys 1997; Corcoran et al. 2004).

$\eta$ Car itself is a colliding wind binary with a period of 2022 days, exhibiting strong variations in emission over a wide wavelength range including X-rays. These are driven by the collision of the dense wind of $\eta$ Car A with the fast, less dense wind of $\eta$ Car B that otherwise remains hidden. It undergoes a deep X-ray minimum every 5.53 yr resulting from its highly eccentric orbital motion and effects on the colliding wind X-ray source and its line-of-sight modulation.

$\eta$ Car is also a source of nonthermal X-ray and $\gamma$-ray emission thought to result from Fermi acceleration at the shock interface of the wind–wind collision and subsequent inverse Compton upscattering of UV photons, demonstrating that colliding wind binaries are a source of energetic particles (Hamaguchi et al. 2018).

A *Chandra* ACIS image of the $\eta$ Carinae region centered on the Homunculus—the hollow, expanding bipolar nebula surrounding $\eta$ Car and corresponding to the shock front originating from the "Great Eruption" in the 1840s—was obtained in 2003 at a time when it was at X-ray minimum and is shown in Figure 4.30. This is the first time the X-ray nebula was detected, as it is normally swamped by the PSF wings of $\eta$ Car itself. Corcoran et al. (2004) found the emission to be characterized by an extremely high temperature in excess of 100 MK and consistent with scattering by the circumstellar environment of the time-delayed X-ray flux associated with the star. A strong Fe K fluorescent line at 6.4 keV was also detected. This line had a width of 4700 km $s^{-1}$, which is much larger than the expansion velocity of the Homunculus itself. Corcoran et al. (2004) interpreted the line widths as likely resulting from reprocessing by fast flows in the lobes of the Homunculus, perhaps with contributions from the companion stellar wind.

The X-ray signal from the central stars of $\eta$ Carinae undergoes dramatic variations on both long and short timescales. On timescales of years, the X-ray emission follows approximately the inverse of the stellar separation, $L_X \sim 1/D_{sep}$, as might be expected for adiabatic colliding wind conditions in a highly eccentric binary. This long-term variability is interrupted by a deep, broad atmospheric eclipse near periastron when the dense inner wind of $\eta$ Car A is in front of the bow shock region of the collision. Superimposed on the long-term trend are shorter "flares," lasting from a couple to 100 days. The origin of these flares remains uncertain. Moffat & Corcoran (2009) considered several mechanisms, including large-scale co-rotating interacting regions in the $\eta$ Car A wind sweeping across the wind collision zone and instabilities intrinsic to the collision zone, but argued that the most likely explanation lies in the largest of the multiscale stochastic wind clumps from $\eta$ Car A entering hard X-ray-emitting wind–wind collision zone.

### 4.4.3 The Role of Magnetism

A fraction amounting to approximately 10% of massive stars are observed to have X-ray emission characteristics that are difficult to explain in terms of pure shocked

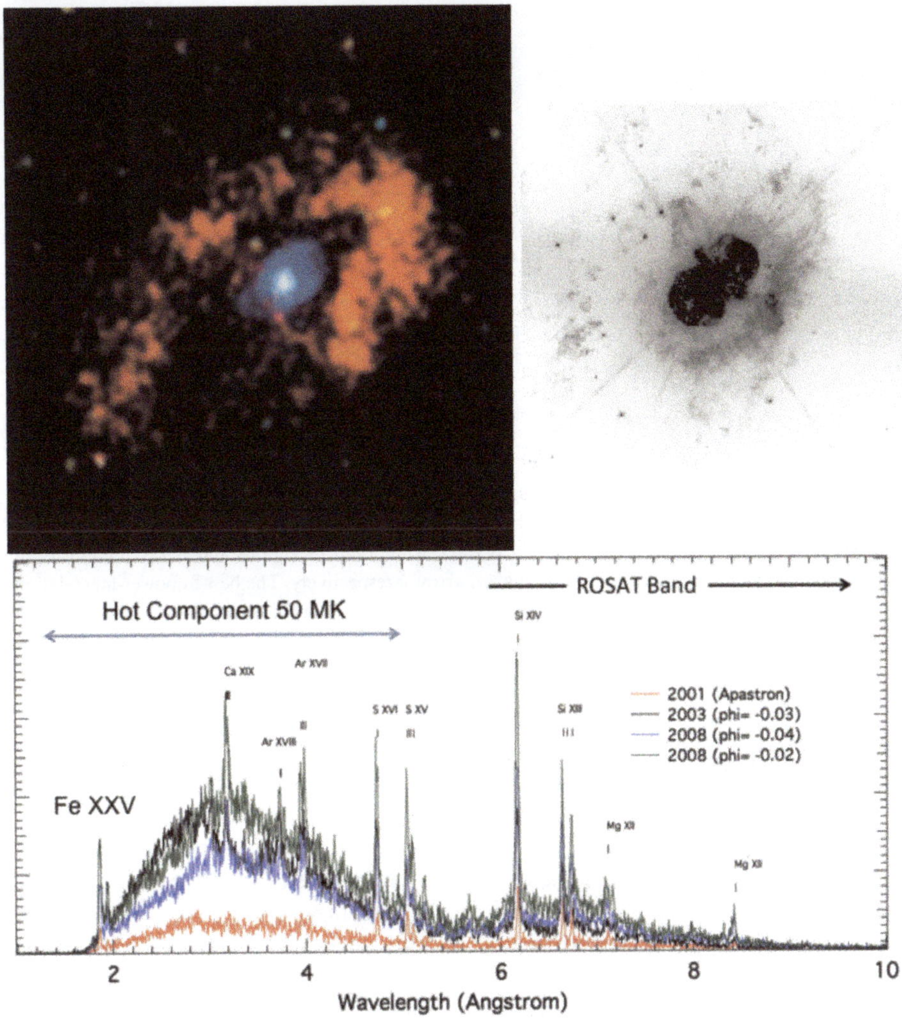


**Figure 4.30.** Top left: a false color *Chandra* ACIS X-ray image of the region centered on $\eta$ Car during X-ray minimum. Colors correspond to different X-ray bands as follows: red, low energy (0.2–1.5 keV); green, medium energy (1.5–3.0 keV); and blue, high energy (3.0–8.0 keV). Emission from $\eta$ Car itself corresponds to the central white point source. A broken elliptical ring of very soft X-ray (shown in red) emission from the shock arising from the "Great Eruption" lies in a partial ring surrounding the star. The bluish patch around the star inside this ring is reflected X-ray emission from the Homunculus nebula. Top right: an *HST* WFPC2 [N II] 6583 Å image of $\eta$ Car with the same plate scale and orientation as the *Chandra* image. North is to the top, and east is to the left. Reproduced from Corcoran et al. (2004) © 2004. The American Astronomical Society. All rights reserved. Bottom: *Chandra* HETG spectra of $\eta$ Car itself at different epochs showing strong X-ray spectral variations that appear to be reasonably well explained by an adiabatic wind–wind shock with X-ray luminosity varying inversely with stellar separation, $L_X \sim 1/D_{sep}$. The spectrum is largely thermal in nature, with a temperature of about 4.5 keV that corresponds to a preshock stellar wind velocity of about 3000 km $s^{-1}$. Reproduced from Corcoran et al. (2015b). The interactive figure shows separately the X-ray emission in low energy (red), medium energy (green), high energy (blue) bands, and the composite image. Additional materials available online at http://iopscience.iop.org/book/978-0-7503-2163-1.

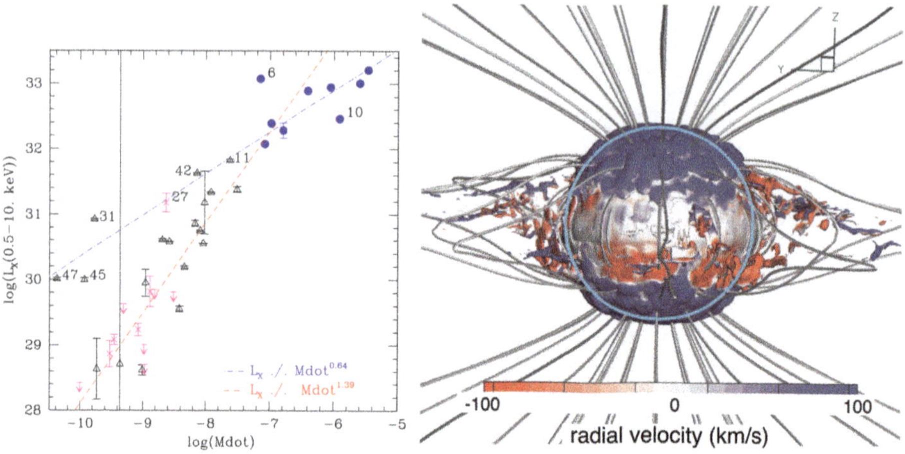


**Figure 4.31.** Left: absorption-corrected X-ray luminosities as a function of mass-loss rate, $\dot{M}$, for magnetic OB stars analyzed by Nazé et al. (2014) © 2014. The American Astronomical Society. All rights reserved. Filled blue dots denote O stars, black empty triangles B stars, while faint detections and X-ray luminosity upper limits are signified by magenta crosses and downward arrows, respectively. The best-fit power-law relation are shown. Right: illustration of a 3D MHD magnetically confined wind shock model for $\theta^1$ Ori C by ud-Doula et al. (2013; by permission of Oxford University Press on behalf of the Royal Astronomical Society) showing an isodensity surface at $\log \rho = 12.5$ g cm$^{-3}$, colored with radial velocity. Some of the gas within the closed loop region falls back on to the stellar surface, while material in the open fields streams out radially.

wind models. These include variability, moderately hard X-ray emission, and in some cases higher plasma densities than expected (e.g., Nazé et al. 2014). It turned out that stars showing such characteristics also tended to have fairly strong magnetic fields thought to either be fossil remnants from star formation or else generated in convective cores. Such magnetic fields channel the winds of massive stars toward the magnetic equator where they collide in a process now known as a "magnetically confined wind." This mechanism was originally proposed by Babel & Montmerle (1997) to explain X-ray emission from the A0p star IQ Aur. Because the magnetically confined wind collides with itself at higher Mach number than the shocks produced by the line de-shadowing instability described in Section 4.4.1, the shocked plasma is expected to be hotter, as observed, and subject to variability, due to instability in the flow (e.g., ud-Doula et al. 2014).

Based on a sample of magnetic stars culled from *XMM-Newton* and *Chandra* archives, Nazé et al. (2014) have shown that the X-ray luminosity is strongly correlated with the stellar wind mass-loss rate and luminosity, with the relations illustrated in Figure 4.31. They determined a power-law dependency that is slightly steeper than linear for the less-luminous B stars with lower mass-loss rates, $\dot{M}$, and that flattens for the more luminous, higher $\dot{M}$ O stars. The observed X-ray luminosities, and their trend with mass-loss rates, were found to be reasonably well reproduced by 2D and 3D magnetohydrodynamic models (ud-Doula et al. 2013, 2014; see Figure 4.31). ud-Doula et al. (2014) found in general that for stars with lower mass-loss rates, X-ray emission is reduced and softened by a "shock retreat" that results from a large postshock cooling length: within the fixed length of a closed magnetic loop, the shock is forced back to lower preshock wind speeds.

This good agreement between 3D models and observations was confirmed in detail by models fitted directly to *Chandra* HETG spectra of the magnetically confined wind Of?p star HD 191612 by Nazé et al. (2016). No significant line shifts were seen, with lines being relatively narrow and formed at a distance of about $2R_{\star}$.

Nazé et al. (2014) still found puzzling exceptions to this picture of "well-behaved" magnetically confined wind stars that still belie explanation, such as the persistence of some X-ray overluminous stars not obviously attributable to colliding winds. Moreover, they found no relation between other X-ray plasma temperature or absorption and stellar or magnetic parameters as might have been expected, indicating that the plasma properties are controlled by different processes from the X-ray luminosity itself.

### 4.4.4 The Weak Winds Problem

One of the outstanding problems in hot star winds is the so-called "weak-wind" problem, in which mass-loss rates modeled from UV line diagnostics for stars that show clear classical P Cygni line profiles can be discrepant by more than an order of magnitude from expected values based on O-star statistical trends and based on theoretical models. The problem then has two aspects: lower mass-loss rates than other stars of similar spectral type; and, on the other hand, other weak-wind stars have lower mass-loss rates than what atmospheric models predict. High X-ray luminosities have sometimes been invoked to explain the weakness of the winds, as they can modify the wind ionization and hence the efficiency of wind acceleration (e.g., Martins et al. 2005).

Huenemoerder et al. (2012) combined *Chandra* LETG+ACIS-S spectroscopy and *Suzaku* observations of the notorious weak wind star $\mu$ Col and found an X-ray emission measure corresponding to an outflow an order of magnitude greater than suggested by UV lines, and comparable with a standard wind luminosity relationship for O stars. The spectrum of $\mu$ Col is soft and lines are broadened, with radiative excitation of the He-like triplets indicating that the bulk of the X-ray emission is formed within 5 stellar radii, all in agreement with expectations of a typical shocked wind.

The "weak-wind" problem identified from cool wind UV and optical spectra is then largely resolved, at least in $\mu$ Col, by accounting for the hot wind only seen in X-rays.

### 4.4.5 The Mysterious X-Rays from Cepheid Variables

Classical Cepheid variables are pulsating yellow supergiant stars with masses in the range 4–20 $M_{\odot}$ and pulsation periods typically ranging from 2 to 60 days. They undergo periodic changes in size, temperature, and brightness as a result of the "$\kappa$-mechanism," so-called because the instability is driven by the sensitivity of the stellar opacity to changes in the stellar structure. In Cepheids, the instability is driven by He ionization. The cooling of gas in an expanding atmosphere leads to lower He ionization and a lower opacity that causes the outer envelope to contract under gravity. Under collapse, the gas is heated and He eventually becomes more ionized and opaque, leading to another expansion and contraction cycle. Cepheids lay the foundations of the cosmic distance scale through the Leavitt law, which describes the correlation between the period of the Cepheid and its luminosity.

X-rays were surprisingly detected from the prototypical classical Cepheid $\delta$ Cep in *XMM-Newton* observations (Engle et al. 2014). Subsequent *Chandra* observations confirmed the hint from the discovery observations that the X-ray flux is correlated with the pulsation phase (Engle et al. 2017). Observations obtained in far-ultraviolet lines with the *Hubble Space Telescope* also demonstrated strong phase-dependent emission. The data for $\delta$ Cep are illustrated in Figure 4.32.

For the majority of the five-day pulsation cycle of $\delta$ Cep, the X-ray luminosity is comparatively low at $\log L_X = 28.5–29$ erg s$^{-1}$, similar to that of nonpulsating yellow supergiants and corresponding to an X-ray surface flux an order of magnitude or more below that of the typical Sun. However, near maximum radius, close to phase 0.5, the X-ray flux was observed to rise rapidly by a factor of 4 or so, and then to fall equally rapidly only 0.10 later in phase. Surprisingly, the X-ray peak occurs at the phase of maximum radius, which is quite different from the beginning of the rise phase at which the peak in the far-UV fluxes is observed.

There are at least two likely mechanisms explaining the X-ray emission. First, a shock wave propagating through the atmosphere might be expected to develop from the pulsation wave gradually propagating into a medium of lower and lower density. X-ray emission from Cepheids was indeed predicted from such a mechanism by Sasselov & Lester (1994). Second, if $\delta$ Cep possessed a sufficiently strong magnetic field, the periodic stressing of the field due to pulsations could drive a periodic type of flaring. Magnetic fields of the order of 1 G have indeed been reported in the Cepheid $\eta$ Aql, as well as in the supergiant $\alpha$ Per (Grunhut et al. 2010).

There are now X-ray detections of several other Cepheids (Engle et al. 2017), although only one of them, $\beta$ Dor, shows strong evidence for similar phase-dependent X-ray emission so far. Progress is hampered because these stars are just generally too distant and too faint in X-rays for a detailed study with *Chandra* and *XMM-Newton*. Thorough examination of the potential of the Cepheids to be a new class of X-ray pulsating sources will likely have to await future missions with larger collecting areas.

## 4.5 Intermediate-mass Stars

We noted previously in Section 4.2 that early observations by the *Einstein Observatory* showed that X-rays from stars are common throughout the Hertzsprung–Russell diagram (Vaiana et al. 1981). Notable exceptions were late-type giants and main-sequence stars from late B type through middle A type. While the bright early A-type stars Vega and Sirius seemed to have been detected by the *Einstein* High Resolution Imager, these were subsequently attributed to UV leaks (Golub et al. 1984; a similar situation noted by Zombeck et al. 1997 also arose with the later *ROSAT* HRI).

The preceding text has detailed two ways in which single, nondegenerate stars produce X-rays: (1) through shock-heated regions in radiatively driven hot stellar

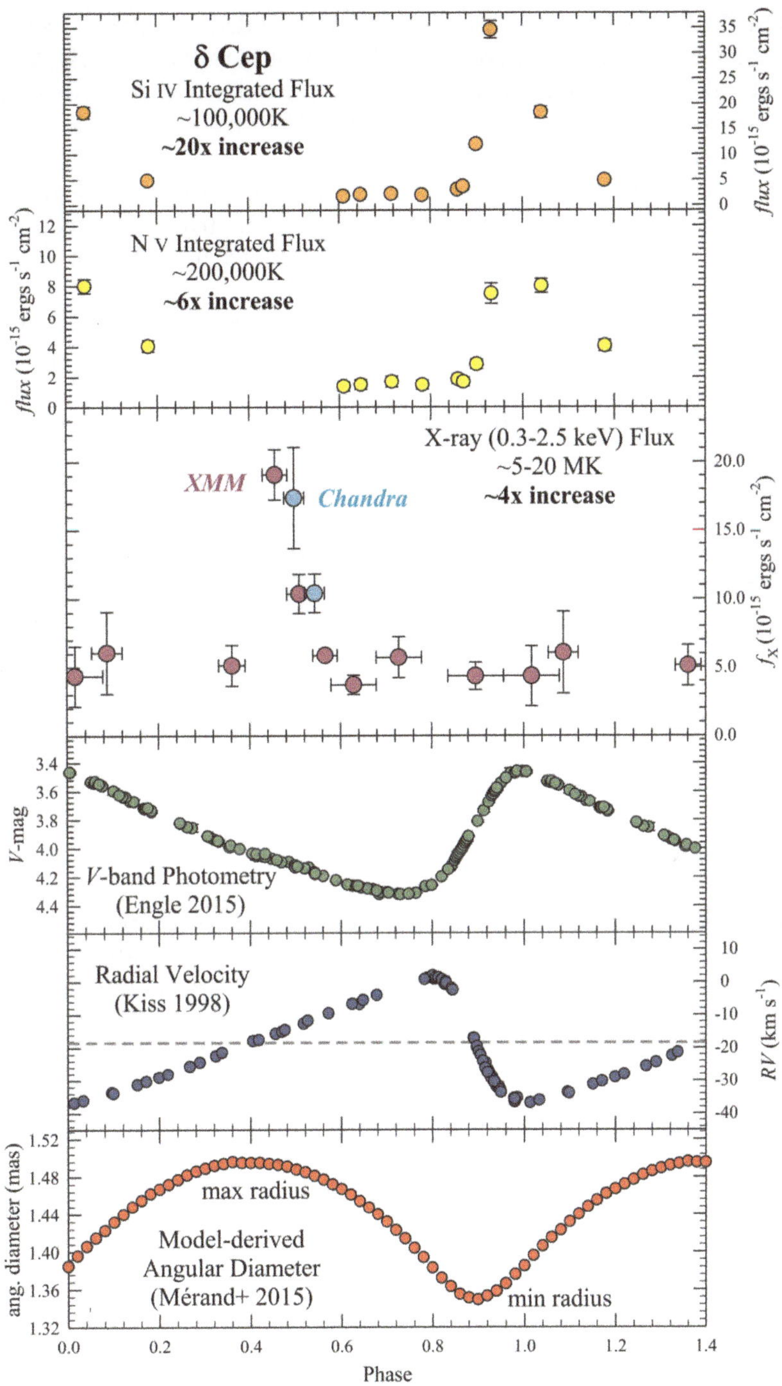


**Figure 4.32.** UV (upper two panels), X-ray (third panel), and photospheric (lower three panels) variations of $\delta$ Cep as a function of pulsation phase. Pink and cyan X-ray data points are from *XMM-Newton* and 2015 *Chandra* observations, respectively. Reproduced from Engle et al. (2017) 

winds and (2) in hot coronae sustained by dynamo-driven magnetic activity that arises in stars with convective envelopes. The lack of X-rays from early A-type stars is then naturally explained by their location in a transition regime between stars that are able to produce X-rays by these mechanisms. Their temperatures are too low and radiation fields too weak to drive strong supersonic winds, but high enough that hydrogen—the dominant source of opacity giving rise to outer convection zones in stars—is largely ionized, rendering their envelopes radiative. The effective temperatures defining the transition regime fall in the range ~8500–12,000 K, with $B - V \sim -0.1$ to 0.2.

One problem arises with the above reasoning: at least some intermediate-mass main-sequence stars are found to be X-ray emitters. Pre-main-sequence Herbig Ae/Be (HAeBe) stars—the intermediate mass nearly fully radiative counterparts to classical T Tauri stars—have routinely been found coincident with X-ray sources with luminosities of a few $10^{31}$ erg $s^{-1}$ down to $\sim 10^{29}$ erg $s^{-1}$ (Damiani et al. 1994; Zinnecker & Preibisch 1994; Stelzer et al. 2006). While unresolved lower mass companions might still be responsible for some of the detections, Stelzer et al. (2006) found an overall detection fraction of 76% for a sample of 17 HAeBes, and only half of these have known unresolved companions.

A plausible scenario raised by the Herbig Ae/Be stars is that an early active magnetic dynamo gives rise to X-ray emission, feeding from the natal shear and differential rotation within the star. This shear is gradually dissipated by the torque exerted by Lorentz forces as a result of magnetic fields generated by the shear operating in concert with the Taylor magnetic instability (Tayler 1973), which is analogous to the pinch instability in plasma physics. Such a dynamo scenario was outlined by Tout & Pringle (1995) and further considered by Spruit (2002) and Braithwaite (2006). The key to test such a scenario is to follow early A-type stars at very young ages to map out their X-ray emission.

Drake et al. (2014b) observed HR 4796A, a nearby (~70 pc) 8 Myr old main-sequence A0 star using the *Chandra* HRC-I, which is more sensitive than the ACIS detector to very soft X-ray emission that might arise from a cooler corona than exhibited by more active stars. HR 4796A possesses a remnant dusty disk that has been scrutinized intensively from both the ground and in space using *Hubble*, which has essentially ruled out the possibility of a stellar mass companion. A 21 ks exposure failed to detect a single photon from HR 4796A! The implied X-ray luminosity upper limit was $L_X \leqslant 1.3 \times 10^{27}$ erg $s^{-1}$.

One other key star of spectral type A0 is Vega, even closer at ~7.8 pc star and also with a dusty remnant disk and thought to be single. Vega is older than HR 4796A, with a rather uncertain age in the range 100–400 Myr. *Chandra* uses Vega for routine monitoring of the detectors for possible UV leaks, and Pease et al. (2006) used accumulated observations to search for an X-ray signal. None was found that could not be attributed to background or a very small residual UV leak and an associated upper limit to the flux of $L_X < 3 \times 10^{25}$ erg $s^{-1}$ was derived, corresponding to a bolometric fraction limit of $L_X/L_{bol} < 9 \times 10^{-11}$ (see also Ayres 2008), or about a million times lower than the Sun near solar minimum. The surface X-ray flux of Vega is in fact at least two orders of magnitude below the "universal minimum"

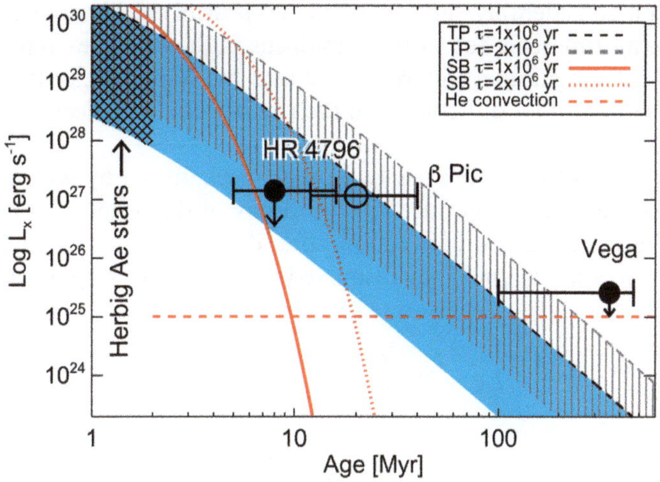


**Figure 4.33.** X-ray luminosity versus time for shear dynamo models compared with observations of single A-type stars from Drake et al. (2014b) © 2014. The American Astronomical Society. All rights reserved. The hashed region to the left represents the range of X-ray luminosities of Herbig Ae stars. The dashed curves correspond to a natal X-ray luminosity of $L_{X_0} = 2 \times 10^{31}$ erg s$^{-1}$ and two different values for the rotation shear decay timescale for the Tout & Pringle (1995, TP) model. The solid and dotted curves correspond to an exponential X-ray luminosity decay corresponding to the rotational shear dissipation found by Spruit (2002, SB; see also Braithwaite 2006) assuming the TP prescription for magnetic flux rising due to buoyancy. The horizontal dashed curve corresponds to the Drake et al. (2014b) estimate of a base-level X-ray luminosity, due to a weak subsurface convection zone. The $\beta$ Pic detection of Günther et al. (2012) is shown with a hollow symbol to denote the likely origin of its X-ray emission from a thin surface convection zone.

surface flux for F–M main-sequence stars found by Schmitt (1997) based on *ROSAT* data, which correspond to the surface flux of solar coronal holes.

The implications of the nondetections of HR 4796A and Vega are illustrated in Figure 4.33 and are compared with typical fluxes observed for Herbig Ae/Be stars and with predictions of the Tout & Pringle (1995), Spruit (2002), and Braithwaite (2006) shear dynamo formalisms. The former predicts a decay of the initial X-ray luminosity, $L_{X_0}$, with time according to

$$L_X(t) = \frac{L_{X_0}}{(1 + t/\tau)^3}, \tag{4.13}$$

where $\tau$ is the decay timescale. Based on reasonable guesses of relevant parameters, Tout & Pringle (1995) estimated $\tau \sim 10^6$yr, which also corresponds to the timescale over which HAeBe stars appear X-ray bright (e.g., Hamaguchi et al. 2005). The analogous formula derived by Drake et al. (2014b) based on Spruit (2002) and Braithwaite (2006) gives

$$L_X(t) = L_{X_0} \exp\left(-\frac{3t}{2\tau}\right), \tag{4.14}$$

although with a considerably shorter timescale, possibly as short as a few hundred years. The decay timescale for both formalisms is, however, extremely uncertain, and by orders of magnitude. Figure 4.33 shows the X-ray decay for timescales of 1 and $2 \times 10^6$ yr. Another difficulty is in the rise of field from deeper layers through diffusive buoyancy, which acts very slowly.

Drake et al. (2014b) also discuss two other mechanisms by which early A-type and late B-type stars might produce X-rays: decay of the primordial magnetic field, and a thin, weak subsurface convection zone driven by opacity bumps from the ionization of iron and helium. They dismissed the former as insignificant and for the latter derived

$$L_X \sim 10^{25}\left(\frac{P}{12\ \mathrm{h}}\right)^{-2} \mathrm{erg\ s^{-1}} \tag{4.15}$$

for a 2 $M_\odot$ star with $L = 25\ L_\odot$, where the approximate power of −2 on the period originates from the same *Ro* relation in Equation (4.7) and $\beta = 2$.

The X-ray luminosity for HR 4796A marginally conflicts with the TP-based model, whereas the X-rays based on Spruit (2002) and Braithwaite (2006) decay too quickly to be seen. Subsurface convection appears only able to produce X-ray luminosities of order $10^{25}$ erg $\mathrm{s}^{-1}$, which lies only moderately below the current detection limit for Vega. Based on these difficulties, Drake et al. (2014b) instead favored an accretion- or jet-based mechanism to explain the X-ray activity of Herbig Ae/Be stars, an idea suggested by Hamaguchi et al. (2005). The lack of photo-excitation of the intercombination lines of H-like O in an *XMM-Newton* observation of the Herbig Ae star HD 163296 (Günther & Schmitt 2009) suggested that the cooler X-ray emission originates away from the star at the base of a jet, whereas hotter emission might be coronal in origin.

Regardless of the exact mechanisms at play, following the Herbig Ae/Be pre-main-sequence phase, early A stars then generally decline in X-ray luminosity at least 100,000-fold in only a few million years.

The spectral type limit earlier than which A stars begin to be plausibly X-ray dark appears to be about A5, based on a detection of $\beta$ Pictoris in a 20 ks *Chandra* HRC-I observation reported by Günther et al. (2012). For later A-type stars, weak X-ray emission can be supported by a thin convection zone.

## 4.6 White Dwarfs and White Dwarf Binary Systems

White dwarfs are the final evolutionary states of low-mass and intermediate-mass stars—stars of less than about 8 $M_\odot$ that are not massive enough to become supernovae and neutron stars, and make up over 97% of stars in the Milky Way (Fontaine et al. 2001). After main-sequence evolution, such a star expands to a red giant in which core He burning through the triple-alpha process builds up a carbon and oxygen core. Helium shell burning proceeds on the asymptotic giant branch (AGB) and the star loses mass at a much higher rate, eventually shedding its envelope. At the end of the AGB phase, the remnant envelope material forms a planetary nebula, inside which the remnant CO core becomes a nascent white dwarf.

Much rarer higher mass progenitors in the 8–10 $M_\odot$ range are expected to have had higher core temperatures on the red giant branch and to have produced cores composed of oxygen, neon, and magnesium (e.g., Iben et al. 1997). However, the dividing line between the production of a white dwarf and a core-collapse supernova is thought to be quite fine, and evidence for remnant ONeMg white dwarfs is fairly scarce.

Stars of less than 0.5 $M_\odot$ have insufficient core temperatures for helium burning and are thought to end their lives as He white dwarfs. Because their evolutionary timescales are longer than the age of the universe, their only appearance is expected to be in binary systems in which their envelopes have been stripped away gravitationally.

A white dwarf is very hot when it forms, with a surface temperature that can easily exceed 100,000 K and therefore having a photosphere with Wien tail visible into the soft X-ray range. Having no internal source of energy, white dwarfs gradually cool with a timescale of millions to hundreds of megayears as this energy is radiated away. The temperature of a single white dwarf is then an age diagnostic, rendering white dwarfs of considerable interest for cosmochronology (e.g., Fontaine et al. 2001).

White dwarfs forming in binary systems can also become X-ray sources by accreting matter captured from their companions, becoming CVs and novae. The average surface gravity of white dwarf stars is $\log g \sim 8$, compared with 4.4 for the Sun. Unlike the accreting protostars discussed in Section 4.2.7, the X-ray spectra of which can show traces of accretion energy, the strong gravitational well of accreting white dwarfs can generate bright, hot, hard X-ray spectra.

*Chandra* has observed white dwarfs stars in all their activity phases, from nascent stars in planetary nebulae (PNe) to single white dwarfs on the cooling track, and binary white dwarfs in CVs and nova explosions.

### 4.6.1 White Dwarf Birth in Planetary Nebulae

PNe are commonly detected as X-ray sources. The X-ray emission is typically in two forms: from the compact point-like sources associated with the central stars, and extended diffuse X-ray emission from shocks and hot bubbles in the nebula itself resulting from the outflow and excitation by the hot ionizing central source.

The most comprehensive study of X-rays from PNe to date has been the *Chandra* Planetary Nebula Survey (ChanPlaNS; Kastner et al. 2012; Freeman et al. 2014) that targeted a total of 59 PNe within ~1.5 kpc. The diffuse X-ray detection rate of the sample was 27%, and the point source detection rate 36%.

Diffuse X-ray emission is expected on the basis of the fast (500–1500 km $s^{-1}$) radiatively driven winds from the pre-white dwarf that collide with the rejected red giant envelope, sweeping the ejecta into a thin shell. This interaction results in shocks that can heat the wind gas to temperatures exceeding $10^6$ K, leading to the formation of a hot bubble of soft X-ray-emitting gas. Freeman et al. (2014) found that diffuse X-ray emission is associated with young ($\lesssim 5 \times 10^3$ yr) and correspondingly compact ($\lesssim 0.15$ pc in radius) PNe with closed morphologies (Figure 4.34).

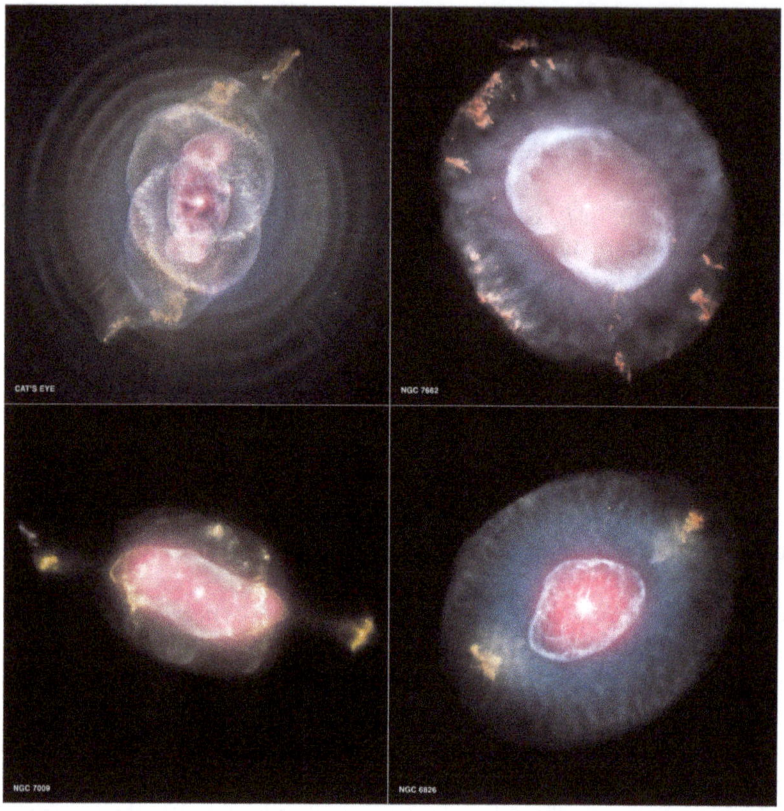


**Figure 4.34.** Four objects from the first systematic X-ray survey of planetary nebulae in the solar neighborhood exhibiting diffuse X-ray emission. Shown here are NGC 6543, also known as the Cat's Eye; NGC 7662; NGC 7009; and NGC 6826. In each case, X-ray emission from *Chandra* is colored purple and optical emission from the *Hubble Space Telescope* is colored red, green, and blue. X-ray emission is produced by the interaction of the fast, radiatively driven outflow from the central nascent pre-white dwarf with itself and with the ejected red giant envelope. From http://chandra.harvard.edu/photo/2012/pne/ and Kastner et al. (2012). X-ray courtesy of NASA/CXC/RIT/J.Kastner et al.; Optical courtesy of NASA/STScI. The interactive figures for each panel: Cat's Eye, NGC7662, NGC7009 and NGC6826, show separately the X-ray emission (purple), optical emission (red, green, blue), and the composite image. Additional materials available online at http://iopscience.iop.org/book/978-0-7503-2163-1.

All five of the PNe in the sample with central stars of the so-called "Wolf–Rayet type" or [WR]-type—so named because of their H deficiencies, high mass-loss rates up to $10^{-6}\ M_{\odot}\ \mathrm{yr}^{-1}$, and general superficial resemblance to true massive Wolf–Rayet stars—were detected in diffuse X-rays. In these cases, Freeman et al. (2014) noted that the diffuse X-ray emission resembles the limb-brightened, wind-blown bubbles blown by massive WR stars.

All of the PNe with diffuse emission detections have relatively high central electron densities of $n_e \gtrsim 1000\ \mathrm{cm}^{-3}$, again reflecting their relatively compact size and the vigor of the outflow of the central star during its early stages. PNe typically last of the order of 20,000 yr, and the *Chandra* diffuse emission detections indicate that beyond the first 5000 years, the systems are either too large and/or the central outflow is too weak to maintain the wind interactions necessary to produce detectable X-ray emission.

X-ray detections of the central stars themselves have proved a bit more difficult to interpret. Twenty of the sample of 59 were detected by *Chandra*. Montez et al. (2015) found the majority of them to be associated with luminous central stars within the relatively young and compact nebulae of the sample. While very soft X-ray emission from the continua of the hot nascent photospheres of these emerging white dwarfs might be expected, the great majority of the detections displayed comparatively hard X-ray emission that Montez et al. (2015) discovered were consistent with optically thin thermal plasma emission. Based on the fitted plasma properties, they identified two classes of central star X-ray emission. One has relatively cool plasma emission characterized by temperatures of $\lesssim 3 \times 10^6$ K and with X-ray to bolometric luminosity ratios $L_X/L_{bol} \sim 10^{-7}$, reminiscent of the radiatively driven wind shocks seen in high-mass stars described in Section 4.4. The other has high-temperature plasma at $10^7$ K or so and with X-ray luminosities apparently uncorrelated with the stellar bolometric luminosities. Montez et al. (2015) identified this latter category with magnetically active binary companions. The binary companion can be spun up in the common envelope phase and by accreting matter from the envelope of its evolving companion, invigorating its magnetic dynamo (see Section 4.2.2).

### 4.6.2 Photospheric Emission

One of the surprises from the *ROSAT* PSPC X-ray all-sky survey was the unexpectedly small number of white dwarf stars detected compared with prelaunch expectations (Fleming et al. 1996). The great majority of the 175 detections (161) are of DA spectral type—essentially photospheric emission from a pure hydrogen envelope. The remainder comprised three helium-dominated DO types, three DAO types (H + He envelopes), and eight PG 1159 stars with He–C–O-dominated photospheres. This sample amounted to only 10% of the white dwarfs thought hot enough to emit in soft X-rays ($T_{eff} > 20,000$ K) and be detected by *ROSAT*.

The reason for the dearth of detections lies in the radiative levitation of metals in the atmospheres of hot white dwarfs that would otherwise sink due to the strong gravitational field. The degree of levitation depends on the white dwarf gravity and on the effective temperature and consequent strength of the photospheric radiation field. The picture can also be complicated by weak mass loss that alters the chemical equilibrium; as the star gradually cools, the photospheric chemical composition is expected to change. For stars with effective temperatures $T_{eff} > 40,000$ K, the atmospheric opacity of metals is very large in the EUV and soft X-ray range, effectively blocking the radiation field and redistributing much of it toward longer wavelengths outside of the X-ray band (e.g., Barstow et al. 1997).

As noted in the introduction to this section, the Wien tail of the photospheric emission from hot white dwarfs extends from the UV and far-UV into the extreme ultraviolet and soft X-ray range—and the longer wavelengths covered by the *Chandra* LETG+HRC-S (1.2–170 Å). This has enabled *Chandra* to study the detailed spectra

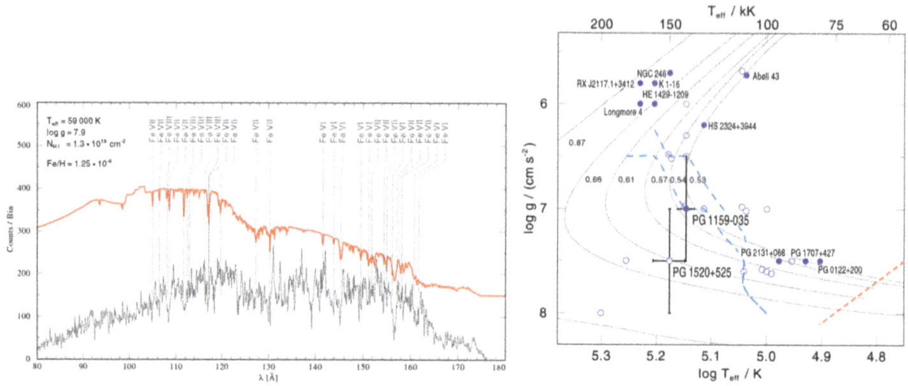


**Figure 4.35.** Left: the *Chandra* LETG+HRC-S spectrum of the DA white dwarf GD 246 (black line) and model spectrum (red; shifted upward by 150 counts for clarity) with some lines due to highly ionized Fe identified. Right: the log $T_{\rm eff}$ – log $g$ diagram illustrating the GW Vir instability strip constrained by the measured effective temperatures of the pulsating star PG 1159–035 and the nonpulsator PG 1520+525. Other pulsating (filled circles) and nonpulsating (empty circles) stars are illustrated. Evolutionary tracks are labeled according to stellar mass. The red edge of the instability region (short dashed line) and two blue edges (long dashed lines) from the models of Quirion et al. (2007) are shown, the latter corresponding to metallicities of $z$ = 0 (upper) and 0.007 (lower), respectively. Both figures from Adamczak et al. (2012), reproduced with permission © ESO.

of a small handful of hot white dwarfs and attempt to unravel the nature of their short-wavelength photospheric spectra.

The well-known DA white dwarfs Sirius B and HZ 43 are observed to have pure H atmospheres in the LETG range, devoid of metal lines. Their smooth continua have been used quite extensively as calibration targets for the *Chandra* LETGS, acting as standard candles providing sources of absolute flux as a function of wavelength through normalization in the UV and application of accurate model atmosphere spectral energy distribution predictions (e.g., Pease et al. 2003).

*Chandra* LETGS observations of the DA white dwarf GD 246, illustrated in Figure 4.35, enabled the first unambiguous identification of lines from highly ionized iron, revealing a slew of features from Fe VI, Fe VII, and Fe VIII in the 100–170 Å range (Vennes & Dupuis 2002; Adamczak et al. 2012). It remains the only white dwarf observed that shows identifiable individual iron lines in the soft X-ray range. A detailed model atmosphere analysis by Adamczak et al. (2012) failed to provide a good match to the observed spectrum with either homogeneous chemical composition or chemically stratified models. The latter models compute the element abundance pattern at each depth point in the atmosphere assuming equilibrium between gravitational and radiative forces. As a result, the atmospheres are no longer chemically homogeneous but vertically stratified in composition. Adamczak et al. (2012) found that the only way to match the GD 246 spectrum with a stratified model was to artificially remove Ni from the atmosphere.

Similar problems were encountered with the spectrum of the DA star LB 1919. This object has a significantly lower metallicity than other DA white dwarfs with otherwise similar atmospheric parameters, and indeed, no metal lines were detected in the LETG spectra. The problem here is that chemically stratified models including radiative levitation predict that significant amounts of light and heavy metals should be accumulated in the atmosphere and readily detectable metal lines should be present. HZ 43 illustrates similar problems, although the situation of LB 1919 is arguably more severe owing to its hotter effective temperature (56,000 K versus 51,000 K). Like HZ 43, Adamczak et al. (2012) found the LETG spectrum of LB 1919 to be well matched with a pure H atmospheric model.

The failure of chemically stratified models to match the soft X-ray spectra of stars like GD 246 and LB 1919 points to missing physics somewhere (see also Barstow et al. 2003). Mass loss, or even accretion of circumstellar material, are not included, while model atoms employed in the models are likely incomplete to some significant extent, though as Adamczak et al. (2012) note, it is difficult to understand how any of these can explain the model's overprediction of metals.

Instead, Adamczak et al. (2012) were able to obtain a good match between observed and model spectra for the pre-white dwarf PG 1159-type star, PG 1520 +525. The PG 1159 stars are hot, post-AGB stars thought to be H deficient because of having their H envelope burned during a late He-shell flash (e.g., Werner & Herwig 2006). Their surface abundances are not expected to be affected by gravitational settling or radiative leviation because a weak radiatively driven wind drives them to a homogeneous equilibrium.

By measuring the effective temperature of PG 1520+525, Adamczak et al. (2012) were able to constrain the blue edge of the GW Vir instability region in the Hertzsprung–Russell diagram illustrated in Figure 4.35. The GW Vir instability results in pulsations with periods of 300–5000 s and is thought to be caused by the $\kappa$ mechanism, similar to the Cepheids discussed in Section 4.4.5 although driven by C and O instead of He (Quirion et al. 2007). PG 1520+525 is not pulsating, whereas the very similar but slightly cooler PG 1159−035 is. The position of the edge constrained in this way is consistent with predictions of the pulsation models of Quirion et al. (2007), providing strong verification for the predictive power of these models and their use in asteroseismological analyses.

*Chandra* LETGS spectroscopy of another PG 1159 star, H1504+65, revealed potential evidence for a rare ONeMg core. H1504+65 is extremely hot, with an effective temperature of approximately 200,000 K. The LETG spectra analyzed by Werner et al. (2004) confirmed that it is not only H deficient, but also He deficient, being primarily composed of carbon and oxygen but also exhibiting lines from highly ionized Ne and Mg. Werner et al. (2004) note that the spectroscopic evidence supports H1504+65 being a naked CO stellar core, but could also be produced by a CO envelope on top of an ONeMg white dwarf.

### 4.6.3 Cataclysmic Variables and Nova Explosions

With of the order of a stellar mass in an object of similar size to Earth, the gravitational fields of white dwarfs lead to a considerable zoo of systems and behavior when combined with other stars in binary systems. Close binaries involving white dwarfs are typically formed due to frictional drag on the secondary companion when the white dwarf progenitor AGB star expands to engulf it.

In the most common case of a late-type dwarf star companion, the same angular momentum loss mechanism noted in Section 4.2.2 that leads to gradual stellar spin-down operates. When the stars are close enough that tidal forces provide spin–orbit coupling, the net loss of angular momentum leads to an inward spiral of the two stars, and eventually to Roche lobe overflow of the companion (see, e.g., Knigge et al. 2011 for a detailed review, and Garraffo et al. 2018 for recent theoretical refinements to CV angular momentum evolution). This typically occurs at separations of a few secondary stellar radii and orbital periods of a few hours. Strong surface gravity then leads to the liberation of considerable amounts of energy from accreting material falling onto the white dwarf. The term CV refers to the rapid and strong variability exhibited by the class as a result of this accretion activity.

A confusing array of subclasses of CVs that divides them based on their different types of behavior has arisen . Novae undergo major outbursts of many magnitudes, appearing as a "new star" in the case of classical novae, and being observed to repeat such outbursts in the case of recurrent novae. Dwarf novae, of which there are several subtypes, undergo frequent outbursts of much smaller amplitude than novae and with typical cycle times of tens of days. Nova-likes have spectra resembling novae spectroscopically but have not been observed to erupt as classical novae. The remaining important subdivision is the magnetic CVs making up 10%–20% of the population with strong-enough fields to influence accretion. They include two subtypes based on the strength of the magnetic field of the white dwarf: the polars, in which the field prevents the formation of an accretion disk, dominating the accretion process within the Roche lobe such that accreting matter follows along magnetic field lines near the inner Lagrangian point and down the magnetic poles of the white dwarf; and intermediate polars in which an accretion disk forms but is truncated at some distance from the white dwarf surface by the magnetic field that again channels accretion onto the magnetic poles.

SS Cygni, a type of dwarf nova, was the first CV to be detected in X-rays, albeit tentatively, based on a rocket flight (Rappaport et al. 1974). The potential for X-ray detection of CVs as a class was subsequently explored by Warner (1974). We shall barely scratch the surface of this now very active subfield of X-ray astronomy and only touch upon novae, dwarf novae, and magnetic CVs here, in addition to the symbiotic binaries that we introduce below. The interested reader is referred to the book by Warner (1995) for a thorough treatise on these fascinating binary systems, and to the review by Mukai (2017) for a recent summary of X-ray properties of CVs.

*Symbiotic Binaries in X-Rays*

Symbiotic binaries are a particular type of CVs, sometimes referred to as Z Andromedae type, comprising a white dwarf star orbiting within the extended envelope or wind of a companion red giant. The white dwarf accretes stellar or stellar wind material via Roche lobe overflow, usually through an accretion disk. Some symbiotics have been observed to display collimated, bipolar outflows, or jets, extending from the white dwarf. The accretion of matter onto the white dwarf is expected to result in X-ray emission, and indeed, this has been observed in several systems.

*Chandra* made the first observation of a two-sided X-ray jet in a symbiotic system in an ACIS-S observation of R Aquarii (R Aqr) in 2000 September (Kellogg et al. 2001; Figure 4.36). R Aqr is a well-studied symbiotic system comprising a mass-losing Mira-like long-period variable with a 387 day period and a $\sim 1\ M_{\odot}$ compact star thought to be a white dwarf. The object is thought to have undergone a nova explosion in the past (see Section 4.6.3 below). The X-ray jets found by *Chandra* were soft, with emission observed only in the energy range <1 keV and are likely shock-heated by interaction with the ambient wind from the red giant star. In addition to the jets, the central, likely white dwarf, source was also detected and found to be consistent with a blackbody spectrum with a temperature of $2 \times 10^6$ K. However, evidence was also present for cold Fe K$\alpha$ and K$\beta$ fluorescence lines at 6.4 and 7.0 keV, together with other weak K lines corresponding to Cr, V, Ca, Ar, and S. Because the production of Fe fluorescence requires a significant ionizing flux at energies of ~7 keV and higher, Kellogg et al. (2001) suggested that a hidden hard source might be associated with an accretion disk.

Observations taken 3.3 years later showed significant changes in X-ray spectral and morphological characteristics. One jet was observed to have moved outward with an apparent projected velocity of $\sim 580$ km s$^{-1}$, while the other faded, likely due to adiabatic expansion and cooling (Kellogg et al. 2007). More centrally, evidence for the formation of a new jet within the inner 500 au of the system was uncovered, and the central object itself hardened considerably, revealing a component consistent with a $T \sim 10^8$ K thermal plasma (Nichols et al. 2007). Similar X-ray emission associated with both a central source and jet activity was seen in the

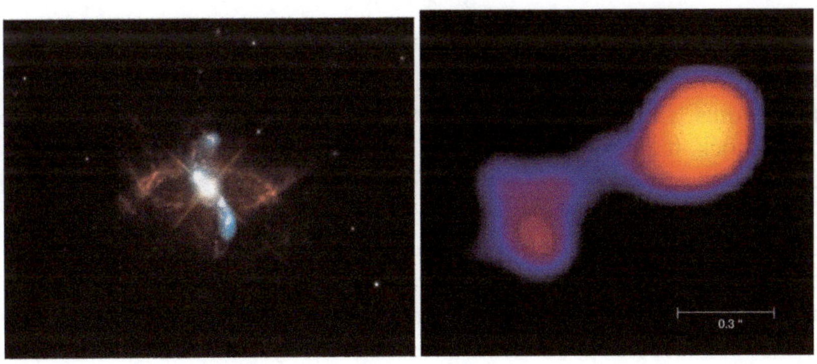


**Figure 4.36.** Left: a composite *Chandra* ACIS (blue) and optical (red) image of the symbiotic system R Aqr. *Chandra* data were obtained in three separate pointings between 2001 September and 2005 October. The ring of optical emission shown in red is thought to have arisen from an earlier nova explosion resulting from thermonuclear runaway of material accreted onto the white dwarf (see Section 4.6.3). The interactive figure shows separately the X-ray emission (blue), the optical emission (red), and the composite image. Image credit: http://chandra.harvard.edu/photo/2017/raqr/. X-ray courtesy of NASA/CXC/SAO/R.Montez et al.; Optical courtesy of Adam Block/Mt. Lemmon SkyCenter/U. Arizona. Right: the Mira system observed and resolved using the *Chandra* ACIS-S detector. The wind-accreting white dwarf is to the left, while the cool AGB star Mira A is to the right. Karovska et al. (2005) speculated that the X-rays from Mira A result from a magnetic reconnection flare. Image from http://chandra.harvard.edu/photo/2005/mira/. Courtesy of NASA/CXC/SAO/M.Karovska et al. Additional materials available online at http://iopscience.iop.org/book/978-0-7503-2163-1.

symbiotic binary CH Cyg by Karovska et al. (2007), with later observations showing the jets to have precessed (Karovska et al. 2010).

Perhaps one of the most surprising discoveries in a symbiotic system came from a *Chandra* observation of Mira—the eponymous star of the pulsating AGB class of luminous variables. Mira is losing mass to a slow wind at a rate of about $10^7\ M_\odot$ $yr^{-1}$, while a companion white dwarf (Mira B) approximately 70 au away (corresponding to 0.6″) accretes from this wind. An ACIS-S observation in 2003 December found not only X-rays from Mira A, but also from B (Figure 4.36). X-rays had never before been detected from an AGB star, and Karovska et al. (2005) speculated that the emission arose from a magnetic reconnection flare similar to those discussed in Section 4.2.4.

*Two Types of Cataclysmic Variable X-Ray Spectra*

It might be expected that the X-ray spectra of different types of CVs can be different based on the characteristics of the accretion flow. In the case of magnetic systems, accretion streams are expected to impact the white dwarf surface vertically, whereas in nonmagnetic systems, gas accretes unimpeded via a disk, forming a boundary layer as it connects to the stellar surface. Mukai et al. (2003) noted that in both cases, the emergent X-ray spectrum is expected to be from plasmas with a continuous range of temperatures, from the temperature of the shock that forms at the white dwarf surface down to its photospheric temperature. For strong shock conditions, the shock temperature, $T_s$, is related to the gravitational potential as follows:

$$T_s = \frac{3}{8}\frac{\mu m_{\rm H}}{k}\frac{GM}{R}, \tag{4.16}$$

where $\mu$ is the mean molecular weight, $m_{\rm H}$ the mass of the hydrogren atom, $k$ the Boltzmann constant, $G$ the gravitational constant, and $M$ and $R$ the white dwarf mass and radius, respectively. Because white dwarf radii decrease for increasing mass, the shock temperature increases significantly with increasing mass, from ~10 keV for a $M = 0.5\ M_\odot$ to ~200 keV for $M = 1.4\ M_\odot$ (e.g., Mukai 2017). Indeed, X-ray shock temperatures have been used quite extensively to infer CV white dwarf masses (e.g., Yu et al. 2018 and references therein).

One of the first breakthroughs enabled by *Chandra* observations of CVs was spearheaded by a study of high-resolution HETG spectra of seven different systems by Mukai et al. (2003), four of which were magnetic systems and three nonmagnetic. They found that the spectra divide into two distinct types (Figure 4.37). The nonmagnetic systems are remarkably well fitted by a simple "cooling flow" model that assumes steady-state isobaric radiative cooling of an optically thin plasma, such as those forming stellar coronae (Section 4.2), between maximum and minimum temperatures. Here, the maximum corresponds to the shock temperature of the impacting flow and the minimum to the white dwarf photospheric temperature. Cooling flow models were originally developed to describe the X-ray-emitting gas in the centers of clusters of galaxies, although the similarities to flows settling onto the surfaces of accreting white dwarfs was noted by Fabian & Nulsen (1977). Mukai

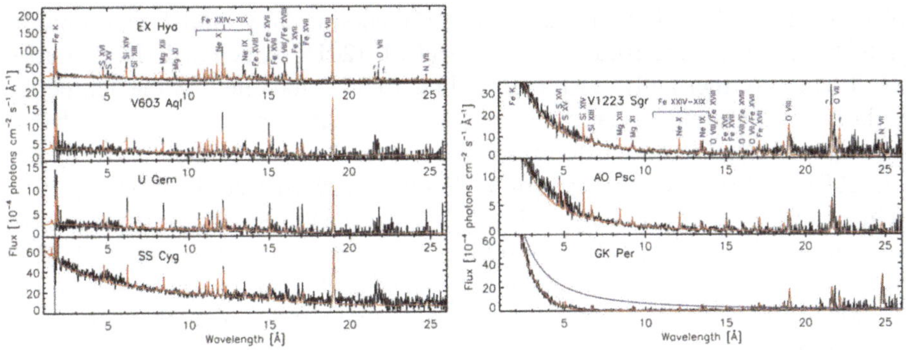


**Figure 4.37.** *Chandra* combined MEG ± first-order data of four CVs that exhibit cooling flow spectra (left) and three showing spectra more characteristic of photoionized plasma (right). Data are shown in black with the model spectra in red. Reproduced from Mukai et al. (2003) © 2003. The American Astronomical Society. All rights reserved.

et al. (2003) note with irony that these models can be used to fit CV spectra, but are now known to be a poor description of galaxy cluster emission!

The spectra of the magnetic systems, all of which are of intermediate polar (IP) type, instead were found to be grossly inconsistent with pure cooling flow models, and exhibited hard continua superimposed by strong H-like and He-like ion emission of abundant elements but little Fe "L-shell" (ions with $n = 2$ ground states) emission. Mukai et al. (2003) found these spectra could be matched with a photoionized plasma model and in some ways were similar to spectra of black hole systems and active galactic nuclei.

The fly in the ointment is that one of the cooling flow-like spectra, that of EX Hya, is also an IP system. What then is the common physics between the different types of spectra? Mukai et al. (2003) suggested that it is the specific accretion rate—the accretion rate per unit area. Accretion on nonmagnetic CVs occurs over a much larger area than in magnetic systems, and so the specific accretion rate is generally expected to be lower than for magnetic systems where accretion is restricted to a small fraction of the stellar surface. The compact "pillbox" geometry of a high specific accretion rate shock renders photon escape from the sides of the postshock region more difficult. EX Hya lies below the CV "period gap" of 2–3 hr where few CVs are found. Below the gap, gravitational radiation tends to dominate the angular momentum loss, which proceeds at quite a low rate, as accretion rates are commensurately lower. EX Hya therefore has a lower accretion rate than most IP systems above the gap, and thus it has a taller shock geometry more similar to nonmagnetic systems and is more amenable to ready escape of X-radiation.

The cooling flow model for EX Hya was qualitatively verified by Luna et al. (2015) based on a much longer 496 ks HETG+ACIS-S observation obtained in 2007 May, but they noted that in detail, the model failed to reproduce certain spectral features such as the ratios of resonance lines of He-like and H-like ions, the former of which were generally stronger than the model prediction. The spectrum suggests that extra heating must be deposited at the base of the accretion column, where cooler ions form. Simple modifications to the cooling flow model were unable to match the observed spectrum, and Luna et al. (2015) concluded that additional physics, such as thermally unstable cooling that would modify the temperature distribution of the emitting gas might be required. A similar problem in which cooler lines were underpredicted was found earlier for the dwarf nova WX Hydri by Perna et al. (2003) and the likely IP system V426 Oph by Homer et al. (2004).

The exquisite quality of the EX Hya *Chandra* spectra has enabled several other important measurements. Based on earlier data obtained in Mauche et al. (2001), Mauche et al. (2003) were able to infer a plasma density of $\sim 10^{14}$ cm$^{-3}$ based on lines of Fe XVII and Fe XXII, respectively—orders of magnitude larger than observed in stellar coronae. By coadding spectral lines, Hoogerwerf et al. (2004) were able to achieve a velocity precision of 15 km s$^{-1}$ and made the first detection of orbital motion using X-ray spectroscopy. The white dwarf mass inferred, $0.49 \pm 0.13\ M_{\odot}$, is consistent with optical and UV measurements. Luna et al. (2010) identified narrow ($\sim$150 km s$^{-1}$) and broad ($\sim$1600 km s$^{-1}$) components in several emission lines and interpreted the broad component as due to photoionization of the preshock flow by radiation from the postshock flow. Because the photoionized region has to be close to the radiation source in order to produce strong photoionized emission lines, the model was able to constrain the height of the standing shock above the white dwarf surface to within the approximate range of 0.2–0.6 white dwarf radii. Thus, EX Hya, while dominated by a cooling flow-type spectrum, shows the link between this thermal emission and the photoionization-dominated sources.

*Nova Explosions*

The term "nova" originates from its use by Tycho Brahe in his book *De Nova Stella* —or *On the New Star*, as we would write today—which featured a description of the supernova SN 1572 now commonly referred to as Tycho's supernova. Unable to see the progenitor of the explosion, from his perspective it appeared like a bright new star where before there was none. Following the accumulation of observational evidence that suggested novae and supernovae are different phenomena, novae were subsequently classified as "classical novae" (CNe). Novae have since been further divided, with the term "recurrent novae" (RNe) being used for objects that have been observed to have undergone more than one outburst. Of course, as we now understand novae to originate in white dwarf binaries, it is interesting to note that "new stars" are actually a manifestation of processes in very old stellar systems!

Novae originate from the explosive detonation of H-rich material that accumulates on the surface of a white dwarf by mass transfer from the stellar companion in CV systems. The bottom layer of the accreted H-rich gas is compressed under the strong white dwarf gravity and becomes hot and degenerate. Once the temperature at the bottom of the accreted layer has reached several million degrees, H fusion can occur through the *p–p* chain, heating the accreted degenerate material even further. At a temperature of about $7 \times 10^7$ K, thermal energy is comparable to the Fermi energy of degeneracy and the gas can begin to expand. At the same time, the energy liberated by nuclear reactions, including the CNO cycle at later times, is increasing rapidly: a thermonuclear runaway occurs and explosively ejects the accreted envelope.

Novae are of special interest not just because of the broad array of astrophysics they encompass, but because they, along with CVs in general, are also the likely progenitors for Type 1a supernovae. For a detailed description of the nova phenomenon, the reader is referred to the reviews of Starrfield et al. (2016) and Williams (1992).

The amount of accreted material required to initiate thermonuclear runaway depends on the temperature and mass of the underlying white dwarf. Heat flow from the white dwarf interior, or from prior outbursts, can be important for heating the accreted layers, while the mass controls the compressional heating and pressure within the layer. A first-order estimate for the ignition mass, $M_{\rm ig}$, is given by the critical pressure for ignition, $P_{\rm crit}$, and white dwarf mass and radius $M_{\rm WD}$ and $R_{\rm WD}$, respectively, by

$$M_{\rm ig} = \frac{4\pi R_{\rm WD}^{4} P_{\rm crit}}{G M_{\rm WD}}, \tag{4.17}$$

where $P_{\rm crit} \sim 10^{20}$ dyne cm$^{-2}$ (see, e.g., Starrfield et al. 2016). Values of $M_{\rm ig}$ based on Equation (4.17) range from $10^{-3}$–$10^{-6} M_{\odot}$ (e.g., Drake et al. 2016). More massive white dwarfs, requiring less accreted material to initiate thermonuclear runaway, tend to be "fast" novae—evolving and fading rapidly.

X-ray emission is important for understanding many of the different aspects of nova explosions, and *Chandra* has made key contributions in all of these areas. Relatively hard X-rays, and even $\gamma$-rays, can be produced in the early stages of the explosion itself when the ejected envelope material interacts violently with circumbinary material inducing particle accleration in the resulting shocks (Cheung et al. 2015). Following the initial blast, nuclear burning continues on the surface of the white dwarf, often leading to a super-Eddington luminosity in which radiation pressure drives an outflow. As the ejecta and the radiatively driven outflow have thinned, the photosphere gradually shrinks, so that the observed effective temperature increases until it reaches several hundred thousand degrees. At this point, analogous to the hot white dwarfs discussed in Section 4.6.2, the Wien tail of the photospheric emission is well into the soft X-ray regime and, provided the intervening interstellar column density is not too high, the nova becomes a "supersoft source" (SSS). At certain accretion rates thought to be close to a few $10^{-7}$ $M_{\odot}$ yr$^{-1}$, steady nuclear burning can also result without thermonuclear runaway, producing a persistent supersoft source, such as CAL 83 and CAL 87 in the Large Magellanic Cloud first seen by *Einstein* (Long et al. 1981; van den Heuvel et al. 1992).

Novae can remain X-ray bright for timescales of days to years. The most fruitful observation strategy has been to employ the *Swift* X-ray satellite (Gehrels et al. 2004) for making short observations to follow X-ray evolution over these long timescales, and *Chandra* and *XMM-Newton* to target key episodes and provide detailed high-resolution X-ray spectra.

**Blast Waves.** Arguably the most spectacular novae in X-rays are the rare events that occur in the types of symbiotic binaries described in Section 4.6.3. In these systems, the explosion occurs inside the wind of the red giant companion and generates an outwardly propagating blast wave that is similar to that of a Type II supernova in which a massive star explodes within the medium of its massive wind. The difference is in the amount of energy involved. Symbiotic nova explosions typically involve energies of the order of $10^{43}$–$10^{44}$ erg, compared with $\sim 10^{51}$ erg liberated in a supernova. As a consequence, symbiotic novae evolve on timescales of weeks instead of millennia, allowing detailed observation of the whole event.

The first symbiotic nova to be observed in detail in the *Chandra* and *XMM-Newton* era occurred on RS Ophiuchi on 2006 February 12, triggering an extensive international multiwavelength campaign. RS Oph is a recurrent nova at a distance of about 1.6 kpc and had been seen in outburst on several previous occasions, with an outburst frequency of approximately once every 20 years. The *Swift* XRT light curve of the 2006 outburst analyzed by Osborne et al. (2011) is illustrated in Figure 4.38 and indicates the times of *Chandra* and *XMM-Newton* observations. The first 30 days are dominated by the blast wave that peaked at a temperature of 10 keV before it gradually faded and cooled as energy was dissipated by the expanding shock in the surrounding medium. The SSS then began to appear during a period of chaotic variability that turned out to be quite common at the beginning of the SSS phase (see Section 4.6.3 below), peaking at an amazing 300 counts/s.

*Chandra* HETG spectra taken on day 13.9 of the outburst (Figure 4.38) revealed a rich thermal plasma spectrum exhibiting Doppler-broadened emission lines with a half-width at zero intensity of $\sim 2400$ km $s^{-1}$ and formed over a wide temperature range from 3–60 million K (Nelson et al. 2008; Drake et al. 2011). Lines were shaped by differential absorption by the remnant itself and indicated a nonspherical, collimated blast expanding largely in the plane of the sky, consistent with the picture provided by radio, and later, *Hubble* observations (O'Brien et al. 2006; Bode et al. 2007). Extended X-ray emission from the blast was detected by *Chandra* one and a half years later, confirming the greatest expansion was close to the plane of the sky and indicating an average expansion velocity of 6000 km $s^{-1}$ (Luna et al. 2009).

Using hydrodynamic simulations, Walder et al. (2008) and Orlando et al. (2009) showed that the blast collimation was due to circumstellar material and a density enhancement in the equatorial plane, a situation that is probably common to all novae (see also the hydrodynamic models of the symbiotic novae V407 Cyg and V745 Sco, Orlando & Drake 2012; Drake et al. 2016; Orlando et al. 2017; and of the recurrent CNe system with an accretion disk, U Sco, Drake & Orlando 2010). The *Chandra* gratings also caught the symbiotic nova V745 Sco that exploded on 2014 February 6. Spectra 17 days after the outburst revealed a similar blast wave spectrum to RS Oph, with line profiles indicating a collimated asymmetric outflow. Hydrodynamic models tuned to match the X-ray observations of symbiotic novae enable estimation of the explosion energy and ejecta mass. Orlando et al. (2009) found that the explosion energy and ejected mass for RS Oph were $E_{ex} \approx 10^{44}$ erg and $M_{ej} \approx 10^{-6}\ M_{\odot}$.

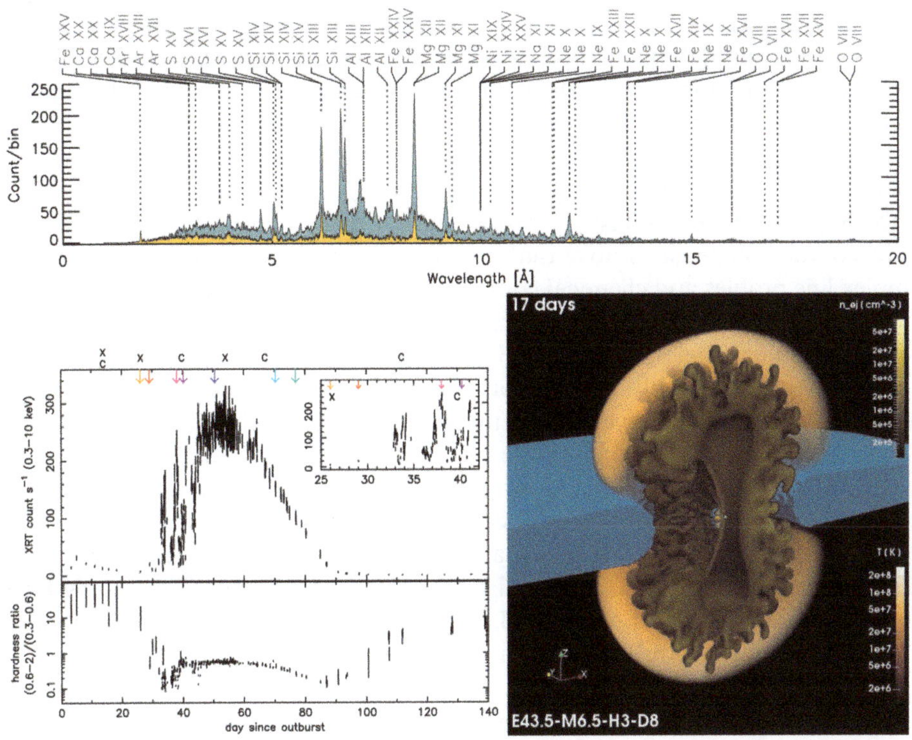


**Figure 4.38.** Top: *Chandra* MEG (background) and HEG (foreground) spectra of RS Oph obtained on day 13.9 of the 2006 outburst showing a hot thermal spectrum from the blast wave. Prominent lines are identified. Reproduced from Drake et al. (2009) © 2009. The American Astronomical Society. All rights reserved. Bottom left: *Swift* XRT light curve and the (0.6–2 keV)/(0.3–0.6 keV) hardness ratio for RS Oph from day 3 to day 135 after outburst discovery. The blast wave emission dominates before day 30, after which violent variability is seen at the onset of SSS emergence. Observations by *Chandra* (C) and *XMM-Newton* (X) are marked. Reproduced from Osborne et al. (2011) © 2011. The American Astronomical Society. All rights reserved. Bottom right: a 3D rendering of the hydrodynamic model of the 2014 February 6 explosion of the symbiotic nova V745 Sco showing the density distribution of the ejecta and the plasma temperature 17 days after the outburst. The blue region is the unshocked equatorial density enhancement isosurface for ejecta densities larger than $10^7$ $cm^{-3}$. The small orange and white spheres at the center represent the red giant and white dwarf, separated by 1.7 au. The explosion is strongly collimated and shaped by the circumbinary material. Reproduced from Orlando et al. (2017), by permission of Oxford University Press on behalf of the Royal Astronomical Society. The video shows a simulation of the V745Sco explosion and subsequent expansion. The status at 17 days is illustrated in the bottom right panel. The 3Dmodel_blast video shows an interactive 3D model of the blast wave from V745Sco 17 days after the outburst on 2014 February 6, as illustrated in the bottom right panel (orange). The 3Dmodel_eject video shows an interactive 3D model of the mass ejected from the explosion of V745Sco 17 days after the outburst on 2014 February 6, as illustrated in the bottom right panel (brown). Additional materials available online at http://iopscience.iop.org/book/978-0-7503-2163-1.

Understanding the ejecta mass in comparison with the inter-outburst accretion rate is important for determining whether the white dwarf is gradually increasing in mass, or whether each outburst reduces its mass. V745 Sco was a “faster” nova than RS Oph, declining by three magnitudes in only nine days, and expectations of a lower outburst energy and ejected mass were borne out by both 1D and 3D hydrodynamic simulations that found $E_{ex} \approx 3 \times 10^{43}$ erg and $M_{ej} \approx 1\text{–}3 \times 10^{-7}\ M_{\odot}$ (Drake et al. 2016; Orlando et al. 2017). The ejected mass is an order of magnitude smaller than required for thermonuclear runaway for a massive white dwarf based on Equation (4.17), indicating that the white dwarf is a mass gainer and likely supernova Ia progenitor.

**The Supersoft Source Phase.** High-resolution *Chandra* and *XMM-Newton* observations of the SSS phase of nova outbursts have been revolutionary in revealing the complex line profiles and chemical compositions of what are often apparently super-Eddington sources. The steady nuclear-burning phase of nova outbursts that produces the characteristic X-ray SSS is thought to begin at the time of outburst, and its appearance simply depends on when the effective photosphere shrinks to yield X-ray-emitting effective temperatures of a few hundred thousand kelvins. The SSS is thought to last as long as the H-rich material, such that its duration is a function of white dwarf mass that controls both the surface pressure and the accreted mass. This picture has been largely confirmed by a mostly *Swift*-based study by Schwarz et al. (2011) of 52 Galactic and Magellanic Cloud novae, for which SSS emission was detected in 26, and by the *Chandra* study of novae in M31 by Henze et al. (2014). Schwarz et al. (2011) did, however, conclude that white dwarf mass is likely not the only controlling factor in SSS turn-on and turn-off times.

The most startling aspect of SSS emission has turned out to be the extreme variability at onset, in which the X-ray count rates are seen to change by factors of 100 or more (e.g., RS Oph in Figure 4.38). This remains largely unexplained, although possible explanations include temporary obscuration by clumpy material, or photospheric radius changes, although neither of these can explain both the X-ray hardness changes and the variability observed. Other dramatic variations include oscillations with a period of the order of 20 minutes thought to be nonradial $g+$-modes on V1494 Aql and V4743 Sgr, the former of which also exhibited a sudden burst of X-rays in which the count rate increased by a factor of 10 for about 1000 s (Drake et al. 2003), and the latter of which showed a very sudden drop in X-rays (Ness et al. 2003). These sudden variations remain unexplained.

SSS X-ray spectra sometimes exhibit P Cygni-like line profiles, with emission features on the redward side of an absorption line, and absorption features redshifted by up to 2000 km $s^{-1}$ or so (e.g., V4743 Sgr, Rauch et al. 2010; nova SMC 2016, Orio et al. 2018), indicating a photosphere formed in rapidly outflowing gas. Ness et al. (2013) found that SSS with emission lines ("SSe" in their notation) were less luminous than those with just absorption features ("SSa") and appear to be in mostly high-inclination systems (Figure 4.39). This suggests the influence of a disk on the spectrum, perhaps as a source of obscuration and the emission-line-producing gas.

The problem facing research into the SSS of nova outbursts is that the theorical modeling side is now lagging behind observational developments. While atmospheric models in which the assumption of local thermodynamic equilibrium is relaxed appear to work well for white dwarfs, and even for SSS spectra to some

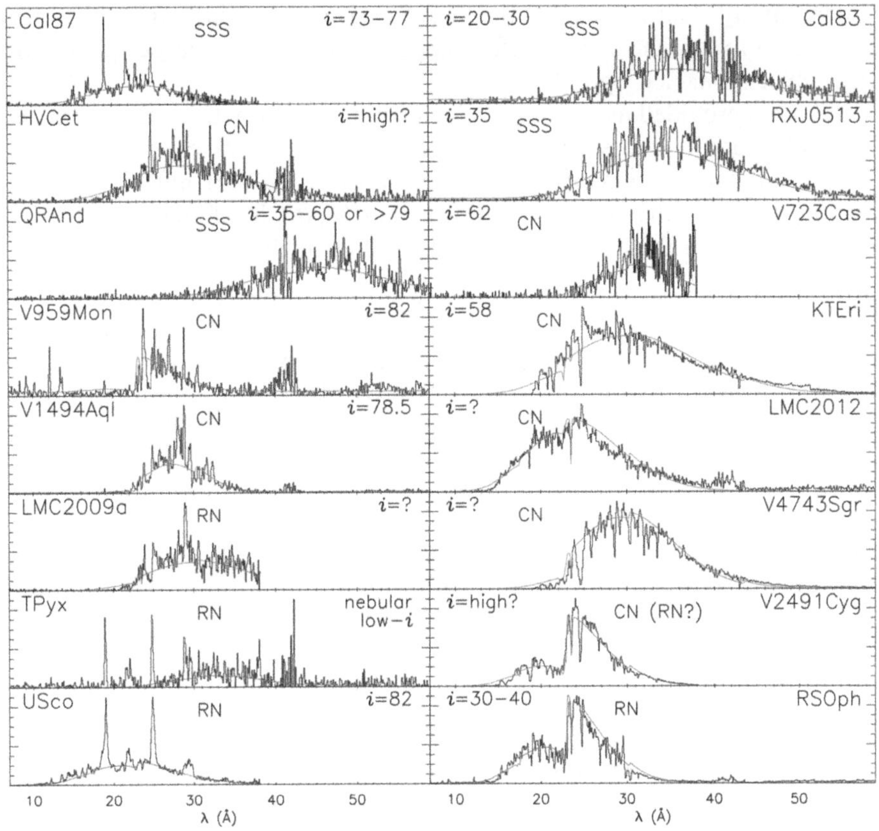


**Figure 4.39.** High-resolution *Chandra* and *XMM-Newton* spectra in arbitrary flux units of SSS exhibiting emission lines without absorption features (SSe, left column) and those that show just continua and absorption lines (SSa, right column), both labeled with inclination angles where known. Thin blue lines represent absorbed blackbody curves. Labels SSS, CN, and RN denote persistent SSS, CNe, and RNe. Note that features near 40 Å in some *Chandra* spectra are due to division by low effective areas near the C K edge. From Ness et al. (2013), reproduced with permission © ESO.

extent (e.g., Rauch et al. 2010), readily available models do not yet incorporate the outflow and associated radiative transfer within the dynamic atmosphere.

## 4.7 Epilogue

Part of the aim of this chapter is to give the reader a sense of the remarkably broad scope of the field of X-rays from stars and planetary systems. Fascinating in its own right, this field encompasses almost all of the physical processes at work in the more distant universe and in systems much less well understood. In this sense, our solar system, the Sun, and stars can be considered true laboratories for X-ray astronomy.

I have given only a brief taste of the astrophysics to be learned from *Chandra* observations of stellar and planetary systems here. By nature, what has been presented is *Chandra*-centric and incomplete, glaringly so in some topics that have been unnaturally compressed by the limited space available, and the reader is encouraged to follow the articles referenced in order to get a deeper picture of the various subjects touched upon here.

*Chandra* has provided us with exquisite mirrors and diffraction gratings to resolve objects both spatially and spectrally for the first time. However, many compelling and important targets remain beyond *Chandra*'s reach and must await a next-generation mission with a larger collecting area. *Chandra* has lit the path that should be followed to make these next-generation breakthroughs.

## References

Adamczak, J., Werner, K., Rauch, T., et al. 2012, A&A, 546, A1
Adler, I., Trombka, J. I., Yin, L. I., et al. 1973, NW, 60, 231
Albacete Colombo, J. F., Drake, J. J., Flaccomio, E., et al. 2018, arXiv:1806.01231
Anderson, K. A. 1958, PhRv, 111, 1397
Argiroffi, C., Caramazza, M., Micela, G., et al. 2016, A&A, 589, A113
Argiroffi, C., Drake, J. J., Bonito, R., et al. 2017, A&A, 607, A14
Argiroffi, C., Drake, J. J., Maggio, A., et al. 2004, ApJ, 609, 925
Argiroffi, C., Maggio, A., & Peres, G. 2007, A&A, 465, L5
Argiroffi, C., Reale, F., Drake, J. J., et al. 2019, NatAs, 3, 742
Armitage, P. J. 2011, ARA&A, 49, 195
Arnaud, M., & Rothenflug, R. 1985, A&AS, 60, 425
Ayres, T. R. 2008, ApJ, 686, 731
Ayres, T. R. 2014, AJ, 147, 59
Babcock, H. W. 1961, ApJ, 133, 572
Babel, J., & Montmerle, T. 1997, A&A, 323, 121
Bai, X.-N. 2016, ApJ, 821, 80
Balbus, S. A. 2003, ARA&A, 41, 555
Balog, Z., Muzerolle, J., Rieke, G. H., et al. 2007, ApJ, 660, 1532
Barstow, M. A., Dobbie, P. D., Holberg, J. B., Hubeny, I., & Lanz, T. 1997, MNRAS, 286, 58
Barstow, M. A., Good, S. A., Burleigh, M. R., et al. 2003, MNRAS, 344, 562
Benz, A. O. 2008, LRSP, 5, 1
Benz, A. O., & Guedel, M. 1994, A&A, 285, 621
Berger, E., Basri, G., Fleming, T. A., et al. 2010, ApJ, 709, 332
Berger, M. J., & Seltzer, S. M. 1972, JATP, 34, 85
Berghoefer, T. W., Schmitt, J. H. M. M., Danner, R., & Cassinelli, J. P. 1997, A&A, 322, 167
Bhardwaj, A., Elsner, R. F., Gladstone, G. R., et al. 2006, JGRA, 111, A11225
Bhardwaj, A., Elsner, R. F., Randall Gladstone, G., et al. 2007, P&SS, 55, 1135
Bhardwaj, A., Elsner, R. F., Waite, J., Hunter, J., et al. 2005a, ApJ, 624, L121
Bhardwaj, A., Elsner, R. F., Waite, J., Hunter, J., et al. 2005b, ApJ, 627, L73
Biermann, L. 1946, NW, 33, 118
Blackman, E. G., & Thomas, J. H. 2015, MNRAS, 446, L51
Bode, M. F., Harman, D. J., O'Brien, T. J., et al. 2007, ApJ, 665, L63
Bonito, R., Orlando, S., Argiroffi, C., et al. 2014, ApJL, 795, L34
Bowyer, S., Drake, J. J., & Vennes, S. 2000, ARA&A, 38, 231
Boyle, J. J., & Pindzola, M. S. 2005, Many-Body Atomic Physics (Cambridge: Cambridge Univ. Press)

Braithwaite, J. 2006, A&A, 449, 451
Branduardi-Raymont, G. 2017, AN, 338, 188
Branduardi-Raymont, G., Bhardwaj, A., Elsner, R. F., & Rodriguez, P. 2010, A&A, 510, A73
Branduardi-Raymont, G., Bhardwaj, A., Elsner, R. F., et al. 2007, P&SS, 55, 1126
Branduardi-Raymont, G., Elsner, R. F., Galand, M., et al. 2008, JGRA, 113, A02202
Branduardi-Raymont, G., Ford, P. G., Hansen, K. C., et al. 2013, JGRA, 118, 2145
Brickhouse, N. S., Cranmer, S. R., Dupree, A. K., Luna, G. J. M., & Wolk, S. 2010, ApJ, 710, 1835
Brinkman, A. C., Behar, E., Güdel, M., et al. 2001, A&A, 365, L324
Brown, C. M., Feldman, U., Doschek, G. A., Seely, J. F., & Lavilla, R. E. 1989, PhRvA, 40, 4089
Bruner, M. E., & McWhirter, R. W. P. 1988, ApJ, 326, 1002
Bryans, P., Landi, E., & Savin, D. W. 2009, ApJ, 691, 1540
Bunce, E. J., Cowley, S. W. H., & Yeoman, T. K. 2004, JGRA, 109, A09S13
Burnight, T. R. 1949, PhRv, 76, 165
Calvet, N., & Gullbring, E. 1998, ApJ, 509, 802
Caramazza, M., Flaccomio, E., Micela, G., et al. 2007, A&A, 471, 645
Cassinelli, J. P., Miller, N. A., Waldron, W. L., MacFarlane, J. J., & Cohen, D. H. 2001, ApJ, 554, L55
Charbonneau, P. 2014, ARA&A, 52, 251
Cherepashchuk, A. M. 1976, SvAL, 2, 138
Cheung, M. C. M., Boerner, P., Schrijver, C. J., et al. 2015, ApJ, 807, 143
Chlebowski, T., & Garmany, C. D. 1991, ApJ, 368, 241
Christian, D. J., Bodewits, D., Lisse, C. M., et al. 2010, ApJS, 187, 447
Cohen, D. H., Cassinelli, J. P., & Waldron, W. L. 1997, ApJ, 488, 397
Cohen, D. H., Li, Z., Gayley, K. G., et al. 2014a, MNRAS, 444, 3729
Cohen, D. H., Wollman, E. E., Leutenegger, M. A., et al. 2014b, MNRAS, 439, 908
Cohen, O., Kashyap, V. L., Drake, J. J., Sokolov, I. V., & Gombosi, T. I. 2011, ApJ, 738, 166
Cook, B. A., Williams, P. K. G., & Berger, E. 2014, ApJ, 785, 10
Corcoran, M. F., Hamaguchi, K., Gull, T., et al. 2004, ApJ, 613, 381
Corcoran, M. F., Nichols, J. S., Pablo, H., et al. 2015a, ApJ, 809, 132
Corcoran, M. F., Hamaguchi, K., Liburd, J. K., et al. 2015b, arXiv:1507.07961
Craig, I. J. D., & Brown, J. C. 1976, A&A, 49, 239
Cravens, T. E. 1997, GeoRL, 24, 105
Cravens, T. E. 2000, AdSpR, 26, 1443
Cravens, T. E., & Maurellis, A. N. 2001, GeoRL, 28, 3043
Cravens, T. E., Robertson, I. P., & Snowden, S. L. 2001, GeoR, 106, 24883
Cravens, T. E., Waite, J. H., Gombosi, T. I., et al. 2003, JGRA, 108, 1465
Damiani, F., & Micela, G. 1995, ApJ, 446, 341
Damiani, F., Micela, G., Sciortino, S., & Harnden, F. R. Jr. 1994, ApJ, 436, 807
Davidson, K., & Humphreys, R. M. 1997, ARA&A, 35, 1
Dennerl, K. 2002, A&A, 394, 1119
Dennerl, K. 2008, P&SS, 56, 1414
Dennerl, K. 2014, in Proc. of Frontier Research in Astrophysics (FRAPWS2014), ed. F. Giovannelli, & L. Sabau-Graziati
Dennerl, K., Burwitz, V., Englhauser, J., Lisse, C., & Wolk, S. 2002, A&A, 386, 319
Dennerl, K., Englhauser, J., & Trümper, J. 1997, Sci, 277, 1625

Dennerl, K., Lisse, C. M., Bhardwaj, A., et al. 2012, AN, 333, 324
Dere, K. P., Landi, E., Mason, H. E., Monsignori Fossi, B. C., & Young, P. R. 1997, A&AS, 125, 149
Dere, K. P., Landi, E., Young, P. R., et al. 2009, A&A, 498, 915
Drake, J. 2001, in ASP Conf. Proc. 234, X-ray Astronomy 2000, ed. R. Giacconi, S. Serio, & S. Luigi (San Francisco, CA: ASP), 53
Drake, J. 2002, in ASP Conf. Proc. 277, Stellar Coronae in the Chandra and XMM-NEWTON Era, ed. F. Favata, & J. J. Drake (San Francisco, CA: ASP), 75
Drake, J., Gladstone, G. R., Stern, S. A., & Wargelin, B. 2004, AAS/DPS Meeting 36, 39.09
Drake, J. J. 1996, in ASP Conf. Ser. 109, Cool Stars, Stellar Systems, and the Sun, ed. R. Pallavicini, & A. K. Dupree (San Francisco, CA: ASP), 203
Drake, J. J. 1999, ApJS, 122, 269
Drake, J. J. 2005, in ESA Special Publication, 13th Cambridge Workshop on Cool Stars, Stellar Systems and the Sun,Vol. 560, ed. F. Favata, G. A. J. Hussain, & B. Battrick, 519
Drake, J. J., Ball, B., Eldridge, J. J., Ness, J.-U., & Stancliffe, R. J. 2011, AJ, 142, 144
Drake, J. J., Brickhouse, N. S., Kashyap, V., et al. 2001, ApJ, 548, L81
Drake, J. J., Braithwaite, J., Kashyap, V., Günther, H. M., & Wright, N. J. 2014b, ApJ, 786, 136
Drake, J. J., Delgado, L., Laming, J. M., et al. 2016, ApJ, 825, 95
Drake, J. J., Laming, J. M., Ness, J.-U., et al. 2009, ApJ, 691, 418
Drake, J. J., Laming, J. M., & Widing, K. G. 1995, ApJ, 443, 393
Drake, J. J., Laming, J. M., & Widing, K. G. 1997, ApJ, 478, 403
Drake, J. J., & Orlando, S. 2010, ApJ, 720, L195
Drake, J. J., Ratzlaff, P., Kashyap, V., et al. 2014a, ApJ, 783, 2
Drake, J. J., Stern, R. A., Stringfellow, G., et al. 1996, ApJ, 469, 828
Drake, J. J., & Testa, P. 2005, Natur, 436, 525
Drake, J. J., Testa, P., & Hartmann, L. 2005, ApJ, 627, L149
Drake, J. J., Wagner, R. M., Starrfield, S., et al. 2003, ApJ, 584, 448
Drake, J. J., Wright, N. J., Guarcello, M. G., et al. 2019, ApJ, in press
Dunn, W. R., Branduardi-Raymont, G., Elsner, R. F., et al. 2016, JGRA, 121, 2274
Dunn, W. R., Branduardi-Raymont, G., Ray, L. C., et al. 2017, NatAs, 1, 758
Durney, B. 1972, Evidence for Changes in the Angular Velocity of the Surface Regions of the Sun and Stars - Comments, Vol. 308, ed. C. P. Sonett, P. J. Coleman, & J. M. Wilcox (Washington, DC: NASA), 282
Elsner, R. F., Ramsey, B. D., Waite, J. H., et al. 2005, Icar, 178, 417
Elsner, R. F., Gladstone, G. R., Waite, J. H., et al. 2002, ApJ, 572, 1077
Engle, S. G., Guinan, E. F., Harper, G. M., et al. 2017, ApJ, 838, 67
Engle, S. G., Guinan, E. F., Harper, G. M., Neilson, H. R., & Remage Evans, N. 2014, ApJ, 794, 80
Ercolano, B., Drake, J. J., Raymond, J. C., & Clarke, C. C. 2008, ApJ, 688, 398
Fabian, A. C., & Nulsen, P. E. J. 1977, MNRAS, 180, 479
Favata, F., Flaccomio, E., Reale, F., et al. 2005, ApJS, 160, 469
Favata, F., Fridlund, C. V. M., Micela, G., Sciortino, S., & Kaas, A. A. 2002, A&A, 386, 204
Feigelson, E. D., Broos, P., Gaffney, J. A. I., et al. 2002, ApJ, 574, 258
Feigelson, E. D., & Decampli, W. M. 1981, ApJ, 243, L89
Feigelson, E. D., Getman, K., Townsley, L., et al. 2005, ApJS, 160, 379
Feldmeier, A., Oskinova, L. M., & Hamann, W.-R. 2003, A&A, 403, 217

Feldmeier, A., Puls, J., & Pauldrach, A. W. A. 1997, A&A, 322, 878
Flaccomio, E., Micela, G., & Sciortino, S. 2003, A&A, 397, 611
Flaccomio, E., Micela, G., & Sciortino, S. 2012, A&A, 548, A85
Flaccomio, E., Albacete-Colombo, J. F., Drake, J. J., et al. 2019, ApJ, in press
Fleming, T. A., Snowden, S. L., Pfeffermann, E., Briel, U., & Greiner, J. 1996, A&A, 316, 147
Fontaine, G., Brassard, P., & Bergeron, P. 2001, PASP, 113, 409
Foster, A. R., Ji, L., Smith, R. K., & Brickhouse, N. S. 2012, ApJ, 756, 128
Foster, A. R., Smith, R. K., Brickhouse, N. S., Kallman, T. R., & Witthoeft, M. C. 2010, SSRv, 157, 135
Frazier, E. N. 1970, SoPh, 14, 89
Freeman, M., Montez, R. J., Kastner, J. H., et al. 2014, ApJ, 794, 99
Freyer, T., Hensler, G., & Yorke, H. W. 2006, ApJ, 638, 262
Friedman, H. 1980, in Oral History Interviews, ed. R. F. Hirsh, https://www.aip.org/history-programs/niels-bohr-library/oral-histories/4613
Friedman, H., Lichtman, S. W., & Byram, E. T. 1951, PhRv, 83, 1025
Gabriel, A. H., & Jordan, C. 1969, MNRAS, 145, 241
Garraffo, C., Drake, J. J., Dotter, A., et al. 2018, ApJ, 862, 90
Gehrels, N., Chincarini, G., Giommi, P., et al. 2004, ApJ, 611, 1005
Getman, K. V., Flaccomio, E., Broos, P. S., et al. 2005, ApJS, 160, 319
Giacconi, R., Gursky, H., Paolini, F. R., & Rossi, B. B. 1962, PhRvL, 9, 439
Giacconi, R., Murray, S., Gursky, H., et al. 1972, ApJ, 178, 281
Seibert, J. 1997, RadioGraphics, 17, 1553
Gilman, D. A., Hurley, K. C., Seward, F. D., et al. 1986, ApJ, 300, 453
Gladstone, G. R., Stern, S. A., Ennico, K., et al. 2016, Science, 351, aad8866
Glassgold, A. E., Najita, J., & Igea, J. 1997, ApJ, 480, 344
Golub, L., Harnden, F. R., Maxson, C. W., et al. 1984, ApJ, 278, 456
Golub, L., & Pasachoff, J. M. 2009, The Solar Corona (Cambridge: Cambridge Univ. Press)
Goode, P. R., Dziembowski, W. A., Korzennik, S. G., & Rhodes, E. J. J. 1991, ApJ, 367, 649
Grunhut, J. H., Wade, G. A., Hanes, D. A., & Alecian, E. 2010, MNRAS, 408, 2290
Guarcello, M. G., Drake, J. J., Wright, N. J., et al. 2013, ApJ, 773, 135
Guarcello, M. G., Drake, J. J., Wright, N. J., et al. 2016, arXiv:1605.01773
Guarcello, M. G., Micela, G., Peres, G., Prisinzano, L., & Sciortino, S. 2010, A&A, 521, A61
Güdel, M. 2004, A&ARv, 12, 71
Güdel, M., Briggs, K. R., Montmerle, T., et al. 2008, Sci, 319, 309
Güdel, M., & Nazé, Y. 2009, A&ARv, 17, 309
Güdel, M., Skinner, S. L., Mel'Nikov, S. Y., et al. 2007, A&A, 468, 529
Günther, H. M., & Schmitt, J. H. M. M. 2009, A&A, 494, 1041
Günther, H. M., Wolk, S. J., Drake, J. J., et al. 2012, ApJ, 750, 78
Hamaguchi, K., Yamauchi, S., & Koyama, K. 2005, ApJ, 618, 360
Hamaguchi, K., Corcoran, M. F., Pittard, J. M., et al. 2018, NatAs, 2, 731
Harnden, F. R. J., Branduardi, G., Elvis, M., et al. 1979, ApJ, 234, L51
Henze, M., Pietsch, W., Haberl, F., et al. 2014, A&A, 563, A2
Herbst, W., & Mundt, R. 2005, ApJ, 633, 967
Hillenbrand, L. A. 1997, AJ, 113, 1733
Homer, L., Szkody, P., Raymond, J. C., et al. 2004, ApJ, 610, 991
Hoogerwerf, R., Brickhouse, N. S., & Mauche, C. W. 2004, ApJ, 610, 411

Huenemoerder, D. P., Oskinova, L. M., Ignace, R., et al. 2012, ApJ, 756, L34
Huenemoerder, D. P., Schulz, N. S., Testa, P., et al. 2010, ApJ, 723, 1558
Hunter, A. 1942, Natur, 150, 756
Iben, I. Jr., Tutukov, A. V., & Fedorova, A. V. 1997, ApJ, 486, 955
Imanishi, K., Tsujimoto, M., & Koyama, K. 2001, ApJ, 563, 361
Jackman, C. M., Knigge, C., Altamirano, D., et al. 2018, JGRA, 123, 9204
Jordan, C. 1970, MNRAS, 148, 17
Jordan, C. 2000, PPCF, 42, 415
Judge, P. G., Solomon, S. C., & Ayres, T. R. 2003, ApJ, 593, 534
Kallman, T. R., & Palmeri, P. 2007, RvMP, 79, 79
Kamio, S., & Mariska, J. T. 2012, SoPh, 279, 419
Karovska, M., Carilli, C. L., Raymond, J. C., & Mattei, J. A. 2007, ApJ, 661, 1048
Karovska, M., Gaetz, T. J., Carilli, C. L., et al. 2010, ApJ, 710, L132
Karovska, M., Schlegel, E., Hack, W., Raymond, J. C., & Wood, B. E. 2005, ApJ, 623, L137
Kashyap, V., & Drake, J. J. 1998, ApJ, 503, 450
Kashyap, V. L., Drake, J. J., & Saar, S. H. 2008, ApJ, 687, 1339
Kastner, J. H., Huenemoerder, D. P., Schulz, N. S., Canizares, C. R., & Weintraub, D. A. 2002, ApJ, 567, 434
Kastner, J. H., Thompson, E. A., Montez, R., et al. 2012, ApJ, 747, L23
Kee, N. D., Owocki, S., & ud-Doula, A. 2014, MNRAS, 438, 3557
Kellogg, E., Anderson, C., Korreck, K., et al. 2007, ApJ, 664, 1079
Kellogg, E., Pedelty, J. A., & Lyon, R. G. 2001, ApJ, 563, L151
Kharchenko, V., Bhardwaj, A., Dalgarno, A., Schultz, D. R., & Stancil, P. C. 2008, JGRA, 113, A08229
Kharchenko, V., & Dalgarno, A. 2000, GeoR, 105, 18351
Kimura, T., Kraft, R. P., Elsner, R. F., et al. 2016, JGRA, 121, 2308
Knigge, C., Baraffe, I., & Patterson, J. 2011, ApJS, 194, 28
Kraft, R. P. 1967, ApJ, 150, 551
Krautter, J., Alcalá, J. M., Wichmann, R., Neuhäuser, R., & Schmitt, J. H. M. M. 1994, RMxAA, 29, 41
Ku, W. H. M., & Chanan, G. A. 1979, ApJ, 234, L59
Lada, C. J. 1987, in IAU Symp. 115, Star-forming Regions, ed. M. Peimbert, & J. Jugaku (Dordrecht: Reidel), 1–17
Laming, J. M. 2015, LRSP, 12, 2
Landi, E., Young, P. R., Dere, K. P., Del Zanna, G., & Mason, H. E. 2013, ApJ, 763, 86
Lisse, C. M., Christian, D. J., Dennerl, K., et al. 2001, Sci, 292, 1343
Lisse, C. M., Dennerl, K., Englhauser, J., et al. 1996, Sci, 274, 205
Lisse, C. M., McNutt, R. L., Wolk, S. J., et al. 2017, Icar, 287, 103
Long, K. S., Helfand, D. J., & Grabelsky, D. A. 1981, ApJ, 248, 925
Long, K. S., & White, R. L. 1980, ApJ, 239, L65
Lucy, L. B., & White, R. L. 1980, ApJ, 241, 300
Luna, G. J. M., Raymond, J. C., Brickhouse, N. S., et al. 2010, ApJ, 711, 1333
Luna, G. J. M., Raymond, J. C., Brickhouse, N. S., Mauche, C. W., & Suleimanov, V. 2015, A&A, 578, A15
Luna, G. J. M., Montez, R., Sokoloski, J. L., Mukai, K., & Kastner, J. H. 2009, ApJ, 707, 1168
Maggio, A., Flaccomio, E., Favata, F., et al. 2007, ApJ, 660, 1462

Maggio, A., Pillitteri, I., Scandariato, G., et al. 2015, ApJ, 811, L2
Mariska, J. T. 1992, The Solar Transition Region, Cambridge Astrophysics Series (New York: Cambridge Univ. Press)
Martins, F., Schaerer, D., Hillier, D. J., et al. 2005, A&A, 441, 735
Mauche, C. W., Liedahl, D. A., & Fournier, K. B. 2001, ApJ, 560, 992
Mauche, C. W., Liedahl, D. A., & Fournier, K. B. 2003, ApJ, 588, L101
Mestel, L., & Spruit, H. C. 1987, MNRAS, 226, 57
Metzger, A. E., Luthey, J. L., Gilman, D. A., et al. 1983, GeoR, 88, 7731
Micela, G., Sciortino, S., Serio, S., et al. 1985, ApJ, 292, 172
Miller, B. P., Gallo, E., Wright, J. T., & Pearson, E. G. 2015, ApJ, 799, 163
Moffat, A. F. J., & Corcoran, M. F. 2009, ApJ, 707, 693
Mohanty, S., & Basri, G. 2003, ApJ, 583, 451
Montesinos, B., Thomas, J. H., Ventura, P., & Mazzitelli, I. 2001, MNRAS, 326, 877
Montez, R. J., Kastner, J. H., Balick, B., et al. 2015, ApJ, 800, 8
Mori, K., Tsunemi, H., Katayama, H., et al. 2004, ApJ, 607, 1065
Moschou, S.-P., Drake, J. J., Cohen, O., et al. 2019, ApJ, 877, 105
Mukai, K. 2017, PASP, 129, 062001
Mukai, K., Kinkhabwala, A., Peterson, J. R., Kahn, S. M., & Paerels, F. 2003, ApJ, 586, L77
Nazé, Y., Broos, P. S., Oskinova, L., et al. 2011, ApJS, 194, 7
Nazé, Y., Petit, V., Rinbrand, M., et al. 2014, ApJS, 215, 10
Nazé, Y., ud-Doula, A., & Zhekov, S. A. 2016, ApJ, 831, 138
Nebot Gómez-Morán, A., & Oskinova, L. M. 2018, A&A, 620, A89
Nelson, T., Orio, M., Cassinelli, J. P., et al. 2008, ApJ, 673, 1067
Ness, J.-U., Brickhouse, N. S., Drake, J. J., & Huenemoerder, D. P. 2003, ApJ, 598, 1277
Ness, J. U., Mewe, R., Schmitt, J. H. M. M., et al. 2001, A&A, 367, 282
Ness, J.-U., Osborne, J. P., Henze, M., et al. 2013, A&A, 559, A50
Ness, J. U., & Schmitt, J. H. M. M. 2000, A&A, 355, 394
Ness, J.-U., Schmitt, J. H. M. M., Burwitz, V., Mewe, R., & Predehl, P. 2002, A&A, 387, 1032
Ness, J. U., Schmitt, J. H. M. M., & Robrade, J. 2004, A&A, 414, L49
Neuhaeuser, R., Sterzik, M. F., Schmitt, J. H. M. M., Wichmann, R., & Krautter, J. 1995, A&A, 295, L5
Neuhäuser, R. 1997, Sci, 276, 1363
Newton, E. R., Irwin, J., Charbonneau, D., Berta-Thompson, Z. K., & Dittmann, J. A. 2016, ApJ, 821, L19
Nichols, J. S., DePasquale, J., Kellogg, E., et al. 2007, ApJ, 660, 651
Noyes, R. W. 1983, in IAU Symp. 102, Solar and Stellar Magnetic Fields: Origins and Coronal Effects, ed. J. O. Stenflo (Dordrecht: Reidel), 133–46
Noyes, R. W., Weiss, N. O., & Vaughan, A. H. 1984, ApJ, 287, 769
O'Brien, D. P., Morbidelli, A., & Levison, H. F. 2006, Icar, 184, 39
Orio, M., Ness, J. U., Dobrotka, A., et al. 2018, ApJ, 862, 164
Orlando, S., & Drake, J. J. 2012, MNRAS, 419, 2329
Orlando, S., Drake, J. J., & Laming, J. M. 2009, A&A, 493, 1049
Orlando, S., Drake, J. J., & Miceli, M. 2017, MNRAS, 464, 5003
Osborne, J. P., Page, K. L., Beardmore, A. P., et al. 2011, ApJ, 727, 124
Ossendrijver, M. 2003, A&ARv, 11, 287
Owen, J. E., & Jackson, A. P. 2012, MNRAS, 425, 2931

Owocki, S. P., Sundqvist, J. O., Cohen, D. H., & Gayley, K. G. 2013, MNRAS, 429, 3379
Pallavicini, R., Golub, L., Rosner, R., et al. 1981, ApJ, 248, 279
Parker, E. N. 1955, ApJ, 122, 293
Pease, D. O., Drake, J. J., & Kashyap, V. L. 2006, ApJ, 636, 426
Pease, D. O., Drake, J. J., Kashyap, V. L., et al. 2003, Proc. SPIE, 4851, 157
Penz, T., Micela, G., & Lammer, H. 2008, A&A, 477, 309
Perna, R., McDowell, J., Menou, K., Raymond, J., & Medvedev, M. V. 2003, ApJ, 598, 545
Pillitteri, I., Wolk, S. J., Lopez-Santiago, J., et al. 2014, ApJ, 785, 145
Pittard, J. M., & Parkin, E. R. 2010, MNRAS, 403, 1657
Pizzolato, N., Maggio, A., Micela, G., Sciortino, S., & Ventura, P. 2003, A&A, 397, 147
Pollock, A. M. T. 1987, ApJ, 320, 283
Poppenhaeger, K., & Schmitt, J. H. M. M. 2011, AN, 332, 1052
Poppenhaeger, K., Schmitt, J. H. M. M., & Wolk, S. J. 2013, ApJ, 773, 62
Porquet, D., Dubau, J., & Grosso, N. 2010, SSRv, 157, 103
Porquet, D., & Dubau, J. 2000, A&AS, 143, 495
Pottasch, S. R. 1963, ApJ, 137, 945
Pravdo, S. H., Feigelson, E. D., Garmire, G., et al. 2001, Natur, 413, 708
Pravdo, S. H., Tsuboi, Y., & Maeda, Y. 2004, ApJ, 605, 259
Preibisch, T., Kim, Y.-C., Favata, F., et al. 2005a, ApJS, 160, 401
Preibisch, T., McCaughrean, M. J., Grosso, N., et al. 2005b, ApJS, 160, 582
Prilutskii, O. F., & Usov, V. V. 1976, SvA, 20, 2
Puls, J., Kudritzki, R. P., Herrero, A., et al. 1996, A&A, 305, 171
Puls, J., Vink, J. S., & Najarro, F. 2008, A&ARv, 16, 209
Quirion, P. O., Fontaine, G., & Brassard, P. 2007, ApJS, 171, 219
Raga, A. C., Noriega-Crespo, A., & Velázquez, P. F. 2002, ApJ, 576, L149
Randich, S., Schmitt, J. H. M. M., & Prosser, C. 1996, A&A, 313, 815
Rappaport, S., Cash, W., Doxsey, R., McClintock, J., & Moore, G. 1974, ApJ, 187, L5
Rauch, T., Orio, M., Gonzales-Riestra, R., et al. 2010, ApJ, 717, 363
Rauw, G., Nazé, Y., Wright, N. J., et al. 2015, ApJS, 221, 1
Rauw, G., & Nazé, Y. 2016, A&A, 594, A82
Reipurth, B., & Bally, J. 2001, ARAA, 39, 403
Ribas, Á., Bouy, H., & Merín, B. 2015, A&A, 576, A52
Ribas, Á., Merín, B., Bouy, H., & Maud, L. T. 2014, A&A, 561, A54
Romine, G., Feigelson, E. D., Getman, K. V., Kuhn, M. A., & Povich, M. S. 2016, ApJ, 833, 193
Rosner, R. 1980, SAOSR, 389, 79
Rosner, R., Tucker, W. H., & Vaiana, G. S. 1978, ApJ, 220, 643
Rosotti, G. P., Dale, J. E., de Juan Ovelar, M., et al. 2014, arXiv:1404.1931
Rutledge, R. E., Basri, G., Martín, E. L., & Bildsten, L. 2000, ApJ, 538, L141
Sanz-Forcada, J., Brickhouse, N. S., & Dupree, A. K. 2002, ApJ, 570, 799
Sanz-Forcada, J., Brickhouse, N. S., & Dupree, A. K. 2003, ApJS, 145, 147
Sanz-Forcada, J., Ribas, I., Micela, G., et al. 2010, A&A, 511, L8
Sasselov, D. D., & Lester, J. B. 1994, ApJ, 423, 795
Schmitt, J. H. M. M. 1997, A&A, 318, 215
Schmitt, J. H. M. M., & Liefke, C. 2004, A&A, 417, 651
Schneider, P. C., & Schmitt, J. H. M. M. 2008, A&A, 488, L13
Schulz, R., Stüwe, J. A., Tozzi, G. P., & Owens, A. 2000, A&A, 361, 359

Schwarz, G. J., Ness, J.-U., Osborne, J. P., et al. 2011, ApJS, 197, 31
Schwarzschild, M. 1948, ApJ, 107, 1
Semenov, D., Wiebe, D., & Henning, T. 2004, A&A, 417, 93
Shibata, K., & Magara, T. 2011, LRSP, 8, 6
Skumanich, A. 1972, ApJ, 171, 565
Snios, B., Kharchenko, V., Lisse, C. M., et al. 2016, ApJ, 818, 19
Snios, B., Lewkow, N., & Kharchenko, V. 2014, A&A, 568, A80
Snios, B., Lichtman, J., & Kharchenko, V. 2018, ApJ, 852, 138
Snios, B., Dunn, W. R., Lisse, C. M., et al. 2019, arXiv:1903.02574
Snow, T. P. J., & Morton, D. C. 1976, ApJS, 32, 429
Snowden, S. L., Freyberg, M. J., Plucinsky, P. P., et al. 1995, ApJ, 454, 643
Spruit, H. C. 2002, A&A, 381, 923
Spruit, H. C. 2011, in IAGA Special Sopron Book Series, Vol. 4, The Sun, the Solar Wind, and the Heliosphere, ed. M.P. Miralles, & J. Sánchez Almeida (Berlin: Springer), 39
Starrfield, S., Iliadis, C., & Hix, W. R. 2016, PASP, 128, 051001
Stelzer, B., Schmitt, J. H. M. M., Micela, G., & Liefke, C. 2006, A&A, 460, L35
Stevens, I. R., Blondin, J. M., & Pollock, A. M. T. 1992, ApJ, 386, 265
Tayler, R. J. 1973, MNRAS, 161, 365
Testa, P., Drake, J. J., & Peres, G. 2004, ApJ, 617, 508
Testa, P., Drake, J. J., Ercolano, B., et al. 2008, ApJ, 675, L97
Testa, P., Reale, F., Garcia-Alvarez, D., & Huenemoerder, D. P. 2007, ApJ, 663, 1232
Tout, C. A., & Pringle, J. E. 1995, MNRAS, 272, 528
Townsley, L. K., Broos, P. S., Chu, Y.-H., et al. 2011a, ApJS, 194, 16
Townsley, L. K., Broos, P. S., Corcoran, M. F., et al. 2011b, ApJS, 194, 1
Uchida, Y., & Shibata, K. 1984, PASJ, 36, 105
ud-Doula, A., Owocki, S., Townsend, R., Petit, V., & Cohen, D. 2014, MNRAS, 441, 3600
ud-Doula, A., Sundqvist, J. O., Owocki, S. P., Petit, V., & Townsend, R. H. D. 2013, MNRAS, 428, 2723
Vaiana, G. S. 1980, SAOSR, 389, 195
Vaiana, G. S., Cassinelli, J. P., Fabbiano, G., et al. 1981, ApJ, 245, 163
Vaiana, G. S., Davis, J. M., Giacconi, R., et al. 1973, ApJ, 185, L47
Vaiana, G. S., & Rosner, R. 1978, ARA&A, 16, 393
van den Heuvel, E. P. J., Bhattacharya, D., Nomoto, K., & Rappaport, S. A. 1992, A&A, 262, 97
Vennes, S., & Dupuis, J. 2002, in ASP Conf. Series 262, The High Energy Universe at Sharp Focus: Chandra Science, ed. E. M. Schlegel, & S. D. Vrtilek (San Francisco, CA: ASP), 57
Vilhu, O. 1984, A&A, 133, 117
Vink, J. S., de Koter, A., & Lamers, H. J. G. L. M. 2000, A&A, 362, 295
Voges, W., Aschenbach, B., Boller, T., et al. 1999, A&A, 349, 389
Walder, R., Folini, D., & Shore, S. N. 2008, A&A, 484, L9
Waldron, W. L., & Cassinelli, J. P. 2001, ApJ, 548, L45
Waldron, W. L., & Cassinelli, J. P. 2007, ApJ, 668, 456
Walsh, C., Nomura, H., Millar, T. J., & Aikawa, Y. 2012, ApJ, 747, 114
Walter, F. M. 1986, ApJ, 306, 573
Walter, F. M., & Bowyer, S. 1981, ApJ, 245, 671
Walter, F. M., & Kuhi, L. V. 1981, ApJ, 250, 254
Wargelin, B. J., Kornbleuth, M., Martin, P. L., & Juda, M. 2014, ApJ, 796, 28

Wargelin, B. J., Markevitch, M., Juda, M., et al. 2004, ApJ, 607, 596
Wargelin, B. J., Saar, S. H., Pojmański, G., Drake, J. J., & Kashyap, V. L. 2017, MNRAS, 464, 3281
Warner, B. 1974, MNRAS, 167, 47P
Warner, B. 1995, Cambridge Astrophysics Series, 28
Warren, H. P. 2014, ApJ, 786, L2
Warren, H. P., Winebarger, A. R., & Brooks, D. H. 2012, ApJ, 759, 141
Weber, E. J., & Davis, L. J. 1967, ApJ, 148, 217
Werner, K., & Herwig, F. 2006, PASP, 118, 183
Werner, K., Rauch, T., Barstow, M. A., & Kruk, J. W. 2004, A&A, 421, 1169
Williams, P. K. G., Cook, B. A., & Berger, E. 2014, ApJ, 785, 9
Williams, R. E. 1992, AJ, 104, 725
Wilson, O. C. 1978, ApJ, 226, 379
Winckler, J. R., Peterson, L., Hoffman, R., & Arnoldy, R. 1959, GeoR, 64, 597
Winston, E., Megeath, S. T., Wolk, S. J., et al. 2007, ApJ, 669, 493
Wolk, S. J., Bourke, T. L., Smith, R. K., Spitzbart, B., & Alves, J. 2002, ApJ, 580, L161
Wood, B. E., Laming, J. M., Warren, H. P., & Poppenhaeger, K. 2018, ApJ, 862, 66
Wright, N. J., Bouy, H., Drew, J. E., et al. 2016, MNRAS, 460, 2593
Wright, N. J., Drake, J. J., Mamajek, E. E., & Henry, G. W. 2011, ApJ, 743, 48
Wright, N. J., Drew, J. E., & Mohr-Smith, M. 2015, MNRAS, 449, 741
Wright, N. J., Newton, E. R., Williams, P. K. G., Drake, J. J., & Yadav, R. K. 2018, MNRAS, 479, 2351
Yu, Z.-l., Xu, X.-j., Li, X.-D., et al. 2018, ApJ, 853, 182
Zinnecker, H., & Preibisch, T. 1994, A&A, 292, 152
Zombeck, M. V., Barbera, M., Collura, A., & Murray, S. S. 1997, ApJ, 487, L69

# Chapter 5

# Supernovae and Their Remnants

**Patrick Slane**

## 5.1 Supernovae

Supernova (SN) explosions play a crucial role in shaping the dynamics and chemical evolution of galaxies, enriching their environments with metals and driving shocks that heat the ISM and trigger new star formation. The release of $\sim 10^{51}$ erg of kinetic energy into a circumstellar environment potentially modified by the mass-loss history of the progenitor system results in $\sim$1–10 $M_{\odot}$ of material moving at initial speeds approaching $10^4$ km s$^{-1}$. The expanding ejecta from the explosion heat the surroundings to high temperatures, producing extremely bright radiation that allows the events to be identified over cosmological distances. For sufficiently high ambient densities, the reverse-shocked ejecta are heated to $T \sim 10^7$ K, resulting in the copious production of X-rays. *Chandra* observations of SNe thus probe the environments immediately surrounding the explosive events, providing direct information on the mass-loss history of the progenitor systems in the late stages leading up to the events.

### 5.1.1 Supernova Types

Classifications of SN types are based on the spectra and time behavior of the optical emission. The details of these classifications are discussed in a multitude of references (e.g., see Branch & Wheeler 2017, which is also a definitive reference on SN explosions in general) and are summarized only briefly here. Broadly, SNe are divided into Types I and II, based on the absence or presence of hydrogen in their optical spectra. Type II events result from the core collapse (CC) of massive stars upon exhaustion of fuel to power nuclear reactions that can provide the energy to support the outer regions of the star against gravity. Upon collapse, an explosion is triggered, which results in the release of $\sim 10^{53}$ erg of energy, about 99% of which is in the form of neutrinos. Roughly $10^{51}$ erg is released in the form of ejected stellar material. The interaction of these rapidly expanding ejecta with the outer layers of the star produces the hydrogen signature. The core collapses to form a neutron star (NS) or, possibly, a black hole (BH).

doi:10.1088/2514-3433/ab43dcch5

The lack of hydrogen in Type I events implies an explosion in an environment that is not surrounded by a typical stellar atmosphere. The events that initially defined the Type I classification occur in all types of galaxies, and their presence in regions of low star formation rates indicate that they are from long-lived (and thus low-mass) progenitors. These were initially recognized as being produced by the thermonuclear detonation of a carbon/oxygen white dwarf (C/O WD), which completely disrupts the core, leaving no stellar remnant, but produces an expanding shell of ejecta with kinetic energy of also $\sim 10^{51}$ erg. The spectra of these SNe also show strong Si features.

Subsequent observations identified subclasses of Type I events that contain little or no Si emission, sometimes with strong He emission and traces of H. These are now recognized as distinctly different types of Type I events, introducing subclassifications of Type Ia for the thermonuclear disruption of a WD, and Type Ib and Type Ic, which are now understood to be CC SNe from massive stars that have shed most or all of their outer H layers—or, in the case of Ic events, He layers—prior to explosion. There are currently major open questions about both thermonuclear (Type Ia) and CC (Type Ib/c and II) SNe, including the nature of the progenitor systems and details of the explosion engines (e.g., Janka 2017; Maoz et al. 2014). As described below, *Chandra* observations of SNe have placed important constraints on models for these events and are continuing to identify new information on the pre-SN evolution of the progenitors.

The nucleosynthesis products from thermonuclear SNe differ considerably from those of CC events. The abundances from Type Ia SNe are dominated by Fe-group elements, with the relative abundances and fraction of neutron-rich isotopes depending on exactly how the burning proceeds. For CC events, the abundances are dominated by the products of the stellar burning stages (C, O, Ne, Mg) with additional contributions of Si-burning and Fe-peak elements from the explosive nucleosynthesis in the hot central regions of the CC. A large O/Fe ratio is a strong signature that differentiates these from SN Ia events. These abundance signatures can become evident as the SN evolves into a young SN remnant (SNR; see Section 5.2.3).

*Type Ia Supernovae*

It is generally agreed that Type Ia SNe are the result of disruption of a C/O WD due to the ignition of C fusion in the electron-degenerate WD core. Because the degeneracy pressure is independent of temperature, the increased thermonuclear energy input cannot be compensated by adiabatic expansion, and an explosive event results. However, the evolutionary scenario that leads to this event is poorly understood. As isolated objects, WD stars are stable. For explosions to occur, the star presumably accretes material from a companion until its mass exceeds the Chandrasekhar mass ($M_{\rm Ch}$), or it merges with another WD. Binary evolution is thus involved, and the two primary scenarios for SNe Ia (Figure 5.1) differ in terms of whether the companion is a normal star (single-degenerate, or SD, scenario) or another WD (double-degenerate, or DD, scenario). The details of the explosions in these two scenarios have considerable differences. In SD models, mass is slowly accreted from a companion—through Roche lobe overflow or a wind—until the

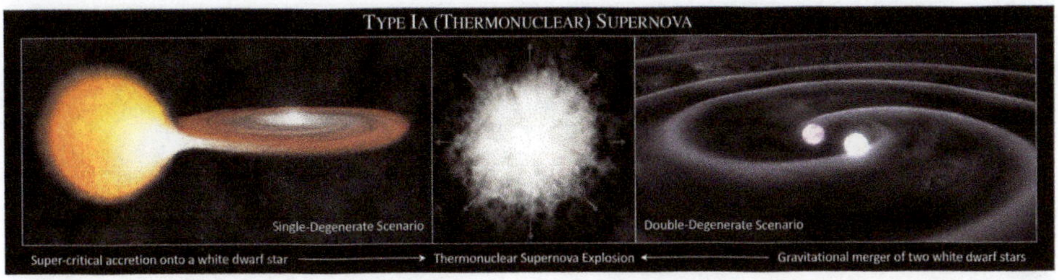


**Figure 5.1.** Illustration of scenarios for thermonuclear supernovae. Left: single-degenerate scenario in which a white dwarf star in a binary captures material from a nondegenerate companion star until Chandrasekhar mass is exceeded. Courtesy of NASA/CXC/M.Weiss. Right: double-generate scenario in which two white dwarf stars in a binary system merge and initiate an explosion. Courtesy of GSFC/D.Berry.

temperature and density are sufficiently high to ignite carbon burning deep in the degenerate core, ultimately triggering a thermonuclear explosion. At the high interior density, efficient electron capture leads to high neutronization, producing enhanced abundances of neutron-rich isotopes such as $^{58}$N and $^{55}$Mn.

In the explosive event, a burning front propagates through the WD, releasing $\sim 10^{51}$ erg that completely disrupts the star. The composition of the resulting $\sim 1.4\ M_\odot$ of rapidly expanding ejecta is determined by $\alpha$-particle captures that produce a layered abundance pattern with peaks around O/Mg/Ne, Si/S, and the iron group (primarily $^{56}$Ni in high-density regions). If extremely dense regions exist, electron capture becomes significant, producing n-rich isotopes with signatures that can potentially be observed in the SNRs that result. The exact manner in which the burning progresses is uncertain. It may be through a pure deflagration (subsonic) wave, a pure detonation, or an initial deflagration that transitions to a detonation. Due to differences in temperature and density structures, these different explosion processes yield different ejecta densities and compositions. Pure detonation models, wherein the burning begins close to the WD center, incinerate most of the WD into Fe-group elements, while pure deflagration models, in which the burning proceeds subsonically, leave some amount of unburned C/O that is ejected in the outer layers. Delayed-detonation models produce an ejecta composition with some of the unburned C/O material processed to Fe-group elements and some to intermediate-mass elements. X-ray observations of the ejecta can potentially differentiate between these different burning scenarios once evolved to the SNR stage (Badenes et al. 2003).

*Core-collapse Supernovae*

The CC of a massive star (Figure 5.2) follows thermonuclear processing of elements that result in an Fe core that cools until supported by electron degeneracy pressure. Subsequent burning in the surrounding shells increase the core mass until it exceeds $M_{\rm Ch}$, and electron degeneracy is incapable of supporting the overlying material against gravitational collapse. The result is the collapse of a $\sim 1.4\ M_\odot$ core with a radius of $\sim 10^4$ km, resulting in the release of gravitational energy, $E_{\rm G} \sim (3GM_{\rm Ch}^2)/5R \approx 3 \times 10^{53}$ erg. The bulk of the energy is eventually carried off in neutrinos, with about 1% being deposited as kinetic energy in the ejecta.

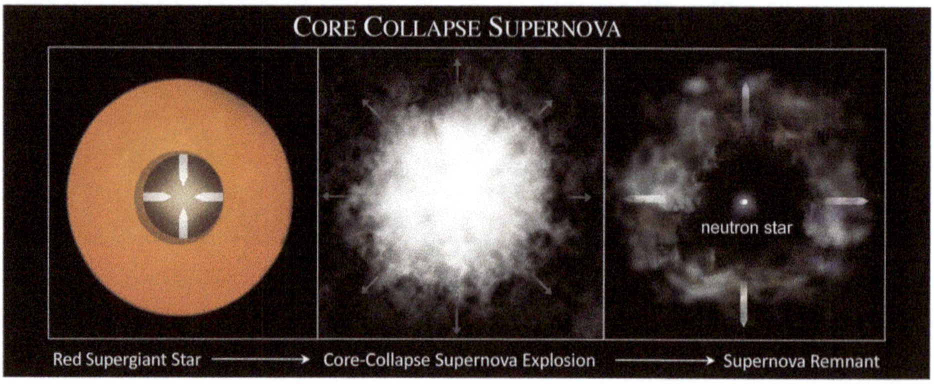


**Figure 5.2.** Illustration of core-collapse supernova. The explosion is initiated by the depletion of viable nuclear fuel at the center of a massive star. As the core collapses, forming a compact NS (or, possibly, a BH), the outer layers of the star are ejected in the supernova event. The expanding ejecta drive a shock into the surrounding CSM, forming an SNR. Details of the explosion process, including the connection between progenitor type and the properties of the compact object, the kick velocity imparted to the compact object, and asymmetries in the ejecta, are not well understood. X-ray observations of SNRs and their associated NSs can provide significant constraints on such explosion models. Courtesy of NASA/CXC.

The explosion announces itself first through the release of neutrinos and gravitational waves (the latter of which have not yet been detected from SNe); the optical signature follows with the so-called shock breakout, when the blast wave (BW) enters the photosphere. The subsequent light curve and spectrum of the SN event depends critically on its composition and on the details of the environment into which it expands.

Because CC SNe can be produced from a broad range of progenitor masses and final stellar configurations, connecting the observed properties to progenitor type provides critical information on the details of these explosive events. Specific areas of importance pertain to how rotation, binarity, and magnetic fields affect the explosions, whether there are large buoyancy-driven instabilities that might imprint themselves on the ejecta distribution, and how (or if) these properties might affect the energetics of the explosions and the character of the associated compact objects left behind. Observations of the SN events themselves provide a wealth of information on some of these elements, while others are best probed when the explosions have evolved to the SNR stage, with the compact objects in full view.

### 5.1.2 Progenitors and Circumstellar Environments

For both thermonuclear and CC SNe, important information about the progenitor systems is contained in the circumstellar environments into which the explosions evolve. While the key classifications and characterizations of SNe are generally carried out in the optical or infrared bands, based on emission from the SN shell, X-ray observations can provide unique and powerful probes of the progenitor through emission of the shocked CSM.

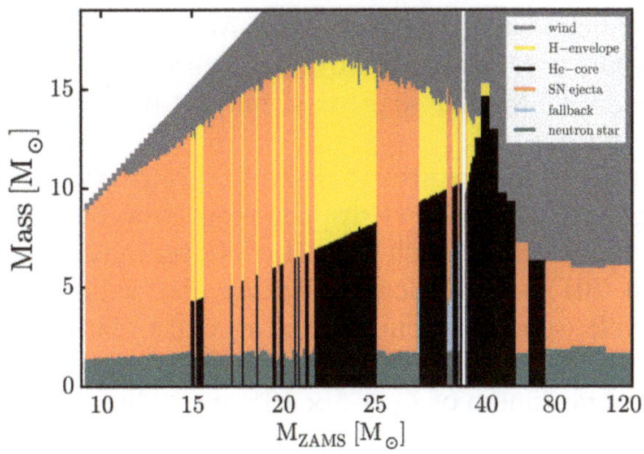


**Figure 5.3.** Mass content of pre-explosion core-collapse progenitors. See text for explanation. Reproduced from Sukhbold et al. (2016). 

*Mass Loss in Core-collapse Progenitors*

The late phases of stellar evolution for massive stars can include episodes of extreme mass loss. This is illustrated in Figure 5.3, which shows the mass distribution just prior to explosion as a function of zero-age main sequence (ZAMS) mass for isolated CC progenitors (Sukhbold et al. 2016). The green regions indicate the mass that collapses to an NS. (Systems without such regions collapse directly to BHs.) The gray regions indicate the mass that has been lost to winds. The remaining orange regions comprise the stellar ejecta that are driven out in the explosion. Systems with large ZAMS mass lose large amounts of their envelopes through winds, leaving smaller amounts of ejecta than systems with lower ZAMS mass. This has important and observable impacts on the properties of the associated SNe and on the SNRs that they produce. The late stages of mass loss may be more complicated than assumed here, including brief episodes of fast (or even explosive) mass loss, the imprint of which on the surrounding medium can have significant effects on the X-ray evolution of these systems. Similarly, evolution in a binary system can significantly modify this picture, again producing results that can be probed in X-ray studies. *Chandra* studies of several such systems illustrate how studies of the X-ray emission can provide detailed information on the late stages of evolution in such systems.

SN 1996cr is an SN in the nearby ($d \sim 4$ Mpc) Circinus galaxy. Originally identified as an ultraluminous X-ray (ULX) source based on a *Chandra* study of the point source population in the Galaxy, the source was observed to brighten in X-rays by a factor of more than 30 between 1997 and 2000. Optical spectroscopy established the object as a Type IIn SN, characterized by narrow H$\alpha$ lines that indicate interaction with dense surroundings (i.e., the emission is from the shock–CSM interaction, not from the SN itself), although this classification is based on post-explosion properties that could reflect later-stage circumstellar interactions (see below). The explosion date was isolated by virtue of radio detection in 1996, and a lack of emission in an earlier 1995 observation (Bauer et al. 2008). Archival data and subsequent observations with *Chandra*, *XMM-Newton*, and *Swift* show dramatic increases in X-ray flux over time, indicative of an encounter between the BW and a large mass of pre-SN CSM that suggests significant mass loss over thousands of years before the explosion.

Circumstellar interactions are also observed for SN 2014C, which was originally classified as a Type Ib event from a H-stripped progenitor, but for which subsequent spectra showed evolution to a H-rich Type IIn SN over the course of roughly a year (Milisavljevic et al. 2015). Pre-explosion *Chandra* observations carried out in 2001 reveal no detectable source within the error circle, but subsequent *Chandra* monitoring observations reveal a sharp rise in the X-ray luminosity for the first 130 days, reaching a maximum of $L_x \sim 5 \times 10^{40}$ erg s$^{-1}$ by an age of 500 days. This, along with an accompanying rise in the radio luminosity, indicates interaction with $\sim 1\ M_\odot$ of H-rich material located at about $6 \times 10^{16}$ cm from the progenitor star (Margutti et al. 2017). These observations suggest that some (perhaps many) observed Type IIn SNe may have initially been Type Ib/c cores that later encountered material from previous mass-loss episodes from the progenitor.

Similar results are obtained for SN 2001em, a Type Ib/c SN, which, after only several years, showed high radio luminosity (Stockdale et al. 2004) and bright X-ray emission (Pooley & Lewin 2004) that was uncharacteristic of Ib/c events of this age. Modeling of the system indicates interaction with a dense, massive CSM shell ($M \sim 3\ M_\odot$) at a distance of $\sim 7 \times 10^{16}$ cm, suggestive of a mass-loss rate $\dot{M} \sim (2 - 10) \times 10^{-3}\ M_\odot$ yr$^{-1}$ for about 1000–2000 years before the explosion (Chugai & Chevalier 2006).

Pre-explosion *Hubble Space Telescope* (*HST*) observations of the Type IIb SN2011dh reveal a yellow supergiant (YSG) progenitor for this event (Van Dyk et al. 2011). X-ray monitoring, including *Chandra* observations ~500 days after the explosion, reveals an X-ray light curve suggestive of interaction with a dense circumstellar environment, consistent with wind-driven mass loss with $\dot{M} \sim 3 \times 10^{-6}\ M_\odot$ yr$^{-1}$ for $v_w \sim 20$ km s$^{-1}$ in the final $\gtrsim$1300 yr before the explosion (Maeda et al. 2014). This is consistent with mass loss for a giant star, which is typically not sufficient to fully remove the H envelope. Thus, the observations are suggestive of binary evolution with large mass transfer ~1300 yr prior to the explosion (Kundu et al. 2019).

*Evidence for Central Engines*

The compact cores created in SN explosions can, in principle, have a significant effect on the SN evolution by supplementing the energy budget. Observations of superluminous SNe, which display luminosities 10 or more times higher than typical SNe, may provide evidence for contributions from such central engines.

The Type Ib SN 20012au is a slowly evolving system with $E \sim 10^{52}$ erg for which optical spectra taken over several years reveal a late-time emission excess that may be associated with power from a central pulsar-driven nebula (Milisavljevic et al. 2018). Observations with *Chandra* reveal no X-ray emission, but this may be consistent with the high optical depth, which, for an O-rich ejecta composition, is $\gg$1 in the soft X-ray band at this stage of evolution.

Observations of two older SNe provide similar evidence for central engines. SN 1970G and SN 1979C are both Type IIL SNe for which monitoring observations reveal recent episodes of X-ray brightening (Figure 5.4, right). X-ray emission from SN 1970G detected in observations with *Chandra* suggest a system dominated by interaction with dense CSM from a progenitor wind (Immler & Kuntz 2005). Subsequent observations reveal a brightening of the X-ray emission by a factor of 3–4, to $L_x \sim 4 \times 10^{37}$ erg s$^{-1}$, with evidence for increased emission above 2 keV. Such an increase is suggestive of a central engine adding energy to the system. For accretion at the Eddington limit, $L_{\rm Edd} = 1.4 \times 10^{38}\ M_{\rm BH,\odot}$ erg s$^{-1}$, where $M_{\rm BH,\odot}$ is the BH mass in units of solar masses. For an accreting BH scenario, the observed luminosity would suggest an unrealistically small BH mass, suggesting that the central engine is likely to be a pulsar-powered nebula (Dittmann et al. 2014). For SN 1979C, on the other hand, for which *Chandra* observations also show a temporal increase in the the X-ray emission, the luminosity is quite high, $L_x \sim 10^{39}$ erg s$^{-1}$, suggesting that the SN may harbor an accreting BH (Patnaude et al. 2011).

*Progenitors of Thermonuclear SNe*

The SD and DD scenarios for producing Type Ia SNe yield different pre-explosion environments that can be probed through both pre- and post-explosion X-ray observations. For the DD scenario, in which the SN results from the merger of two WD stars, the expectation is that the ambient medium experiences little modification from any sort of wind and that pre-explosion X-ray emission will be very faint. For the SD scenario, however, significant modification may occur through accretion

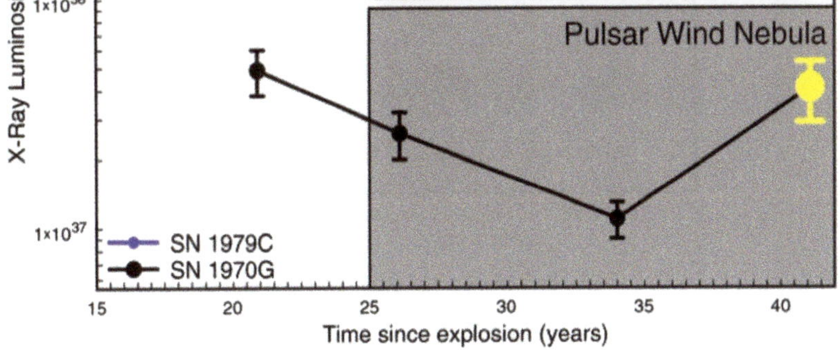


**Figure 5.4.** X-ray light curves for SN 1979C and SN 1970G, showing late-time brightening indicative of energy input from a central engine. The upper and gray box corresponds to the viable luminosity range for an accreting BH left behind in the explosion, while the lower box corresponds to emission from a pulsar-powered nebula. Reproduced from Dittmann et al. (2014). © 2014. The American Astronomical Society. All rights reserved.

winds, and the interaction of the SN with this CSM can produce detectable X-ray emission for sufficiently large mass-loss rates. Thus, probing these surroundings with the expanding SN (or SNR) can place significant constraints on the progenitor system. This is illustrated in Figure 5.5 (from Patnaude & Badenes 2017), which plots density profiles for different wind-driven mass-loss scenarios and compares these with observations of SNe Ia and SNRs. Note that the SN measurements probe mass loss at times just before the explosion, while SNR measurements probe mass loss over much larger pre-explosion timescales.

SN 2011fe is a Type Ia SN in the nearby and well-studied galaxy M101. Pre-explosion observations of the galaxy with *HST* and *Chandra* reveal no evidence of emission from a progenitor system. The X-ray flux limits rule out steady nuclear burning from an $M = M_{\rm Ch}$ WD, but other SD scenarios that include emission from an extended photosphere with $kT < 60$ eV or sub-Chandrasekhar WDs undergoing quasi-stationary burning are still permitted (Liu et al. 2012). Post-explosion X-ray observations, including a 50 ks *Chandra* just four days after the event, place stronger constraints on SD scenarios, with the flux upper limits constraining any pre-SN mass loss to $\dot{M}/v_{\rm w} < (2 \times 10^{-9}\ M_\odot\ {\rm yr}^{-1})/(100\ {\rm km\ s}^{-1})$. The X-ray results are consistent with, but do not fully require, a DD scenario for this WD system.

A similar constraint results from *Chandra* and *Swift* observations of SN 2014J, a Type Ia SN in the nearby starburst galaxy M82, where X-ray nondetections rule out SD progenitor systems that have undergone mass loss up until the explosion time (Margutti et al. 2014).

*Chandra* studies of nine other Type Ia SNe also yield nondetections that argue against accreting, nuclear-burning WDs (typically observed as supersoft X-ray sources) as the progenitors, or indicate that the sources are significantly obscured (Nielsen et al. 2012). SN 2012ca provides a potentially different result, however.

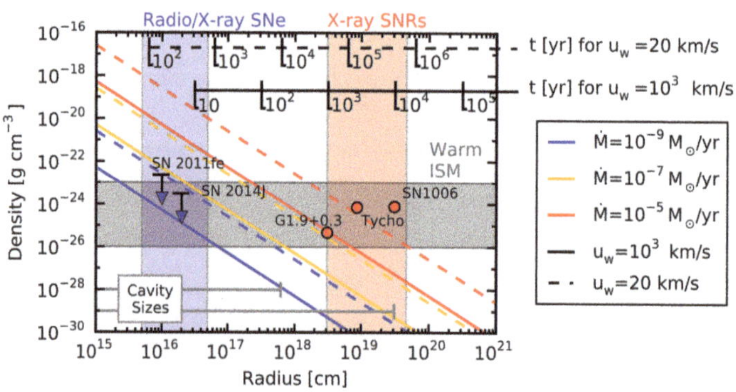


**Figure 5.5.** Circumstellar density profiles for isotropic SN Ia progenitor outflows, compared with regions probed by studies of SNe and SNRs. The colored curves correspond to different mass-loss rates and wind speeds, and the gray box illustrates typical density ranges in the warm ISM. Rulers on the bottom show sizes of typical cavities blown by outflows, and upper rulers indicate the number of years before explosion that material at the indicated speeds was ejected by the progenitor. Purple and pink bands illustrate the radius/age ranges probed by studies of SNe and SNRs, and measurements for several SNe and SNRs are indicated on the plot. Reprinted/adapted by permission from Patnaude and Badenes (2017). © Springer International Publishing AG (2017).

While designated as a Type Ia event, there is some controversy as to whether or not this is actually a CC SN. *Chandra* observations carried out ~500–800 days after the explosion yield $L_x \sim 2 \times 10^{40}$ erg s$^{-1}$, making this the first X-ray detection of a Type Ia SN if that designation is correct (Bochenek et al. 2018). The data provide some evidence for expansion into an asymmetric CSM for which $n_0 > 10^8$ cm$^{-3}$ in the higher density region. This would seem to imply a clumpy wind or some sort of dense disk or torus which, if associated with a Type Ia SN, would be consistent with an SD scenario for this system.

### 5.1.3 Supernova 1987A

On 1987 February 23, the brightest SN since Kepler was discovered in the LMC. Observations across the electromagnetic spectrum, as well as in neutrinos, have provided an unprecedented view of the evolution of this system, SN 1987A, now transitioning from the SN to the SNR phase. For a thorough recent review, the reader is referred to McCray (2017).

SN 1987A was classified as a Type II event, and identification of neutrinos associated with the event confirmed its CC nature. Pre-SN images of the field identified a blue supergiant (BSG) progenitor star—Sanduleak −69 202—which explains the unusual light curve that continued to brighten for ~3 months after outburst. Early optical observations revealed an hour-glass-like structure composed of three rings (Figure 5.6, upper right), possibly created in the late stages of binary

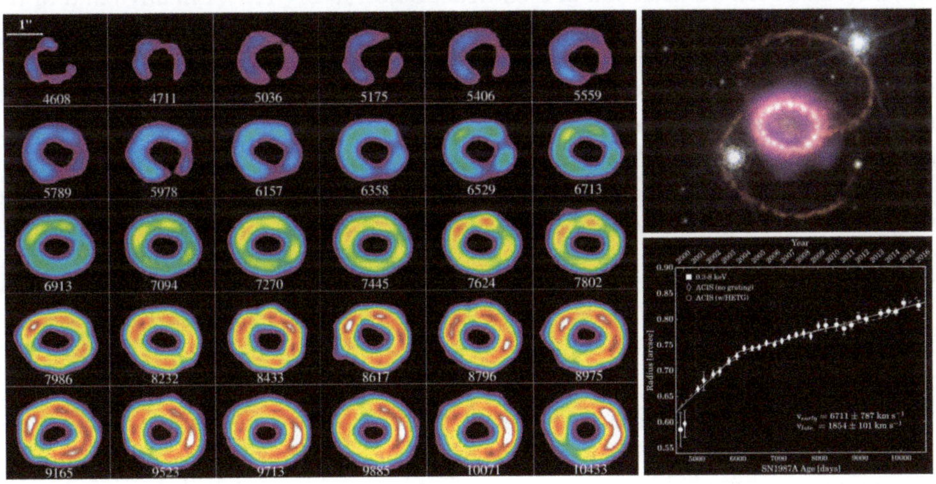


**Figure 5.6.** Left: brightening and expansion of SN 1987A as measured in *Chandra* observations covering 16 years. Numbers in each panel correspond to days after explosion. The video shows a time-lapse movie of X-ray emission from SN1987A observed with *Chandra* from 1999–2013, encompassing most of the data shown in the left-hand panel. The interactive figure shows separately the X-ray emission (blue), optical emission (green), and millimeter emission (red), and the composite image. Right (upper): optical and X-ray image of SN 1987A illustrating the three-ring structure with bright knots in the equatorial ring. Right (lower): radius as a function of age from *Chandra* observations of SN 1987A. The simulation shows a simulated visualization of the initial explosion of SN 1987A, and the evolution of the resulting supernova remnant up until the present day. Reproduced figure compiled from Frank et al. (2016; © 2016. The American Astronomical Society. All rights reserved) and CXC/NASA. Additional materials available online at http://iopscience.iop.org/book/978-0-7503-2163-1.

evolution that might explain why the star exploded as a BSG. The bright central ring, thought to be an equatorial structure, has a radius of $\sim 6.5 \times 10^{17}$ cm and an expansion speed of $\sim 10^3$ km s$^{-1}$, indicating that it was ejected roughly 20,000 yr prior to the explosion (assuming undecelerated expansion). As noted in previous sections, recent observations of other SNe provide additional evidence of such episodic mass-loss events in the final stages of evolution.

X-rays from SN 1987A were first detected ~1400 days after the explosion (Hasinger et al. 1996). After ~4000 days, the BW began interacting with dense clumps at the inner edge of the equatorial ring, producing bright spots observed in both optical and X-ray observations. The BW entered the bulk of the ring material after ~6000 days, resulting in a significant increase in the X-ray brightness. The evolution of the X-ray properties of SN 1987A is shown in Figure 5.6, where images from *Chandra* monitoring observations are shown at the left (with the day number since the explosion at the bottom of each panel), and the radial expansion shown at the lower right, where a clear deceleration is observed at the time the BW entered the main ring (Frank et al. 2016). The X-ray flux has continued to rise as the BW progresses through the ring and is expected to flatten when the BW exits the ring. *Chandra* LETG and HETG spectra reveal emission from both slow and fast shocks, with the temperature of the latter decreasing with age (Zhekov et al. 2009). Line profiles indicate the presence of gas that has been shocked by the BW and then again from shocks reflected by the inner ring.

Despite the clear signature of NS formation by virtue of neutrino production, to date there has been no detection of a compact object in SN 1987A in any band of the electromagnetic spectrum. The upper limit for the X-ray luminosity from *Chandra* observations is $L_x < 10^{36}$ erg s$^{-1}$ (Alp et al. 2018), which rules out the presence of a pulsar or nebula as bright as the Crab (see Section 5.3.1). However, due to the presence of significant absorbing material from the foreground column of ejecta gas and dust, the limit on the surface temperature of a cooling NS ($\log T \lesssim 6.9$) does not rule out typical cooling models and is still roughly four times higher than that observed for the young NS in Cas A (see Section 5.2.3). The emergence of a central NS may occur eventually, as the ejecta continue to expand, reducing the column density.

## 5.2 Supernova Remnants

The very fast shocks formed in young and middle-aged SNRs act to heat matter to temperatures exceeding many millions of degrees. As a result, these systems are copious emitters of thermal X-rays. This emission, characterized by bremsstrahlung continuum accompanied by line emission from recombination and de-excitation in the ionized gas, can originate from three distinct regions of the SNR: (1) behind the forward shock (FS), where interstellar or circumstellar material is swept up and heated; (2) interior to the SNR boundary, where cold ejecta are heated by the reverse shock (RS); and (3) outside a central pulsar wind nebula (PWN), if one exists, where the slow-moving central ejecta are heated by the expanding nebula. Thermal X-ray spectra in an SNR thus provide crucial information on the surrounding environment, which may have been modified by strong stellar winds from the progenitor, as well as the ejecta that bear the imprint of the stellar and explosive nucleosynthesis in the progenitor star. With capabilities for studying the X-ray emission with high angular resolution, *Chandra* has produced breakthrough results in the study of SNRs.

Studies of the thermal X-ray emission from SNRs reveal details on the temperature, density, composition, and ionization state of the shocked plasma. These, in turn, can provide specific information on the density structure for both the ejecta and the surrounding circumstellar material, the nature of the SN explosion (CC versus Type Ia), the age and explosion energy of the SNR, the mass of the progenitor star, and the distribution of metals in the post-explosion SN. For an excellent review of X-ray emission from SNRs, including a detailed discussion of the thermal X-ray emission, the reader is referred to Vink (2012).

The fast shocks in SNRs are capable of accelerating charged particles to extremely high energies, and X-ray studies are of particular importance in revealing synchrotron emission from energetic electrons at and behind these shock fronts.

### 5.2.1 Properties of SNR Shocks

As the ejecta from an SN explosion expand into the surrounding CSM/ISM, the BW compresses and heats the ambient material. The sound speed in the ambient material is

$$c_s = \sqrt{\frac{\gamma P}{\rho}}, \tag{5.1}$$

where $\gamma$ is the adiabatic index of the gas ($\gamma = 5/3$ for an ideal monatomic gas), $P$ is the pressure, and $\rho$ is the density. For typical values in the warm ISM, $c_s \approx 10$ km s$^{-1}$, which is much lower than the SNR expansion speed. The BW thus moves supersonically, forming a so-called FS. An important characteristic of this and other shocks in SNRs is that the size scale over which the density changes is very small. The collisional mean free path for the gas particles is

$$\lambda_{\rm col} = \frac{1}{n\pi a_0^2}, \tag{5.2}$$

where $n$ is the number density and $a_0$ is the collision radius. For Coulomb collisions, the collision radius is roughly the separation at which the kinetic energy of one particle is equal to the electrostatic force from the other:

$$\frac{1}{2} m_1 v_1^2 = \frac{Z_1 Z_2 e^2}{a_0}, \tag{5.3}$$

where $Z_1 e$ and $Z_2 e$ are the charges of the colliding particles. Assuming

$$v_1 = \left(\frac{3kT}{2m_1}\right)^{1/2} \tag{5.4}$$

and that the velocity of the more massive particle is negligible, typical postshock velocities of $\gtrsim 10^3$ km s$^{-1}$ (see Section 5.2.2) yield $\lambda_{\rm col} > 3 \times 10^{19}/n$ cm $\approx 10/n$ pc. This is much larger than the observed shock widths in SNRs (and, indeed, larger than the entire SNR for most cases), demonstrating that the shocks are mediated by something other than collisions. These so-called "collisionless" shocks are apparently the results of plasma waves, although the exact character of the waves is still the subject of considerable study.

The properties of the gas flowing through the shock are illustrated in Figure 5.7 (left). Here, the shock is viewed as being stationary, and material with initial density $\rho_1$, velocity $v_1$, and pressure $P_1$ flows in from the right. The values of the downstream quantities are determined by the conservation of mass, momentum, and energy:

$$\begin{aligned} \rho_1 v_1 &= \rho_1 v_2 \\ \rho_1 v_1^2 + P_1 &= \rho_2 v_2^2 + P_2 \\ e_1 + \frac{1}{2}v_1^2 + P_1/\rho_1 &= e_2 + \frac{1}{2}v_2^2 + P_2/\rho_2, \end{aligned} \tag{5.5}$$

where $e$ is the specific internal energy. For high Mach number hydrodynamic shocks ($\mathcal{M} \equiv v_1/c_s$), the so-called "shock jump conditions" yield

$$\frac{\rho_2}{\rho_1} = \frac{v_1}{v_2} = \frac{\gamma + 1}{\gamma - 1} \tag{5.6}$$

and

$$P_2 = \frac{2\rho_1 v_1^2}{\gamma + 1}. \tag{5.7}$$

For $\gamma = 5/3$, the shock compression ratio is $r' \equiv \rho_2/\rho_1 = 4$. Shifting to a frame where the shock is propagating with a speed $v_s$ into a stationary medium, the downstream gas speed is $v_2 = 3v_s/4$. In addition,

$$T_2 = \frac{3\mu m_H v_s^2}{16k}, \tag{5.8}$$

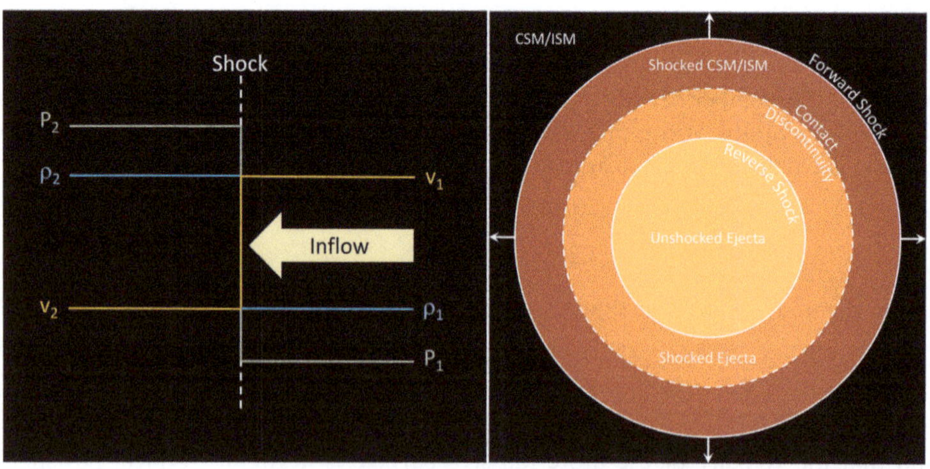


**Figure 5.7.** Left: shock jump conditions for velocity, density, and pressure. Right: schematic of the SNR structure identifying FS, RS, and CD along with locations of shocked/unshocked material.

where $k$ is Boltzman's constant and $\mu$ is the mean molecular weight of the gas ($\mu \approx 0.6$ for ionized gas of solar composition). For typical shock velocities, the postshock temperature is thus of order $10^7$ K, yielding plasmas sufficiently hot to produce significant X-ray emission.

### 5.2.2 SNR Structure

The shock driven by the expanding ejecta sweeps up a shell of shocked CSM/ISM (see Figure 5.7). As the shock decelerates due to the increasing mass of swept-up material, a pressure wave propagates supersonically back into the ejecta that are driving the expansion, forming a shock—the so-called RS. The two shocked fluids are separated by a "contact discontinuity" (CD) that is subject to Rayleigh–Taylor instabilities, resulting in a turbulent interface where mixing can occur. Studies of emission from the FS thus probe the properties of the ambient material into which the SNR is evolving, while emission from the RS probes the ejecta formed in the SN explosion.

Figure 5.8 illustrates these basic SNR structures with a *Chandra* observation of Tycho's SNR, the remnant of SN 1572. A three-color image in the left panel identifies clumpy thermal emission primarily associated with shocked ejecta, with a distinct thin outer boundary of high-energy (blue) emission defining the FS. The X-rays from the FS region are dominated by synchrotron emission from cosmic-ray electrons accelerated at the shock (see Section 5.2.4). The center panel identifies the CD with a green contour. The observed corrugations are associated with Rayleigh–Taylor instabilities at the interface. The right panel identifies Fe–K from the inner regions of the SNR, with the innermost contour identifying the position of the RS.

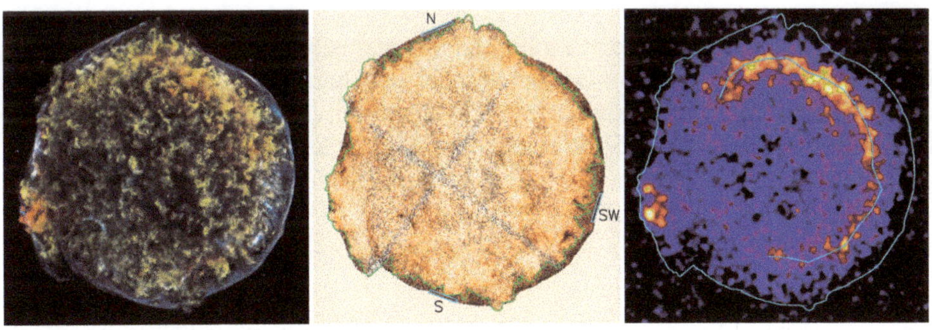


**Figure 5.8.** Left: *Chandra* image of Tycho's SNR. The red (0.95–1.26 keV) and green (1.63–2.26 keV) emission are thermal and predominantly ejecta. The blue (4.1–6.1 keV) outer boundary demarcates the FS and corresponds to nonthermal emission from accelerating particles (see Section 5.2.4). Center: Tycho image separating line-rich emission (light regions) from featureless emission (dark regions) using a principal component analysis. The CD is identified with a green contour. Right: Tycho 4–6 keV image highlighting the interior Fe–K emission. The inner contour identifies the position of the RS, and the outer contour corresponds to the FS. Reproduced from Warren et al. (2005) © 2005. The American Astronomical Society. All rights reserved. The interactive figure shows separately the X-ray emission in energy bands: 1.6–2.0 keV, red; 2.2–2.6 keV, green; 4–6 keV, blue, and the composite image, shown in the left hand panel. Additional materials available online at http://iopscience.iop.org/book/978-0-7503-2163-1.

In the earliest stages, the SNR expands freely with a velocity

$$v_{fe} \approx \sqrt{\frac{2E}{M_{ej}}} = 10^4 \left(\frac{E_{51}}{M_{ej,\odot}}\right)^{1/2} \text{ km s}^{-1}, \tag{5.9}$$

where $E_{51}$ is the kinetic energy released in the SN explosion, in units of $10^{51}$ erg, and $M_{ej}$ is the ejecta mass in solar masses. As the remnant sweeps up the surrounding medium, the RS heats the ejecta, first propagating outward but more slowly than the ejecta, but eventually propagating inward to the SNR center. In the process, the ejecta (which have initially cooled due to adiabatic expansion) are heated to X-ray-emitting temperatures by the RS.

As the SNR evolves to a characteristic age of

$$t_{ch} = E^{-1/2} M_{ej}^{5/6} \rho_0^{-1/3}, \tag{5.10}$$

where $\rho_0$ is the density of the ambient medium, the evolution transitions from a phase in which the ejecta dominate the total mass to one in which the swept-up gas dominates, and the expansion law transitions from $R_s \propto t$ associated with free expansion, to $R_s \propto t^{2/5}$. During this evolutionary phase, the SNR properties can be described by the Sedov–Taylor similarity solution (Taylor 1950; Sedov 1959), which gives

$$R_s = 1.15 \left(\frac{E}{\rho_0}\right)^{1/5} t^{2/5} = 5.9 \left(\frac{E_{51}}{n_0}\right)^{1/5} t_3^{2/5} \text{ pc}, \tag{5.11}$$

where $n_0$ is the number density of the ambient medium and $t_3$ is the SNR age in units of $10^3$ yr. (For a more complete treatment of the evolution, see Truelove & McKee 1999).

Many young remnants are at evolutionary stages between the free-expansion and Sedov–Taylor phases, with an expansion index $m$ (where $R_s \propto t^m$) that falls between 1 and 2/5. For the general case where the ejecta density is described by a power law in radius $\rho_{ej} \propto r^{-n}$ and the SNR is expanding into the ambient density $\rho_{amb} \propto r^{-s}$ (e.g., $s = 0$ for a uniform medium, and $r = 2$ for a bubble blown by a progenitor wind), the BW radius evolves as

$$R_s \propto t^{\frac{(n-3)}{(n-s)}}.$$

A measurement of the expansion index thus provides important constraints on the properties and evolutionary state of the SNR.

The rapid shock speeds described above correspond to observable expansion on the sky, with an angular rate of $\dot{\theta} = 0.21 v_3 d_{kpc}^{-1}$ arcsec yr$^{-1}$, where $v_3$ is the expansion velocity in units of $10^3$ km s$^{-1}$ and $d_{kpc}$ is the distance in kiloparsecs. With its exceptional angular resolution, *Chandra* is able to measure the expansion rates for a number of young SNRs. Observations of G1.9+0.3, the youngest SNR known in the Galaxy, show a mean expansion rate of ~0.6% yr$^{-1}$ (Carlton et al. 2011), but also show evidence for large spatial variations, with motions varying by a factor of 5 throughout the SNR (Borkowski et al. 2017). Measurements for Tycho's SNR reveal proper motions of 0.2″–0.4″ $yr^{-1}$ for the FS, giving an expansion index of $m = 0.33 - 0.65$, indicating that the SNR is approaching the Sedov–Taylor phase of evolution (Katsuda et al. 2010). Figure 5.9 (left) shows the brightness profile of the FS region in the western region of Tycho's SNR from observations taken in 2000 (red) and 2015 (blue) from expansion measurements made by Williams et al. (2016), who found azimuthal variations that indicate a density gradient in the ambient medium.

Velocity measurements along the line of sight can also be made with *Chandra* for sufficiently rapid motions, through Doppler shifts of spectral lines. For Tycho's SNR, direct ejecta velocity measurements reveal both redshifted and blueshifted clumps with velocities of $\lesssim 7800$ km $s^{-1}$ and $\lesssim 5000$ km $s^{-1}$, respectively, as illustrated in Figure 5.9 (right; Sato & Hughes 2017a). Here, the centroid energies of the Si–He $\alpha$ lines (corresponding to the $n = 3{-}2$ transition in He-like Si, i.e., Si with all but two electrons stripped) are shown for positions around the SNR shell. The shifts are large at the SNR center and decrease with radius as expected for Doppler shifts from an expanding shell.

*Electron–Ion Equilibration*

As described above, in the simplest picture, all charged particles that go through the SNR shock attain a velocity $v = 3v_s/4$, where $v_s$ is the shock velocity. Because the associated kinetic temperature scales with mass (Equation (5.8)), the result is that

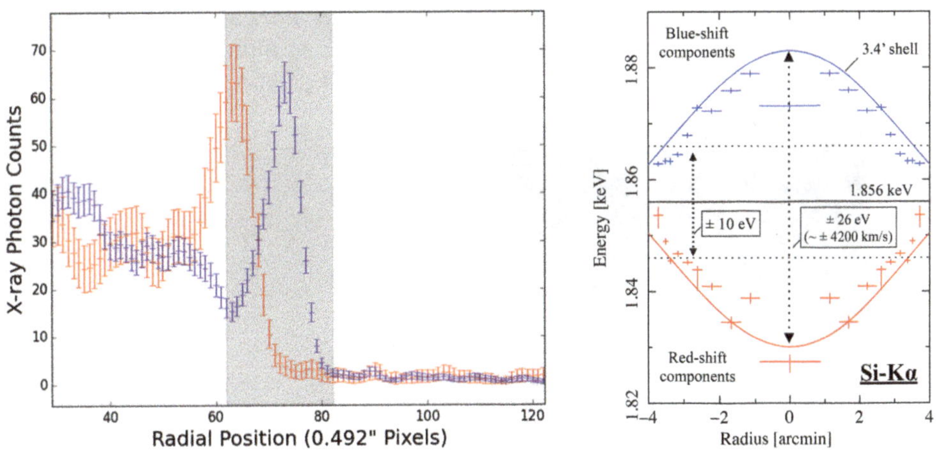


**Figure 5.9.** Left: radial profiles of the emission from a western region in Tycho's SNR. The red (blue) profile is from an observation in 2000 (2015). Reproduced from Williams et al. (2016). © 2016. The American Astronomical Society. All rights reserved. Right: radial dependence of the Si–He $\alpha$ energies of the red- and blueshifted shell components of Tycho. Reproduced from Sato & Hughes (2017a). © 2017. The American Astronomical Society. All rights reserved. The video shows a timelapse movie of the X-ray emission (color-coded by energy band: 0.95–1.26 keV, red; 1.63–2.26 keV, green; 4.1–6.1 keV, blue) from the Tycho supernova, observed with *Chandra* from 2000–2015. The start and end correspond to the radial profiles shown in the left hand panel. A background optical star field is also included in the video. Additional materials available online at http://iopscience.iop.org/book/978-0-7503-2163-1.

the electron temperature is initially much lower than that of the ions. Because the ions comprise the bulk of the mass swept up by the SNR, it is the ion temperature that characterizes the dynamical evolution. However, the temperature determined from X-ray measurements is that of the electrons.

On the slowest scales, Coulomb collisions between electrons and ions will bring the populations into temperature equilibrium. More rapid plasma processes may result in faster equilibration, and often the two extreme cases of instantaneous temperature equilibration and Coulomb equilibration are considered in order to investigate the boundaries of the problem. SNR expansion velocities can be used to estimate the ion temperatures for some remnants, and the estimated temperatures are much higher than the observed electron temperatures, suggesting slow temperature equilibration.

For example, expansion measurements of Cas A give a shock velocity of $\sim$5000 km s$^{-1}$ (DeLaney et al. 2004), corresponding to a mean ion temperature $T_i = 3.4 \times 10^8$ K. Assuming no additional heating of the electrons, the expected temperature (replacing $\mu m_H$ with $m_e$ in Equation (5.8)) is $T_e = 3.5 \times 10^5$ K. Based upon spectra from *Chandra* observations, the temperature of the CSM material behind the FS (Hwang & Laming 2012) is $T_e \approx 2.6 \times 10^7$ K($kT_e \approx 2.2$ keV, demonstrating that while significant heating of the electrons has occurred, full electron–ion temperature equilibration has not yet been reached. In connecting X-ray measurements to the SNR evolution, it is thus clear that an understanding of the electron–ion temperature ratio is required.

In a small number of cases, measurements of Balmer lines at the FS can provide the proton temperature and the degree of electron–ion equilibration, and there is a significant trend showing larger $T_e/T_p$ values for lower shock velocities. This is

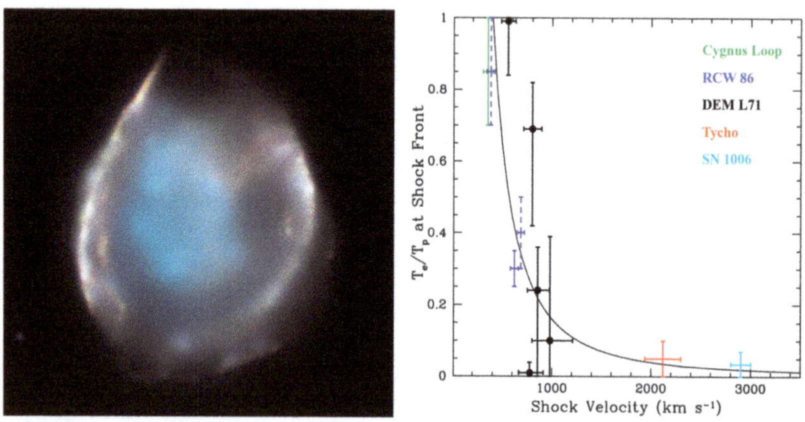


**Figure 5.10.** Left: *Chandra* image of DEM L71, a Type Ia SNR in the LMC. The energy bands for the image are: 0.3–0.7 keV (red), 0.7–1.1 keV (green), 1.1–4.2 keV (blue). The outer rim is dominated by swept-up ISM material, while the central regions are dominated by ejecta enriched in Si and Fe. Right: electron to proton temperature ratio at the shock front as a function of shock velocity. Black dots correspond to measurements from different rim locations in DEM L71. Reproduced from Ghavamian et al. (2007). © 2007. The American Astronomical Society. All rights reserved. The interactive figure shows separately the X-ray emission at low energy (0.3–0.7 keV, red), medium energy (0.7–1.1 keV, green), high energy (1.1–4.2 keV, blue) bands, and the composite image, as shown in the left hand panel. Additional materials available online at http://iopscience.iop.org/book/978-0-7503-2163-1.

demonstrated in studies of DEM L71 (Figure 5.10, left), for which variations in the FS radius at different azimuths correspond to different shock velocities. The $T_e/T_i$ ratios derived from values of $T_i$ based on optical spectra and of $T_e$ based on spectra from *Chandra* (Rakowski et al. 2003) are shown in Figure 5.10 (left), along with measurements based on several other SNRs (Ghavamian et al. 2007).

An additional prediction of Equation (5.8) is that different ion species will have different postshock temperatures, with the timescale for temperature equilibration depending on the exact heating process. The width $\sigma_D$ of the emission lines from such ions will be Doppler broadened, providing a way of determining the individual temperatures:

$$\sigma_D = \frac{E}{c}\sqrt{\frac{kT_i}{m_i}}, \tag{5.12}$$

where $E$ is the natural line energy, and $T_i$ and $m_i$ are the temperature and mass of the ion. Miceli et al. (2019) used high-resolution spectra from SN 1987A taken with the *Chandra* HETG to show that the ratio of the ion temperature to the proton temperature exceeds one and increases with ion mass, showing that equilibration between ions has not yet been reached in the shocked plasma.

Future high-resolution X-ray measurements of thermal line broadening offer the promise of direct measurements of both $T_e$ and $T_p$, as well as that for different ion species, in a larger sample of SNRs, providing stronger constraints on models for the heating process.

*Nonequilibrium Ionization*

Due to the high temperatures in postshock gas, high ionization states are produced. However, because the ionization proceeds through collisions, and because the gas density is relatively low, it can take a considerable amount of time before the gas reaches the collisional equilibrium ionization (CIE) state expected for its temperature. In the immediate postshock region, and for young SNRs, the gas is in a nonequilibrium ionization (NEI) state. The approach to ionization equilibrium is characterized by the ionization timescale parameter $\tau = n_e t$. The plasma reaches CIE for $\tau \gtrsim 10^{12.5}$ cm$^{-3}$ s; for smaller values, the plasma is in an underionized NEI state. This is illustrated in Figure 5.11. The left panel, from Hughes & Helfand (1985), shows the ionization fraction for different ion states of oxygen as a function of $\tau$, for a temperature of $10^{6.5}$ K. In equilibrium, the gas is largely in the fully ionized state, while for $\tau = 10^{11}$ cm$^{-3}$ s it is dominated by O VII and O VIII, the He-like and H-like states with all but two or one electrons removed. The right panel shows a portion of the soft X-ray spectrum (folded through the ACIS-S detector response) for this plasma for different values of $n_e t$. It is clear from this that determination of the plasma temperature based on relative line strengths must thus account for NEI conditions in the plasma. While the limited spectral resolution provided by CCD detectors obviously obscures details of specific line ratios, spectral fits to the spectrum can easily identify gross differences of the ionization state.

The progression of the ionization state in the evolving postshock gas in SNRs is readily observed. Early *Chandra* observations of 1E 0102.2–7219, for example, show

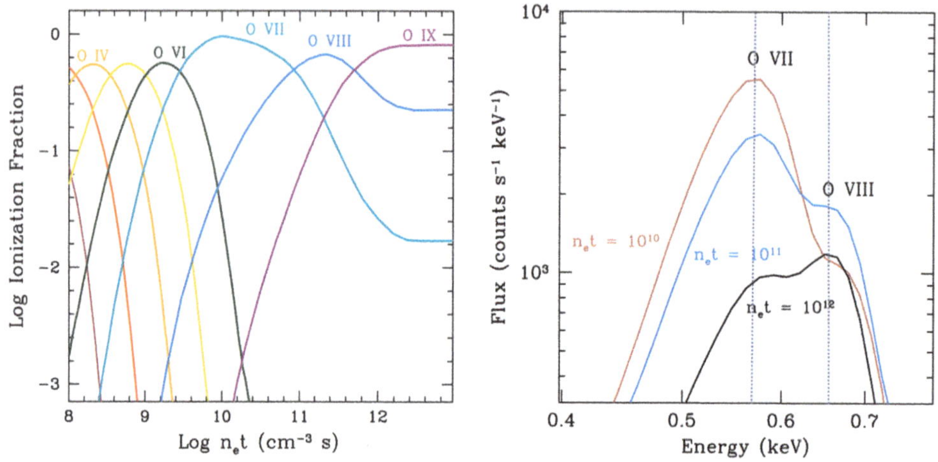


**Figure 5.11.** Left: ionization fraction for different ion states of oxygen as a function of $\tau = n_e t$, for a temperature of $10^{6.5}$ K (reproduced from Hughes & Helfand 1985 © 1985. The American Astronomical Society. All rights reserved). Right: evolution of X-ray spectrum (folded through *Chandra* ACIS-S detector response) with ionization age, emphasizing lines from O VII and O VIII (the centroids of which are indicated by the dashed vertical lines). The plasma temperature is $10^{6.5}$ K, and the units for $n_e t$ are $cm^{-3}\,s^{-1}$.

that the peak of the O VII emission is found at a smaller radius than that of O VIII, consistent with the expectation that the ionization state of ejecta most recently encountered by the inward-propagating RS lags behind that of the ejecta that have been shocked earlier (Gaetz et al. 2000). The same effect is evident in shocked circumstellar material in G292.0+1.8, where the ionization state of the emission directly behind the FS is observed to be considerably lower than that for regions farther downstream (Lee et al. 2010). Such measurements thus provide constraints on the thermal history of the shocked material as well as on variations in the density structure.

While conditions of underionization are found in many SNRs, recent X-ray observations have identified exactly the opposite situation in a growing number of remnants. In W49B, for example, Kawasaki et al. (2005) used the relative intensity of H-like and He-like lines of Ar and Ca from *ASCA* observations to determine an ionization temperature that is higher than $T_e$, indicating an overionized plasma. Ozawa et al. (2009) used *Suzaku* observations of W49B to identify a distinct radiative recombination continuum (RRC) feature that confirms this overionized state. Using *Chandra* observations, Lopez et al. (2013) mapped the distribution of the overionized plasma and concluded that these are associated with regions where adiabatic expansion into regions of reduced density allows the gas to cool rapidly, lowering the plasma temperature to a value below its ionization state in equilibrium. Similar spectral signatures are seen in other SNRs and appear to be particularly evident in so-called mixed-morphology SNRs with central regions that outshine the outer shells. The exact cause of such morphology is still under investigation, although models invoking expansion into a medium infused with cold clouds that slowly evaporate in the SNR interior after being overtaken by the FS have had some success in reproducing some of the overall properties (White & Long 1991). MHD

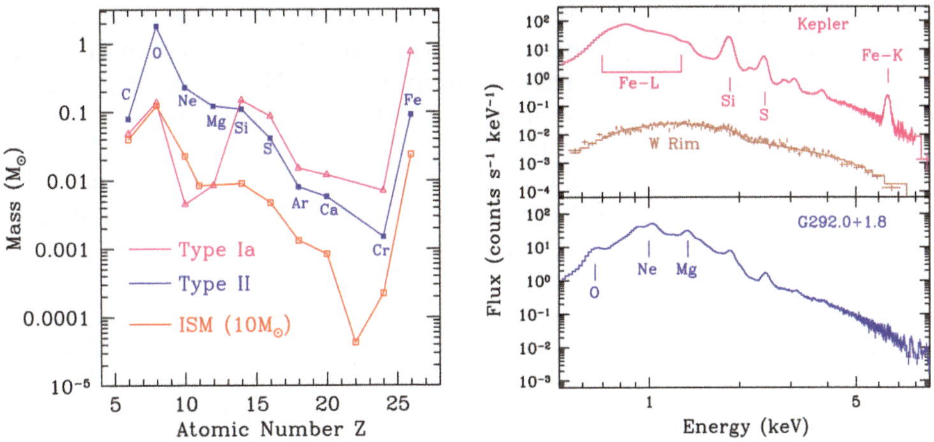


**Figure 5.12.** Left: mass of ejecta components, by atomic number, in representative CC and Type Ia SNe. For comparison, the mass contained in 10 $M_\odot$ of solar-abundance material is also shown. Right: thermal X-ray spectra from a Type Ia remnant (Kepler's SNR, top) and CC remnant (G292.0+1.8, bottom), illustrating the significant difference in Fe and O/Ne content. The lower spectrum in the Kepler panel is extracted from a portion of the western rim, and the histogram represents a pure power-law fit to the data, indicating synchrotron radiation from highly relativistic electrons.

models indicate that thermal conduction is important in such a scenario (Slavin et al. 2017) and the transport of heat in the SNR interior may also play a role in producing an overionized plasma.

### 5.2.3 SNR Ejecta and Constraints on Progenitors

The very different stellar evolution histories and explosion processes for Type Ia and CC SNe result in distinct signatures in the shock-heated ejecta of SNRs. Type Ia events, corresponding to the complete disruption of a C/O WD star, produce more than 0.5 $M_\odot$ of Fe-group elements (including Cr, Mn, Fe, and Ni), accompanied by a significant contribution of intermediate-mass elements (e.g., Si, S, Ar, Ca). CC SNe, on the other hand, are dominated by materials synthesized during the stellar evolution of the massive progenitor—particularly O, and also Ne—with additional products from explosive nucleosynthesis in the innermost regions surrounding the collapsed core. As illustrated in Figure 5.12 (left), where we plot the mass distributions for key nucleosynthesis products for characteristic Type Ia and CC events (Iwamoto et al. 1999), the former are dominated by Fe while the latter contain much larger amounts of O. For comparison, the total mass of these elements contained in 10 $M_\odot$ of swept-up material with solar abundances is also shown. Particularly at young ages when the total amount of mass swept up by the FS is not exceedingly high, the thermal X-ray spectra from such remnants provide rich information about SN ejecta and thus on the nature of the progenitor.

*Remnants of Type Ia Supernovae*

In Figure 5.12 (right), we compare the spectrum from Kepler's SNR (top), a Type Ia SNR (see below), with that from the CC SNR G292.0+1.8 (bottom). The dominant flux just below 1 keV in Kepler is largely from Fe–L emission, characteristic of the large amount of Fe created in such events, while the spectrum from G292.0 +1.8 shows strong emission features from O and Ne (Park et al. 2004). Identification and modeling of such spectral features identified in *Chandra* observations have been used to identify the SN type that produced numerous SNRs. For example, the LMC remnant DEM L71 (Figure 5.10) shows a double-shock morphology with an outer BW dominated by LMC abundance material and a central (blue) region rich in Si and Fe ejecta, with mass estimates $M_{\rm Si} \sim 0.12\ M_\odot$ and $M_{\rm Fe} \sim 0.8\ M_\odot$, consistent with expectations for a Type Ia event (Hughes et al. 2003).

The LMC SNR 0509−67.5 presents a similar case (Warren & Hughes 2004), and comparisons of the abundances and ionization properties with SN Ia models appear to favor a delayed-detonation explosion for this remnant, with a progenitor similar to that of the bright, highly energetic SN 1991T (Badenes et al. 2008), a result subsequently confirmed with light-echo spectra from the original event (Rest et al. 2008). Expansion measurements, possible with the superb angular resolution of *Chandra*, show an expansion speed of $7500 \pm 1700$ km s$^{-1}$, consistent with optical measurements (Roper et al. 2018). These, along with SNRs 0534−69.9 and 0544 −70.4 (Hendrick et al. 2003), demonstrate that Fe-rich ejecta from SN Ia events are measurable in their SNRs at ages as large as $10^4$ yr and distances as large as those of the Magellanic Clouds, and confirm our ability to characterize properties of SNe through X-ray studies of their remnants even thousands of years after their explosions.

*Chandra* studies that classify Kepler's SNR (the remnant of SN 1604) as a Type Ia remnant based on the dominant Si, S, and Fe emission are particularly important (Reynolds et al. 2007). The remnant displays a distinct enhancement of shocked CSM in the north (Figure 5.13, left), suggesting interaction with mass lost from a fairly massive progenitor—either that of the WD or a companion star. This may provide evidence for SD progenitor for this system. *Chandra* proper motion measurements for particular knots within the SNR show angular motions of 0.11″–0.14″ yr$^{-1}$ (Sato & Hughes 2017b). In addition, spectra show Doppler shifts indicating radial velocities of ~9000–10,000 km s$^{-1}$ (Figure 5.13, right). Moreover, measurements of the Ni-to-Fe line flux ratio indicates a progenitor with supersolar metallicity, suggesting that the SN resulted from a WD that exploded through a relatively prompt channel, corresponding to a progenitor lifetime much shorter than that for a typical single star with a WD evolutionary state (Park et al. 2013).

Tycho's SNR (Figure 5.8) is another Type Ia remnant showing rich ejecta structure surrounded by a mostly circular BW. We discuss details of several important characteristics of its structure below.

Additional important constraints on the nature of Type Ia SNe come from LMC remnants DEM L238 and DEM L249, the ionization structure of which indicates Fe ejecta densities expected only from higher mass progenitors (Borkowski et al. 2006), and the SMC SNR 0104−72.3, for which dense clouds to the east also suggest a young Ia progenitor. Similarly, the Galactic SNR G272.2−3.2 shows a clumpy CSM shell from a complex environment with dense, dusty molecular clouds (McEntaffer et al. 2013), providing additional evidence for a prompt Ia population.

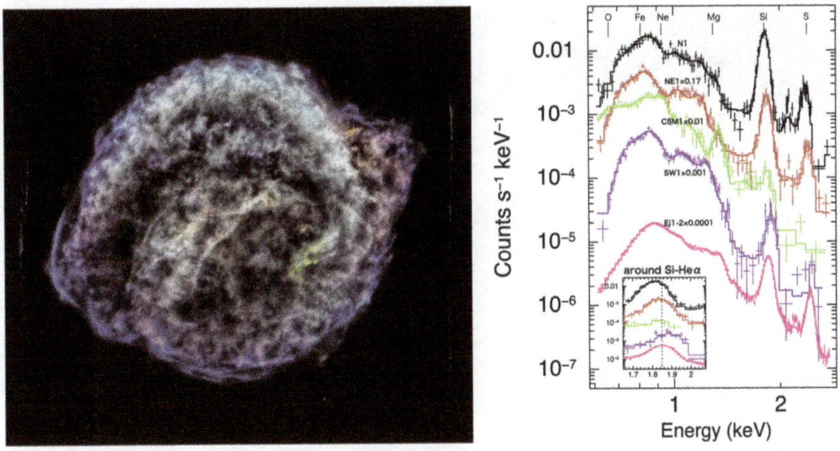


**Figure 5.13.** Left: *Chandra* image of Kepler's SNR. The enhanced emission in the north originates from swept-up circumstellar material, while the dark blue rims are associated with nonthermal emission from electrons accelerated at the FS. Right: spectra for a collection of knots from Kepler's SNR (reproduced from Sato & Hughes 2017b © 2017. The American Astronomical Society. All rights reserved). The spectrum shown in green corresponds to a CSM-dominated knot, while the others are ejecta dominated. The inset shows the Si–He $\alpha$ line, illustrating different Doppler shifts for different knots.

*Remnants of Core-collapse Supernovae*

The early development of SN explosions can imprint signatures on the SNRs that they form. Studies of the spatial distribution of ejecta can thus provide evidence of asymmetries and mixing in these events. X-ray studies of Cas A provide a particularly important example. Cas A (Figure 5.14) is a ~340 yr old SNR for which the O-rich optical spectra establish a CC origin—a Type IIb SN based on studies of light-echo spectra (Krause et al. 2008). X-ray observations show distinct evidence of Fe ejecta in the outermost regions of the remnant (Hughes et al. 2000), despite the expectation that Fe is produced in the regions closest to the remnant core. This large-scale disruption of the ejecta layers has been studied in detail by Hwang & Laming (2012), who performed fits to over 6000 X-ray spectra in Cas A and found distinct examples of regions where Fe is accompanied by other products of incomplete Si burning, along with others that are nearly pure Fe, presumably produced in regions of $\alpha$-rich freezeout during complete Si burning. They conclude that nearly all of the Fe in Cas A is found outside the central regions of the remnant, apparently the result of hydrodynamic instabilities in the explosion. Figure 5.14 (left) shows the *Chandra* image of Cas A, with emission from different elements displayed in different colors (right). Plumes of Fe (purple) extending to the southeast are evident, as are general differences in the distributions of other elements, providing constraints on the ejecta kinematics.

Curiously, observations with *NuSTAR* reveal the presence of $^{44}$Ti in the central regions of Cas A (Grefenstette 2017), but with a different spatial distribution than that of the Fe emission, despite the expectation that these elements should be formed in the same regions during explosive nucleosynthesis. The observed X-rays from

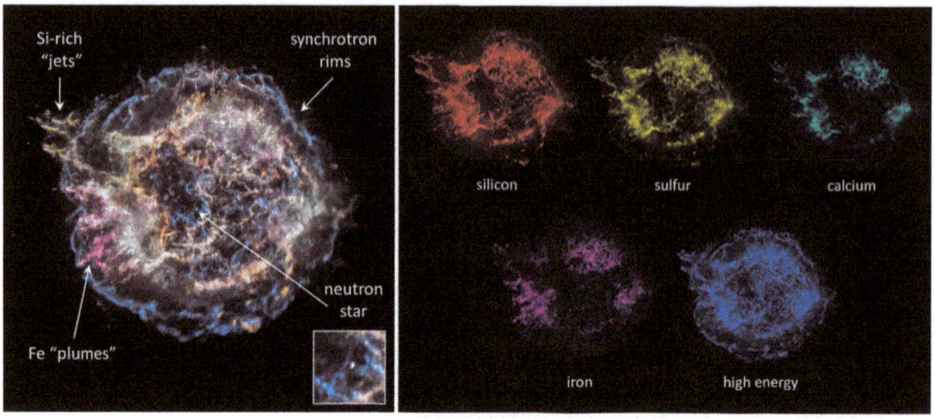


**Figure 5.14.** *Chandra* image of Cas A (left) with emission from different elements presented in different colors. Regions with plumes of Fe extending outside of the lighter elements and with Si-rich jet-like structures are identified, as is the central NS and an example of the synchrotron-emitting rims associated with electrons accelerated to cosmic-ray energies by the FS. The individual element images (selected from energy bands centered on emission lines of the identified elements, but also containing some emission from other elements, due to the finite spectral resolution of the detector) are shown at the right. Courtesy of NASA/CXC/SAO. The interactive figure shows the separate elements in the Cassiopeia A supernova remnant observed in the X-rays with *Chandra* (silicon, red; sulfur, yellow; calcium, green; iron, purple), the high energy emission (blue), as shown in the right hand panel, and the composite image, as shown in the left hand panel. The 3D model of Cas A based on Doppler shift measurements from *Chandra* HETG measurements of Si VIII (Lazendic et al. 2006) along with *Chandra* ACIS measurement of Fe–K and *Spitzer* infrared spectra from [Ne II], [Ar II], and [Si II] DeLaney et al. (2010). The 3D model of the Cassiopeia A supernova remnant can be rotated to view the SNR from any angle and different aspects of the object, such as elemental concentrations and morphological features, can be toggled to explore the complex properties of the remnant. Additional materials available online at http://iopscience.iop.org/book/978-0-7503-2163-1.

$^{44}$Ti are from radioactive decay, meaning that both shocked and unshocked $^{44}$Ti can be observed, while X-ray emission probes only shocked Fe. It is thus possible that $^{44}$Ti regions lacking Fe signatures reside interior to the RS. In addition, the temperature dependence for $^{44}$Ti and Fe yields differ, meaning that some regions showing Fe without detected $^{44}$Ti may have originated from explosion regions where the Fe yields were higher.

A census of the ejecta in Cas A indicates a total mass of about 3.5 $M_{\odot}$, of which $\sim$0.1 $M_{\odot}$ is Fe. Distinct "jet/counterjet" structures enriched in Si are evident in optical and X-ray observations. While not true jets of relativistic material, these structures suggest significant asymmetries that may be related to rotation or binarity. Spatially resolved X-ray spectra from the outermost regions of Cas A, obtained with *Chandra*, show that the density profile of the ambient CSM is consistent with $\rho \propto r^{-2}$, indicative of expansion into a steady progenitor wind with $\dot{M} \gtrsim 10^{-5}\, M_{\odot}^{-1}\,\mathrm{yr}^{-1}$ and $v_{\mathrm{w}} \approx 10\ \mathrm{km\ s^{-1}}$ (Lee et al. 2014).

In the centermost regions, *Chandra* observations of Cas A reveal an NS. Unlike the bright pulsar that powers the Crab Nebula (see Section 5.3.1), the NS in Cas A is decidedly weak, shows no evidence of a surrounding wind nebula, and has no counterpart in any other wavelength band. The X-ray luminosity is only $L_{\mathrm{x}} \sim 4 \times 10^{33}\ \mathrm{erg\ s^{-1}}$, with a characteristic temperature of $\log T \sim 6.3$ (Posselt et al. 2013). Modeling of the *Chandra* spectrum suggests a low-magnetic-field NS covered with an atmosphere of carbon, with emission emerging from the entire surface, potentially explaining the lack of any detected pulsations that might be expected from hot polar caps (Ho & Heinke 2009). Following its discovery, similar such "central compact objects" have been observed in ~10 other SNRs, identifying a class of young NSs with apparently low magnetic fields, possibly being powered in part by SN fall-back accretion.

A second exceptional example of a CC SNR is the O-rich remnant G292.0+1.8. Its age is ~3000 yr, based on expansion rates of fast-moving ejecta knots seen in the optical band. *Chandra* observations (Figure 5.15) reveal X-rays dominated by O, Ne, and Mg ejecta in the central regions, with a complex distribution of filaments and knots, along with a central bar of circumstellar material that may be associated with an equatorial ring of material ejected in earlier stages of the progenitor's life. The outer rim of shocked material has CSM composition as well, and the density profile inferred from *Chandra* measurements is consistent with $\rho \propto r^{-2}$, as expected from expansion into the progenitor wind (Lee et al. 2010).

Slightly offset from the dynamical center of the SNR is a young pulsar (PSR J1124−5916) for which *Chandra* observations reveal a compact jet/torus structure inside a larger extended PWN. The offset from the SNR center is presumably associated with a kick produced in the SN explosion and corresponds to a velocity of

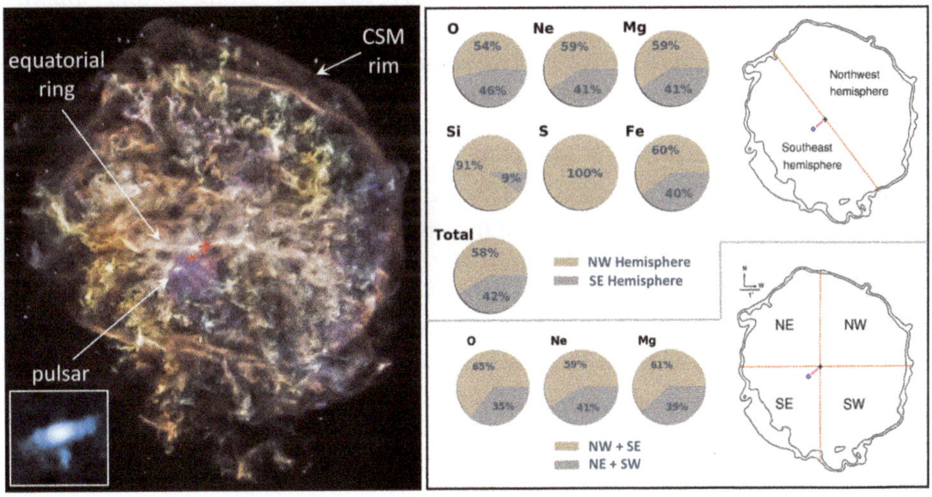


**Figure 5.15.** Left: *Chandra* image of G292.0+1.8. The energy bands for the different colors, with the primary emission contributions, are red = 0.58–0.95 keV (O), orange = 0.98–1.1 keV (Ne), green = 1.28–1.43 keV (Mg), and blue = 1.81–2.62 keV (Si/S/synchrotron). Compare with the integrated spectrum shown in Figure 5.12. Arrows identify the central equatorial ring, a rim of CSM in the NW, and the central pulsar and its surrounding jet/torus structure. The red cross marks the dynamical center of the SNR based on kinematics of optical ejecta, and the red arrow points in the direction of the NS. Right: fractional distribution of different elements by hemisphere (upper) and opposing quadrants (lower), illustrating an asymmetry with larger amounts of ejecta opposite the inferred direction of the NS kick. Adapted from Bhalerao et al. (2019).  Additional materials available online at http://iopscience.iop.org/book/978-0-7503-2163-1.

$\sim$440 km s$^{-1}$ (Hughes et al. 2001). The jet axis, which defines the pulsar rotation axis, is oriented somewhat east of the north–south direction, providing evidence against the jet/kick alignments inferred for some other pulsar systems.

Measurements of the ejecta mass distribution (Bhalerao et al. 2019) show significant variations between elements, with the O, Ne, and Mg material associated with hydrostatic nucleosynthesis during the evolution of the progenitor being distributed broadly across the SNR, though enhanced in the NW and SE quadrants, with the Si, S, and Fe produced in the explosive nucleosynthesis process during the explosion being concentrated primarily in the NW (Figure 5.15, right). Both distributions suggest an association with the kick-velocity axis, with the explosive products concentrated largely in a direction opposite the NS kick.

An example of an O-rich SNR outside of our galaxy is 1E 0102.2–7219, the brightest remnant in the SMC. *Chandra* images (Figure 5.16, left) show a roughly circular outer shell with a radius of 22″ corresponding to 6.3 pc at a distance of 60 kpc. Interior to the BW is a bright region dominated by O, Ne, and Mg ejecta. The progression of the ionization state (see Section 5.2.2) suggests that the RS may be propagating into a density gradient (Gaetz et al. 2000). HETG observations (Figure 5.16, right) show H-like and He-like lines from O, Ne, Mg, and Si, with very weak emission from Fe (Flanagan et al. 2004). The dispersed images indicate Doppler shifts of $\sim$1000 km s$^{-1}$ and suggest a ring-like or cylindrical structure inclined to the line of sight. Redshifted and blueshifted regions are spatially segregated, inconsistent with the morphology expected from simple spherical expansion.

A census of the ejecta from deep *Chandra* observations shows abundance ratios of O/Ne, O/Mg, and Ne/Mg consistent with a $\sim$40 $M_\odot$ progenitor (Alan et al. 2019) and a spatially resolved ejecta structure that again argues for an asymmetric distribution. Measurements of the SNR expansion by comparing images taken over the course of 17 years show a BW expansion speed of $v_{\rm BW} = (1.6 \pm 0.4) \times 10^3$ km s$^{-1}$, and modeling of the overall dynamical structure indicates a total swept-up mass of 26–52 $M_\odot$ (22–40 $M_\odot$) and $M_{\rm ej}$ = 2–6 $M_\odot$ (2–3.5 $M_\odot$), for cases with constant (wind-like) ambient density profiles (Xi et al. 2019). These models indicate that the RS has not yet reached the remnant center, leaving enough unshocked material to potentially explain the unusual lack of detected Fe in 1E 0102.2–7219,

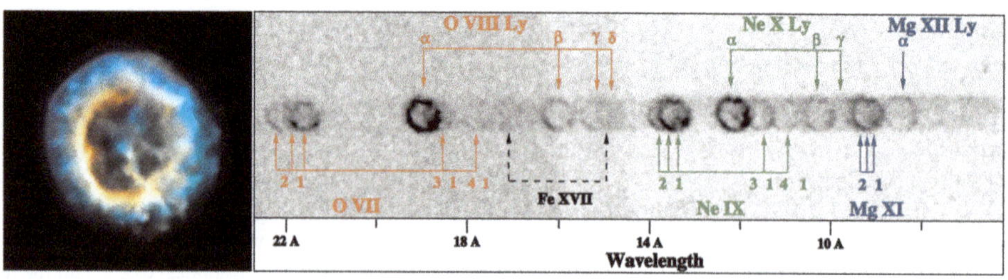


**Figure 5.16.** Left: *Chandra* image of 1E 0102.2–7219 revealing the outer shell of the shocked CSM with a bright inner ring of shocked ejecta. Right: HETG spectrum of 1E 0102.2–7219 showing dispersed images in H-like and He-like lines of O, Ne, and Mg, along with other line features. Reproduced from Flanagan et al. (2004). 

although the models, being spherically symmetric, are only approximations to the actual structure of this SNR.

The large-scale distribution of SNR ejecta has been investigated for a large number of other SNRs by Lopez et al. (2011), who found that the thermal X-ray emission from CC remnants shows a lower degree of spherical symmetry and mirror symmetry than for remnants of Type Ia SNe (Figure 5.17, left). This may indicate that CC SNe evolve in more asymmetric environments, or perhaps that the events themselves are asymmetric. The "jet"-like structures in Cas A and the distribution of the ejecta in G292.0+1.9 provide evidence for some degree of asymmetry in the explosions themselves, which is important for understanding the underlying explosion models.

The ionization structure of an SNR can also provide clues to the nature of the progenitor systems. Figure 5.17 (right) shows the Fe–K centroid energy and luminosity for a collection of Milky Way and LMC remnants (Yamaguchi et al. 2014). A clear bifurcation is observed between CC SNRs (blue) and Type Ia events (red), with the CC remnants showing more highly ionized ejecta, displaying Fe–K centroids higher than 6.55 keV. Models indicate that this is primarily a result of the ambient medium density, with Type Ia SNRs evolving into regions that have not significantly modified their surroundings (Yamaguchi et al. 2014), while CC SNR evolve into regions strongly modified by mass loss, resulting in stronger RSs that produce more highly ionized ejecta (Patnaude et al. 2015).

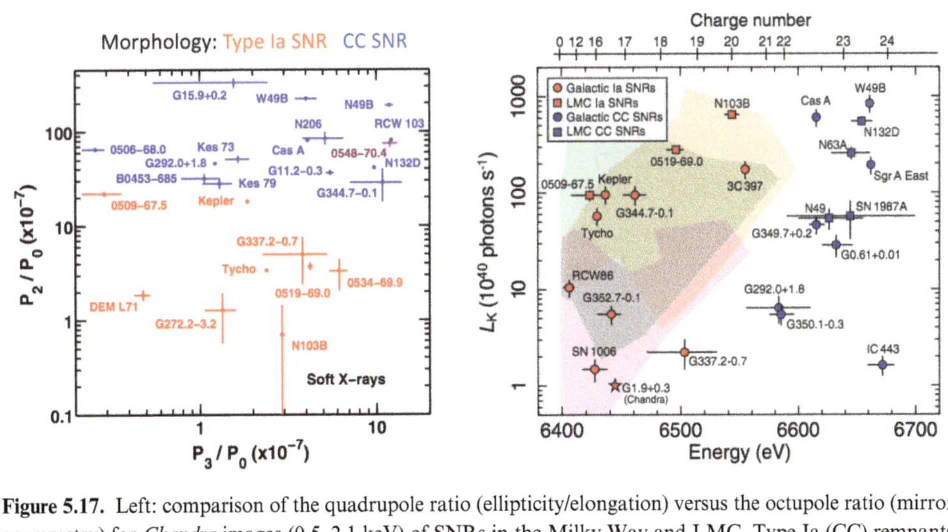


**Figure 5.17.** Left: comparison of the quadrupole ratio (ellipticity/elongation) versus the octupole ratio (mirror asymmetry) for *Chandra* images (0.5–2.1 keV) of SNRs in the Milky Way and LMC. Type Ia (CC) remnants are shown in red (blue). Reproduced from Lopez et al. (2011). © 2011. The American Astronomical Society. All rights reserved. Right: Fe–K luminosity versus line centroid for Type Ia (red) and CC (blue) SNRs in Milky Way and LMC, from *Suzaku* and *Chandra* spectra. Shaded regions correspond to theoretical Type Ia SNR models (DDTa: green, DDTg: magenta, PDD: orange) expanding into uniform-density environments. Reproduced from Yamaguchi et al. (2014). © 2014. The American Astronomical Society. All rights reserved.

### 5.2.4 Cosmic-Ray Acceleration in SNRs

In addition to thermal heating of the swept-up gas, some fraction of the SNR shock energy density can also go into the production of relativistic particles. This is readily observed through the radio emission from SNRs, which is synchrotron radiation from electrons with $\gtrsim$GeV energies. However, in the past two decades, it has become evident that many young SNRs produce X-ray synchrotron radiation from electrons with energies exceeding tens of TeV. The emission is readily observed at high energies, above the thermal emission from the remnants, and is generally concentrated in thin rims at the FS. Examples include Tycho's SNR (Figure 5.8, left), Kepler's SNR (Figure 5.13, right), and Cas A (Figure 5.14, left).

The particle acceleration most likely proceeds through diffusive shock acceleration (DSA), wherein energetic particles streaming away from the shock form turbulent waves which act to scatter other particles back toward the shock. Subsequent reacceleration builds up a nonthermal population of high-energy particles, with the maximum energy being limited by radiative losses, the age of the SNR, or particle escape. Such particle acceleration by SNR shocks has long been suggested as a process by which cosmic rays are produced, at least up to the "knee" of the cosmic-ray spectrum at $\sim 10^{14}$–$10^{15}$ eV.

In DSA, particles scatter off of MHD waves in the background plasma that are either pre-existing or generated by the streaming ions themselves. The scattering mean free path is

$$\lambda_{\rm mfp} = \eta r_{\rm g} = \eta\left(\frac{E}{eB}\right), \tag{5.13}$$

where $r_{\rm g}$ is the particle gyroradius and $\eta = (\delta B/B)^2 \geqslant 1$ is the so-called gyrofactor. The associated diffusion coefficient is

$$D = \frac{\lambda_{\rm mfp} c}{3} = \frac{\eta E c}{3eB}. \tag{5.14}$$

The case of Bohm diffusion corresponds to $\eta = 1$, where the magnetic field is totally random on the scale of $r_{\rm g}$. The energy gain per shock crossing in DSA is

$$\frac{\Delta E}{E} \approx \frac{4}{3}\left(\frac{r' - 1}{r'}\right)\frac{v_{\rm s}}{c}, \tag{5.15}$$

where $r'$ is the shock compression ratio. This leads to a nonthermal particle distribution function,

$$N(E) \propto E^{-\frac{r'+2}{r'-1}}. \tag{5.16}$$

For a compression ratio $r' = 4$, the particle spectrum thus goes as $E^{-2}$, which is close to the observed spectrum for Galactic cosmic rays.

At the very high shock speeds in SNRs, particle acceleration can proceed to extremely high energies. The above description holds in the so-called "test particle" limit where the energy of the accelerated particles is dynamically unimportant. If the relativistic particle component of the energy density becomes comparable to that of the thermal component, the shock acceleration process can become highly nonlinear. The gas becomes more compressible, which results in higher magnetic fields and enhanced acceleration. The resulting efficient production of nonthermal particles has a significant impact on the dynamical evolution of the shock. In particular, efficient particle acceleration robs energy from the thermal gas, resulting in a slower expansion and, thus, a lower temperature. This is illustrated in Figure 5.18 (left), from Ellison et al. (2007), which shows the SNR temperature as a function of radius at an age of 500 yr, from simulations assuming expansion into a uniform density $n_0 = 0.1\ \mathrm{cm}^{-3}$ with an ambient magnetic field strength $B_0 = 15\ \mu\mathrm{G}$. The short-dashed curve (labeled TP) corresponds to the test particle case in which the accelerated particles are energetically unimportant in terms of their effects on the shock. The regions corresponding to the FS, RS, and CD are indicated. Also shown are temperature plots for cases of moderate ($\varepsilon = 36\%$) and efficient ($\varepsilon = 63\%$) particle acceleration. (Here, $\varepsilon$ refers to the fraction of the energy flux crossing the shock that ends up in relativistic particles.)

Two distinct observable effects are immediately evident: at a given age, the separation between the FS and the RS (or CD) decreases considerably with increased particle acceleration, and the temperature at the FS is reduced for cases of high acceleration efficiency. The former effect is observed in Tycho's SNR, for

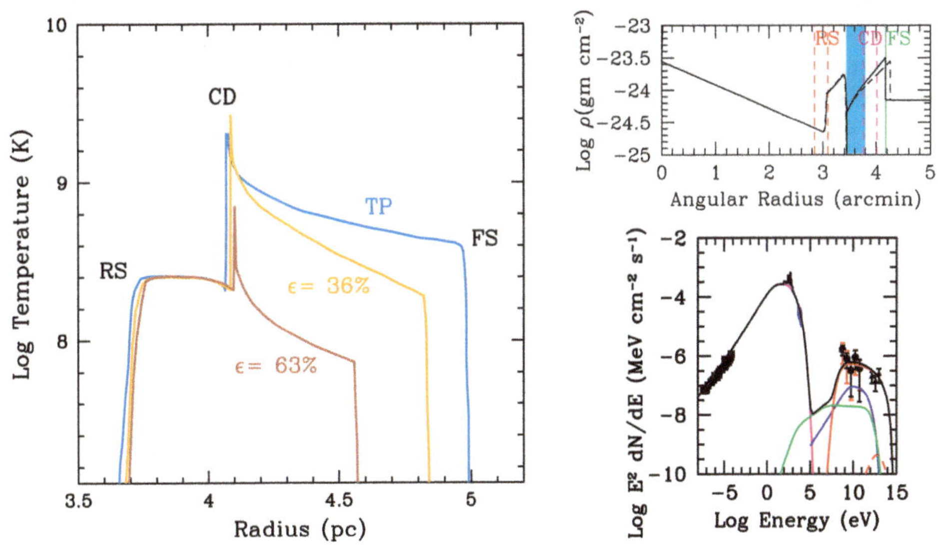


**Figure 5.18.** Left: effects of CR acceleration on FS, CD, and RS radius. See text for discussion. Reproduced from Ellison et al. (2007).  Right (upper): results from the model of Tycho's SNR with efficient particle acceleration. The dashed curve represents the case with no acceleration. The measured positions of the FS, CD, and RS are indicated. The cyan region indicates the range over which Rayleigh–Taylor instabilities may extend from the CD to the FS. Right (lower): predicted broadband emission for Tycho based on the model above—synchrotron emission (magenta), nonthermal bremsstrahlung (cyan), inverse-Compton emission (blue), and $\pi^0$-decay emission (red). Radio, X-ray, and $\gamma$-ray measurements are shown for comparison. Reproduced from Slane et al. (2014). 

which *Chandra* observations reveal that the ratios of the FS, RS, and CD radii are inconsistent with models unless the thermal energy is lower than expected—presumably due to energy going into particle acceleration (Warren et al. 2005). Hydrodynamical models for the evolution of Tycho's SNR, in which cosmic-ray acceleration of both electrons and ions is included, are consistent with this interpretation (Slane et al. 2014), as illustrated in Figure 5.18, where the plot of density as a function of radius identifies the FS, CD, and RS positions. Here the solid curve corresponds to a model in which 26% of the energy crossing the shock goes into particle acceleration, and the dashed curve corresponds to the case with no acceleration. The positions of the FS, CD, and RS shown in Figure 5.8 are identified.

For most of the known young SNRs, thin rims of nonthermal emission surround the remnants directly along their FSs. These observations make it clear that SNRs are capable of accelerating electrons to very high energies. It is assumed that ions are accelerated by the same process, although the evidence for this is less direct. Because ions dominate the cosmic-ray energy density, however, it is exactly this evidence that is of particular importance for our understanding of cosmic-ray production. For the model described above for Tycho's SNR, the predicted broadband emission is shown in Figure 5.18. The X-ray emission is well described by synchrotron emission from relativistic electrons, while the observed $\gamma$-ray emission arises primarily from the decay of neutral pions created in collisions of accelerated protons with ambient gas, showing that efficient acceleration of both electrons and ions is occurring.

Efficient particle acceleration via DSA requires strong magnetic fields in order to produce sufficiently small gyroradii to keep the particles within the acceleration region. The observed thin X-ray synchrotron rims can be used to estimate the field strength. The observed rim thickness corresponds to either the diffusion length,

$$l_{\rm diff} = \frac{D}{v}, \tag{5.17}$$

or to the advection length of the plasma corresponding to the synchrotron loss time,

$$l_{\rm ad} = v\tau_{\rm syn}. \tag{5.18}$$

Here, $v = v_s/r'$, where $r'$ is the shock compression factor, and

$$\tau_{\rm syn} \approx 820 E_{\rm e,100}^{-1} B_{10}^{-2}\ {\rm yr}, \tag{5.19}$$

where $B_{10}$ is the magnetic field strength in units of 10 μG and $E_{\rm e,100}$ is the energy of the electrons producing the observed synchrotron radiation, in units of 100 TeV. This energy is related to the observed synchrotron energy by

$$h\nu_s \approx 2.2 E_{\rm e,100}^2 B_{10}\ {\rm keV}. \tag{5.20}$$

The observed rim widths in these young remnants are typically 3″–5″—resolvable only with *Chandra*—and lead to field strengths ranging from ~100–500 μG, demonstrating significant amplification of the typical 3 μG upstream field in the ISM. In addition to thin synchrotron rims, *Chandra* observations of Tycho also

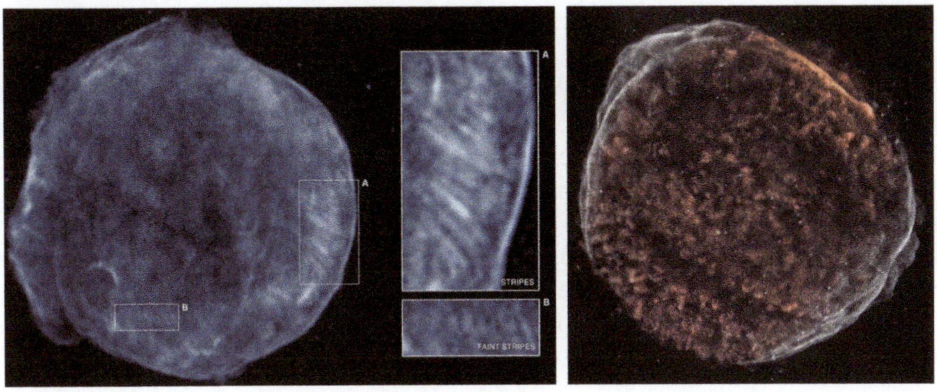


**Figure 5.19.** Left: Tycho's SNR (4–6 keV) showing the synchrotron rim as well as discrete stripe-like features in the SNR interior. Courtesy of NASA/CXC/Rutgers/K.Eriksen et al. Right: *Chandra* image of SN 1006 showing thermal X-rays from ejecta in the interior (red) and synchrotron emission (blue) concentrated along the NE and SW rims. Courtesy of NASA/CXC/Middlebury College/F.Winkler.

reveal a series of stripe-like structures (Figure 5.19) the separations of which may be associated with the gyroradii of ions accelerated to energies at high as $10^{14}$–$10^{15}$ eV (Eriksen et al. 2011).

In several cases—e.g., SN 1006 (Koyama et al. 1995), G347.3−0.5 (Koyama et al. 1997; Slane et al. 1999), and G266.2−1.2 (Slane et al. 2001)—the nonthermal emission components completely dominate the thermal components, and the X-ray spectra from the shells are featureless. The structure of SN 1006 is of particular interest (Figure 5.19), with thermal X-rays from hot ejecta filling the interior, and synchrotron emission along only the NE and SW rims. In DSA, particle acceleration is most efficient for quasi-parallel shocks, in which the velocity vector has a substantial component along the magnetic field. Radial fields are thus anticipated in regions of efficient acceleration. Radio polarization measurements for Tycho (Reynoso et al. 1997) and Cas A (Dubner & Giacani 2015) do indeed show primarily radial fields, while for SN 1006 the fields appear radial in regions where the bright synchrotron rims reside and tangential in the NW and SE regions.

An additional observation of particular importance regarding particle acceleration in SNRs is evidence from *Chandra* observations of Cas A for particle acceleration at the RS (Helder & Vink 2008). This is a surprising result, given that magnetic field strengths in the vicinity of the RS are expected to be extremely low. The observation of synchrotron X-rays from the RS may suggest the creation and amplification of fields at the RS, which could conceivably lead to episodes early in the SNR evolution where particles are accelerated to energies approaching the knee of the cosmic-ray spectrum. Additional searches for synchrotron X-rays from the RS regions of other young SNRs are clearly of importance.

## 5.3 Pulsar Wind Nebulae

The structure of a PWN is determined by both the properties of the host pulsar and the environment into which the nebula expands. Observations across the electromagnetic spectrum allow us to constrain the nature of the pulsar wind, including both its magnetization and geometry, and the global properties of the PWN allow us to constrain the evolutionary history as it evolves through the ejecta of the SNR in which it was born. Spectroscopic observations yield information on the mass and composition of shocked ejecta into which the nebula expands, and on the expansion velocity. Measurements of the broadband spectrum provide determinations of the nebular magnetic field and the maximum energy of the particles injected into the PWN. X-ray observations, in particular, reveal structures within PWNe that define the overall system geometry and identify sites of particle injection. These observations continue to inform theoretical models of relativistic shocks, which, in turn, have broad importance across the realm of high-energy astrophysics.

### 5.3.1 Pulsars

The discovery and basic theory of pulsars have been summarized in many places. First discovered by their radio pulsations, it was quickly hypothesized that these objects are rapidly rotating, highly magnetic NSs. Observations show that the spin period $P$ of a given pulsar increases with time, indicating a gradual decrease in rotational kinetic energy:

$$\dot{E} = I\Omega\dot{\Omega}, \tag{5.21}$$

where $\Omega = 2\pi/P$ and $I$ is the moment of inertia of the NS (nominally $I = \frac{2}{5}MR^2$, where $M$ and $R$ are the mass and radius of the star; $I \approx 10^{45}$ g cm$^2$ for $M = 1.4\ M_\odot$ and $R = 10$ km). This spin-down energy loss is understood to be the result of a magnetized particle wind produced by the rotating magnetic star. Treated as a simple rotating magnetic dipole, the energy loss rate is

$$\dot{E} = -\frac{B_p R^6 \Omega^4}{6c^3}\sin^2\chi, \tag{5.22}$$

where $B_p$ is the magnetic dipole strength at the pole, and $\chi$ is the angle between the magnetic field and the pulsar rotation axis. Typical values for $P$ range from ~0.03–3 s, with period derivatives of $10^{-17}$–$10^{-13}$ s s$^{-1}$ (though values outside these ranges are also observed, particularly for so-called magnetars and millisecond pulsars). This leads to inferred magnetic field strengths of order $10^{11}$–$10^{13}$ G for typical pulsars.

As the pulsar rotates, a charge-filled magnetosphere is created, with particle acceleration occurring in charge-separated gaps in regions near the polar cap or in the outer magnetosphere, which extends to the so-called light cylinder where $R_{LC} = c/\Omega$. The maximum potential generated by the rotating pulsar field under the assumption of coalignment of the magnetic and spin axes is

$$\Phi = \left(\frac{\dot{E}}{c}\right) \approx 6 \times 10^{13}\left(\frac{\dot{E}}{10^{38}\ \mathrm{erg\ s^{-1}}}\right)^{1/2} \mathrm{V}, \tag{5.23}$$

providing conditions for acceleration of charged particles to extremely high energies. The minimum particle current required to sustain the charge density in the magnetosphere is

$$\dot{N}_{GJ} = \frac{c\Phi}{e} \approx 4 \times 10^{33} \left( \frac{\dot{E}}{10^{38} \text{ erg s}^{-1}} \right)^{1/2} \text{s}^{-1}, \tag{5.24}$$

where $e$ is the electron charge (Goldreich & Julian 1969). As the particles comprising this current are accelerated, they produce curvature radiation that initiates an electron–positron pair cascade. Based on observations of PWNe, values approaching $\dot{N} = 10^{40}$ s$^{-1}$ are required to explain the radio synchrotron emission. The implied multiplicity (i.e., the number of pairs created per primary particle) of $\sim 10^5$–$10^7$ appears difficult to obtain from pair production in the acceleration regions within pulsar magnetospheres (Timokhin & Harding 2015), suggesting that a relic population of low-energy electrons created by some other mechanism early in the formation of the PWN may be required (Atoyan & Aharonian 1996, e.g.). Studies of the particle content of PWNe thus provide crucial information for understanding the details by which the pulsar spin-down power is converted into outflows of relativistic particles.

### 5.3.2 Pulsar Wind Structure

For pulsars with a magnetic axis that is inclined relative to the rotation axis, the result of the above is a striped wind, with an alternating poloidal magnetic field component separated by a current sheet (Bogovalov 1999). The magnetization of the wind, $\sigma$, is defined as the ratio between the Poynting flux and the particle energy flux:

$$\sigma = \frac{B^2}{4\pi m n \gamma_0 c^2}, \tag{5.25}$$

where $B$, $n$, and $\gamma_0$ are the magnetic field, number density of particles of mass $m$, and bulk Lorentz factor in the wind, respectively. The energy density of the wind is expected to be dominated by the Poynting flux as it leaves the magnetosphere, with $\sigma \sim 10^4$. Ultimately, the wind is confined by ambient material (slow-moving ejecta in the host SNR at early times; the ISM once the pulsar has exited the SNR), forming an expanding magnetic bubble of relativistic particles—the PWN. As the fast wind entering the nebula decelerates to meet the boundary condition imposed by the much slower expansion of the PWN, a wind termination shock (TS) is formed at a radius $R_{\rm TS}$, where the ram pressure of the wind is balanced by the pressure within the nebula:

$$R_{\rm TS} = \sqrt{\dot{E}/(4\pi\omega c P_{\rm PWN})}, \tag{5.26}$$

where $\omega$ is the equivalent filling factor for an isotropic wind and $P_{\rm PWN}$ is the total pressure in the nebula. The geometry of the pulsar system results in an axisymmetric wind (Lyubarsky 2002), forming a torus-like structure in the equatorial plane, along with collimated jets along the rotation axis. The higher magnetization at low latitudes confines the expansion here to a higher degree, resulting in an elongated shape along the pulsar spin axis for the large-scale nebula (Begelman & Li 1992;

van der Swaluw 2003). This structure is evident in Figure 5.20 (left), where X-ray and optical observations of the Crab Nebula clearly reveal the jet/torus structure surrounded by the elongated wind nebula bounded by filaments of swept-up ejecta. The X-ray nebula, which is shown enlarged at the right, is smaller than the optical (and radio) nebula because the energetic particles that produce the X-ray emission have undergone significant synchrotron losses before reaching the edge of the nebula; their synchrotron lifetimes (Equation (5.19)) are shorter than the time required to diffuse to the edge of the nebula. The innermost ring in the X-ray image corresponds to the TS, and its radius is well described by Equation (5.26). The inclination angle of the ring provides the orientation of the pulsar spin axis, which is aligned with the jet axis, and this orientation also explains the brighter NW portion of the torus, which is Doppler brightened. MHD models of the jet/torus structure in pulsar winds reproduce many of the observed details of these systems (see Bucciantini 2011 for a review).

While the Crab Pulsar and its nebula are far from representative of the population, being much brighter than most other young systems, jet/torus structures are observed in numerous PWNe, including that within G292.8+1.8 (Figure 5.15), as discussed earlier. The Vela Pulsar shows a double-arc structure (Figure 5.21, left) accompanied by a jet/counterjet for which monitoring observations reveal complex dynamical motions that may be associated with kink instabilities in the outflow (Pavlov et al. 2003). The pulsar in G54.1+0.3 (Figure 5.21, right) also shows a well-defined torus around the pulsar, along with elongated jets. Doppler brightening is presumably responsible for the observed asymmetric X-ray intensity in these systems, as with the Crab.

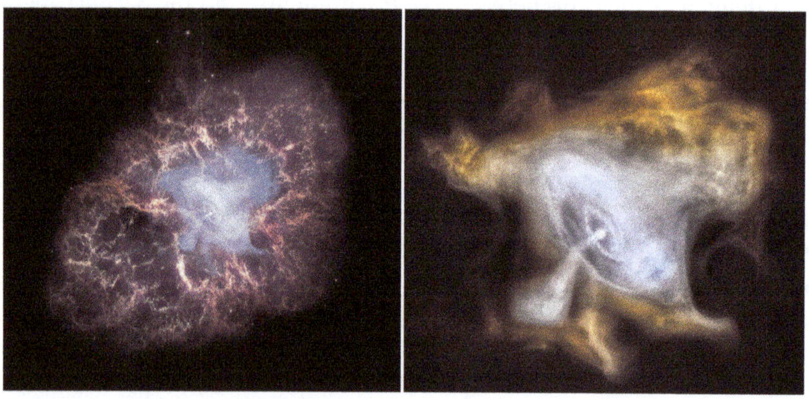

**Figure 5.20.** Left: composite image of the Crab Nebula with X-ray emission (blue) from *Chandra*, optical emission (red and yellow) from *HST*, and IR emission (purple) from *Spitzer*. X-ray courtesy of NASA/CXC/SAO/F.Seward; Optical courtesy of NASA/ESA/ASU/J.Hester & A.Loll; Infrared courtesy of NASA/JPL-Caltech/Univ. Minn./R.Gehrz. Right: *Chandra* image of the Crab Nebula, illustrating the inner termination shock ring, the surrounding torus, and the jet. Courtesy of NASA/CXC/SAO. The interactive figure shows separately the X-ray emission (purple), ultraviolet emission (blue), optical emission (green), infrared emission (yellow-green), radio emission (red) and the composite image, extending the image in the left hand panel to cooler material in the outskirts of the nebula. The video shows a timelapse movie of the X-ray emission from the Crab Nebula shown in the right hand panel, including 7 *Chandra* images taken from 2000 November to 2001 April. Additional materials available online at http://iopscience.iop.org/book/978-0-7503-2163-1.

**Figure 5.21.** Left: *Chandra* image of the Vela Pulsar, illustrating the surrounding jet and double-arc structure. Courtesy of NASA/CXC/Univ of Toronto/M.Durant et al. Right: composite image of G54.1+0.3 (Temim et al. 2010) with X-ray emission (blue) from *Chandra* and IR emission from *Spitzer* (red–yellow, 24 μm; green, 8 μm). X-ray courtesy of NASA/CXC/SAO/T.Temim et al.; IR courtesy of NASA/JPL-Caltech. The video shows a timelapse movie of the X-ray emission from the Vela Pulsar, as shown in the left hand panel, including eight *Chandra* images taken from 2010 June to September. This movie shows a fast moving jet of particles produced by a rapidly rotating neutron star. The interactive figure of G54.1 shows separately the X-ray emission (blue), infrared emission at 24μm (red-yellow), infrared at 8μm (green), and the composite image shown in the right hand panel. Additional materials available online at http://iopscience.iop.org/book/978-0-7503-2163-1.

Importantly, models of the dynamical structure and emission properties of the Crab Nebula require $\sigma \sim 10^{-3}$ just upstream of the termination shock (Kennel & Coroniti 1984). Apparently, somewhere between the pulsar magnetosphere and the termination shock, the wind converts from being Poynting dominated to being particle dominated. Magnetic reconnection in the current sheet has been suggested as a mechanism for dissipating the magnetic field, transferring its energy into that of the particles (e.g., Lyubarsky 2003). Recent particle-in-cell simulations of relativistic shocks show that shock compression of the wind flow can drive regions of opposing magnetic fields together, causing the reconnection (Sironi & Spitkovsky 2011). However, the maximum Lorentz factor that appears achievable is limited by the requirement that the diffusion length of the particles be smaller than the termination shock radius,; $\gamma_{\rm max} \sim 8.3 \times 10^6 \dot{E}_{38}^{3/4} \dot{N}_{40}^{-1/2}$. This is insufficient to explain the observed X-ray synchrotron emission in PWNe, suggesting that an alternative picture for acceleration of the highest energy particles in PWNe is required (Sironi et al. 2013), possibly associated with turbulent magnetic reconnection in the outer nebula (Olmi et al. 2014; Lyutikov et al. 2019). *Chandra* observations of 3C 58 (Figure 5.22) show complex loop-like features in the nebula surrounding the pulsar, torus, and jet, possibly corresponding to turbulent magnetic structures within the PWN (Slane et al. 2004).

### 5.3.3 PWN Evolution

The evolution of a PWN within the confines of its host SNR is determined by both the rate at which energy is injected by the pulsar and by the density structure of the ejecta material into which the nebula expands. The location of the pulsar itself,

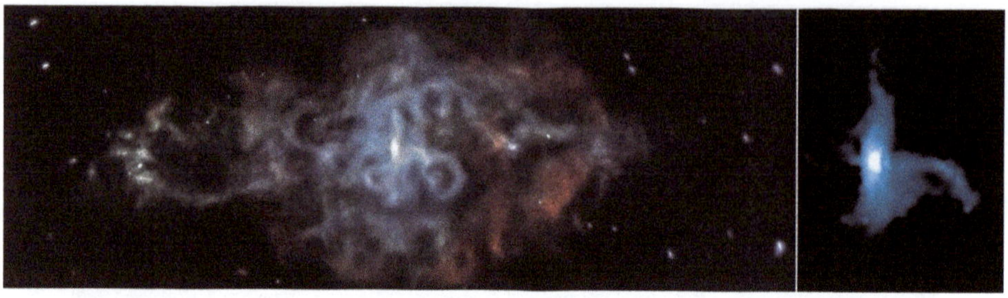

**Figure 5.22.** Left: *Chandra* image of 3C 58. Low (high) energy X-rays are shown in red (blue). Courtesy of NASA/CXC/SAO. Right: expanded view of the central region of 3C 58 showing the toroidal structure and jet associated with the central pulsar. Courtesy of NASA/CXC/SAO/P.Slane et al.

relative to the SNR center, depends upon any motion given to the pulsar in the form of a kick velocity during the explosion, as well as on the density distribution of the ambient medium into which the SNR expands. As discussed in Section 5.2.2, the SNR BW initially expands freely at a speed of $\sim(5 - 10) \times 10^3$ km s$^{-1}$, much higher than typical pulsar velocities of ~200–1500 km s$^{-1}$. As a result, for young systems the pulsar will be located near the SNR center unless the SNR surroundings have a large density gradient (which is quite possible, given that CC SNRs evolve in complex surroundings containing molecular clouds and modified by stellar winds).

*Early Evolution*

The energetic pulsar wind is injected into the SNR interior, forming a high-pressure bubble that expands supersonically into the surrounding ejecta, forming a shock. The input luminosity is generally assumed to have the form (e.g., Pacini & Salvati 1973)

$$\dot{E} = \dot{E}_0\left(1 + \frac{t}{\tau_0}\right)^{-\frac{(n+1)}{(n-1)}}, \tag{5.27}$$

where $\tau_0$ is the initial spin-down timescale of the pulsar. Here $\dot{E}_0$ is the initial spin-down power, $P_0$ and $\dot{P}_0$ are the initial spin period and its time derivative, and $n$ is the so-called "braking index" of the pulsar ($n = 3$ for magnetic dipole spin-down). The pulsar has roughly constant energy output until a time $\tau_0$, beyond which the output declines fairly rapidly with time.

Because the ejecta can be expanding rapidly, the shock velocity $v_s = v_{\rm PWN} - v_{\rm ej}$ (where $v_{\rm PWN}$ is the expansion velocity of the PWN) can be modest. Indeed, while *Chandra* observations of 3C 58 reveal a shell of thermal emission with enhanced abundances (Figure 5.22; Slane et al. 2004), suggesting a high shock velocity, other studies of PWNe show much lower velocities. For G21.5–0.9, for example, which is a young SNR containing an energetic pulsar and PWN (Figure 5.23; Slane et al. 2000; Safi-Harb et al. 2001), IR observations reveal a shell of [Fe II] surrounding the PWN, indicating a low shock velocity (Zajczyk et al. 2012), a result confirmed by *Herschel* observations that also reveal a shell of [C II] (Temim & Slane 2017).

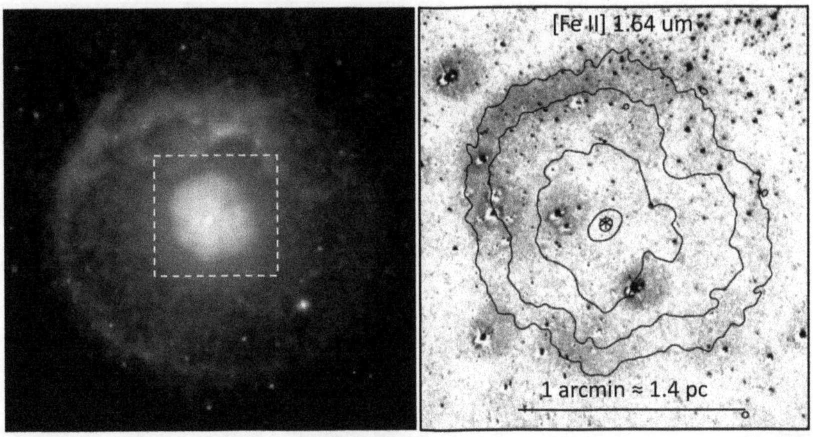


**Figure 5.23.** Left: *Chandra* image of G21.5–0.9. The pulsar is located at the center and is surrounded by a PWN. The faint outer shell is the SNR, and a portion of the emission between the PWN and the outer shell is scattered flux from the PWN. Image CXC/NASA. Right: infrared image at 1.64 μm showing a shell of [Fe II] emission from ejecta that has been swept up and shocked by the expanding PWN. (Reproduced from Zajczyk et al. (2012), reproduced with permission © ESO.)

*Herschel* observations of Kes 75, for which direct expansion measurements with *Chandra* show $v_{\rm PWN} \sim 1000$ km s$^{-1}$ (Reynolds et al. 2018), reveal broadened emission lines from [O II], [O III], and [C II] (Temim et al. 2019), indicating a shock velocity of only ∼50–200 km s$^{-1}$.

Figure 5.24 (Slane 2017) illustrates the evolution of a PWN within its host SNR. The left panel shows a hydrodynamical simulation of an SNR evolving into a nonuniform medium, with a density gradient increasing from left to right. The pulsar is moving upward in the simulation. The SNR FS, RS and CD separating the shocked CSM and shocked ejecta are identified, as is the PWN shock driven by expansion into the cold ejecta. The right panel illustrates the radial density distribution, highlighting the PWN TS as well as the SNR FS, CD, and RS.

*PWN/SNR Interaction*

As the SNR BW sweeps up increasing amounts of material, the RS propagates back toward the SNR center. In the absence of a central PWN, it reaches the center at a time $t_c \approx 7(M_{\rm ej}/10\ M_\odot)^{5/6} E_{51}^{-1/2} n_0^{-1/3}$ kyr, where $E_{51}$ is the explosion energy, $M_{\rm ej}$ is the ejecta mass, and $n_0$ is the number density of ambient gas (Reynolds & Chevalier 1984). When a PWN is present, however, the RS interacts with the nebula before it can reach the center (Figure 5.24). The shock compresses the PWN, increasing the magnetic field strength and resulting in enhanced synchrotron radiation that burns off the highest energy particles. If the ambient CSM/ISM is significantly nonuniform, the FS expands more (less) rapidly in regions of lower (higher) density. This has two significant effects. First, it changes the morphology of the SNR to a distorted shell for which the associated pulsar is no longer at the center. Second, the RS also propagates asymmetrically, reaching the center more quickly from the direction of the higher density medium (Blondin et al. 2001).

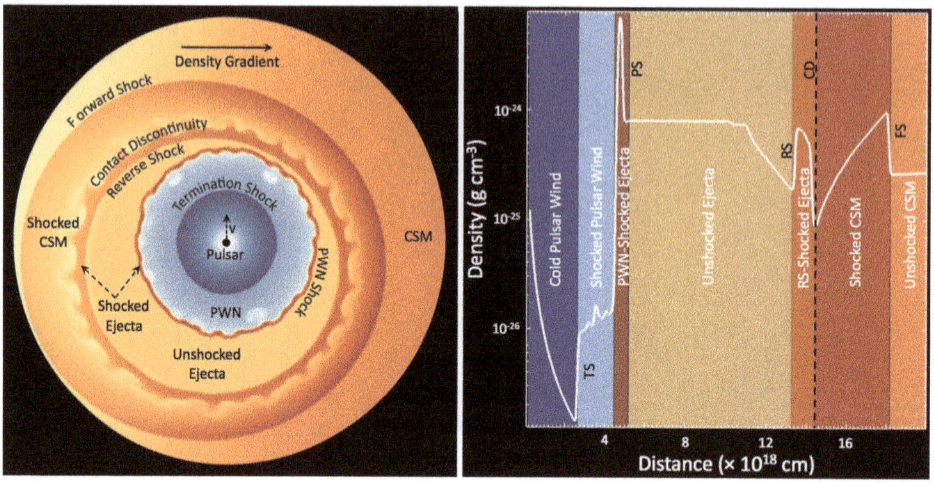


**Figure 5.24.** Left: density image from a hydrodynamical simulation of a PWN expanding into an SNR that is evolving into a medium with a CSM density gradient increasing to the right. The pulsar itself is moving upward. The reverse shock is propagating inward, approaching the PWN preferentially from the upper right due to the combined effects of the pulsar motion and the CSM density gradient. Right: density profile for a radial slice through the simulated composite SNR. Colored regions correspond to different physical regions identified in the SNR image. Reproduced from Slane (2017). © Springer International Publishing AG 2017. With permission of Springer.

The return of the RS ultimately creates a collision with the PWN. Simulations of this interaction show that the PWN is compressed until the pressure in the nebula is sufficiently high to rebound, and again expand into the ejecta. During the compression phase, the magnetic field of the nebula increases, resulting in enhanced synchrotron radiation and significant radiative losses from the highest energy particles. The PWN/RS interface is Rayleigh–Taylor (R–T) unstable, and is subject to the formation of filamentary structure where the dense ejecta material is mixed into the relativistic fluid. If the SNR has evolved in a nonuniform medium, an asymmetric RS will form, disrupting the PWN and displacing it in the direction of lower density (Figure 5.24). The nebula subsequently re-forms as the pulsar injects fresh particles into its surroundings, but a significant relic nebula of mixed ejecta and relativistic gas will persist.

Because the SNR RS typically reaches the central PWN on a timescale that is relatively short compared with the SNR lifetime, all but the youngest PWNe that we observe have undergone an RS interaction. This has significant impact on the large-scale geometry of the PWN, as well as on its spectrum and dynamical evolution. Remnants such as MSH 15−56 (Temim et al. 2013), G327.1−1.1 (Temim et al. 2015), and Vela (Slane et al. 2018) all show complex structure indicative of RS/PWN interactions, and observations of extended sources of very high-energy (VHE) $\gamma$-rays indicate that many of these objects correspond to PWNe that have evolved beyond the RS-crushing stage.

An example of such an RS-interaction stage is presented in Figure 5.25, where we show the composite SNR G327.1−1.1 (Temim et al. 2015). Radio observations (a) show a complete SNR shell surrounding an extended flat-spectrum PWN in the

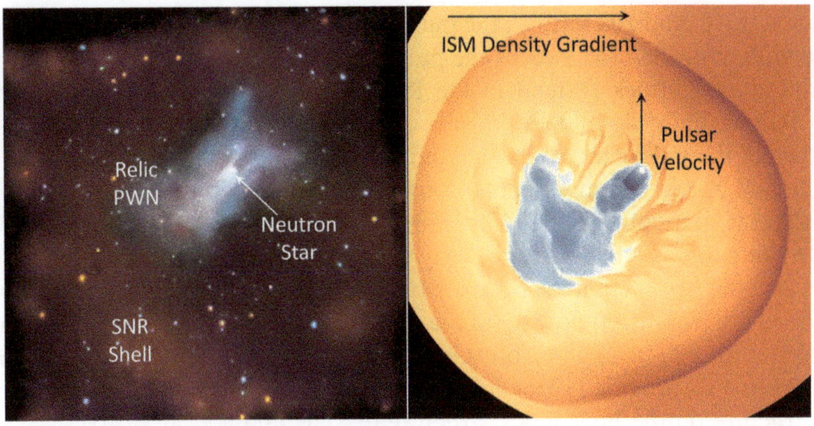


**Figure 5.25.** Left: *Chandra* image of G327.1−1.1 showing the faint X-ray SNR shell, the NS encased in a compact cometary nebula, and the relic PWN. Right: hydrodynamical simulation of an evolved composite SNR with properties similar to G327.1−1.1. (See text for details.) Reproduced from Temim et al. (2015) © 2015. The American Astronomical Society. All rights reserved.

remnant interior, accompanied by a finger-like structure extending to the northwest. *Chandra* observations (b) show faint emission from the SNR shell along with a central compact source located at the tip of the radio finger, accompanied by a tail of emission extending back into the radio PWN. The X-ray properties of the compact source are consistent with emission from a pulsar (though, to date, pulsations have not yet been detected) which, based on its position relative to the geometric center of the SNR, appears to have a northward motion. Spectra from the SNR shell indicate a density gradient in the surrounding medium, increasing from east to west, and the presence of ejecta-rich material in the PWN region.

Results from hydrodynamical modeling of the evolution of such a system using these measurements as constraints, along with an estimate for the spin-down power of the pulsar based upon the observed X-ray emission of its PWN (see Section 3.1), are shown in Figure 5.25 (right) where we show the density (compare with Figure 5.24). The RS has approached rapidly from the west, sweeping past the pulsar, disrupting the PWN and mixing ejecta into the regions around the relic PWN. The result is a trail of emission swept back into the relic PWN, in excellent agreement with the radio morphology. The X-ray spectrum of the tail shows a distinct steepening with distance from the pulsar, consistent with synchrotron cooling of the electrons based on the estimated magnetic field and age of the injected particles tracked in the hydro simulation. *Chandra* images show that the central source is embedded in a cometary structure produced by a combination of the northward motion of the pulsar and the interaction with the RS propagating from the west. Extended prong-like structures are also observed in X-rays, the origin of which is currently not understood.

*Bow Shock PWNe*

Late in the evolution of a PWN, the pulsar will exit its host SNR and begin traveling through the ISM. Because the sound speed for the cold, warm, and hot phases of the

ISM is $v_s \sim 1, 10$, and $100 \text{ km s}^{-1}$, the pulsar motion will typically be supersonic. The relative motion of the ISM sweeps the pulsar wind back into a bow shock structure. The nonradiative shock formed in the ISM interaction results in the emission of optical Balmer lines, dominated by H$\alpha$, providing a distinct signature from which properties of the pulsar motion and wind luminosity can be inferred.

Radio and X-ray measurements of bow shock nebulae probe the shocked pulsar wind. Observations of PSR J1747–2958 and its associated nebula G359.23–0.82 reveal a long radio tail and an X-ray morphology that reveals both a highly magnetized tail from wind shocked from the forward direction, and a weakly magnetized tail from wind flowing in the direction opposite that of the pulsar motion (Gaensler et al. 2004). High-resolution measurements of the emission near several pulsars have also provided evidence for asymmetric pulsar winds imprinting additional structure on the bow shock structure (e.g., Romani et al. 2010).

While the X-ray emission from the bow shock can be quite faint, deep *Chandra* observations have been used to uncover important elements of their structure. Observations of PSR J1714–2054 taken 3.2 yr apart reveal proper motion of the pulsar, $\mu = 109 \pm 10 \text{ mas yr}^{-1}$, in a direction along the symmetry axis of the observed H$\alpha$ nebula, along with a tail of extended X-ray emission that shows no evidence for cooling, suggesting a low magnetic field (Auchettl et al. 2015). Deep *Chandra* observations of PSR B0355+54 show a compact nebula surrounding the pulsar, with an extended tail. The morphology of the system suggests a small angle between the spin axis and the line of sight, consistent with inferences based on the radio pulse profile (Klingler et al. 2016).

Deep *Chandra* observations of the Geminga pulsar, for which previous observations had revealed two lateral trails (Caraveo et al. 2003; Pavlov et al. 2006), show that these structures are most likely swept-back jets rather than limb-brightened edges of an extended bow shock (Posselt et al. 2017). The geometry indicates that the pulsar spin axis lies very nearly in the plane of the sky (Figure 5.26), consistent with the lack of observed radio pulses, which originate near the polar axis, but with distinct $\gamma$-ray pulsations which originate closer to the equatorial regions of the pulsar

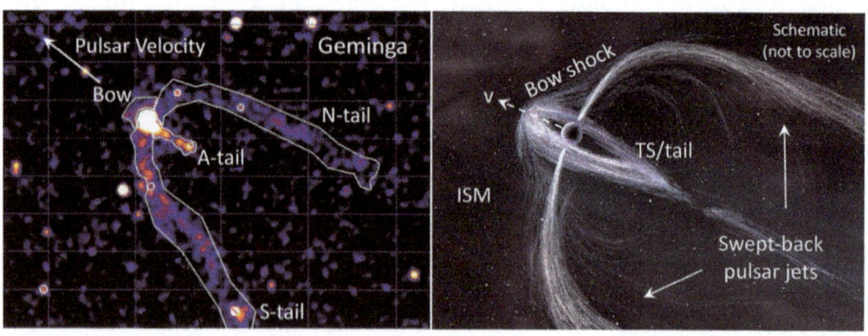


**Figure 5.26.** Left: *Chandra* image of the Geminga PWN. Reproduced from Posselt et al. (2017). © 2017. The American Astronomical Society. All rights reserved. Right: schematic diagram of the Geminga-like PWN, with swept-back pulsar jets.

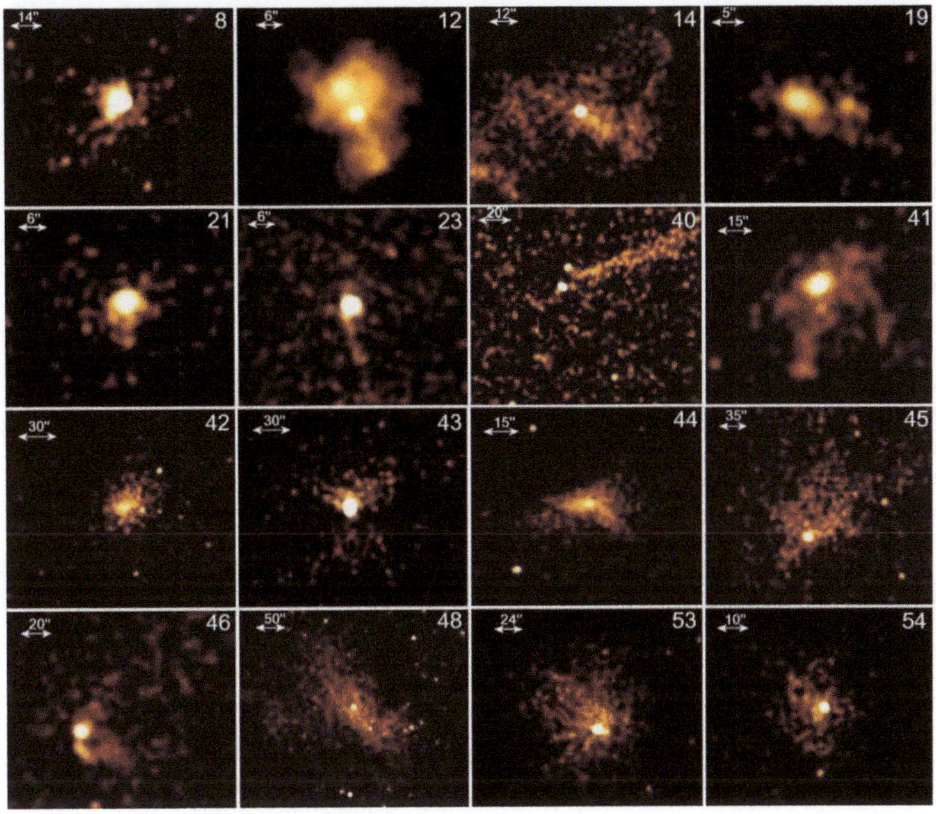


**Figure 5.27.** Incomplete collection of other pulsars and PWNe for which *Chandra* observations reveal evidence of jet, torus, or bow shock structures. Angular sizes are indicated in upper left of each panel. Deeper observations are required to fully explore these structures, with angular resolution at least as high as that provided by *Chandra*. Reproduced from Kargaltsev & Pavlov (2008), with the permission of AIP Publishing. See publication for source identification.

magnetosphere. The images also show an axial tail that appears to be the extended TS region behind the known direction of the pulsar motion. The results indicate a distinct misalignment between the proper motion and the pulsar spin axis, which is important for constraints on the nature of pulsar kicks.

While *Chandra* observations of PWNe have revolutionized our understanding of their structure and evolution, many of these systems are quite faint. A large number of PWNe for which jet/torus and other extended structures are evident have been identified with *Chandra*, a subset of which are shown in Figure 5.27 (Kargaltsev & Pavlov 2008). Deeper *Chandra* observations and sensitive observations from future X-ray telescopes with larger effective area and *Chandra*-like angular resolution (or better) are required to begin probing the larger part of the population and establish more detailed pictures for the long-term evolution and eventual particle escape from pulsar-driven nebulae.

## References

Alp, D., Larsson, J., Fransson, C., et al. 2018, ApJ, 864, 174
Alan, N., Park, S., & Bilir, S. 2019, ApJ, 873, 53
Atoyan, A. M., & Aharonian, F. A. 1996, A&AS, 120, 453
Auchettl, K., Slane, P., Romani, R. W., et al. 2015, ApJ, 802, 68
Badenes, C., Bravo, E., Borkowski, K. J., & Domínguez, I. 2003, ApJ, 593, 358
Badenes, C., Hughes, J. P., Cassam-Chenaï, G., & Bravo, E. 2008, ApJ, 680, 1149
Bauer, F. E., Dwarkadas, V. V., Brandt, W. N., et al. 2008, ApJ, 688, 1210
Begelman, M. C., & Li, Z.-Y. 1992, ApJ, 397, 187
Bhalerao, J., Park, S., Schenck, A., Post, S., & Hughes, J. P. 2019, ApJ, 872, 31
Blondin, J. M., Chevalier, R. A., & Frierson, D. M. 2001, ApJ, 563, 806
Bochenek, C. D., Dwarkadas, V. V., Silverman, J. M., et al. 2018, MNRAS, 473, 336
Bogovalov, S. V. 1999, A&A, 349, 1017
Borkowski, K. J., Hendrick, S. P., & Reynolds, S. P. 2006, ApJ, 652, 1259
Borkowski, K. J., Gwynne, P., Reynolds, S. P., et al. 2017, ApJ, 837, L7
Branch, D., & Wheeler, J. C. 2017, Supernova Explosions (Berlin: Springer)
Bucciantini, N. 2011, ApSSP, 21, 473
Caraveo, P. A., Bignami, G. F., De Luca, A., et al. 2003, Sci., 301, 1345
Carlton, A. K., Borkowski, K. J., Reynolds, S. P., et al. 2011, AAS Meeting 217, 256.12
Chugai, N. N., & Chevalier, R. A. 2006, ApJ, 641, 1051
DeLaney, T., Rudnick, L., Fesen, R. A., et al. 2004, ApJ, 613, 343
DeLaney, T., Rudnick, L., Stage, M. D., et al. 2010, ApJ, 725, 2038
Dittmann, J. A., Soderberg, A. M., Chomiuk, L., et al. 2014, ApJ, 788, 38
Dubner, G., & Giacani, E. 2015, A&ARv, 23, 3
Ellison, D. C., Patnaude, D. J., Slane, P., Blasi, P., & Gabici, S. 2007, ApJ, 661, 879
Eriksen, K. A., Hughes, J. P., Badenes, C., et al. 2011, ApJ, 728, L28
Flanagan, K. A., Canizares, C. R., Dewey, D., et al. 2004, ApJ, 605, 230
Frank, K. A., Zhekov, S. A., Park, S., et al. 2016, ApJ, 829, 40
Gaensler, B. M., van der Swaluw, E., Camilo, F., et al. 2004, ApJ, 616, 383
Gaetz, T. J., Butt, Y. M., Edgar, R. J., et al. 2000, ApJ, 534, L47
Ghavamian, P., Laming, J. M., & Rakowski, C. E. 2007, ApJ, 654, L69
Goldreich, P., & Julian, W. H. 1969, ApJ, 157, 869
Grefenstette, B. W., Fryer, C. L., Harrison, F. A., et al. 2017, ApJ, 834, 19
Hasinger, G., Aschenbach, B., & Truemper, J. 1996, A&A, 312, L9
Helder, E. A., & Vink, J. 2008, ApJ, 686, 1094
Hendrick, S. P., Borkowski, K. J., & Reynolds, S. P. 2003, ApJ, 593, 370
Ho, W. C. G., & Heinke, C. O. 2009, Natur, 462, 71
Hughes, J. P., Ghavamian, P., Rakowski, C. E., & Slane, P. O. 2003, ApJ, 582, L95
Hughes, J. P., & Helfand, D. J. 1985, ApJ, 291, 544
Hughes, J. P., Rakowski, C. E., Burrows, D. N., & Slane, P. O. 2000, ApJ, 528, L109
Hughes, J. P., Slane, P. O., Burrows, D. N., et al. 2001, ApJ, 559, L153
Hwang, U., & Laming, J. M. 2012, ApJ, 746, 130
Immler, S., & Kuntz, K. D. 2005, ApJ, 632, L99
Iwamoto, K., Brachwitz, F., Nomoto, K., et al. 1999, ApJS, 125, 439
Janka, H.-T. 2017, in Handbook of Supernovae, Neutrino-Driven Explosions, ed. A. W. Alsabti, & P. Murdin (Berlin: Springer), 1095

Kargaltsev, O., & Pavlov, G. G. 2008, in AIP Conf. Ser. 983, 40 Years of Pulsars: Millisecond Pulsars, Magnetars and More, ed. C. Bassa, Z. Wang, A. Cumming, & V. M. Kaspi (Melville, NY: AIP), 171–85
Katsuda, S., Petre, R., Hughes, J. P., et al. 2010, ApJ, 709, 1387
Kawasaki, M., Ozaki, M., Nagase, F., Inoue, H., & Petre, R. 2005, ApJ, 631, 935
Kennel, C. F., & Coroniti, F. V. 1984, ApJ, 283, 710
Klingler, N., Rangelov, B., Kargaltsev, O., et al. 2016, ApJ, 833, 253
Koyama, K., Kinugasa, K., Matsuzaki, K., et al. 1997, PASJ, 49, L7
Koyama, K., Petre, R., Gotthelf, E. V., et al. 1995, Natur, 378, 255
Krause, O., Birkmann, S. M., Usuda, T., et al. 2008, Sci, 320, 1195
Kundu, E., Lundqvist, P., Sorokina, E., et al. 2019, ApJ, 875, 17
Lazendic, J. S., Dewey, D., Schulz, N. S., & Canizares, C. R. 2006, ApJ, 651, 250
Lee, J.-J., Park, S., Hughes, J. P., & Slane, P. O. 2014, ApJ, 789, 7
Lee, J.-J., Park, S., Hughes, J. P., et al. 2010, ApJ, 711, 861
Liu, J., Di Stefano, R., Wang, T., & Moe, M. 2012, ApJ, 749, 141
Lopez, L. A., Pearson, S., Ramirez-Ruiz, E., et al. 2013, ApJ, 777, 145
Lopez, L. A., Ramirez-Ruiz, E., Huppenkothen, D., Badenes, C., & Pooley, D. A. 2011, ApJ, 732, 114
Lyubarsky, Y. E. 2002, MNRAS, 329, L34
Lyubarsky, Y. E. 2003, MNRAS, 345, 153
Lyutikov, M., Temim, T., Komissarov, S., et al. 2019, MNRAS, 489, 2403
Maeda, K., Katsuda, S., Bamba, A., Terada, Y., & Fukazawa, Y. 2014, ApJ, 785, 95
Maoz, D., Mannucci, F., & Nelemans, G. 2014, ARA&A, 52, 107
Margutti, R., Kamble, A., Milisavljevic, D., et al. 2017, ApJ, 835, 140
Margutti, R., Parrent, J., Kamble, A., et al. 2014, ApJ, 790, 52
McCray, R. 2017, in Handbook of Supernovae, The Physics of Supernova 1987A, ed. A. W. Alsabti, & P. Murdin (Berlin: Springer), 2181
McEntaffer, R. L., Grieves, N., DeRoo, C., & Brantseg, T. 2013, ApJ, 774, 120
Miceli, M., Orlando, S., Burrows, D. N., et al. 2019, NatAs, 3, 236
Milisavljevic, D., Margutti, R., Kamble, A., et al. 2015, ApJ, 815, 120
Milisavljevic, D., Patnaude, D. J., Chevalier, R. A., et al. 2018, ApJ, 864, L36
Nielsen, M. T. B., Voss, R., & Nelemans, G. 2012, MNRAS, 426, 2668
Olmi, B., Del Zanna, L., Amato, E., Bandiera, R., & Bucciantini, N. 2014, MNRAS, 438, 1518
Ozawa, M., Koyama, K., Yamaguchi, H., Masai, K., & Tamagawa, T. 2009, ApJ, 706, L71
Pacini, F., & Salvati, M. 1973, ApJ, 186, 249
Park, S., Badenes, C., Mori, K., et al. 2013, ApJ, 767, L10
Park, S., Hughes, J. P., Slane, P. O., et al. 2004, ApJ, 602, L33
Patnaude, D., & Badenes, C. 2017, in Handbook of Supernovae, Supernova Remnants as Clues to Their Progenitors, ed. A. W. Alsabti, & P. Murdin (Berlin: Springer), 2233
Patnaude, D. J., Lee, S.-H., Slane, P. O., et al. 2015, ApJ, 803, 101
Patnaude, D. J., Loeb, A., & Jones, C. 2011, NewA, 16, 187
Pavlov, G. G., Sanwal, D., & Zavlin, V. E. 2006, ApJ, 643, 1146
Pavlov, G. G., Teter, M. A., Kargaltsev, O., & Sanwal, D. 2003, ApJ, 591, 1157
Pooley, D., & Lewin, W. H. G. 2004, IAUC, 8323
Posselt, B., Pavlov, G. G., Slane, P. O., et al. 2017, ApJ, 835, 66
Posselt, B., Pavlov, G. G., Suleimanov, V., & Kargaltsev, O. 2013, ApJ, 779, 186

Rakowski, C. E., Ghavamian, P., & Hughes, J. P. 2003, ApJ, 590, 846
Rest, A., Matheson, T., Blondin, S., et al. 2008, ApJ, 680, 1137
Reynolds, S. P., Borkowski, K. J., & Gwynne, P. H. 2018, ApJ, 856, 133
Reynolds, S. P., Borkowski, K. J., Hwang, U., et al. 2007, ApJ, 668, L135
Reynolds, S. P., & Chevalier, R. A. 1984, ApJ, 278, 630
Reynoso, E. M., Moffett, D. A., Goss, W. M., et al. 1997, ApJ, 491, 816
Romani, R. W., Shaw, M. S., Camilo, F., Cotter, G., & Sivakoff, G. R. 2010, ApJ, 724, 908
Roper, Q., Filipovic, M., Allen, G. E., et al. 2018, MNRAS, 479, 1800
Safi-Harb, S., Harrus, I. M., Petre, R., et al. 2001, ApJ, 561, 308
Sato, T., & Hughes, J. P. 2017a, ApJ, 840, 112
Sato, T., & Hughes, J. P. 2017b, ApJ, 845, 167
Sedov, L. I. 1959, Similarity and Dimensional Methods in Mechanics (New York: Academic)
Sironi, L., & Spitkovsky, A. 2011, ApJ, 741, 39
Sironi, L., Spitkovsky, A., & Arons, J. 2013, ApJ, 771, 54
Slane, P. 2017, in Handbook of Supernovae, Pulsar Wind Nebulae, ed. A. W. Alsabti, & P. Murdin (Berlin: Springer), 2159
Slane, P., Chen, Y., Schulz, N. S., et al. 2000, ApJ, 533, L29
Slane, P., Gaensler, B. M., Dame, T. M., et al. 1999, ApJ, 525, 357
Slane, P., Helfand, D. J., van der Swaluw, E., & Murray, S. S. 2004, ApJ, 616, 403
Slane, P., Hughes, J. P., Edgar, R. J., et al. 2001, ApJ, 548, 814
Slane, P., Lee, S.-H., Ellison, D. C., et al. 2014, ApJ, 783, 33
Slane, P., Lovchinsky, I., Kolb, C., et al. 2018, ApJ, 865, 86
Slavin, J. D., Smith, R. K., Foster, A., et al. 2017, ApJ, 846, 77
Stockdale, C. J., Van Dyk, S. D., Sramek, R. A., et al. 2004, IAUC, 8282
Sukhbold, T., Ertl, T., Woosley, S. E., Brown, J. M., & Janka, H.-T. 2016, ApJ, 821, 38
Taylor, G. 1950, RSPTA, 201, 159
Temim, T., & Slane, P. 2017, in Astrophysics and Space Science Library, Modelling Pulsar Wind Nebulae,Vol. 446, ed. D. F. Torres, 29
Temim, T., Slane, P., Reynolds, S. P., et al. 2010, ApJ, 710, 309
Temim, T., Slane, P., Castro, D., et al. 2013, ApJ, 768, 61
Temim, T., Slane, P., Kolb, C., et al. 2015, ApJ, 808, 100
Temim, T., Slane, P., Sukhbold, T., et al. 2019, arXiv:1905.02849
Timokhin, A. N., & Harding, A. K. 2015, ApJ, 810, 144
Truelove, J. K., & McKee, C. F. 1999, ApJS, 120, 299
van der Swaluw, E. 2003, A&A, 404, 939
Van Dyk, S. D., Li, W., Cenko, S. B., et al. 2011, ApJ, 741, L28
Vink, J. 2012, A&ARv, 20, 49
Warren, J. S., & Hughes, J. P. 2004, ApJ, 608, 261
Warren, J. S., Hughes, J. P., Badenes, C., et al. 2005, ApJ, 634, 376
White, R. L., & Long, K. S. 1991, ApJ, 373, 543
Williams, B. J., Chomiuk, L., Hewitt, J. W., et al. 2016, ApJ, 823, L32
Xi, L., Gaetz, T. J., Plucinsky, P. P., Hughes, J. P., & Patnaude, D. J. 2019, ApJ, 874, 14
Yamaguchi, H., Badenes, C., Petre, R., et al. 2014, ApJ, 785, L27
Zajczyk, A., Gallant, Y. A., Slane, P., et al. 2012, A&A, 542, A12
Zhekov, S. A., McCray, R., Dewey, D., et al. 2009, ApJ, 692, 1190

IOP Astronomy

The Chandra X-ray Observatory
Exploring the high energy universe
Belinda Wilkes and Wallace Tucker

# Chapter 6

# X-Ray Binaries

**Michael A Nowak and Dominic J Walton**

## 6.1 Introduction

### 6.1.1 X-Ray Binaries at the Extremes of Flux

In the study of X-ray binaries, the *Chandra* X-ray Observatory brings to bear three unique attributes. The first two are commonly appreciated, namely its spatial resolution of $\lesssim 1''$ and its spectral resolution of $E/\Delta E \gtrsim 1000$ at 1 keV when using the High Energy Transmission Gratings Spectrometer (HETGS). The third attribute is less commonly recognized but also has proven to be a powerful tool. Namely, by alternating between the use of the transmission gratings and the imaging detectors, *Chandra* can achieve perhaps the widest range in observed flux of any single satellite mission ever flown. In terms of observed 0.5–8 keV flux, *Chandra* has observed sources as bright as the $\approx 10^{-8}$ erg cm$^{-2}$ s$^{-1}$ of the black hole (BH) candidate V404 Cyg in outburst, and as faint as $10^{-18}$ erg cm$^{-2}$ s$^{-1}$ for sources in the *Chandra* deep field surveys (see Chapter 8). This is 10 orders of magnitude, or in terms of astronomical magnitudes, 25 mag! This is essentially the equivalent of going from the faintest sources in the *Hubble* Deep Field to the faintest stars visible with the naked eye.

What makes this latter fact so important for observations of X-ray binaries is that they can be incredibly dynamic objects in terms of their luminous output. X-ray binaries over timescales of months or years can indeed vary by these many orders of magnitude in flux, with *Chandra* giving us the ability to follow an outburst of a single source from rise, to peak, and then decline into quiescence with the same observatory. There are two important examples that we will return to throughout this chapter: V404 Cyg and Cir X-1. V404 Cyg is a BH that had last been observed to go into outburst in 1989 (Kitamoto et al. 1989) and was in quiescence at the start of the *Chandra* mission. Early *Chandra* observations found a 0.5–8 keV quiescent level of $\approx 3 \times 10^{-13}$ erg cm$^{-2}$ s$^{-1}$ (Hynes et al. 2004, 2009), but HETGS observations of a recent outburst of V404 Cyg found a peak luminosity of $\approx 10^{-8}$ erg cm$^{-2}$ s$^{-1}$ (King et al. 2015), with *Chandra* then following V404 Cyg into a quiescent level

doi:10.1088/2514-3433/ab43dcch6

comparable to that previously observed (Plotkin et al. 2017). This is a remarkable 4.5 orders of magnitude (11.8 optical magnitudes) with *Chandra* observations alone, which allowed *Chandra* to serve as the linchpin in multisatellite/multiwavelength campaigns of V404 Cyg.

A second object that we shall return to a number of times throughout this chapter is Cir X-1, which has recently been recognized as perhaps the youngest known neutron star (NS) in the galaxy (Heinz et al. 2013). In contrast to V404 Cyg, at the start of the *Chandra* mission, Cir X-1 was bright with a 2–8 keV flux of $1.8 \times 10^{-8}$ erg cm$^{-2}$ s$^{-1}$ (Brandt & Schulz 2000). More recent observations have found it with a 0.5–8 keV flux as low as $10^{-11}$ erg cm$^{-2}$ s$^{-1}$ (Sell et al. 2010). This is 7.5 astronomical magnitudes, which has allowed *Chandra* to observe detailed features in the Cir X-1 high-resolution X-ray spectra (Brandt & Schulz 2000; Schulz & Brandt 2002) at bright epochs, and then image faint X-ray structures surrounding or in front of Cir X-1 at faint epochs (Sell et al. 2010; Heinz et al. 2013, 2015).

As will be discussed for these two examples and others, both *Chandra*'s high-resolution spectroscopic and imaging capabilities have not only allowed us to study the physics of these objects themselves, but have also allowed us to enhance our knowledge of the environment surrounding these exotic objects. X-rays from the vicinity of the compact object act as probes of the stellar winds of their companion stars. Additionally, these X-rays act as probes of the interstellar medium (ISM) between us and these objects, and have been used to map this medium both in the plane and in the halo of our Galaxy.

In this chapter, we first begin with a consideration of compact object sources in quiescence and discuss one of the first scientific debates for which *Chandra* played a crucial role: what are the observational differences between quiescent BHs and NSs, and what does this say about the underlying physics governing these systems? We then discuss *Chandra*'s contribution to the study of these systems in outburst, and how *Chandra* has been vital for understanding coronae, winds, and jets, and the transition among these processes, in compact object systems. We then turn to the role of *Chandra* observations in understanding the environments surrounding and in between these systems. Finally, we consider *Chandra* studies of "extreme" compact object systems: ultraluminous X-ray sources (ULXs; extreme apparent luminosities), and the recently observed electromagnetic counterpart of a merging binary NS system, GW 170817.

## 6.2 X-Ray Binaries in Quiescence

### 6.2.1 Black Hole Candidates versus Neutron Stars

Shortly prior to the launch of *Chandra*, there was debate as to whether there were significant observed X-ray flux differences between quiescent binary systems containing a BH or an NS (Menou et al. 1999). Observations suggested that quiescent systems containing BHs were fainter, for a given orbital period, than those that contained NSs (Garcia et al. 2001). Similarly, there were suggestions that BHs showed greater ratios between peak flux in outburst and minimum flux in quiescence as compared to NSs (Campana & Stella 2000). Debate centered around the question of whether BH systems are fainter because they lack a surface, or whether NS systems are brighter because they have a surface. This question was not merely tautological! It was suggested that comparable orbital periods, regardless of whether the compact object was a BH or NS, implied similar quiescent accretion rates, leading to two hypotheses. BHs, for a given orbital period, are fainter because their quiescent emission arises from radiatively inefficient accretion flows (RIAFs), with the bulk of the gravitational energy released via accretion being lost as thermal energy, either falling through the event horizon (Narayan et al. 1996) or escaping the system as an outflowing wind (Blandford & Begelman 1999). Conversely, the emission processes and radiative efficiency of the accretion flow are the same in BHs and NSs, except that for the latter we are additionally seeing residual thermal radiation from the NS surface (see, for example, Brown et al. 1998).

*Chandra* spectroscopic observations, in conjunction with coordinated observations in other wavelength bands, have pushed the field beyond this simple debate over possible flux differences among these systems. Some form of radiatively inefficient flow is generally accepted as being relevant to BH systems in quiescence, and modeling thermal radiation from NS surfaces is seen as crucial in understanding their quiescent states. However, both theory and observations provided by *Chandra* and other facilities have progressed to the point that detailed models are being applied to multiwavelength spectra of these systems. Debate now centers around the contributions of different physical components in the accretion flows of these systems (e.g., corona versus jet versus disk), and how observations constrain the underlying physics of these systems. This is especially true for quiescent NS systems, where X-ray observations are used to constrain models of NS cooling and NS equations of state. Below we consider separately the implications of *Chandra* observations for quiescent BH and NS models.

### 6.2.2 Jet and Advection-dominated States of Black Hole Candidates

Prior to the launch of *Chandra*, it was not even clear what components of the binary system were responsible for the observed X-ray fluxes of quiescent BH systems. For example, Bildsten & Rutledge (2000) suggested that for the very faintest BH systems observed, X-ray fluxes were consistent with coronal emission from the mass-donating main-sequence secondary *star*. Thanks to the superb spatial resolution of *Chandra*, which essentially eliminates background concerns in these quiescent systems, coupled with its CCD-quality spectra, astronomers were able to obtain $\approx$0.5–8 keV spectra over a wide range of quiescent X-ray flux levels. One of the earliest *Chandra* studies of quiescent BHs (Kong et al. 2002) demonstrated that four different sources (A0620–00, GRO J1655–40, XTE J1550–564, V404 Cyg) showed similar power-law spectra with the brighter luminosities being too high for the coronal model. Kong et al. (2002) argued that these power laws were broadly consistent with the expectations of RIAF models, and further suggested that the lack of any detectable thermal emission (from a boundary layer or surface emission) was evidence of an event horizon in these systems. Similar arguments were made for

*Chandra* observations of XTE J1118+480 in quiescence by McClintock et al. (2004). It should be noted, however, that it is not the lack of thermal emission per se, but rather the lack of surface emission that is key. Even in the standard "Shakura–Sunyaev" thermal disk model (Shakura & Sunyaev 1973), the lack of additional boundary layer emission—contrary to the extra thermal component sometimes seen even in bright NS X-ray sources—could have just as readily been argued as evidence for an event horizon.

The earliest RIAF models (Ichimaru 1977; Narayan et al. 1996) considered gravitational potential energy released by the accretion flow as being lost as thermal energy being swallowed by the event horizon. Later models instead considered the role of outflows transporting this energy away from the system (e.g., Blandford & Begelman 1999). The role of outflows in BH systems, specifically jets, had already been under serious consideration owing to the discovered correlation between X-ray and radio fluxes in the bright, spectrally hard states of BH systems (see Hannikainen et al. 1998, who discovered this correlation in GX 339–4). Later observations showed that the radio/X-ray correlations in GX 339–4 extended over a range of four magnitudes in X-ray flux, with a characteristic relationship of radio to X-ray fluxes of $F_R \propto F_X^{0.7}$ (Corbel et al. 2003). We show the current observational status of this correlation in Figure 6.1 (Bahramian et al. 2018), which includes data for BH and

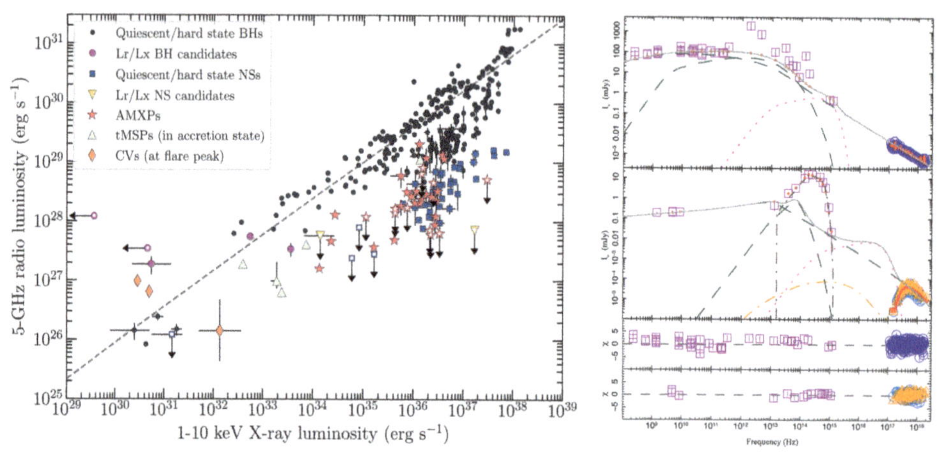


**Figure 6.1.** Left: observed correlation between radio flux and X-ray flux in various types of compact object systems (reproduced from Bahramian et al. 2018). The dashed line shows $F_R \propto F_X^{0.7}$, as appropriate for black hole systems. Right: a synchrotron-emission-dominated jet model of the low-luminosity active galactic nucleus M81* (top) and the quiescent black hole system V404 Cyg (middle), reproduced from Markoff et al. (2015) © 2015. The American Astronomical Society. All rights reserved. (The bottom two panels show the $\Delta\chi$ residuals from the fits presented, respectively, in the top two panels.) In both cases, the X-ray observations are from *Chandra* observations, and radio/IR/optical data have been obtained from a variety of facilities. The gray and red lines and points show the full fitted model, while the dashed and dotted lines show individual model components. (Model components include synchrotron emission, synchrotron self-Compton, disk emission, optical emission from either the stellar companion or the galaxy, line emission, and absorption; see Markoff et al. 2015 for a full description.) The model ties numerous parameters together between the fit to M81* and V404 Cyg to demonstrate that the quiescent accretion flow physics and emission mechanisms are fundamentally the same in each of these two systems, when one accounts for mass scaling of the relevant model parameters.

various types of NS systems. This nonlinear relationship between radio and X-ray luminosity even holds true for supermassive BHs, and in fact, these more massive systems can be placed on the stellar-mass BH relationship if one includes a nonlinear dependence upon BH mass, $M$. This leads to what has been dubbed the "fundamental plane" of BH radio/X-ray activity (Gallo et al. 2003; Merloni et al. 2003; Falcke et al. 2004). In the scheme described by Merloni et al. (2003), $\log L_R = (0.6 \pm 0.11)\log L_X + (0.78^{+0.11}_{-0.09})\log M$.

Fender et al. (2003) pointed out that the existence of the nonlinear relationship for stellar-mass BHs allows for the transition to quiescence, not being one of advective loss of energy through the event horizon, but rather one of a transition to a jet-dominated state. Specifically, they suggested that total accretion power scales with accretion rate, $\dot{m}$, but that X-ray luminosity $L_X \propto \dot{m}$ above the "transition" to quiescence and $\propto \dot{m}^2$ below the transition to quiescence, while the jet power $L_j \propto \dot{m}^{0.5}$ above the transition and $\propto \dot{m}$ below it. This would reproduce the $F_R \propto F_X^{0.7}$ in simple optically thick jet models (e.g., Markoff et al. 2001). The ability of *Chandra* to obtain background-free spectra down to low flux levels has become crucial in developing these concepts further. Because the *Chandra* spectral imaging capabilities are so well matched to both those of radio and optical, these correlated studies were extended to include optical spectra (e.g., Hynes et al. 2004) and carried to extremely low flux levels in both the X-ray and radio. See the *Chandra* observations of A0620–00 discussed by Gallo et al. (2006), who, with coordinated VLA radio observations, showed the correlation extending down to a 2–10 keV X-ray flux $(4^{+0.5}_{-1.4}) \times 10^{-14}$ erg cm$^{-2}$ s$^{-1}$ and a radio flux of 51 $\pm$ 7 $\mu$Jy.

*Chandra* spectroscopy helped demonstrate that, as sources approached quiescence, their power-law spectra (i.e., $F_\gamma \equiv$ photons/area/time/energy $\propto E^{-\Gamma}$, where $E$ is the photon energy) softened to a photon index of $\Gamma \approx 2.1$ (Corbel et al. 2006), in contrast to the $\Gamma \approx 1.7$ index typically seen as a BH source first transitions from a spectrally soft state to a bright, spectrally hard state. Hynes et al. (2009) conducted a study of V404 Cyg with *Chandra* that included radio, optical, and UV coverage, and demonstrated that there was a mid-IR excess more consistent with an extrapolation of the radio spectrum, as opposed to an unmodeled excess from the optical companion, further strengthening arguments for a jet model underlying observed quiescent BH spectra. Results such as these gave impetus for researchers to use *Chandra* observations in conjunction with those in other wavelength bands to constrain more sophisticated jet models.

One set of examples are the jet models of Markoff et al. (2015), who, highlighting the existence of the fundamental plane, considered models jointly applied to the stellar-mass BH V404 Cyg and the supermassive BH M81*. They analyzed a set of spectra where each source had a similar X-ray luminosity scaled relative to its Eddington luminosity, $\ell_X \equiv L_X/L_{Edd} \approx 10^{-6}$, where the Eddington luminosity $L_{Edd} \equiv 4\pi G M m_p c/\sigma_T$ is the luminosity at which the inward pull of gravity by an object of mass $M$ is balanced by the outward push of radiation pressure. (Here, $G$ is the gravitational constant, $c$ is the speed of light, $m_p$ is the mass of the proton, and $\sigma_T$ is the Thompson cross section.) As much as possible, jet model parameters were cast in terms independent of mass, and accretion rates were described relative to the system's Eddington luminosity, so that the core parameters of the model could be tied together for the models of V404 Cyg and M81*. Additionally, these models considered multiple emission process: thermal synchrotron, nonthermal synchrotron (from a postshock regime, where electrons have been accelerated into a power-law distribution), and synchrotron self-Compton (SSC) all from a jet; disk emission; and line emission (e.g., from fluorescence from the disk). Figure 6.1 shows a relatively successful multiwavelength fit using this approach for a model where the observed emission is dominated by synchrotron processes. The most successful fits further suggested that each jet's magnetic field energy density was roughly in equipartition with the thermal pressure, and that the electron acceleration in the jet was either inefficient or that the jet was in a cooling-dominated regime.

Connors et al. (2017) employed a similar analysis approach, but consider *Chandra* observations of even fainter sources: A0620–00, a quiescent stellar-mass BH with $\ell_X \sim 10^{-8.5}$, and Sgr A*, the supermassive BH at the center of our Galaxy, in a flaring state with $\ell_X \sim 10^{-9}$. Using an updated version of the jet model of Markoff et al. (2015), these joint radio/optical/ *Chandra* X-ray spectra were instead best described with SSC-dominated emission and a jet that had a subequipartition magnetic field, but still implied either inefficient particle acceleration, or a jet in a synchrotron-cooling-dominated regime. Plotkin et al. (2017), using *Chandra*, the Karl G. Jansky Very Large Array (JVLA), and the Very Large Baseline Array (VLBA), also modeled joint observations of the 2015 decay into quiescence of V404 Cyg with an SSC-dominated, radiatively cooled jet model. They also note that in this decay to quiescence, V404 Cyg makes a fairly rapid transition, over the course of only three days, from a photon index $\Gamma \approx 1.7$ to a softer index of $\Gamma \approx 2$ at a fractional Eddington luminosity of $\ell_X \sim 10^{-5}$. They identify this spectral evolution as a "state change" to quiescence.

Which jet model is the definitive description for quiescent BH emission is still subject to debate. *Chandra* observations, however, are seen to be crucial to these studies. The specific ability of *Chandra* to determine the spectral shape over a very wide range of very faint luminosities ranging from $\ell_X \sim 10^{-9}$–$10^{-4}$, free from the worry of background, when applied in conjunction with comparably sensitive measurements in radio, IR, and optical wavelength bands, offers the most promising path forward.

### 6.2.3 Quiescent and Cooling Neutron Stars

Studies of quiescent NS spectra with *Chandra* have provided insights into the structure of NSs and the equation of state of nuclear matter. Modern thermal atmosphere models are parameterized by the NS distance, mass, radius, and surface temperature observed at infinity (see Heinke et al. 2006). If the quiescent NS distance is known, fits to its spectra, typically dominated by the emission of low-temperature thermal radiation, would allow one to constrain the NS mass–radius relationship, and therefore its equation of state. Similarly, if there is a good estimate of the long-term average heating of the NS via accretion onto its surface during bright phases, one can learn about cooling processes in the NS, and hence NS structure, by studying the long-term evolution of the cooling spectrum during quiescent phases.

The success of such studies, however, relies on either knowing that all accretion has ceased, or being able to clearly identify the thermal surface emission component of the spectrum separate from any other spectral components. A number of low-flux NS systems exhibit a hard X-ray tail (Rutledge et al. 2001; Jonker & Nelemans 2004). Theories for this component have included atmosphere effects (see Ho & Heinke 2009), synchrotron radiation from either a jet or an outflowing pulsar wind interacting with an incoming accretion stream (Campana et al. 1998; Campana & Stella 2000), or emission from an optically thin, hot thermal plasma, possibly associated with an outflowing RIAF (Narayan & Yi 1995). Figure 6.2 shows 4U 2129+47, which when first observed by *Chandra* in 2001 exhibited a hard tail and furthermore showed sinusoidal modulation of its soft X-ray flux (as well as a rapid, total periodic eclipse by the secondary) over the source's 5.24 hr period (Nowak et al. 2002). Both facts taken together were interpreted as being due to active, albeit low-level, accretion in this system. This interpretation was bolstered by subsequent *Chandra* observations (see Figure 6.2 and Lin et al. 2009), as both the hard X-ray tail and sinusoidal modulation disappeared together. Furthermore, the soft X-ray flux dropped by roughly one-quarter, with much smaller drops in flux in all subsequent observations. Joint *XMM-Newton* and *NuSTAR* observations discovered a cutoff in the hard X-ray tail of the highly variable, quiescent NS Cen X-4. Coupled with historical 0.2–10 keV observations, including from *Chandra*, this was also used to argue that such hard tails are indicative of active ongoing accretion (Chakrabarty et al. 2014).

This points out the importance of being able to verify that active accretion does not occur when studying quiescent NS systems. If there is no active accretion, then

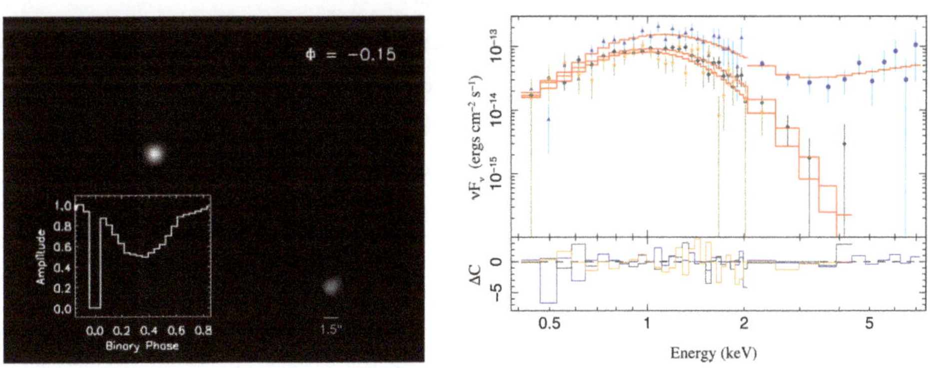


**Figure 6.2.** Left: an animated GIF of *Chandra* images of the quiescent NS 4U 2129+47 observed over the course of its 5.24 hr binary orbit (Nowak et al. 2002). The inset shows the normalized, averaged 0.5–2 keV phase-dependent light curve. (The source in the lower right is a foreground star.) Right: *Chandra* spectra (integrated over out of eclipse phases) of 4U 2129+47 taken in 2001 (blue), 2007 (green), and 2012 (orange), fit with absorbed, thermal emission from the neutron star surface plus a power law. (Note the power-law disappearance of the latter two spectra, and the slight cooling between 2007 and 2012.) Additional materials available online at http://iopscience.iop.org/book/978-0-7503-2163-1.

soft X-ray measurements of the spectrum can constrain NS cooling models and physics. There are three possible types of cooling NS systems to consider, depending upon the mechanism responsible for heating the compact object. The NS could have been heated in the initial supernova explosion (as for the case of the central compact object in the Cas A supernova remnant; see Ho & Heinke 2009; Ho et al. 2019), or the NS surface could have been heated by a (relatively) brief episode of active accretion, or the deep interior of the NS could have been heated by prolonged accretion characterized by a long-term average rate (see Figure 6.3, and the review by Wijnands et al. 2017). *Chandra* observations have made important contributions to the study of each of these three cases.

Residual heat after the formation of the NS in a supernova explosion is a rare, but important, example. As discussed in the review by Ho et al. (2019), the newly formed, hot neutron star will initially cool by a combination of thermal surface emission and neutrino emission in three observable stages. On timescales of order one hundred years, the NS surface temperature remains roughly constant as the NS interior cools predominantly via neutrino emission. Thermal conductivity, however, delays the appearance of this cooling front in the surface temperature. This is then followed by a rapid cooling phase, also lasting on the order of only a few hundred years, as the cooling front from the core reaches the surface. The neutrino emission then continues to dominate cooling for on the order of 100,000 years, with the interior temperature scaling with time, $t$, as $t^{-1/6}$, before photon cooling from the surface becomes the dominant mechanism and, the interior temperature then scales as $t^{-1/2}$. (See the discussion and references in Ho et al. 2019.)

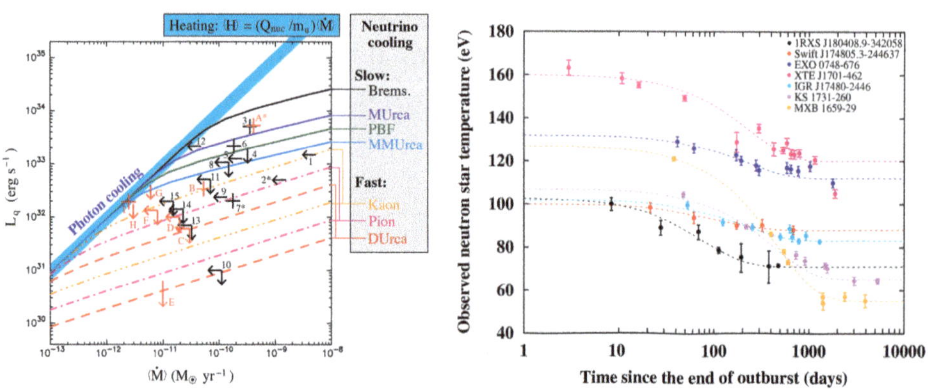


**Figure 6.3.** Left: observed X-ray luminosity for various quiescent NS systems versus estimated long-term average accretion rate during outburst (reproduced from Wijnands et al. 2017 © Indian Academy of Sciences 2017). The lines are for different theoretical models of NS core cooling, with "slow cooling" models represented by solid lines toward the top of the figure, and "fast cooling" models represented by dashed/dotted lines toward the bottom of the figure. The "photon cooling" line is where the observed luminosity is equal to the presumed long-term averaged heating rate. (The source 4U 2129+47 shown in Figure 6.2 is the leftward-pointing arrow labeled 1, while 1H 1905+000 is represented by the leftward-/downward-pointing arrows labeled 10, and SAX J1808.4−3658 is represented by the downward-pointing arrow labeled E.) Right: fitted surface temperatures for various NSs observed over time, post cessation of active accretion, with curves showing model fits using various crustal cooling models (reproduced from Wijnands et al. 2017 © Indian Academy of Sciences 2017).

The details of and spectral emission associated with the thermal emission phase of the cooling will depend upon the composition of the NS atmosphere. Any residual material from accreted fallback material potentially will undergo nuclear burning to higher Z elements. For example, it has been suggested that the main component of the atmosphere for the NS in the Cas A supernova remnant is carbon (Ho & Heinke 2009). This supernova remnant is likely only 340 years old (based upon observations of expansion of the supernova remnant; Fesen et al. 2006), and hence the NS is in an early phase of cool down, possibly the rapid cool-down phase after the cooling front has reached the surface, after the initial formation of the NS. *Chandra* observations have been extremely important to the study of this NS as, prior to the advent of its arcsecond X-ray imaging capability, the NS remained undiscovered. Its emission could not be separated from that of the rest of the SN remnant. *Chandra* has performed multiple observations of this system over the past two decades, allowing for an estimate of its temperature drop over this timescale. Because this system has no known binary companion providing accreted material, it should behave purely as an isolated, cooling NS. Models of these observations presented by Ho & Heinke (2009) and Ho et al. (2019) suggest that the observed emission is from a carbon-rich atmosphere, and further that the surface temperature of the NS has dropped by $\sim$3% (from $1.87 \times 10^6$ K to $1.80 \times 10^6$ K) over the lifetime of *Chandra*. Other analyses of a subset of these data, using different atmosphere models, do not find a significant temperature drop (Posselt & Pavlov 2018). The observations of Cas A are particularly challenging to analyze because they occur at the very lowest energy end of the *Chandra* bandpass, where degradation of the instrument response due to contamination buildup is an issue. As calibration of the soft X-ray response is improved, certainty about whether or not Cas A is cooling will increase. Regardless, even attempting such measurements as described by Ho & Heinke (2009) would be impossible with any other current X-ray instrument.

The heating of the NS in the Cas A supernova remnant occurs during its initial formation. However, if the NS star is in an X-ray binary and periodically undergoes episodes of accretion, then there is the potential to heat the interior of the NS. If the accretion rate is high enough such that the accreted material achieves a temperature higher than the interior of the NS, then heat will be driven inward by thermal conduction. The details of this heating are complicated, with the heating thought to occur at a deep level within the crust, which is hotter then the surface temperature of the NS (allowing for the possibility of some heat flowing outward), but still hotter than the NS interior (sending net heat inward on a timescale of order 10,000 years; see Brown et al. 1998).

Under the assumption that the periods of heating are episodic, shorter in duration than the quiescent periods in between heating episodes, and shorter than the characteristic cooling timescale of the NS, then the surface temperature will come into equilibrium with the internal temperature of the NS. This temperature in turn will be determined by the long-term (10,000 year) average mass accretion rate onto the NS as well as by the cooling mechanisms at work in the NS (see the review by Wijnands et al. 2017 and references therein). If the cooling of the NS were solely due to emission from the NS surface, such cooling would be rather slow and would drive the NS emission toward a high flux and temperature for a given average accretion rate. This is illustrated on the left side in Figure 6.3 (compiled by Wijnands et al. 2017, based upon the references given in that review), which shows the observed quiescent luminosity of a number of quiescent NS systems relative to their estimated long-term average accretion rate. The long-term average rate is based upon the observed outburst history over the past several decades coupled with theoretical models of outburst recurrence timescales.

All systems are seen to be at a lower temperature than would be the case solely for photon emission; however, this is expected, due to additional cooling from neutrino processes. Very broadly speaking, these processes can either be "slow" or "fast." A description of these processes are beyond the scope of this chapter, but a broad range of predictions for both slow and fast processes are shown in Figure 6.3 (see Wijnands et al. 2017 for references to the specific mechanisms labeled in this figure). In general, however, a fair fraction of the observed systems require fast cooling mechanisms. This was noted with the *Chandra* observations of KS 1731−260 by Wijnands et al. (2001). After a decade-long outburst, this source settled into quiescence at a flux level of $2 \times 10^{33}$ erg cm$^{-2}$ s$^{-1}$, which would be consistent with slow cooling processes only if the outburst recurrence timescale in this system were significantly longer than 100 years.

Even more extreme examples were provided by *Chandra* observations of SAX J1808.4−3658 (Heinke et al. 2009) and 1H 1905+000 (Jonker et al. 2006, 2007). for which only upper limits on the flux were obtained. No photons were detected with 325 ks of *Chandra* observations of 1H 1905+000. Given the superb imaging capabilities and low background of *Chandra*, these limits are extremely constraining, as seen in Figure 6.3. Given the possibility of ongoing, low-level active accretion, the luminosities depicted in Figure 6.3 are upper limits, which means that for the most extreme low-luminosity systems, unless their recurrence timescales are extremely long (and hence their average accretion rates are extremely low), fast cooling processes are dominant. The brighter systems, however, could be consistent with slow cooling processes. Differences between these two extremes of populations could be due to the underlying mass of the NS, with higher mass systems being subject to "fast" cooling processes (see the discussion and references in Wijnands et al. 2017).

For sufficiently long timescales past outburst (years to decades), the observed NS is expected to come into equilibrium with the internal temperature. However, for systems that undergo recurrent outbursts, or are observed only shortly after outburst, crustal cooling mechanisms come into play instead. During the outburst, pycnonuclear reactions occurring at a few hundred meters depth in the NS atmosphere release on the order of 2 MeV/nucleon, heating the crust to a temperature higher than that of the interior and bringing the surface out of equilibrium. After accretion stops, the crust then cools on a timescale determined by its composition and structure. If the atmosphere structure is highly ordered, heat conduction is efficient and cooling can be rapid (Shternin et al. 2007; Brown & Cumming 2009). *Chandra* observations of EXO 0748−676 in quiescence, after it completed a 24 year outburst, suggested that cooling is indeed rapid, although this interpretation is complicated by the fact that a variable, hard X-ray power-law tail is also seen in quiescence in this system (Degenaar et al. 2011b). Cooling curves for a number of post-outburst systems observed by *Chandra* (and other X-ray instruments), along with model fits, are shown in Figure 6.3 (from Wijnands et al. 2017).

In some cases, modeling the *Chandra* observations of these systems not only constrains the atmospheric structure of the NS, but also indicates that there may be extra heating beyond the expected 2 MeV/nucleon. Degenaar et al. (2011a), modeling *Chandra* observations of the post-outburst light curve of IGR J17480 −2446, suggested that both efficient cooling and extra heating were required. On the other hand, Degenaar et al. (2015) found that no extra heating was required to model the *Chandra*-observed light curve of Swift J174805.3−244637. Studies of these cooling NS systems continues to be an active area of *Chandra* research.

### 6.2.4 Transitional Neutron Star Systems

Observations with the *Rossi X-ray Timing Explorer* led to the discovery of millisecond X-ray pulsations in a class of NS systems now dubbed accreting millisecond X-ray pulsars (AMXPs), confirming the previously long-held hypothesis that low-mass X-ray binary systems (LMXBs) are the progenitors of millisecond radio pulsars (MSPs) via accretion spin up of these systems (Wijnands and van der Klis 1998). Radio pulsations are not expected to be observable in AMXPs during active phases, as the radio field is "buried" by the accretion. Once accretion ceases, then the radio pulse can turn on. Thus, some fraction of the quiescent systems discussed above should be observable as MSP once they enter a truly "quiescent" phase. Furthermore, given the variable X-ray activity recorded by *Chandra* and other X-ray instruments for these systems, one might expect a fraction of these systems to transition between MSP and LMXB phases.

Such transitional millisecond X-ray pulsars (TMXPs) indeed have been discovered in just the past decade. The first such discovered system was PSR J1023+0038, which was seen to transition from an LMXB phase (indicated by the presence of an accretion disk determined from optical observations, among other observations) to an MSP phase (Archibald et al. 2009). A similar transition from LMXB to MSP was also observed for PSR J1227−4853 (Roy et al. 2015). Subsequent X-ray observations of PSR J1023+0038 showed it transitioning back into an LMXB phase (Takata et al. 2014; Patruno et al. 2014; Stappers et al. 2014), showing that such quiescent sources can transition back and forth between the two phases on short timescales.

In fact, the first such source to show repeated transitions between LMXB and MSP phases was PSR J1824−0038 (Papitto et al. 2013). *Chandra* observations were not only used to precisely localize the X-ray source, but also provided historical measurements of its X-ray flux variability. This allowed one, with hindsight, to see that the source likely made multiple transitions between these two phases over a few-year timescale.

*Chandra* observations will continue to be an important component of this new field of study. Its imaging capability will be used for precise localization of the X-ray counterpart (transient AMXPs and LMXBs are often found in extremely crowded regions of the sky), variability signatures will be searched for with the High Resolution Camera (HRC), and the (typically faint) spectra will be studied with ACIS.

## 6.3 X-Ray Binaries in Action

In the previous section, we discussed how the imaging spectroscopy of *Chandra*, partly relying on its superb angular resolution, allowed the study of X-ray binaries in potentially extreme quiescent states. Here we consider *Chandra* studies of "bright" X-ray binaries that are clearly actively accreting. There is no formal definition of "active" versus "quiescent," but roughly speaking, here we are considering sources emitting at or above $10^{-4}$ of their Eddington luminosity (perhaps substantially so). For nearly all X-ray binaries observed within the galaxy, this would lead to a *Chandra* spectrum that is "piled up" (i.e., more than one photon landing in the same pixel region within a single CCD readout frame, leading to the event either being dismissed as a particle event, or resulting in an incorrectly assigned energy; see Davis 2001 for a description of this process). Both of necessity, and for many of the interesting scientific questions that *Chandra* addresses in the study of X-ray binaries, this means that the High Energy Transmission Gratings Spectrometer instrument is typically employed. High- resolution X-ray spectroscopy has therefore become a prime area of focus for *Chandra* studies of actively accreting binaries.

### 6.3.1 Probes of the Inner, Relativistic Accretion Flow

If an X-ray spectrum extends to high energies (often in a power-law spectrum with $\Gamma \lesssim 2$, followed by an exponential cutoff at energies at $\gtrsim$50–100 keV), and illuminates optically thick, cold material, one expects a "reflection spectrum" from the cold material, which is a modified form of the incident power law that exhibits a "Compton hump" from backscattered radiation, edge structure (from absorption, e.g., by inner K shell and M shell electrons being liberated), and fluorescence emission lines (from recombinations culminating, e.g., in an L to K transition), together yielding a characteristic spectrum (George & Fabian 1991). Based upon expected abundances, cross sections, and fluorescence yields, the 6.4 keV line of Fe is expected to be the most prominent line in this process. If this (or any other identifiable) fluorescence line originates close to the surface of an NS or the event horizon of a BH, then the relativistic motions of the accreting material, along with the strong gravity of the compact object, should distort this line into a broadened shape. This profile typically has a prominent extended tail redward of the expected energy (due to special relativistic Doppler shifts and gravitational redshifting) and a sharper cutoff blueward (due to special relativistic Doppler shifts) of the expected energy (see the review by Reynolds & Nowak 2003). Modeling the profile of this line for a given system potentially can lead to measurements of the underlying parameters of the compact object, namely, the spin of the BH.

Evidence for such lines existed long before the launch of *Chandra* (see, for example, Fabian et al. 1989); however, the spectral resolution of these instruments made it very difficult to distinguish between narrow and broad Fe line profiles, or at the very least, made it difficult to separate narrow components (due to reflection or emission from distant, slowly moving material), from broad-line components. Although *Chandra* is capable of measuring broadened line profiles, its chief contribution to this area of study has been to definitively measure, or strongly constrain, the *narrow* component of the line, allowing for more confident measurements of broad-line profiles.

Figure 6.4 shows a *Chandra*-HETGS measurement of the Fe line region of the BH Cyg X-1 (Miller et al. 2002), where both broad and narrow components are present, but the former dominates the line flux. Similar measurements were made for other BH systems, e.g., GX 339–4; however, that system only exhibited a broad component of the line (Miller et al. 2004). Surveys of the Fe line region were also conducted for NS systems (Cackett et al. 2009), with two systems, GX 17+2 and GX 349+2, clearly exhibiting a broad-line component. The line region for the latter system is also shown in Figure 6.4 and shows excellent agreement between *Chandra* and lower (CCD) spectral resolution *Suzaku* measurements. An additional example of a broad line in an NS system is the case of Serpens X-1, where an extremely deep 300 ks *Chandra*-HETGS observation only showed a broad-line component, well described by the relativistic model, without any indications of complications due to narrower features (Chiang et al. 2016).

For the case of broad lines in NS systems, there has been some debate as to whether a well-formed disk can exist so close to the NS surface (e.g., due to disruption by the NS magnetic field, or X-ray irradiation from emission from the surface of the NS), in which case the broadened line would not be due to general relativistic effects, but instead might be the result of broadening by, e.g., bulk motion Comptonization. *Chandra*-HETGS observations were also used to address this question by studying the absorption spectra of NS systems exhibiting broad Fe line

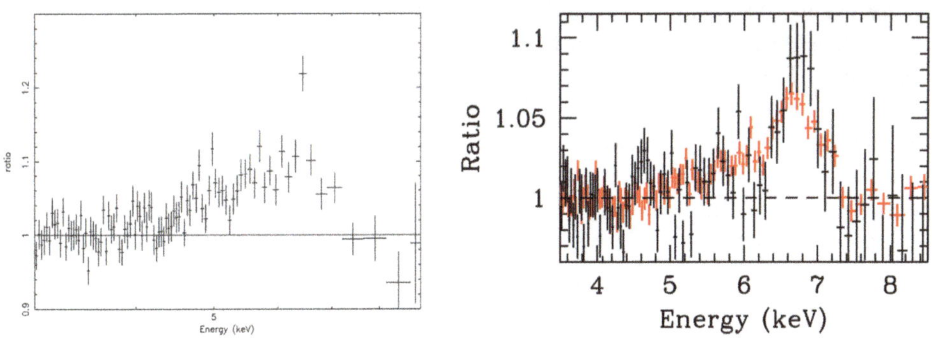


**Figure 6.4.** Left: ratio residuals in the Fe K$\alpha$ region for a power-law fit to *Chandra*-HETGS spectra of Cyg X-1 (reproduced from Miller et al. 2002 © 2002. The American Astronomical Society. All rights reserved). The residuals show the presence of a broad line, consistent with relativistic broadening, but a narrow component can also be seen near 6.4 keV. (See also Figure 6.8.) Right: ratio residuals in the Fe K$\alpha$ region for a power-law fit to simultaneous *Chandra*-HETGS (black) and *Suzaku* (red) spectra of GX 339–4 (reproduced from Cackett et al. 2009 © 2009. The American Astronomical Society. All rights reserved). Again, the presence of a broad line is indicated (with little evidence for a narrow core), and consistency is found between the HETGS and *Suzaku* spectra.

profiles (Cackett & Miller 2013). If bulk motion scattering in an infalling wind were the dominant mechanism for broadening the line profile, one might expect a correlation between broad-line properties and the depth and width of (local) absorption-line profiles. None were observed, leaving the relativistic line broadening interpretation as the more favored one (Cackett & Miller 2013). We shall see below that *Chandra*-HETGS studies of absorption in X-ray binary systems has become an important tool in understanding the physical processes occurring in these systems.

### 6.3.2 Outer Disk Structure

Just as *Chandra* observations help clarify the structure of the inner, relativistic regions of the accretion flow by clearly delineating the broad and narrow components of the line region, they have also been used to elucidate outer disk structure. This has been primarily, but not exclusively, via observations of high-inclination (i.e., nearly edge-on) binary systems. Many of these have fallen under the classification of "accretion disk corona" (ADC) or "dipper" sources, as we discuss below.

Returning to the example of Cir X-1, this was one of the first high-inclination sources for which *Chandra*-HETGS observations revealed evidence of an outflow in the form of P Cygni profiles associated with hydrogen-like and helium-like ions from a number of different elements (Ne, Mg, Si, S, Ar, Ca, and Fe; Brandt & Schulz 2000; Schulz & Brandt 2002). A P Cygni profile is characterized by redshifted emission and blueshifted absorption, as would be expected for a radial, near equatorial flow, in a highly inclined system. In this case, the velocity difference between the absorption and emission components was seen to be variable and ranged from 200–1900 km $s^{-1}$. It was argued that these winds were describable with the thermally driven wind models of Begelman et al. (1983), or possibly radiatively driven wind models (see Proga & Kallman 2002 and references therein). Here, and for the other outflows that we discuss in this section, it might be appropriate to distinguish these winds from those discussed below in Section 6.3.3, where there is still debate as to the driving mechanism. This distinction is not firm, as we discuss below, and continues to be an extremely active area of research for *Chandra*. In Section 6.3.3, we specifically consider the relationship between the presence of wind outflows versus jet outflows as a function of significant variations in both the binary's flux and the emergent X-ray spectral shape. In this section, we concentrate on structures that are more clearly associated with heating/radiative driving in the outer regions of the disk.

*Chandra*-HETGS generally has found spectroscopic signatures of wind absorption in the so-called "dipper" sources. These tend to be bright binary systems with an NS central compact object, viewed at inclination angles of $\approx 60°$–75 °, where soft X-ray dips are observed over the course of the system's orbital period. (The dips, however, are not strictly periodic.) A typical example is 4U 1916–05, where absorption from H-like ions of Ne, Mg, Si, and S, and from both Fe XXV and Fe XXVI, is observed with essentially no velocity shifts, and narrow widths (Juett & Chakrabarty 2006). Modest velocities and narrow widths for the absorption lines are consistent with the X-ray source being much smaller than the absorber, which is presumed tied to the rotation of the outer disk rather than outflowing, such that the motion of the absorber is predominantly perpendicular to our line of sight. Comparable results were found for the dipper source 1A 1744−361 (velocity widths and offsets $\lesssim$200 km s$^{-1}$; Gavriil et al. 2012). For the case of 4U 1624–490, velocity widths of the absorption lines again were modest, but photoionization models of the absorber suggested two zones: one with a radius of $\approx 3 \times 10^{10}$ cm, and one coincident with the outer edge of the disk (Xiang et al. 2009 and see also Iaria et al. 2007).

For sufficiently high inclination angles, enough of the central compact object X-ray source can be obscured from our line of sight such that (typically weak) line emission from the outer disk can be seen in contrast to the weak, likely scattered into our line of sight, continuum. A prominent example of this is for *Chandra*-HETGS observations of EXO 0748−676 (before it went into quiescence; see Section 6.2.3 above), where not only absorption is seen, but line emission and radiative recombination continua (emission from thermal free electrons recombining with ionized material) are also observed (Jimenez-Garate et al. 2003). The relative strength of these lines become more pronounced during dips, again indicating both that the central source is obscured and that the line emission is predominantly associated with the heated outer rim of the accretion disk.

For inclination angles $\approx 85°$, the central X-ray source is obscured, and we are likely observing continuum radiation indirectly, e.g., via scattering in the extended heated atmosphere. Thus, in this case, the X-ray source is extended. A further consequence of the high inclination angle, and extended nature of the continuum X-ray source, is that we observe broad, partial eclipses of the X-ray source on the orbital period of the system. Such partially eclipsing, extended X-ray sources are referred to as ADC sources, with the classic example being X1822−371 (see Heinz & Nowak 2001 and references therein). In fact, this was also the likely model for 4U 2129+47 (see Figure 6.2) when it was in outburst decades before the launch of *Chandra* (McClintock et al. 1982). For the case of X1822−371, *Chandra*-HETGS observations reveal the orbital phase-dependent emission lines associated with the X-ray heating of the outer disk rim, as illustrated in Figure 6.5 (from the analysis of Ji et al. 2011; see also Iaria et al. 2013). The outer disk rim extends to the tidal truncation radius in the binary system and is raised due to the interaction of the incoming accretion stream with the disk. This disk rim itself also leads to a periodic modulation of the X-ray continuum. We see the most prominent line emission when the disk/accretion stream is within our line of sight, and the secondary position is perpendicular to our line of sight (Ji et al. 2011, and Figure 6.5).

### 6.3.3 Wind Structure, Driving Mechanisms, and Transitions to Jet-dominated States

The presence of wind absorption lines in X-ray binary systems was known prior to the launch of *Chandra*, e.g., from observations of the BH GRO J1655–40 (Ueda et al. 1998) and GRS 1915+105 (Kotani et al. 2000) by the ASCA observatory. However, these studies were only performed at CCD resolution with an instrument that was difficult to use with such bright sources (in these cases, a substantial fraction of the X-ray flux of the Crab Nebula and pulsar, which is among one of the brightest X-ray sources in the sky). Furthermore, these CCD-quality spectra only provided

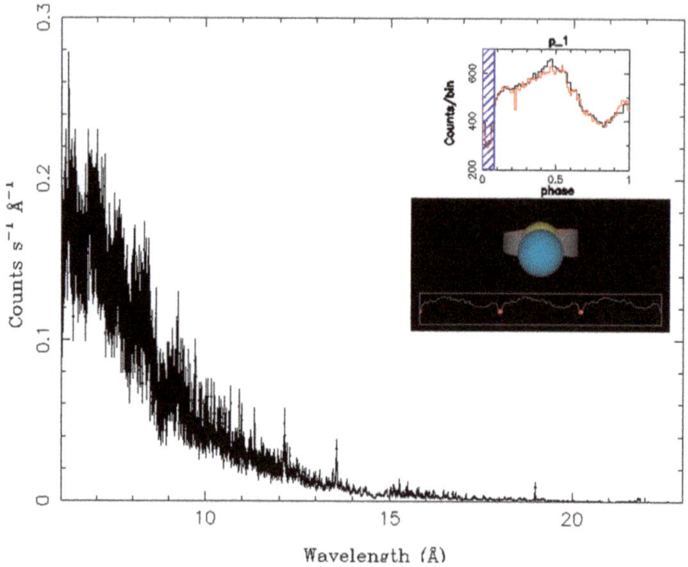


**Figure 6.5.** Animated GIF of *Chandra*-HETGS spectra of the low-mass X-ray binary (LMXB) X1822−371 varying as a function of phase over its 5.6 hr orbit (Ji et al. 2011). The insets show the X-ray light curve as a function of orbital phase (as well as the phase range used to create the spectrum in each animation frame), and a figure, drawn to proper relative scale, of the binary geometry. The latter shows the secondary orbiting the compact object, which is likely surrounded by both a disk with a flared edge and an inner "accretion disk corona," which is an extended atmosphere due to the X-ray heating of the disk (Heinz & Nowak 2001). Additional materials available online at http://iopscience.iop.org/book/978-0-7503-2163-1.

modestly adequate spectral resolution near the Fe K$\alpha$ line region and were even more difficult to use at lower energies. (For CCD spectra, resolution $E/\Delta E$ approximately scales as $E^{1/2}$, and line regions tend to be more crowded at lower energies.) *Chandra*-HETGS observations bring to bear higher resolution that improves toward lower energy, allowing a more precise measurement of wind properties in these systems. One of the first *Chandra* spectroscopic studies of wind outflows in a BH system was for GRS 1915+105 (Lee et al. 2002). The spectra were extremely bright (2–25 keV flux $\approx 2 \times 10^{-8}$ erg cm$^{-2}$ s$^{-1}$, i.e., roughly 0.4 times the flux of the Crab), and describable as a heavily absorbed (equivalent neutral column $N_{\rm H} = 5 \times 10^{22}$ cm$^{-2}$) disk spectrum (with a peak temperature of $kT \approx 1.4$ keV) with a significant power-law tail of photon index $\Gamma = 2.4$. This would be classified as a moderately "hard" (i.e., dominated by high-energy X-rays) spectrum of the type that in this source is also associated with radio emission from an outflowing jet. In this particular case, however, absorption lines similar to those observed by Kotani et al. (2000) using ASCA were observed, indicating the presence of an outflowing wind along with a radio jet. Depending upon the solid angle subtended by the wind, it was found that its mass outflow rate could have been a substantial fraction of the inflowing accretion rate that is ultimately responsible for producing the X-rays.

Understanding the driving mechanisms behind winds in such BH systems and their relationship to the presence of jets became, and continues to be, an important

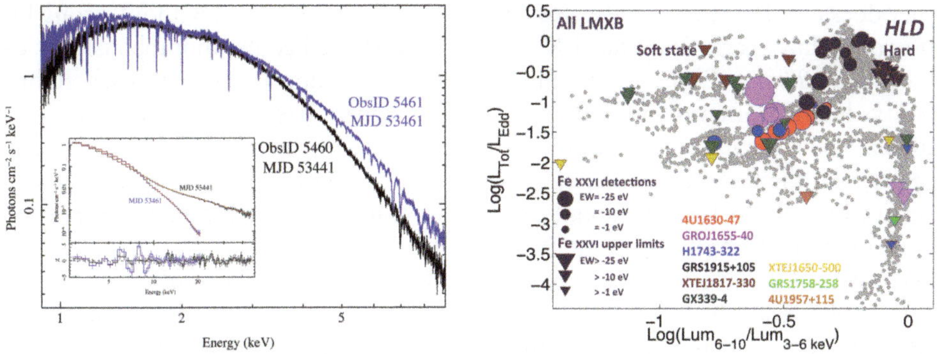


**Figure 6.6.** Left: *Chandra*-HETGS spectra of the BH system GRO J1655–40 (reproduced from Neilsen & Homan 2012 © 2012. The American Astronomical Society. All rights reserved, using the same observations discussed in Miller et al. 2006) at two different epochs separated by only 25 days. The observation taken on Modified Julian Date (MJD) 53461 (blue) shows copious line absorption, whereas the observation taken only 25 days earlier (black), despite similar continuum spectrum in the 1–9 keV band observed by *Chandra*, does not. The inset shows contemporaneous 3–50 keV spectra obtained by the *Rossi X-ray Timing Explorer* on MJD 53441 (black) and MJD 53461 (blue). The former shows a significant power-law tail extending to 50 keV, whereas the latter does not (Neilsen & Homan 2012). Right: the measured equivalent widths (colored circles) or upper limits on magnitudes of equivalent widths (colored triangles) for wind outflow absorption lines seen in BH binary systems (reproduced from Ponti et al. 2012, by permission of Oxford University Press on behalf of the Royal Astronomical Society). The sample BH systems are listed in the figure. The gray points represent the color–intensity diagram for all of these systems, with color being the ratio of the 6–10 keV to 3–6 keV flux (from *Rossi X-ray Timing Explorer* data), while the intensity is an estimate of the source's total luminosity relative to its Eddington luminosity.

focus of *Chandra* studies of these systems. Miller et al. (2006) discussed two *Chandra*-HETGS observations of the BH GRO J1655–40, taken only a few weeks apart from one another, exhibiting broadly similar continuum spectra within the HETGS bandpass, as seen in Figure 6.6. The second of the two spectra, however, was seen to exhibit significantly more absorption lines, consistent with an outflowing wind with velocities ranging from 300–1600 km $s^{-1}$. In contrast to the spectra of GRS 1915 +105 discussed above, here the spectra are dominated by the lower-energy X-ray disk component with the source emitting at ≈4% of its Eddington luminosity. Miller et al. (2006) considered photoionization models of the wind in order to constrain its launching radius, and based upon the lack of observed expected lines (which would arise from higher density winds at smaller radii from the BH) and the inability of the continuum to sufficiently ionize the wind, due to dilution of the continuum, at large radii, they placed the launching radius within the range $10^{7.5}$–$10^{9.5}$ cm. This is too close to the central BH to be effectively launched by thermal driving mechanisms. Similarly, they argued that the wind was too ionized to be driven by radiative forces. They thus argued that the wind driving must instead be accomplished by magnetic forces. They later used refined photoionization models to reinforce this conclusion (Miller et al. 2008; Luketic et al. 2010).

One of the fascinating differences between the two GRO J1655–40 spectra shown in Figure 6.6 is that despite the continuum X-ray spectra looking so similar to each other in the HETGS band, one spectrum shows strong line absorption, while the other does not. However, as also seen in that figure, the spectrum without strong line absorption has a significant hard X-ray tail (as observed contemporaneously by the *Rossi X-ray Timing Explorer*) absent in the other spectrum, which was observed only 20 days later (Neilsen & Homan 2012). It was argued, based upon photoionization models of the second observation that were then re-evaluated with the continuum spectrum of the first observation, that the differences between the two spectra were not attributable to the additional power-law component in the first observation. Thus, the differences between the two must be due to bulk property changes in the wind between the two observations. Neilsen & Homan (2012) argued that what was observed between the two observations was in fact a hybrid thermal/magnetically driven wind, with the latter component coming to dominate in the second observation. It is important to note that the first observation occurred closer to the transition from a bright, spectrally hard state (associated with the presence of a jet) to a softer, disk-dominated state. We shall return to this concept again below.

Miller et al. (2006b) noted that for *Chandra*-HETGS observations of the BH system H1743−322 there was a dichotomy between radio-quiet soft spectral states that exhibit evidence of a wind outflow and radio-loud (jetted) hard spectral states that show no spectral evidence of an outflow. Neilsen & Lee (2009) used *Chandra* observations to highlight similar behavior in GRS 1915+105, but went further in noting that the spectral hard, jetted states showed a broad line. These Fe line region fit residuals for these observations are shown in Figure 6.7. Estimates of the mass loss in the wind states were shown to be comparable to expectations of the inner accretion flow rate in the jetted state. Neilsen & Lee (2009) therefore hypothesized that jet illumination of the disk was responsible for the broad line and that the mass loss in the wind explicitly was responsible for quenching the jet, thus leading to the dichotomy between radio-loud/broad-line emission and radio-quiet/absorption spectra.

Ponti et al. (2012) conducted a survey of results from high X-ray spectral resolution studies of X-ray binaries that further lent support for this dichotomy between jet-dominated states and wind-dominated states. The basic results from this survey are shown in Figure 6.6. This figure shows a standard "hardness–intensity" diagram for broadband X-ray continuum observations of low-mass X-ray binary BH systems (see Fender et al. 2004). The $x$-axis shows a ratio of 6–10 keV to 3–6 keV luminosity (hardness), while the $y$-axis shows source luminosity as a fraction of its Eddington luminosity (intensity). Transient systems in their rise from quiescence start in the lower right-hand corner of the diagram, and rise in luminosity in a spectrally hard, power-law-dominated state while remaining radio bright. At high Eddington fraction, they then soften and move leftward in the diagram and become radio quiet. They then fall in fractional luminosity, and eventually move rightward on the diagram, hardening and again becoming radio loud. Ponti et al. (2012) shows this diagram for a number of LMXB BH sources superimposed on top of one another, but additionally they label detections of, or upper limits on, wind-absorption signatures (concentrating on absorption by Fe XXV and Fe XXVI). Generally, radio-loud sources on the right of the diagram did not show wind-absorption signatures, while radio-quiet, softer X-ray sources potentially did, in agreement with the behavior noted by Miller et al. (2006b) and Neilsen & Lee

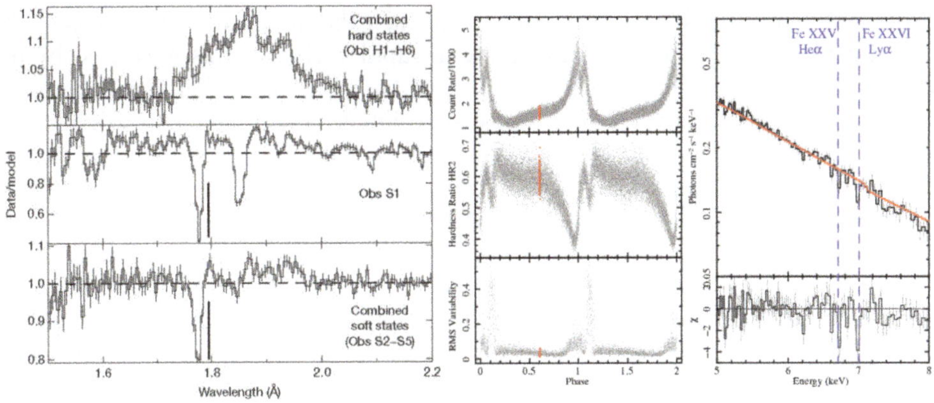


**Figure 6.7.** Left side: residuals in the Fe K$\alpha$ line region for a continuum fit, absent any emission- or absorption-line components, to *Chandra*-HETGS observations of the BH GRS 1915+105 (reprinted by permission from Springer Nature from Neilsen & Lee 2009 © Springer Nature 2009). The top panel is for observations from spectrally hard states, while the middle and bottom panels are for observations from spectrally soft states. The line indicates the expected position of an Fe XXVI line. Right side: an animated GIF of time-dependent HETGS spectroscopy of GRS 1915+105 in the "heartbeat" state, an exotic ~50 s oscillation likely driven by radiation pressure and the high accretion rate onto the black hole. Top left of GIF: phase-folded *Rossi X-ray Timing Explorer* X-ray light curve showing hundreds of individual cycles stacked (1 s resolution). Middle left of GIF: "hardness" ratios of 9–30 keV to 4.5–9 keV count rates, demonstrating strong spectral evolution during the cycle. Bottom left of GIF: root mean square (rms) variability of the X-ray light curve. Right of GIF: zoom-in of the 5–8 keV spectrum, with rest energies of Fe XXV and Fe XXVI shown as dashed blue lines. As the luminosity and spectral shape vary, these iron absorption lines appear and disappear. Photoionization analysis indicates that variability originates in a pulsating wind from the disk that also experiences variable ionization. Residuals show the best-fit model with the line flux set to zero in order to highlight the absorption lines alone. Adapted from Neilsen et al. (2011). Additional materials available online at http://iopscience.iop.org/book/978-0-7503-2163-1.

(2009). This dichotomy became even more pronounced, with a greater fraction of soft X-ray observations revealing evidence of wind absorption, when the diagram was limited to sources known to be viewed at high inclination (i.e., closer to edge-on).

This separation of jet states and wind states, however, is not absolute. Further observations with *Chandra*-HETGS and other X-ray observatories, in conjunction with simultaneous radio observations, have revealed cases where both winds and jets are observed simultaneously in the same system, with examples seen for both NSs and BHs (Homan et al. 2016). As pointed out by these authors, even if the onset of a wind is directly responsible for the turn off of a jet, one might expect a time delay such that a given observation might catch an intermediate phase when both are active. X-ray binary systems are highly dynamic on a broad range of timescales, which can potentially lead to complex jet and/or wind behavior related to this variability. An example of such complexity is shown in Figure 6.7. Here we see *Chandra*-HETGS observations of GRS 1915+105 in its exotic "heartbeat" state (Neilsen et al. 2011). In this state, the X-ray light curve goes through quasi-regular (but not strictly periodic) double-peaked oscillations in its light curve, on ≈50 s timescales, reminiscent of the shape of an echocardiogram, as shown in Figure 6.7.

In the first peak, the light curve softens, and then the light curve rehardens in the second peak, accompanied by an increase in rapid variability (as measured by the fractional root mean square, rms, variability of the light curve; Figure 6.7). Spectral fits to the continuum show an increasing disk radius approaching the first light-curve peak, and then a rapid drop in disk radius and increase in disk temperature during the first peak, and then a Compton corona component rising in the second peak. With regard to the absorption features observed by HETGS, Fe XXV absorption is seen outside the peak, and Fe XXVI is seen prominently during the peak of the light curve. This line behavior can be seen in Figure 6.6. Neilsen et al. (2011) suggested that the heartbeat behavior can be attributed to a thermal-viscous instability in the disk leading to an evaporation or ejection of the inner disk region, possibly into a transient jet. The behavior then repeats in a limit cycle behavior. More recent simultaneous *Chandra*-HETGS and *NuSTAR* observations do not fundamentally alter this interpretation, but do show that the details are even more complex (Zoghbi et al. 2016). These observations suggest that rather than the disk radius changing in the outburst, its density is undergoing drastic changes instead. Further, Zoghbi et al. (2016) used the HETGS observations to identify multiple wind outflow components in this state.

Again, we see that these X-ray binary systems are highly dynamic and complex. Another example is the previously cited (Section 6.1.1) 2015 June outburst of V404 Cyg. *Chandra*-HETGS observations showed variations between strong, highly ionized emission-line profiles, plus the addition of a broad Fe K$\alpha$ structure, and significant P Cygni line profiles, with wind velocities as high as 4000 km $s^{-1}$, in the same 25 ks observation (King et al. 2015). (This is reminiscent of, but a more extreme example of the behavior seen in, GRS 1915+105; Neilsen & Lee 2009.) Continued studies by *Chandra*-HETGS of outbursts of these systems as they recur, or as new systems are discovered, will be crucial for elucidating their nature.

## 6.4 Circumbinary and Interstellar Medium

As *Chandra* performs sensitive imaging and spectroscopic observations of X-ray binaries in our Galaxy, it is not only probing signatures of the compact object and of the material close to its innermost (and potentially highly relativistic) regions—it is also probing the physics of the surrounding and intervening environment. *Chandra*, most often through its spectroscopic observations with HETG and LETG, is providing crucial information about the atmosphere and wind structure of the binary companion donating mass to the compact object (Section 6.4.1), and is revealing the composition and structure of the interstellar medium (ISM) along our line of sight to the compact object (Section 6.4.2 and 6.4.3). The latter studies especially make heavy use of both *Chandra*'s unique spectroscopic (Section 6.4.2) and imaging (Section 6.4.3) capabilities. Below, each of these three areas—circumbinary material, ISM spectroscopy, and dust-scattering halos—is considered in turn.

### 6.4.1 Probes of Stellar Wind Structure

LMXBs are expected to accrete solely via Roche lobe overflow. That is, if the radius of the star exceeds the equipotential surface where material at rest in the rotating frame of the binary falls under the influence of the gravity of the compact object primary, rather than the gravity of the secondary, then an accretion disk can form surrounding the compact object (see Frank et al. 2002). In many cases, this accretion is transient, and the LMXB system will not be persistently bright. High-mass X-ray binaries (HMXBs), however, can be more persistently accreting sources as their secondary stars can have significant winds that launch material beyond the Roche lobe surface, allowing accretion even if the secondary does not otherwise fill its Roche lobe. It is worth noting that some of the most prominent, persistently bright BH sources—LMC X-1, LMC X-3, and Cyg X-1—are in HMXBs. To the extent that the secondary star in a system comes close to filling its Roche lobe, the resulting accretion flow may look more or less like "standard" Roche lobe accretion flow with a well-formed disk. LMC X-3 likely has a very well-formed disk (and in fact is likely truly Roche lobe fed; Orosz et al. 2014), whereas Cyg X-1 and LMC X-1 more likely are accreting via what is known as "focused wind accretion," where a disk forms but the wind can strongly affect the dynamics of the accretion flow (see Nowak et al. 1999, 2001, Wilms et al. 2001 for discussions of these three specific systems).

Winds and outflows in these systems can intersect our line of sight to the compact object even for a broad range of binary inclination angles, with our line of sight passing through different regions within the wind as a function of binary orbital phase. This is illustrated in Figure 6.8. Even for observations that occur outside of times where the wind significantly affects our view, indications of its presence can be manifested in scattered and re-emitted light. Effects of both absorption and emission can become prominent in X-ray spectra. Thus, *Chandra*-HETGS observations become a powerful probe of the atmospheric and wind structure of the secondary stars in these systems.

Torrejón et al. (2010) presented a *Chandra*-HETGS survey of 41 binaries, 10 HMXBs, and 31 LMXBs, where they specifically modeled the narrow component of the Fe K$\alpha$ fluorescence line, as well as the soft X-ray absorption in the system. All HMXBs showed the presence of a narrow Fe K$\alpha$ line, whereas only 10% of the LMXBs exhibited this feature. Furthermore, the equivalent width of the Fe K$\alpha$ line was seen to be correlated with the equivalent neutral column in these systems. This indicates that some substantial fraction of the overall column in these systems is local, and not due to the intervening interstellar medium along our line of sight, and we are seeing fluorescence/emission from the wind and atmosphere of the secondary.

*Chandra*-HETGS spectroscopy of the Cyg X-1 system shows significant orbital phase-dependent absorption due to the secondary's wind, as seen in Figure 6.8 (Nowak et al. 2011). This figure shows the spectrum as measured by HETG and as measured simultaneously with the CCD detectors on board the *Suzaku* X-ray observatory. Although the presence of the absorption is detectable with CCD detectors, its structure is far too complex and narrow to describe without the gratings resolution of HETG. The absorption shown in Figure 6.8 is predominantly

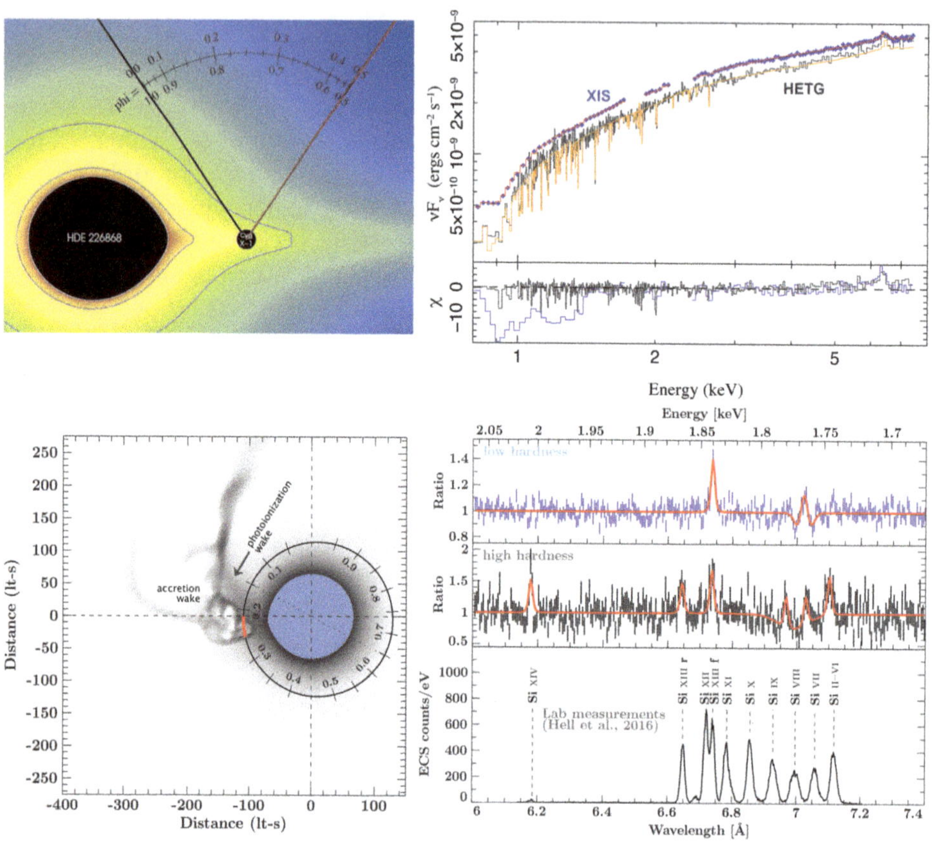


**Figure 6.8.** Top left: schematic showing the geometry/viewing angle of the HDE 226568/Cyg X-1 system (Hanke et al. 2009). The companion star size and component separation are drawn to scale, and the colored contours are based upon models of the expected stellar wind density. The lines and grid show our viewing angle to the compact object as a function of orbital phase, with phase 0 being superior conjunction and our view passing through the densest part of the wind from the secondary. Top right: simultaneous *Suzaku*-XIS and *Chandra*-HETG observations of Cyg X-1 at orbital phase 0 (reproduced from Nowak et al. 2011 © 2011. The American Astronomical Society. All rights reserved). The model is an absorbed power law, with numerous absorption lines from the ionized component of the stellar wind, as well as both a relativistically broadened and narrow fluorescent Fe K$\alpha$ line. Residuals are for the ionized absorption and Fe K$\alpha$ lines removed from the model. Bottom left: schematic of the hypothesized geometry for the Vela X-1 system, as viewed perpendicular to and above the orbital plane and in the orbital frame of the system (reproduced from Grinberg et al. 2017, reproduced with permission © ESO). The schematic shows to scale the size of the secondary, the orbit of the neutron star, the density of the ionized wind, and our viewing angle as a function of orbital phase (labeled along the neutron star orbit). Bottom right: *Chandra*-HETGS observations of the Si region in the Vela X-1 spectrum near orbital phase 0.25 (reproduced from Grinberg et al. 2017, reproduced with permission © ESO). The spectra are for brighter (upper) and fainter (middle) states of the Vela X-1 spectrum, with the lower panel showing laboratory measurements with an electron beam ion trap (EBIT) of emission wavelengths for various different ionized species of Si (Hell et al. 2016). Until these recent EBIT measurements, *Chandra*-HETGS observations actually provided more accurate wavelength measurements for these species than were known from either theoretical or laboratory work.

from hydrogen-like and helium-like ions of various "metal" species in the secondary's wind. (Here we are using the astronomer's definition of metal: any element heavier than hydrogen or helium!) Detailed absorption fits to the spectra shown in Figure 6.8 are presented by Hanke et al. (2009), who observed the system near superior conjunction (orbital phase 0, i.e., where we are viewing through the densest part of the secondary wind, with the secondary in the foreground of the compact object). *Chandra*-HETGS is able to constrain the velocities of the observed lines, and they are in fact somewhat modest, typically being $\lesssim$200 km s$^{-1}$. This is in fact consistent with simulations of the expected properties of the wind as viewed from our perspective of the system. The wind is seen to be almost completely ionized, and the metals appear to be overabundant with respect to solar abundance values.

As expected, the depths of these absorption lines vary with orbital phase (Miškovičová et al. 2016), becoming less pronounced toward inferior conjunction (orbital phase 0.5, i.e., the compact object in the foreground), and showing a clear orbital modulation. Near orbital phase 0.5, P Cygni profiles (see the description above for Cir X-1) are also seen, with velocity differences between the peak of emission and the depths of absorption being on the order of $\lesssim$1000 km s$^{-1}$. An interesting point to make about the spectral studies of Hanke et al. (2009) and Miškovičová et al. (2016) is that they were performed on the least absorbed spectra at a given orbital phase. Specifically, as a function of orbital phase, Cyg X-1 shows intense dipping in the light curve on timescales ranging from seconds to tens of minutes, with the low-energy ($\lesssim$3 keV) X-ray flux dipping to lower flux values than the high-energy X-ray flux (see Hirsch et al. 2019 and references therein). This is attributed to the wind being a multiphase medium, being composed of hot, low-density regions and cool, high-density clumps. The spectra shown in Figure 6.8 are solely for the hot (nearly totally ionized), low-density absorption by the wind.

The frequency, duration, and depth of dips are all at a maximum near superior conjunction and all at a minimum near inferior conjunction (Hirsch et al. 2019). *Chandra*-HETGS observations of this behavior become a probe of the structure of the secondary wind. During the dips, absorption lines from lower ionization stages of the metals are observed with velocities comparable to the highly ionized component of the wind, indicating that structure in the wind facing toward the BH shields material behind it from the ionizing effects of the emission arising from near the compact object. It is also interesting to note that prior to recent laboratory measurements (Hell et al. 2016) with electron beam ion traps (EBITs), *Chandra*-HETGS observations actually provided the most accurate measurements of the wavelengths of many lines from these intermediate ionization stage metals!

Such studies of wind structure are being extended to HMXBs that contain an NS compact object. Figure 6.8 shows a schematic of the geometry of the Vela X-1 X-ray pulsar system (with hypothesized wind/accretion structures, secondary size, and orbital separation all drawn to relative scale). Also shown is a portion of a *Chandra*-HETGS spectrum, taken at orbital phase 0.25, where we are seeing emission lines from a "photoionization wake" in the accretion stream (Grinberg et al. 2017). The accretion geometry in this system is more complex than that for Cyg X-1. The wind from the secondary is likely more significant, and there is also a strong pulsar magnetic field to be included in the models. Similar to Cyg X-1, however, *Chandra*-HETGS observations reveal that the wind structure in this system is also a highly complex, multiphase medium, with nearly completely ionized, hot, low-density regions, and colder, denser clumps. Note the numerous intermediate ionization stage emission lines shown in Figure 6.8. *Chandra*-HETGS observations are opening up a rich treasure trove of observational data that can now be applied to theoretical models of stellar winds and wind-fed accretion flows (see, e.g., the theoretical models of El Mellah et al. 2018, 2019).

### 6.4.2 Interstellar Medium

Bright X-ray sources, particularly binaries, are especially important for the study of the ISM, because the absorption and scattering of the X-rays that they emit probe all phases of matter in the medium. The ISM comprises a cold phase (atomic gas and dust), a warm phase (mildly ionized material), and a hot phase (highly ionized metals). High-resolution X-ray spectroscopic studies of the ISM suggest that by mass, the cold, warm, and hot phases comprise 89%, 8%, and 3%, respectively, of the ISM, at least within the plane of the galaxy (Gatuzz & Churazov 2018). There is significant variation along different lines of sight, with a prime motivation of future *Chandra* ISM observations being mapping of this spatial structure. This sensitivity to all phases of the ISM is unique to X-rays: radio observations of emission at the hydrogen fine structure transition at 21 cm only probe the neutral medium, IR mainly tracks the dust distribution (with only weak constraints on its composition), while optical/UV absorption studies can also trace the warm and hot phases, but are insensitive to solids (depletion of metals into dust is inferred by comparing line-of-sight gas-phase columns with total columns expected from overall Galactic abundances). *Chandra*-HETGS studies, as well as *XMM-Newton*-RGS studies, have also addressed the geometric structure of the ISM, using high Galactic latitude targets to probe the hot phase of our own Galactic halo (Wang et al. 2005, Yao et al. 2006), as we discuss further below.

Although ISM absorption models are typically parameterized in terms of an equivalent neutral hydrogen column, $N_H$, in the X-rays most ISM absorption is due to metals: e.g., C, N, O, Ne, Mg, Si, S, and Fe (Wilms et al. 2000). Prior to the launch of *Chandra*, X-ray absorption models (e.g., the `phabs` model; Balucinska-Church & McCammon 1992) typically were extremely simple, using a Henke model for the absorption optical depth near metal ionization energies, $E_{\rm edge}$: $\tau_{\rm edge} = \tau_0 (E_{\rm edge}/E)^3$ for $E \geqslant E_{\rm edge}$, and $\tau_{\rm edge} = 0$ for $E < E_{\rm edge}$. The modern era of X-ray absorption studies started shortly after the launch of *Chandra* with improved models of X-ray absorption (e.g., `TBabs`; Wilms et al. 2000). For example, the `TBabs` model included revised ISM abundances, updated atomic cross sections, the inclusion of complex structure in absorption edge structure, absorption due to resonance lines of atomic species, absorption by dust, and depletion of metals in dust grains.

These models were further refined and informed by *Chandra*-HETGS (as well as *XMM-Newton*-RGS) observations of X-ray binaries. It was quickly noted, for

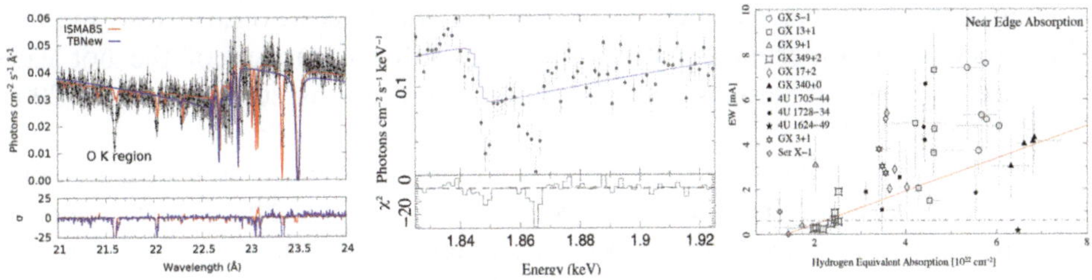


**Figure 6.9.** Left: *Chandra*-HETGS observations of a bright compact object, highlighting interstellar medium (ISM) absorption in the oxygen edge region (reproduced from Gatuzz et al. 2015, by permission of Oxford University Press on behalf of the Royal Astronomical Society). Two different models for ISM absorption are shown (`TBnew`, blue, based upon Wilms et al. 2000, and `ismabs`, red, from Gatuzz et al. 2015), but both contain absorption from neutral and more highly ionized metal species in the ISM. Middle: HETG observations of a high-mass X-ray binary (HMXB), showing the Si edge region (reproduced from Schulz et al. 2016 © 2016. The American Astronomical Society. All rights reserved). The observed absorption features are due to a combination of neutral and ionized Si in the gaseous phase of the ISM, silicates in dust, and ionized winds local to the compact object. Right: the correlation of near-edge Si features with a fitted equivalent neutral column for a number of different HMXBs (reproduced from Schulz et al. 2016 © 2016. The American Astronomical Society. All rights reserved). In general, the two values are correlated, but the near-edge features are variable, likely indicating that dust local to the system is sublimated by the X-ray flux from the compact object, and then re-forms.

example, that absorption structure around the oxygen edge included K$\alpha$ resonance lines from O I and O II (Juett et al. 2004) with *Chandra*-HETGS being used to verify the benchmark wavelengths of these lines (Gorczyca et al. 2013; Liao et al. 2013; Gatuzz et al. 2014). *Chandra* observations were then used to refine the absorption models's structure for the Fe L edge region near 0.7 keV and the neon edge region near 0.9 keV (Juett et al. 2006). In observations of X-ray binaries, the latter region was noted to exhibit resonant K$\alpha$ absorption from Ne I and Ne II (the warm phase of the ISM) and Ne IX (the hot phase of the ISM).

This detailed structure of ISM absorption features as observed in *Chandra*-HETGS observations of X-ray binaries is shown in Figure 6.9, where the `TBabs` and `ISMabs` models are compared. The latter model is especially concerned with describing warm-phase and hot-phase absorption, whereas the `TBabs` model chooses to let many of those features be independently fit. This latter philosophy is partly driven by the fact that one expects the relative contribution of the various ISM phases to vary depending upon the line of sight (e.g., lines of sight that lie in the Galactic plane versus those that pass through the halo of our Galaxy). Furthermore, there could be abundance and composition gradients along different lines of sight. *Chandra*-HETGS spectra have been used to explore each of these issues. Within the plane of the galaxy, *Chandra* X-ray spectra provide little evidence for abundance differences with distance (Gatuzz et al. 2016); however, for sources within the Large Magellanic Cloud (LMC), it was argued that different abundances should be used with *Chandra* observations (Hanke et al. 2010). As discussed above, high-resolution X-ray spectroscopy suggests that the cold, warm, and hot phases of the ISM are 89%, 8%, and 3% by mass in the plane of the galaxy (Gatuzz & Churazov 2018).

Looking along sight lines through the halo of the galaxy, however, the ISM requires a higher contribution from the hot phase, and informs models of the hot gas's distribution within the halo (Wang et al. 2005; Yao & Wang 2005; Yao et al. 2006; Yao & Wang 2006).

Moving to higher energies, *Chandra*-HETG has made important measurements of the Si edge structure near 1.85 keV Schulz et al. (2016). The Si edge region is especially important in ISM studies because, given sufficient spectral resolution and exposure, this region is expected to reveal evidence of the solid-state structure of the ISM, i.e., absorption and scattering by dust grains, which are expected to be composed of different possible combinations of silicates and graphites, with the ratio of compositions, crystalline structure, etc., influencing the effects on the absorption edge (Lee & Ravel 2005; Lee et al. 2009; Corrales et al. 2016). These resonant effects in the X-ray spectra can in principle be modeled in *Chandra*-HETGS spectra of absorbed X-ray binaries (Lee et al. 2009; Zeegers et al. 2017).

As shown in Figure 6.9, *Chandra* spectra do indeed show complex spectra near the Si edge. In this X-ray spectrum of the binary GX 5−1, an absorption feature is seen with an energy of ≈1.85 keV, at just above the Si edge, while an additional absorption feature is seen near the Si XIII resonance line energy of 1.865 keV. Such features are in fact commonly seen in a survey of moderately absorbed Galactic X-ray binaries, with the former line being attributed to a solid-state dust absorption feature (Schulz et al. 2016). These spectra were modeled with both a phenomenological approach to the Si edge, as well as a more physical approach using the edge models of Corrales et al. (2016). Schulz et al. (2016) noted a rough correlation between fitted equivalent neutral column and the equivalent width of the near Si edge feature, and an even weaker correlation between column and the observed Si XIII absorption. The latter line could very well be local to the system and is observed with Doppler velocities too large to reasonably be associated with the ISM. Variations in the correlation between equivalent neutral column and near Si edge equivalent width, however, could be due to compositional differences along different lines of sight, abundance gradients in the ISM, etc. However, Schulz et al. (2016) noted that for some sources, the correlation between column and near Si edge feature equivalent width statistically significantly varies from observation to observation. They hypothesize that this is due to a large fraction of both the absorption and spectroscopic dust features being local to the sources, with dust being destroyed and re-formed in response to the local radiative environment of the X-ray binary.

### 6.4.3 Dust-scattering Halos

The presence of dust in the ISM not only manifests in *Chandra* spectra of X-ray binaries, but also in *Chandra* images of these systems. This is due to X-ray scattering off of foreground dust clouds. As discussed by Xu et al. (1986) and illustrated in Figure 6.10 (see Heinz et al. 2015), observed X-rays from a binary in the galaxy take multiple paths to reach our detectors. The bulk of the X-rays come to us along our direct line of sight to the system. A fraction of soft X-rays are absorbed by the dust, but a further fraction is also scattered out of our line of sight. However, X-rays that initially

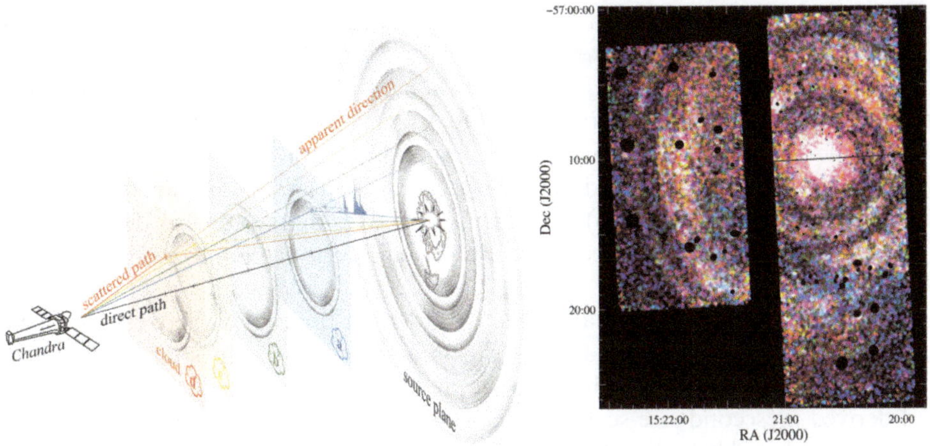


**Figure 6.10.** The geometry (right) for *Chandra* observations of the dust-scattering halos surrounding the image (left) of Cir X-1 (reproduced from Heinz et al. 2015 © 2015. The American Astronomical Society. All rights reserved). The light curve from the central object comes directly to us, with some losses due to dust scattering a fraction of the soft X-rays out of our line of sight. Emission leaving at small angles to our line of sight scatters back into our view, albeit with a time delay, due to scattering off of intervening dust clouds (in this case, four clouds between us and the source). These are then observed as four separate rings, as seen in the *Chandra* image.

travel at small angles away from our line of sight can be scattered into our field of view, albeit with a time delay relative to the direct X-rays as they travel a longer path. This scattered radiation will appear as a soft X-ray "halo" around the point source, typically on arcminute spatial scales. The spatial scale of the halo, its spectrum, and the time delays between the direct and scattered radiation will be functions of the dust locations (dust clouds are often associated with, e.g., the spiral arm structure between us and the source) as well as dust composition, morphology, and size distributions.

Dust-scattering X-ray halos were first recognized as spatially extended soft X-ray excesses above the detector point-spread function (PSF) in *Einstein* (Rolf 1983) and *ROSAT* (Predehl & Schmitt 1995) observations of Galactic X-ray binaries. The first *Chandra* observations of dust halos were performed in the same manner, by modeling excesses above the PSF (see, e.g., Predehl et al. 2000 and Smith et al. 2006 who modeled the dust halos in front of Cyg X-3 and GX 5–1, respectively). In *Chandra* images, this excess emission can in fact appear on size scales as small as arcseconds for dust clouds close to very distant sources (see Corrales et al. 2017).

This highlights a major difference between *Chandra* and many previous X-ray instruments, namely the vastly improved PSF which much more cleanly separates the scattered flux from direct flux. To lowest order, X-ray emission scattered out of the line of sight is scattered back into the line of sight, albeit with a time delay (Corrales et al. 2016). Thus, for X-ray detectors with arcminute scale or poorer spatial resolution (and minimal temporal variation in the source flux and spectrum), dust scattering has been ignored in the spectrum. This is typically not the case for *Chandra*-HETGS spectra, where dust scattering must be included as an additional "loss term" at soft X-ray energies.

This is in fact included in the modeling of the Cyg X-1 spectra shown in Figure 6.8. The absorption lines due to the wind of the companion are included in the models of both the *Suzaku* and *Chandra*-HETGS spectra, but the latter models also include the additional loss due to dust X-ray scattering (Nowak et al. 2011). The differences expected between spatially resolved *Chandra* spectra and more poorly spatially resolved spectra (e.g., *Suzaku, NuSTAR*) are described in detail by Corrales et al. (2016). As discussed above, this work also describes how dust scattering affects the shape of the observed Si edge region, as well as the Fe L edge region.

*Chandra* observations also have directly shown the time delays between direct and scattered photons, as expected from dust-scattering halos. In sources exhibiting significant obscuration events local to the source, the dips in the gratings dispersed, and derived arcsecond-size-scale light curves are seen time delayed (by roughly 10 ks), weakened, and broadened in the light curve derived from arcminute scales in the zeroth-order image of the *Chandra* gratings. For the case of the LMXB 4U 1624 −490, as described in Section 6.3.2, we are viewing the central X-ray source through the atmosphere of the outer disk, and our view is occasionally obscured by the disk wind and (geometric) edge structure. Models of dips in the *Chandra*-observed arcminute-scale light curve suggest that most of the absorption seen in this system is local to the source, and the source distance is $15^{+2.9}_{-2.6}$ kpc (Xiang et al. 2007). A similar analysis has been performed with *Chandra*-HETG observations of the BH Cyg X-1, where the dipping events are instead caused by obscuration by clumps in the stellar wind from the companion star seen near orbital phase 0 (i.e., colder, denser material seen near the same orbital phase as probed by the spectra shown in Figure 6.8). Analysis of both the halo light curve (relative to the directly observed light curve) and the radial halo profile suggested dust clouds significantly comprised of graphite grains at distances 88% and 94% of the distance to Cyg X-1, with the distance to this BH being between 1.72–1.90 kpc (Xiang et al. 2011), consistent with radio-parallax measurements of $1.86 \pm 0.12$ kpc (Reid et al. 2011).

Even more dramatic and distinct examples of dust-scattering echoes are shown in Figure 6.10, which shows the X-ray scattering on 10′–20′ spatial scales in front of the NS LMXB Cir X-1 (see Heinz et al. 2015). In this case, the scattered radiation is not in response to simple obscuration events but rather are the manifestation of large-scale, relatively short-timescale (in relation to the scattering delays) outbursts of Cir X-1. The scattered X-rays are predominantly due to two main outburst periods, lasting ≈10 days and ≈30 days, with significantly shorter-timescale substructure, that occurred ≈50–100 days prior to the *Chandra* observations (Heinz et al. 2015).

## 6.5 Extreme Physics Systems

In the previous sections, we discussed *Chandra* observations of X-ray binaries within our own Galaxy. However, thanks to its superb spatial resolution, *Chandra* has also been crucial for studies of extragalactic binary sources. In this section, we consider two prominent examples, each of which present "extreme" behavior: "ultraluminous" X-ray sources (ULXs), so called because of their extreme X-ray luminosities, and merging NS binary sources and the discovery of electromagnetic counterparts to gravitational radiation detections. Here we describe the *Chandra* contributions to both of these areas.

### 6.5.1 Ultraluminous X-Ray Sources

"Ultraluminous" X-ray sources (ULXs) are so labeled because, if one assumes isotropic emission, their luminosities exceed $10^{39}$ erg $s^{-1}$, roughly $L_{Edd}$ for the typical $\sim 10\ M_{\odot}$ stellar-remnant black holes seen in Galactic BH X-ray binaries. The extreme luminosities exhibited by ULXs have resulted in significant debate over their nature. This has centered around whether these objects have such high isotropic luminosities because they are powered by genuine super-Eddington accretion onto otherwise normal X-ray binaries (e.g., Poutanen et al. 2007; Middleton et al. 2015), or whether they might host a population of "intermediate-mass" BHs (IMBHs; $\sim 10^{3-4} M_{\odot}$) and thus do not violate the source's Eddington luminosity (e.g., Miller et al. 2003; Strohmayer & Mushotzky 2009). For a recent review focusing on ULXs, see Kaaret et al. (2017).

Although ULXs are relatively rare, *Chandra* has played a key role in efforts to significantly build the known population of these sources (Swartz et al. 2004; Liu 2011), which now numbers well into the hundreds (Walton et al. 2011; Earnshaw et al. 2019). Indeed, despite their rarity, the outstanding spatial resolution of *Chandra* has enabled it to reveal remarkable individual galaxies (or galaxy mergers) that host large numbers of ULXs, for example, the Cartwheel galaxy (ESO 350−G040; Gao et al. 2003; Wolter & Trinchieri 2004), the Antennae galaxies (NGC 4038/39; Fabbiano et al. 2001), the Whirlpool galaxy (M51a/b; Terashima & Wilson 2004), and the Cigar galaxy (M82; Matsumoto et al. 2001). One thing all of these galaxies have in common is a high rate of galaxy-scale star formation, implying a potential connection between this process and the presence of ULXs. However, the precise source positions provided by *Chandra*'s outstanding imaging capabilities combined with the sizable sample of ULXs *Chandra* has observed allows us to go further and robustly investigate the immediate environments in which these sources typically reside within their host galaxies. Such work has further helped to establish the spatial connection between ULXs and regions of recent star formation in a statistical sense (Swartz et al. 2009). This association between ULXs and recent star formation strongly implies that, at least on average, ULXs are themselves fairly young systems and are likely analogous to high-mass X-ray binaries, which *Chandra* has helped demonstrate are also strongly connected to star formation (Grimm et al. 2003; Mineo et al. 2012).

The most spectacular example of this association with star formation comes from the Cartwheel galaxy mentioned above (see Figure 6.11). This is the archetypal example of a "ring" galaxy, in which a smaller satellite galaxy has plunged through the plane of a larger galaxy (e.g., Amram et al. 1998), resulting in a wave of star formation sweeping progressively outwards from the point of impact. *Chandra* imaging shows that the large population of ULXs present in this galaxy are clearly associated with this star-forming ring, with a dearth of such sources in the interior (Gao et al. 2003; Wolter & Trinchieri 2004). The lack of ULXs interior to the star-forming ring can be

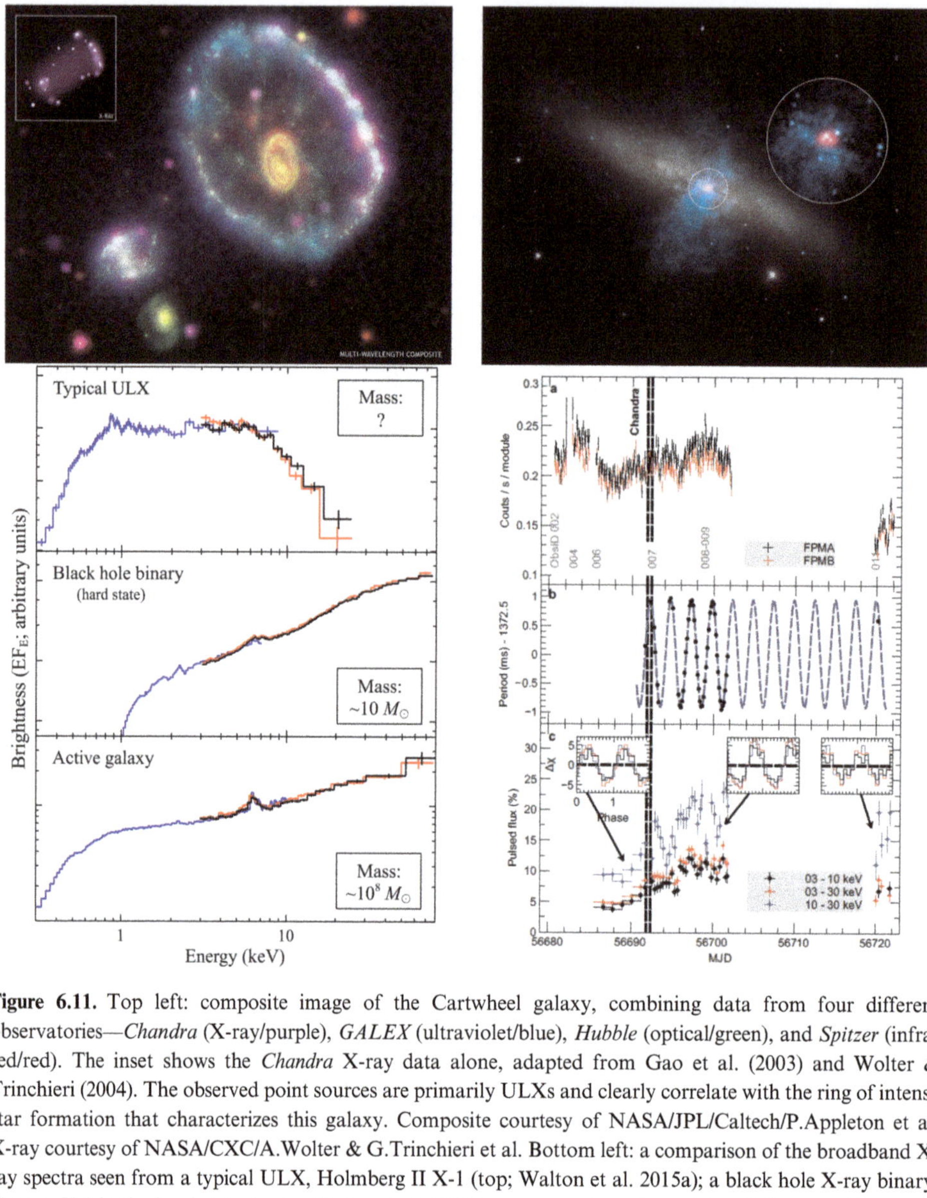


**Figure 6.11.** Top left: composite image of the Cartwheel galaxy, combining data from four different observatories—*Chandra* (X-ray/purple), *GALEX* (ultraviolet/blue), *Hubble* (optical/green), and *Spitzer* (infrared/red). The inset shows the *Chandra* X-ray data alone, adapted from Gao et al. (2003) and Wolter & Trinchieri (2004). The observed point sources are primarily ULXs and clearly correlate with the ring of intense star formation that characterizes this galaxy. Composite courtesy of NASA/JPL/Caltech/P.Appleton et al, X-ray courtesy of NASA/CXC/A.Wolter & G.Trinchieri et al. Bottom left: a comparison of the broadband X-ray spectra seen from a typical ULX, Holmberg II X-1 (top; Walton et al. 2015a); a black hole X-ray binary, Cygnus X-1 in the hard state (center; Parker et al. 2015); and an active galaxy, Ark 120 (bottom; Matt et al. 2014). In each case, the soft X-ray coverage (either *XMM-Newton* or *Suzaku*) is shown in blue, and the *NuSTAR* data are shown in red/black. The latter two sources have been selected to exhibit the approximate spectral form expected should ULXs be powered by sub-Eddington accretion onto IMBHs. Instead, ULXs show clearly distinct high-energy spectra, supporting a super-Eddington interpretation (note that the differences at low energies are related to differing amounts of line-of-sight absorption). Top right: composite image of the M82 galaxy, with X-ray data from *Chandra* (blue) overlaid on an optical image (taken from the ground-based Kitt Peak Observatory). The inset shows a zoom-in (~30″ radius) on M82 X−2, highlighting the crowded field, with the *NuSTAR* difference image (i.e., pulse-on—pulse-off) overlaid in magenta. Courtesy of NASA/JPL-Caltech/SAO/NOAO. Bottom right: timing results for M82 X−2 from Bachetti et al. (2014). The top panel shows the integrated light curve from M82 seen by *NuSTAR*, the middle panel shows the evolution

used to place an approximate upper limit of $\sim 10^7$ yr on the age of these systems. King (2004) estimates this would imply an unrealistic fraction of the star-forming mass in the Cartwheel would have ended up in IMBHs if its ULX population was primarily powered by such objects and thus argues that its ULX population is, instead, primarily made up of typical stellar remnants. In turn, this would imply that most ULXs are in an extreme, super-Eddington accretion regime.

However, such arguments only hold for the population as a whole and do not exclude the possibility that individual sources could still harbor IMBHs. Indeed, these *Chandra* observations also helped to reveal the most extreme members of the ULX population, with isotropic luminosities exceeding even $10^{41}$ erg s$^{-1}$ (e.g., Kaaret et al. 2001; Gao et al. 2003). This latter subset of ULXs are often referred to as "hyperluminous" X-ray sources (HLXs), and, owing to their truly extreme luminosities, are often considered the strongest candidates for hosting IMBHs. Further potential evidence for IMBHs in some ULXs came from X-ray spectroscopy; the CCD spectra obtained by *XMM-Newton*, *Suzaku*, and *Chandra* over the $\sim$0.5–10.0 keV band revealed evidence for two continuum components that could be interpreted within the disk-corona framework seen in sub-Eddington X-ray binaries (and also sub-Eddington active galactic nuclei). This implied that ULXs have low-temperature accretion disks ($T \sim 0.2$ keV), and, in turn, larger black hole masses ($M \sim 1000\ M_\odot$), assuming that the disk extended into the innermost stable circular orbit (e.g., Miller et al. 2003; Cropper et al. 2004, Roberts et al. 2005), as the standard Shakura & Sunyaev (1973) model predicts that the disk temperature should scale as $T \propto M^{-1/4}$.

The launch of the *NuSTAR* observatory in 2012, carrying the first focusing X-ray optics to operate in the hard X-ray band ($E > 10$ keV), marked a major step in our ability to study the spectra of ULXs by providing the first detections of these sources above 10 keV; prior hard X-ray detectors did not have sufficient angular resolution or sensitivity to isolate and study the ULX population. A major effort was undertaken to obtain broadband X-ray spectra in the first few years of *NuSTAR* operations, with a sample of ULXs observed in coordination with *XMM-Newton*, *Suzaku*, and/or *Chandra*, typically resulting in spectral coverage across the $\sim$0.5–30 keV band. These observations revealed broadband spectra that were clearly distinct from those seen in sub-Eddington X-ray binaries and active galactic nuclei (e.g., Bachetti et al. 2013; Walton et al. 2014; Figure 6.11), which would also have been expected should these sources host sub-Eddington IMBHs, given that the basics of BH accretion appear to be independent of mass. These observations instead

of the pulse period, which reveals clear variations related to the 2.5 day orbit of the binary system, and the bottom panel shows the evolution of the pulsed fraction, with pulse profiles shown as insets (which can be seen to be $\sim$sinusoidal). The timing of the *Chandra* observation is also indicated. The Cartwheel interactive figure shows separately the X-ray emission (purple), ultraviolet emission (blue), optical emission (green), infrared emission (red), a composite image, and a composite image with an inset (as shown in the top, left hand panel), to better identify the ULXs correlated with the ring of star formation. The M82 interactive figure shows the separate X-ray emission (*Chandra*, 0.5-7 keV, blue, and *NuSTAR*, high energies, purple) and optical wavelength images, a composite image, and a composite image with an inset zoomed on the X-ray emission from M82 X-2. Additional materials available online at http://iopscience.iop.org/book/978-0-7503-2163-1.

suggested these sources were exhibiting a distinct, and therefore likely super-Eddington, accretion regime, although even in this scenario BH accretors were still almost universally assumed.

However, our understanding of ULXs experienced a major (and unexpected) paradigm shift in 2014, when coherent X-ray pulsations with a period of ~1.4 s were detected from the ULX M82 X-2 (Bachetti et al. 2014), unambiguously demonstrating that the accretor in this system is not a BH at all, but a magnetized NS. This was a remarkable discovery, as NSs are known to have a tight range of masses ($\sim 1 - 2\,M_{\odot}$), yet M82 X-2 exhibits a peak luminosity of $\sim 2 \times 10^{40}$ erg s$^{-1}$. For a typical NS mass of 1.4 $M_{\odot}$, this corresponds to an astonishing ~100 $L_{\rm Edd}$! Although the pulsations were detected by *NuSTAR*, the *Chandra* observatory also made a significant contribution to this discovery, providing high-resolution imaging at the time the pulsations were detected. *NuSTAR* only has an angular resolution of ~1′, meaning it integrates over a large fraction of the M82 X-ray binary population (Matsumoto et al. 2001; Figure 6.11) and could not determine from which source in M82 the pulsations were originating by itself. The *Chandra* imaging demonstrated that there were only two sources that were bright enough to provide the pulsations detected in the integrated flux seen by *NuSTAR*, M82 X-1 and X-2. Then, through a combination of difference imaging over the pulse cycle with *NuSTAR* and continued monitoring with *Swift*, the pulsating source could be firmly identified as X-2.

In addition to confirming the nature of the accretor in this system, the discovery of X-ray pulsations suddenly revealed a wealth of information about the binary system (Bachetti et al. 2014). The extreme accretion onto the NS can be seen to be spinning the star up with an average rate of $\dot{P} \sim -2 \times 10^{-10}$ s s$^{-1}$, as the infalling material applies a significant torque. Periodic variations in the pulse period are also imprinted on top of this secular spin up (Figure 6.11), allowing the orbital dynamics of the system to be determined, as these variations result from Doppler shifts experienced by the NS as it orbits its companion. Although the inclination of the system is still not known, the orbital period of ~2.5 days revealed by these variations still places a lower limit on the mass of the stellar companion of >5.2 $M_{\odot}$ (assuming a typical 1.4 $M_{\odot}$ NS), further confirming the connection between ULXs and high-mass X-ray binaries.

Since this initial discovery, three more ULX pulsars have been reported (although several more are likely to be reported in the near future): NGC 7793 P13 (Fürst et al. 2016; Israel et al. 2017b), NGC 5907 ULX1 (Israel et al. 2017a), and NGC300 ULX1 (Carpano et al. 2018). Nevertheless, despite this remarkable progress, significant questions still remain regarding how these NS are able to reach such extreme luminosities. The current record holder, NGC 5907 ULX1, exhibits peak isotropic luminosities of $\sim 10^{41}$ erg s$^{-1}$, or ~500 $L_{\rm Edd}$ (Israel et al. 2017a; Fürst et al. 2017)! The magnetic nature of the accretion required to produce the pulsations means that there must be some degree of anisotropy to the emission, as the magnetic field of the NS funnels the infalling material onto its magnetic poles (e.g., Basko & Sunyaev 1976). If the radiation is strongly collimated, it may be possible that we are significantly overpredicting the total luminosity by assuming isotropy. However, it is difficult to reconcile the roughly sinusoidal pulse profiles, seen in all ULX pulsars to date (Figure 6.11), with a scenario in which the radiation is highly beamed. Instead, the debate has focused on the effects of the magnetic field of the central NS. Some authors have suggested this may be related to the presence of extreme, magnetar-level magnetic fields (up to $B \sim 10^{15}$ G; e.g., Dall'Osso et al. 2015; Mushtukov et al. 2015), as fields this strong can suppress the electron-scattering cross section (Herold 1979). However, others have argued that such a strong magnetic field would prevent the accreting material from providing the lever arm required to explain the observed spin-up rates these sources are experiencing, and that the fields are actually much lower (as low as $B \sim 10^9$ G; e.g., Kluźniak & Lasota 2015). These theoretical estimates span an astonishing six orders of magnitude!

Significant efforts have therefore recently been made to place observational constraints on the magnetic field strengths in these systems, and *Chandra* has played a key role in these preliminary results. In principle, the *B*-field can be estimated from the spin-up rate (assuming the source is close to spin equilibrium). Early results here imply fairly modest fields of $\sim 10^{12}$ G (Bachetti et al. 2014; Fürst et al. 2016), typical of the field strengths seen in Galactic X-ray pulsars (Caballero & Wilms 2012). Long-term monitoring with *Swift*, *Chandra*, and *XMM-Newton* has also shown that the known ULX pulsars show "off" states, in which their fluxes drop by factors of ~50 or more (e.g., Motch et al. 2014; Walton et al. 2015b; Brightman et al. 2016a). These have been proposed to be related to transitions to the "propeller" regime, in which the rotation of the magnetic field anchored to the neutron star suddenly shuts off the accretion (Illarionov & Sunyaev 1975), potentially providing a means to estimate the *B*-field. Early results here instead support magnetar-level fields (Tsygankov et al. 2016). However, the most robust method for estimating NS magnetic field strengths is through the detection of cyclotron resonant scattering features (CRSFs), absorption-like features that are produced when strong magnetic fields quantize particle energy levels (referred to as Landau levels), resulting in resonant scattering of photons at specific energies. Provided the nature of the scattering particle is known, the energy of the CRSF provides a direct measure of the magnetic field strength. However, here, too, the early results paint a mixed picture. Based on a long *Chandra* observation, Brightman et al. (2018) reported the likely detection of a proton CRSF at 4.5 keV in the spectrum of M51 ULX-7 (note that pulsations have not yet been detected from this source, but the presence of a CRSF would confirm it as another NS ULX). If correct, this would imply a truly extreme magnetic field of $B \sim 10^{15}$ G (although see Middleton et al. 2019 for an alternative interpretation). In contrast, based on a deep *XMM-Newton* + *NuSTAR* observation, Walton et al. (2018a) reported the potential presence of an electron CRSF at ~13 keV in the ULX pulsar NGC 300 ULX1, which would imply a more standard field of $B \sim 10^{12}$ G. Further observational and theoretical effort is clearly required to understand the physics of these ULX pulsars.

There is also still substantial uncertainty over the contribution of NSs versus BHs to the broader ULX population. Although pulsations have only currently been seen in a handful of systems, all of the ULXs observed by *NuSTAR* to date (both pulsing and nonpulsing) show qualitatively similar broadband X-ray spectra, raising the possibility that NSs dominate the ULX population (Pintore et al. 2017; Koliopanos et al. 2017; Walton et al. 2018b). Efforts to identify further NS ULXs are ongoing and include searches for more ULXs that could be exhibiting propeller phase transitions (Earnshaw et al. 2018); the sensitivity of *Chandra* to faint sources is particularly well suited for this work. However, ultimately confirming the presence of an additional BH ULX population will require dynamical constraints on the binary parameters, which in turn requires the study of the stellar companions in these systems. In most cases, this will require the next generation of ground-based facilities (e.g., the Thirty Meter Telescope), because these counterparts are very faint and extremely challenging to study with current facilities (e.g., Gladstone et al. 2013; Heida et al. 2014; 2019, López et al. 2017, although there are a few rare exceptions that are bright enough to study now, e.g., Liu et al. 2013; Motch et al. 2014; Heida et al. 2015). However, here, too, the outstanding imaging provided by *Chandra* will continue to play a critical role, as the identification of the correct counterpart often requires a precision in the source position that only *Chandra* can provide.

Finally, although these recent developments have helped provide a strong case that the majority of ULXs are powered by super-Eddington accretion onto standard stellar remnants, it is also worth noting that there are still a small number of individual sources that remain good candidates for hosting IMBHs. Most notable among these is ESO 243–49 HLX1, which has an astonishing peak luminosity of $\sim 10^{42}$ erg s$^{-1}$ (Farrell et al. 2009)! Beyond its extreme luminosity, this source stands out from the rest of the ULX population as observations with *XMM-Newton*, *Chandra*, and *Swift* (e.g., Servillat et al. 2011) have revealed that it exhibits repeated outbursts across which it shows the same pattern of state transitions as seen in LMXBs with BH accretors (discussed in Section 6.3.3; see Figure 6.6). Although no dynamical mass constraints have been possible to date, essentially all aspects of its behavior consistently point to the presence of a $\sim 10^4\ M_\odot$ BH in this system (see also Webb et al. 2012; Straub et al. 2014). In addition, M82 X−1 is also often considered to be a promising IMBH candidate (peak luminosity $\sim 10^{41}$ erg s$^{-1}$). Based on the apparent presence of specific temporal frequencies in its variability properties, which are thought to scale directly with BH mass, Pasham et al. (2014) proposed that M82 X−1 hosts a $\sim 400\ M_\odot$ black hole. However, M82 X−1 is difficult to study owing to the crowded field (Figure 6.11; X−1 is only $\sim 5''$ northwest of X−2), and in this case, the available spectral results paint a mixed picture for the mass of the accretor, with estimates ranging from $\sim 10 - 1000\ M_\odot$ (e.g., Feng & Kaaret 2010; Brightman et al. 2016b). The most extreme ULXs nevertheless remain a promising population in which to search for potential IMBH candidates (although we note again that the ULX pulsar NGC 5907 ULX1 can also reach luminosities of $\sim 10^{41}$ erg s$^{-1}$). Further expanding the known ULX population to help unearth these rare objects is therefore of significant interest, and efforts to do so with *Chandra* are ongoing.

### 6.5.2 Multimessenger Systems—GW 170817

The discovery of gravitational radiation from merging BHs in a binary system (Abbott et al. 2016) is one of the most important astrophysical discoveries of this decade. Numerous observatories operating in various different wavebands (ranging from radio through optical to X-ray and gamma-ray energies) have searched for electromagnetic counterparts to this event and to the many subsequent gravitational wave events discovered by these detectors. Until recently, all of these events have been associated with BH–BH mergers. With the exception of a statistically marginal gamma-ray burst detected coincident with GW 150914 (Connaughton et al. 2016; Greiner et al. 2016), no electromagnetic counterparts have been detected associated with these gravitational wave events. However, there are no strong theoretical reasons to expect electromagnetic counterparts to a binary BH merger. Conversely, an NS–NS binary merger/gravitational wave event is theoretically expected to produce a potentially observable electromagnetic counterpart.

This expectation was borne out on 2017 August 17 with the detection of the gravitational wave event GW 170817 (Abbott et al. 2017; see Figure 6.12), which was identified as an NS–NS binary merger. This event was followed only 2 s later by the *Fermi*-Gamma Ray Burst Monitor (GBM) detection of the short gamma-ray burst GRB 170817A (Goldstein et al. 2017), which fell within the (large) gravitational wave positional error region determined by the LIGO and Virgo detectors.

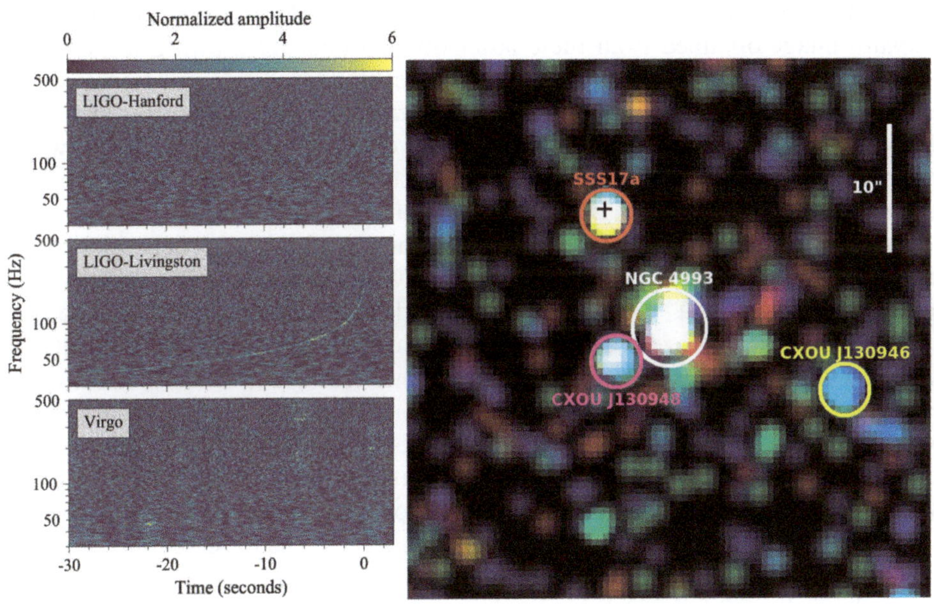


**Figure 6.12.** Left: the gravity wave chirp signal associated with GW 170817, interpreted as arising from an NS–NS merger, as observed by the LIGO-Hanford, LIGO-Livingston, and Virgo sites (reproduced from Abbott et al. 2017. CC 3.0.). Right: *Chandra* three-color X-ray image (red = 0.5–1.2 keV, green = 1.2–2.0 keV, blue = 2.0–7.0 keV), made from observations taken 15–16 days past the gravity wave event (reproduced from Haggard et al. 2017 © 2017. The American Astronomical Society. All rights reserved). The position of the GW 170817 optical counterpart is labeled as SSS17a (Coulter et al. 2017). The location of the probable host galaxy, NGC 4413, for the merging NS binary is also indicated, along with positions of other *Chandra*-detected X-ray sources within the field of view. The vertical line indicates an angular distance of 10″, the approximate separation between SSS17a and NGC 4413.

The INTEGRAL SPI-ACS instrument also detected GRB 170817A ~2 s after the end of the GW event (Savchenko et al 2017). The positional error regions provided by the combination of the three gravitational wave detectors and *Fermi*-GBM allowed for the rapid optical identification of a counterpart by the Swopes Sky Survey, dubbed SSS17a, the spectrum and light curve of which were consistent with a "kilonova" event that is typically associated with short gamma-ray bursts (Coulter et al. 2017). This counterpart was seen to lie just 10″ away from the core of galaxy, NGC 4413. This positional determination quickly led to follow-up observations with other facilities, including X-ray campaigns with the *Swift* and *Chandra* X-ray observatories.

*Swift* observations, occurring only 0.6 days after the gravitational wave event, failed to detect any X-ray counterpart, placing an isotropic luminosity upper limit of $\approx 10^{39}$ erg s$^{-1}$ on the counterpart (Evans et al. 2017). Similarly, *Chandra* observations two days after the event also failed to detect an X-ray counterpart (Margutti et al. 2017). It was not until nine days past the event, using *Chandra*, that an X-ray counterpart was first detected (Troja et al. 2017). This was also the time of the first detection of a radio counterpart. *Chandra* observations occurring 15–16 days past the event provided the first X-ray spectrum, which was found to be consistent with a power-law spectrum with photon index $\Gamma = 2.4 \pm 0.8$ (Haggard et al. 2017). The *Chandra* image obtained from these latter observations are shown in Figure 6.12. Note the proximity of the X-ray source to the brighter X-ray emission from the nearby galaxy nucleus, NGC 4413; the *Chandra* imaging resolution was crucial for these observations. (*XMM-Newton* successfully observed GRB 170817A. However, separating the source emission from that of the nearby nucleus is much more difficult using these observations; see D'Avanzo et al. 2018).

The first interpretations of these observations attributed the delayed onset of radio and X-ray emission to our viewing a relativistic jet at an off-axis angle of $\gtrsim 20°$ (presuming the jet has a "top hat" velocity and density profile, i.e., constant across a cylindrical profile, and zero outside of it), with our line of sight entering into the widening cone of beamed emission as the jet slows. The evolution of the X-ray light curve, however, could not be followed in the short term as the Sun's angular position became too close to that of GRB 170817A. X-ray observations were not feasible again until over 100 days past the initial gravitational wave event. When the source was observed again by *Chandra* 108–109 days past the initial event, the X-ray flux was found to be four times higher (Ruan et al. 2018). The observed flux was only slightly higher at 154–164 days past the event (Margutti et al. 2018) and was finally seen to be in decay a full 260 days past the event (Nynka et al. 2018). The X-ray light curve obtained with *Chandra* observations (Nynka et al. 2018) is shown in Figure 6.13. The spectral evolution of the source from radio through X-ray bands (Margutti et al. 2018) is also shown in this figure, where aside from IR/optical spectra consistent with kilonova emission, the spectrum is roughly consistent with a power law from radio to X-ray.

This light curve is not consistent with the simplest top-hat profile jet model. Combining *Chandra* observations of the X-ray light curve with radio light-curve observations (which mostly track the X-ray light-curve profile), alternative models

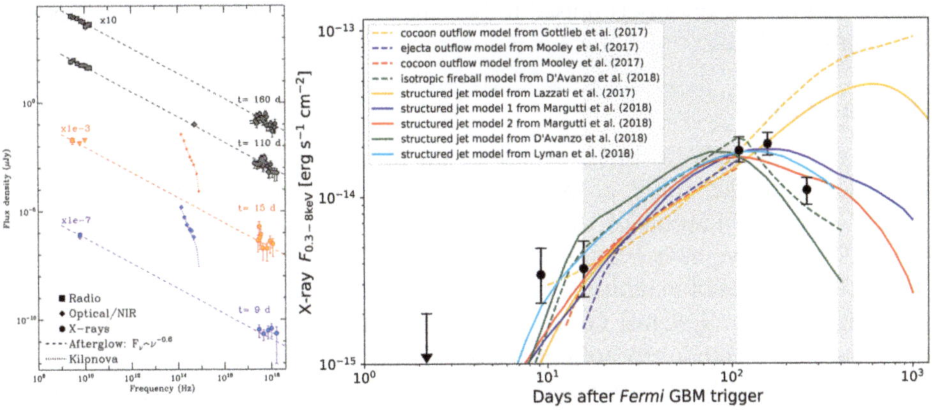


**Figure 6.13.** Left: broadband spectral energy distributions (SEDs) for GW 170817 as a function of time (reproduced from Margutti et al. 2018 © 2018. The American Astronomical Society. All rights reserved). Radio (squares), IR/optical (diamonds), and *Chandra* X-ray (circles) spectra are shown for times 9 days (blue, offset by a factor of $10^{-7}$), 15 days (red, offset by a factor of $10^{-3}$), 110 days, and 160 days (black, offset by a factor of 100) past the initial detection of the gravity wave event. The spectra are modeled with a combination of a kilonova model (for the IR/optical) and an afterglow/cocoon (see Margutti et al. 2018). Right: *Chandra* light curves (circles) as a function of days past the gravity wave event (reproduced from Nynka et al. 2018 © 2018. The American Astronomical Society. All rights reserved). Lines are various models that can be broken up into two classes: a structured jet (solid lines) and an afterglow from a cocoon/ejecta/fireball (dashed lines). The gray shaded areas are periods where *Chandra* observations were not possible due to the source's angular position on the sky being too close to that of the Sun.

have been suggested. These models, broadly speaking, fall into two classes: jets with a fast inner core and a slower outer sheath, or "cocoon"-type models, where the jet is propagating through an expanding wavefront of ejecta and creating an afterglow via the interaction with this ejecta. Both classes of models allow for the continued brightening of the flux for an observed 100 days past the initial event, followed by a few month plateau, and then decline. Predictions of a number of these models are shown overplotted on the *Chandra* light curve in Figure 6.13 (Nynka et al. 2018). It should be noted that most of these model light curves were fit to the *Chandra* data before the X-ray decline had been observed and therefore have not had their parameters optimized. As of this point, the *Chandra* observations do not uniquely distinguish between fast-core jet models and afterglow models; however, many models plausibly predict that the afterglow could be observable for up to 1000 days past the initial event (Alexander et al. 2018).

As of the writing of this chapter (2019 July), this is the only fully convincing binary NS merger detected by LIGO and Virgo to be published in the refereed literature,[1] and the only gravitational wave event with a clear electromagnetic counterpart. However, many more events, including NS binary mergers, are expected to be detected over the coming years, with further opportunities for follow-up observations in the next decade (or two!) of *Chandra* observations.

[1] Candidate BH–NS and NS–NS star mergers, however, have been detected in 2019 April.

Multimessenger astronomy is poised to become an ever more important component of *Chandra* studies.

## 6.6 Summary

We have given only a brief overview of the achievements by the *Chandra* observatory for the study of X-ray binaries, primarily those within our own Galaxy. The combination of the best-ever X-ray spatial and spectral resolution, and the fact that we have been able to study given systems, such as V404 Cyg and Cir X-1, over orders of magnitude in flux, has proven to be crucial for advancing our knowledge of these systems. Given that these systems are highly dynamic, and in many cases go through repeated episodes of outburst and quiescence, means that *Chandra* undoubtedly will observe many of these objects again in its next decades of observations. This will allow us to understand long-term behavior (e.g., as in the cooling of NSs, such as in Cas A), and compare predictions of models for repeated outbursts. Additionally, new sources will continue to be discovered. Previously unknown transient NSs and BHs continue to be discovered at the rate of a few per year, with *Chandra* playing an important role in initial localization, high-resolution spectroscopy in bright phases, and then following the evolution back into quiescence. Likewise, existing and newly discovered ULX sources will continue to be a major area of research with *Chandra* observations. Finally, the era of studies of counterparts to gravitational radiation events has only just begun. A few such events per year should be detectable over the coming decade, with *Chandra* being the X-ray observatory most capable of separating their X-ray emission from the background or nearby (e.g., nuclear) X-ray emission.

## References


Abbott, B. P., Abbott, R., Abbott, T. D., et al. 2016, PhRvL, 116, 061102

Abbott, B. P., Abbott, R., Abbott, T. D., et al. 2017, PhRvL, 119, 161101

Alexander, K. D., Margutti, R., Blanchard, P. K., et al. 2018, ApJ, 863, L18

Amram, P., Mendes de Oliveira, C., Boulesteix, J., & Balkowski, C. 1998, A&A, 330, 881

Archibald, A. M., Stairs, I. H., Ransom, S. M., et al. 2009, Sci, 324, 1411

Bachetti, M., Harrison, F. A., Walton, D. J., et al. 2014, Natur, 514, 202

Bachetti, M., Rana, V., Walton, D. J., et al. 2013, ApJ, 778, 163

Bahramian, A., Miller-Jones, J., Strader, J., et al. 2018, https://github.com/bersavosh/XRB-LrLx_pub

Balucinska-Church, M., & McCammon, D. 1992, ApJ, 400, 699

Basko, M. M., & Sunyaev, R. A. 1976, MNRAS, 175, 395

Begelman, M. C., McKee, C. F., & Shields, G. A. 1983, ApJ, 271, 70

Bildsten, L., & Rutledge, R. E. 2000, ApJ, 541, 908

Blandford, R. D., & Begelman, M. C. 1999, MNRAS, 303, L1

Brandt, W. N., & Schulz, N. S. 2000, ApJ, 544, L123

Brightman, M., Harrison, F. A., Barret, D., et al. 2016b, ApJ, 829, 28

Brightman, M., Harrison, F. A., Fürst, F., et al. 2018, NatAs, 2, 312

Brightman, M., Harrison, F., Walton, D. J., et al. 2016a, ApJ, 816, 60

Brown, E. F., Bildsten, L., & Rutledge, R. E. 1998, ApJ, 504, L95

Brown, E. F., & Cumming, A. 2009, ApJ, 698, 1020
Caballero, I., & Wilms, J. 2012, MmSAI, 83, 230
Cackett, E. M., & Miller, J. M. 2013, ApJ, 777, 47
Cackett, E. M., Miller, J. M., Homan, J., et al. 2009, ApJ, 690, 1847
Campana, S., & Stella, L. 2000, ApJ, 541, 849
Campana, S., Stella, L., Mereghetti, S., et al. 1998, ApJ, 499, L65
Carpano, S., Haberl, F., Maitra, C., & Vasilopoulos, G. 2018, MNRAS, 476, L45
Chakrabarty, D., Tomsick, J. A., Grefenstette, B. W., et al. 2014, ApJ, 797, 92
Chiang, C. -Y., Cackett, E. M., Miller, J. M., et al. 2016, ApJ, 821, 105
Connaughton, V., Burns, E., Goldstein, A., et al. 2016, ApJ, 826, L6
Connors, R. M. T., Markoff, S., Nowak, M. A., et al. 2017, MNRAS, 466, 4121
Corbel, S., Nowak, M. A., Fender, R. P., Tzioumis, A. K., & Markoff, S. 2003, A&A, 400, 1007
Corbel, S., Tomsick, J. A., & Kaaret, P. 2006, ApJ, 636, 971
Corrales, L. R., García, J., Wilms, J., & Baganoff, F. 2016, MNRAS, 458, 1345
Corrales, L. R., Mon, B., Haggard, D., et al. 2017, ApJ, 839, 76
Coulter, D. A., Foley, R. J., Kilpatrick, C. D., et al. 2017, Sci, 358, 1556
Cropper, M., Soria, R., Mushotzky, R. F., et al. 2004, MNRAS, 349, 39
Dall'Osso, S., Perna, R., & Stella, L. 2015, MNRAS, 449, 2144
D'Avanzo, P., Campana, S., Salafia, O. S., et al. 2018, A&A, 613, L1
Davis, J. E. 2001, ApJ, 562, 575
Degenaar, N., Brown, E. F., & Wijnands, R. 2011a, MNRAS, 418, L152
Degenaar, N., Wijnands, R., Bahramian, A., et al. 2015, MNRAS, 451, 2071
Degenaar, N., Wolff, M. T., Ray, P. S., et al. 2011b, MNRAS, 412, 1409
Earnshaw, H. P., Roberts, T. P., Middleton, M. J., Walton, D. J., & Mateos, S. 2019, MNRAS, 483, 5554
Earnshaw, H. P., Roberts, T. P., & Sathyaprakash, R. 2018, MNRAS, 476, 4272
El Mellah, I., Sander, A. A. C., Sundqvist, J. O., & Keppens, R. 2019, A&A, 622, A189
El Mellah, I., Sundqvist, J. O., & Keppens, R. 2018, MNRAS, 475, 3240
Evans, P. A., Cenko, S. B., Kennea, J. A., et al. 2017, Sci, 358, 1565
Fabbiano, G., Zezas, A., & Murray, S. S. 2001, ApJ, 554, 1035
Fabian, A. C., Rees, M. J., Stella, L., & White, N. E. 1989, MNRAS, 238, 729
Falcke, H., Körding, E., & Markoff, S. 2004, A&A, 414, 895
Farrell, S. A., Webb, N. A., Barret, D., Godet, O., & Rodrigues, J. M. 2009, Natur, 460, 73
Fender, R. P., Belloni, T. M., & Gallo, E. 2004, MNRAS, 355, 1105
Fender, R. P., Gallo, E., & Jonker, P. G. 2003, MNRAS, 343, L99
Feng, H., & Kaaret, P. 2010, ApJ, 712, L169
Fesen, R. A., Hammell, M. C., Morse, J., et al. 2006, ApJ, 645, 283
Frank, J., King, A., & Raine, D. J. 2002, Accretion Power in Astrophysics (3rd ed.; Cambridge: Cambridge Univ. Press)
Fürst, F., Walton, D. J., Stern, D., et al. 2017, ApJ, 834, 77
Fürst, F., Walton, D. J., Harrison, F. A., et al. 2016, ApJ, 831, L14
Gallo, E., Fender, R. P., Miller-Jones, J. C. A., et al. 2006, MNRAS, 370, 1351
Gallo, E., Fender, R. P., & Pooley, G. G. 2003, MNRAS, 344, 60
Gao, Y., Wang, Q. D., Appleton, P. N., & Lucas, R. A. 2003, ApJ, 596, L171
Garcia, M. R., McClintock, J. E., Narayan, R., et al. 2001, ApJ, 553, L47
Gatuzz, E., & Churazov, E. 2018, MNRAS, 474, 696

Gatuzz, E., García, J., Kallman, T. R., Mendoza, C., & Gorczyca, T. W. 2015, ApJ, 800, 29
Gatuzz, E., García, J., Mendoza, C., et al. 2014, ApJ, 790, 131
Gatuzz, E., García, J. A., Kallman, T. R., & Mendoza, C. 2016, A&A, 588, A111
Gavriil, F. P., Strohmayer, T. E., & Bhattacharyya, S. 2012, ApJ, 753, 2
George, I. M., & Fabian, A. C. 1991, MNRAS, 249, 352
Gladstone, J. C., Copperwheat, C., Heinke, C. O., et al. 2013, ApJS, 206, 14
Goldstein, A., Veres, P., Burns, E., et al. 2017, ApJ, 848, L14
Gorczyca, T. W., Bautista, M. A., Hasoglu, M. F., et al. 2013, ApJ, 779, 78
Greiner, J., Burgess, J. M., Savchenko, V., & Yu, H. -F. 2016, ApJ, 827, L38
Grimm, H. -J., Gilfanov, M., & Sunyaev, R. 2003, MNRAS, 339, 793
Grinberg, V., Hell, N., El Mellah, I., et al. 2017, A&A, 608, A143
Haggard, D., Nynka, M., Ruan, J. J., et al. 2017, ApJ, 848, L25
Hanke, M., Wilms, J., Nowak, M. A., Barragán, L., & Schulz, N. S. 2010, A&A, 509, L8
Hanke, M., Wilms, J., Nowak, M. A., et al. 2009, ApJ, 690, 330
Hannikainen, D. C., Hunstead, R. W., Campbell-Wilson, D., & Sood, R. K. 1998, A&A, 337, 460
Heida, M., Harrison, F. A., Brightman, M., et al. 2019, ApJ, 871, 231
Heida, M., Jonker, P. G., Torres, M. A. P., et al. 2014, MNRAS, 442, 1054
Heida, M., Torres, M. A. P., Jonker, P. G., et al. 2015, MNRAS, 453, 3510
Heinke, C. O., Jonker, P. G., Wijnands, R., Deloye, C. J., & Taam, R. E. 2009, ApJ, 691, 1035
Heinke, C. O., Rybicki, G. B., Narayan, R., & Grindlay, J. E. 2006, ApJ, 644, 1090
Heinz, S., Burton, M., Braiding, C., et al. 2015, ApJ, 806, 265
Heinz, S., & Nowak, M. A. 2001, MNRAS, 320, 249
Heinz, S., Sell, P., Fender, R. P., et al. 2013, ApJ, 779, 171
Hell, N., Brown, G. V., Wilms, J., et al. 2016, ApJ, 830, 26
Herold, H. 1979, PhRvD, 19, 2868
Hirsch, M., Hell, N., Grinberg, V., et al. 2019, A&A, 626, A64
Ho, W. C. G., & Heinke, C. O. 2009, Natur, 462, 71
Ho, W. C. G., Wijngaarden, M. J. P., Chang, P., et al. 2019, in AIP Conf. Proc. 2127, Xiamen-CUSTIPEN Workshop on the EOS of Dense Neutron-Rich Matter in the Era of Gravitational Wave Astronomy, 020007
Homan, J., Neilsen, J., Allen, J. L., et al. 2016, ApJ, 830, L5
Hynes, R. I., Bradley, C. K., Rupen, M., et al. 2009, MNRAS, 399, 2239
Hynes, R. I., Charles, P. A., Garcia, M. R., et al. 2004, ApJ, 611, L125
Iaria, R., Di Salvo, T., D'Aì, A., et al. 2013, A&A, 549, A33
Iaria, R., Lavagetto, G., D'Aí, A., di Salvo, T., & Robba, N. R. 2007, A&A, 463, 289
Ichimaru, S. 1977, ApJ, 214, 840
Illarionov, A. F., & Sunyaev, R. A. 1975, A&A, 39, 185
Israel, G. L., Belfiore, A., Stella, L., et al. 2017a, Sci, 355, 817
Israel, G. L., Papitto, A., Esposito, P., et al. 2017b, MNRAS, 466, L48
Ji, L., Schulz, N. S., Nowak, M. A., & Canizares, C. R. 2011, ApJ, 729, 102
Jimenez-Garate, M. A., Schulz, N. S., & Marshall, H. L. 2003, ApJ, 590, 432
Jonker, P. G., Bassa, C. G., Nelemans, G., et al. 2006, MNRAS, 368, 1803
Jonker, P. G., & Nelemans, G. 2004, MNRAS, 354, 355
Jonker, P. G., Steeghs, D., Chakrabarty, D., & Juett, A. M. 2007, ApJ, 665, L147
Juett, A. M., & Chakrabarty, D. 2006, ApJ, 646, 493
Juett, A. M., Schulz, N. S., & Chakrabarty, D. 2004, ApJ, 612, 308

Juett, A. M., Schulz, N. S., Chakrabarty, D., & Gorczyca, T. W. 2006, ApJ, 648, 1066
Kaaret, P., Feng, H., & Roberts, T. P. 2017, ARA&A, 55, 303
Kaaret, P., Prestwich, A. H., Zezas, A., et al. 2001, MNRAS, 321, L29
King, A. L., Miller, J. M., Raymond, J., Reynolds, M. T., & Morningstar, W. 2015, ApJ, 813, L37
King, A. R. 2004, MNRAS, 347, L18
Kitamoto, S., Tsunemi, H., Miyamoto, S., Yamashita, K., & Mizobuchi, S. 1989, Natur, 342, 518
Kluźniak, W., & Lasota, J. -P. 2015, MNRAS, 448, L43
Koliopanos, F., Vasilopoulos, G., Godet, O., et al. 2017, A&A, 608, A47
Kong, A. K. H., McClintock, J. E., Garcia, M. R., Murray, S. S., & Barret, D. 2002, ApJ, 570, 277
Kotani, T., Ebisawa, K., Dotani, T., et al. 2000, ApJ, 539, 413
Lee, J. C., & Ravel, B. 2005, ApJ, 622, 970
Lee, J. C., Reynolds, C. S., Remillard, R., et al. 2002, ApJ, 567, 1102
Lee, J. C., Xiang, J., Ravel, B., Kortright, J., & Flanagan, K. 2009, ApJ, 702, 970
Liao, J. -Y., Zhang, S. -N., & Yao, Y. 2013, ApJ, 774, 116
Lin, J., Nowak, M. A., & Chakrabarty, D. 2009, ApJ, 706, 1069
Liu, J. 2011, ApJS, 192, 10
Liu, J. -F., Bregman, J. N., Bai, Y., Justham, S., & Crowther, P. 2013, Natur, 503, 500
López, K. M., Heida, M., Jonker, P. G., et al. 2017, MNRAS, 469, 671
Luketic, S., Proga, D., Kallman, T. R., Raymond, J. C., & Miller, J. M. 2010, ApJ, 719, 515
Margutti, R., Berger, E., Fong, W., et al. 2017, ApJ, 848, L20
Margutti, R., Alexander, K. D., Xie, X., et al. 2018, ApJ, 856, L18
Markoff, S., Falcke, H., & Fender, R. 2001, A&A, 372, L25
Markoff, S., Nowak, M. A., Gallo, E., et al. 2015, ApJ, 812, L25
Matsumoto, H., Tsuru, T. G., Koyama, K., et al. 2001, ApJ, 547, L25
Matt, G., Marinucci, A., Guainazzi, M., et al. 2014, MNRAS, 439, 3016
McClintock, J. E., London, R. A., Bond, H. E., & Grauer, A. D. 1982, ApJ, 258, 245
McClintock, J. E., Narayan, R., & Rybicki, G. B. 2004, ApJ, 615, 402
Menou, K., Esin, A. A., Narayan, R., et al. 1999, ApJ, 520, 276
Merloni, A., Heinz, S., & di Matteo, T. 2003, MNRAS, 345, 1057
Middleton, M. J., Brightman, M., Pintore, F., et al. 2019, MNRAS, 486, 2
Middleton, M. J., Heil, L., Pintore, F., Walton, D. J., & Roberts, T. P. 2015, MNRAS, 447, 3243
Miller, J. M., Fabbiano, G., Miller, M. C., & Fabian, A. C. 2003, ApJ, 585, L37
Miller, J. M., Raymond, J., Fabian, A., et al. 2006, Natur, 441, 953
Miller, J. M., Raymond, J., Reynolds, C. S., et al. 2008, ApJ, 680, 1359
Miller, J. M., Fabian, A. C., Wijnands, R., et al. 2002, ApJ, 578, 348
Miller, J. M., Raymond, J., Fabian, A. C., et al. 2004, ApJ, 601, 450
Miller, J. M., Raymond, J., Homan, J., et al. 2006b, ApJ, 646, 394
Mineo, S., Gilfanov, M., & Sunyaev, R. 2012, MNRAS, 419, 2095
Miškovičová, I., Hell, N., Hanke, M., et al. 2016, A&A, 590, A114
Motch, C., Pakull, M. W., Soria, R., Grisé, F., & Pietrzyński, G. 2014, Natur, 514, 198
Mushtukov, A. A., Suleimanov, V. F., Tsygankov, S. S., & Poutanen, J. 2015, MNRAS, 454, 2539
Narayan, R., McClintock, J. E., & Yi, I. 1996, ApJ, 457, 821
Narayan, R., & Yi, I. 1995, ApJ, 452, 710

Neilsen, J., & Homan, J. 2012, ApJ, 750, 27
Neilsen, J., & Lee, J. C. 2009, Natur, 458, 481
Neilsen, J., Remillard, R. A., & Lee, J. C. 2011, ApJ, 737, 69
Nowak, M. A., Heinz, S., & Begelman, M. C. 2002, ApJ, 573, 778
Nowak, M. A., Vaughan, B. A., Wilms, J., Dove, J. B., & Begelman, M. C. 1999, ApJ, 510, 874
Nowak, M. A., Wilms, J., Heindl, W. A., et al. 2001, MNRAS, 320, 316
Nowak, M. A., Hanke, M., Trowbridge, S. N., et al. 2011, ApJ, 728, 13
Nynka, M., Ruan, J. J., Haggard, D., & Evans, P. A. 2018, ApJ, 862, L19
Orosz, J. A., Steiner, J. F., McClintock, J. E., et al. 2014, ApJ, 794, 154
Papitto, A., Ferrigno, C., Bozzo, E., et al. 2013, Natur, 501, 517
Parker, M. L., Tomsick, J. A., Miller, J. M., et al. 2015, ApJ, 808, 9
Pasham, D. R., Strohmayer, T. E., & Mushotzky, R. F. 2014, Natur, 513, 74
Patruno, A., Archibald, A. M., Hessels, J. W. T., et al. 2014, ApJ, 781, L3
Pintore, F., Zampieri, L., Stella, L., et al. 2017, ApJ, 836, 113
Plotkin, R. M., Miller-Jones, J. C. A., Gallo, E., et al. 2017, ApJ, 834, 104
Ponti, G., Fender, R. P., Begelman, M. C., et al. 2012, MNRAS, 422, L11
Posselt, B., & Pavlov, G. G. 2018, ApJ, 864, 135
Poutanen, J., Lipunova, G., Fabrika, S., Butkevich, A. G., & Abolmasov, P. 2007, MNRAS, 377, 1187
Predehl, P., Burwitz, V., Paerels, F., & Trümper, J. 2000, A&A, 357, L25
Predehl, P., & Schmitt, J. H. M. M. 1995, A&A, 293, 889
Proga, D., & Kallman, T. R. 2002, ApJ, 565, 455
Reid, M. J., McClintock, J. E., Narayan, R., et al. 2011, ApJ, 742, 83
Reynolds, C. S., & Nowak, M. A. 2003, PhR, 377, 389
Roberts, T. P., Warwick, R. S., Ward, M. J., Goad, M. R., & Jenkins, L. P. 2005, MNRAS, 357, 1363
Rolf, D. P. 1983, Natur, 302, 46
Roy, J., Ray, P. S., Bhattacharyya, B., et al. 2015, ApJ, 800, L12
Ruan, J. J., Nynka, M., Haggard, D., Kalogera, V., & Evans, P. 2018, ApJ, 853, L4
Rutledge, R. E., Bildsten, L., Brown, E. F., Pavlov, G. G., & Zavlin, V. E. 2001, ApJ, 551, 921
Savchenko, V., Ferrigno, C., Kuulkers, E., et al. 2017, ApJ, 848, L15
Schulz, N. S., & Brandt, W. N. 2002, ApJ, 572, 971
Schulz, N. S., Corrales, L., & Canizares, C. R. 2016, ApJ, 827, 49
Sell, P. H., Heinz, S., Calvelo, D. E., et al. 2010, ApJ, 719, L194
Servillat, M., Farrell, S. A., Lin, D., et al. 2011, ApJ, 743, 6
Shakura, N. I., & Sunyaev, R. A. 1973, A&A, 24, 337
Shternin, P. S., Yakovlev, D. G., Haensel, P., & Potekhin, A. Y. 2007, MNRAS, 382, L43
Smith, R. K., Dame, T. M., Costantini, E., & Predehl, P. 2006, ApJ, 648, 452
Stappers, B. W., Archibald, A. M., Hessels, J. W. T., et al. 2014, ApJ, 790, 39
Straub, O., Godet, O., Webb, N., Servillat, M., & Barret, D. 2014, A&A, 569, A116
Strohmayer, T. E., & Mushotzky, R. F. 2009, ApJ, 703, 1386
Swartz, D. A., Ghosh, K. K., Tennant, A. F., & Wu, K. 2004, ApJS, 154, 519
Swartz, D. A., Tennant, A. F., & Soria, R. 2009, ApJ, 703, 159
Takata, J., Li, K. L., Leung, G. C. K., et al. 2014, ApJ, 785, 131
Terashima, Y., & Wilson, A. S. 2004, ApJ, 601, 735
Torrejón, J. M., Schulz, N. S., Nowak, M. A., & Kallman, T. R. 2010, ApJ, 715, 947

Troja, E., Piro, L., van Eerten, H., et al. 2017, Natur, 551, 71
Tsygankov, S. S., Mushtukov, A. A., Suleimanov, V. F., & Poutanen, J. 2016, MNRAS, 457, 1101
Ueda, Y., Inoue, H., Tanaka, Y., et al. 1998, ApJ, 492, 782
Walton, D. J., Bachetti, M., Fürst, F., et al. 2018a, ApJ, 857, L3
Walton, D. J., Fürst, F., Heida, M., et al. 2018b, ApJ, 856, 128
Walton, D. J., Harrison, F. A., Bachetti, M., et al. 2015b, ApJ, 799, 122
Walton, D. J., Harrison, F. A., Grefenstette, B. W., et al. 2014, ApJ, 793, 21
Walton, D. J., Middleton, M. J., Rana, V., et al. 2015a, ApJ, 806, 65
Walton, D. J., Roberts, T. P., Mateos, S., & Heard, V. 2011, MNRAS, 416, 1844
Wang, Q. D., Yao, Y., Tripp, T. M., et al. 2005, ApJ, 635, 386
Webb, N., Cseh, D., Lenc, E., et al. 2012, Sci, 337, 554
Wijnands, R., Degenaar, N., & Page, D. 2017, JApA, 38, 49
Wijnands, R., Miller, J. M., Markwardt, C., Lewin, W. H. G., & van der Klis, M. 2001, ApJ, 560, L159
Wijnands, R., & van der Klis, M. 1998, Natur, 394, 344
Wilms, J., Allen, A., & McCray, R. 2000, ApJ, 542, 914
Wilms, J., Nowak, M. A., Pottschmidt, K., et al. 2001, MNRAS, 320, 327
Wolter, A., & Trinchieri, G. 2004, A&A, 426, 787
Xiang, J., Lee, J. C., & Nowak, M. A. 2007, ApJ, 660, 1309
Xiang, J., Lee, J. C., Nowak, M. A., & Wilms, J. 2011, ApJ, 738, 78
Xiang, J., Lee, J. C., Nowak, M. A., Wilms, J., & Schulz, N. S. 2009, ApJ, 701, 984
Xu, Y., McCray, R., & Kelley, R. 1986, Natur, 319, 652
Yao, Y., Schulz, N., Wang, Q. D., & Nowak, M. 2006, ApJ, 653, L121
Yao, Y., & Wang, Q. D. 2005, ApJ, 624, 751
Yao, Y., & Wang, Q. D. 2006, ApJ, 641, 930
Zeegers, S. T., Costantini, E., de Vries, C. P., et al. 2017, A&A, 599, A117
Zoghbi, A., Miller, J. M., King, A. L., et al. 2016, ApJ, 833, 165

AAS | IOP Astronomy

The Chandra X-ray Observatory
Exploring the high energy universe
**Belinda Wilkes and Wallace Tucker**

# Chapter 7

# X-Rays from Galaxies

**Giuseppina Fabbiano**

## 7.1 Introduction

Galaxies, of which our own Milky Way is the closest example, are complex systems, composed of stars, gas, dust, nonbaryonic (dark) matter, and black holes. The deepest images from the *Hubble Space Telescope* (*Hubble*) show that galaxies populate the universe out to $z > 8.5$ (Ellis et al. 2013). *Chandra* has allowed the detection and study of the different components of the X-ray emission of galaxies in the local universe out to about 100 Mpc and of the overall X-ray output of galaxies at higher redshift.

In the X-rays, we can uniquely probe the nonthermal emission of stars and the interaction of stellar winds with the surrounding interstellar medium; the end products of stellar evolution, including both gaseous supernova remnants and compact stellar remnants (white dwarfs, neutron stars); the hot ~10 million-degree gas in the interstellar medium, gaseous outflows, and galaxy halos; and a wide range of black holes, from those originating from stellar evolution to the supermassive black holes at the nuclei of galaxies. These results are setting important constraints on the energy input (feedback) from stellar and nuclear sources during galaxy evolution. During the merging of gas-rich disk galaxies, the X-ray emission is enhanced with increased stellar formation because of both the birth of luminous X-ray binaries and feedback onto the interstellar medium (ISM), which is also enriched with the elements produced by stars and supernovae. Accretion onto the nuclear supermassive black hole is responsible for intense X-ray emission, leading to the formation of a strong active galactic nucleus (AGN), and to galaxy-scale X-rays from the interaction of the AGN-emitted photons with the ISM. Hot, extended, X-ray-emitting gas in massive elliptical galaxies points to the presence of massive dark matter halos.

The types of X-ray sources found in galaxies and the physical processes responsible for the X-ray emission are discussed in detail in other chapters of this

doi:10.1088/2514-3433/ab43dcch7  

book. Here we concentrate on the collective behavior of these sources in their native environment. We will discuss:

(1) The populations of X-ray binaries that compose the larger fraction of the X-ray emission of normal galaxies at energies > ~2 keV (Section 7.2);
(2) The hot gaseous component, prevalent at energies < 1 keV, and its evolution in both star-forming and old stellar population galaxies (Section 7.3);
(3) The discovery of low-luminosity, "hidden" active nuclear sources in normal galaxies that are expanding the range of nuclear black holes down to "intermediate" masses, bridging the gap between stellar black holes (Mass up to ~100 $M_{\odot}$) and supermassive nuclear black holes ($M \sim 10^6$–$10^{10}$ $M_{\odot}$; Section 7.4); and finally,
(4) The interaction of supermassive black holes with their host galaxy, which pose constraints both on nuclear feedback and on our understanding of the nature of AGNs (Section 7.5).

The results discussed in this chapter could not have been achieved without the subarcsecond resolution of the *Chandra* X-ray telescope and the ability of the ACIS detector to record, simultaneously, the position, energy, and time of the incoming photons (Chapter 2). However, the small collecting area of *Chandra* requires very long exposures to pursue this science. All of this makes an excellent scientific case for a large-area, next-generation X-ray telescope that retains the angular resolution of *Chandra* and is equipped with focal plane instruments capable of providing higher resolution in spectral and time domains.

The following discussion also demonstrates the importance of multiwavelength observations of comparable quality, provided by *Hubble* and *Spitzer* in space, and the JVLA and ALMA from the ground, for achieving a full understanding of the processes in action. It is important that this multiwavelength capability be retained by the astronomical community in the years to come, if we do not want the type of science possible in the NASA Great Observatories era to became an example of an irretrievable golden past (Section 7.6).

## 7.2 X-Ray Binary Populations

X-ray astronomy began with the unexpected discovery of a very luminous source, Sco X-1 (Giacconi et al. 1962). Sco X-1 was the first Galactic X-ray binary (XRB) ever to be observed. The first X-ray survey of the sky with the NASA *Uhuru* satellite (also conceived and led by Giacconi and collaborators) showed that XRBs are the most common luminous X-ray sources in the Milky Way. XRBs are binary systems composed of an evolved stellar remnant (neutron star—NS, black hole—BH, or white dwarf—WD), and a stellar companion. The X-rays are produced by the gravitational accretion of the atmosphere of the companion onto the compact remnant (for reviews of XRBs, see *X-ray Binaries,* ed. Lewin et al. 1995, and *Compact Stellar X-ray Sources,* ed. Lewin & van der Klis 2006).

If the companion is a massive star (mass > 10 $M_\odot$), the XRB is called a high-mass X-ray binary (HMXB); HMXBs are short-lived X-ray sources, with lifetimes ~10 million years, regulated by the evolution of the massive companion, and therefore are associated with young stellar populations. If the companion is a low-mass star (mass $\leqslant$ 1 $M_\odot$), the XRB is called a low-mass X-ray binary (LMXB). There are two types of LMXBs: (1) those resulting from the evolution of native stellar binary systems in the parent galaxy stellar field (field LMXBs; see Verbunt & van den Heuvel 1995), and (2) those formed from dynamical interactions in globular clusters (GC-LMXBs; Clark 1975; Grindlay 1984). Field LMXBs evolve at a slow pace, with lifetimes of ~$10^8$–$10^9$ yr, due to the long time needed to produce a close binary and start the accretion process, so these LMXBs are generally old systems. On the other hand, GC-LMXBs can be formed virtually at any time, and some of them may be short-lived systems (e.g., the NS–WD ultracompact binaries; Bildsten & Deloye 2004).

Ever since the discovery of Sco X-1, the study of Galactic XRBs has continued to be a thriving area of research. This body of work has investigated the physical processes related to the gravitational accretion onto a compact object in a binary system and has set constraints on the mass and nature of the compact stellar remnants. In the late 1970s and 1980s, following the advent of imaging X-ray astronomy with the *Einstein Observatory*, XRBs were detected in a number of nearby galaxies (see the review by Fabbiano 1989). While observations with other X-ray telescopes (*ROSAT*, *ASCA*, *XMM-Newton*) have contributed to the study of extragalactic XRBs, the subarcsecond resolution of the *Chandra X-ray Observatory* (Weisskopf et al. 2000) has revolutionized this field (see review by Fabbiano 2006). With *Chandra*, populations of luminous point-like sources (typically $L_X > 1 \times 10^{37}$ erg $s^{-1}$ in the *Chandra* energy band ~0.5–7 keV) have been detected in all galaxies within 50 Mpc of the Milky Way and even farther away (Figure 7.1).

A large body of work convincingly associates these sources with XRBs (we refer the reader to Fabbiano 2006). In particular, the X-ray colors of the majority of these sources are consistent with the spectra of Galactic LMXBs and HMXBs. Classes of softer emission sources (super-soft sources and quasi-soft sources), possibly associated with nuclear-burning WD binaries, were also detected. Monitoring observations of some of these systems have uncovered the widespread source variability characteristic of XRBs, pointing to compact accreting objects. Spectral variability patterns consistent with those of BH LMXBs were found in sources detected in the nearby elliptical galaxy NGC 3379 (Brassington et al. 2008). In this galaxy, and in NGC 4478, the spectra of most luminous LMXBs are consistent with the emission of accretion disks of black holes (BHs) with masses of 5–20 $M_\odot$ (Brassington et al. 2010; Fabbiano et al. 2010). These masses are in the range of those measured in Milky Way BH binaries (Remillard & McCLintock 2006).

*Chandra* observations of galaxies provide a new tool for constraining the formation and evolution of XRBs. Given that the XRBs in a particular galaxy are all at the same distance, with *Chandra* we can derive their relative luminosities accurately (which is not the case for Galactic XRBs); the absolute luminosities will be affected by the uncertainty on the distance of the parent galaxy. Moreover, we

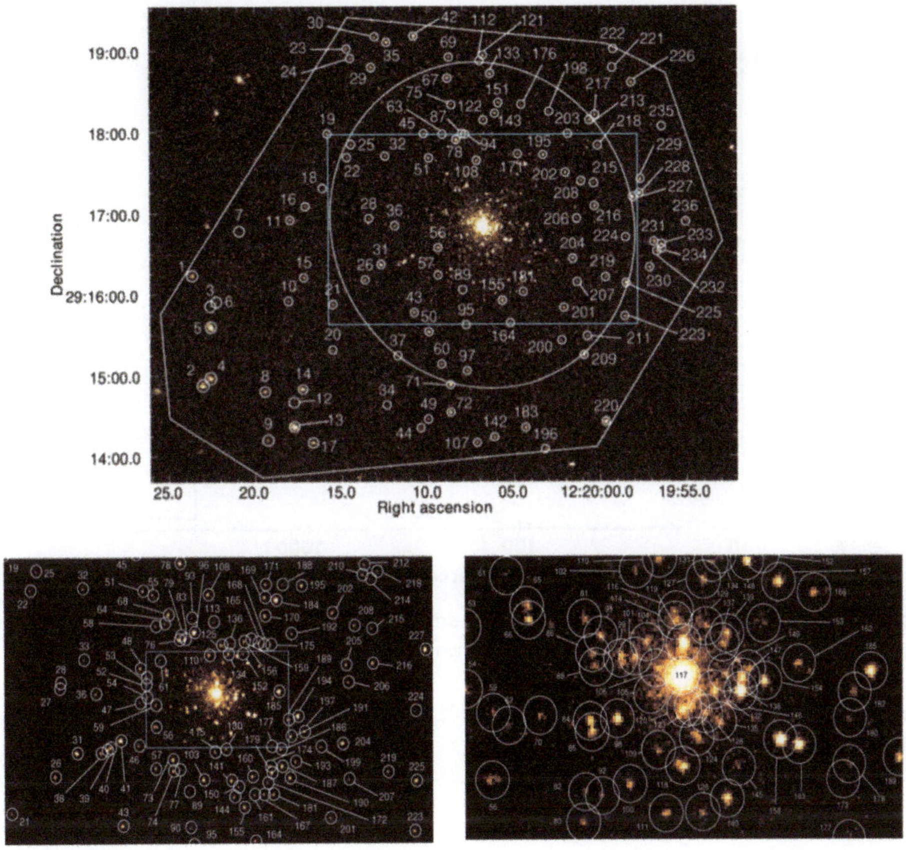


**Figure 7.1.** The LMXB population of the elliptical galaxy NGC 4278 (reproduced from Brassington et al. 2009 © 2009. The American Astronomical Society. All rights reserved). The top panel shows the sources detected at larger radii, within the footprint of the *Chandra* fields (polygon). The ellipse in the top panel represents the $D_{25}$ isophote of the optical stellar light. The bottom-left panel expands the region within the blue rectangle in the top panel, and the bottom-right panel expands the region within the blue rectangle in the left bottom panel. NGC 4278 was observed with *Chandra* ACIS-S in six separate pointings, resulting in a coadded exposure of 458 ks. From this deep observation, 236 sources have been detected within the region overlapped by all observations, 180 of which lie within the $D_{25}$ ellipse of the galaxy. These 236 sources (of which only 29 are expected to be background AGNs) range in $L_X$ from $3.5 \times 10^{36}$ erg s$^{-1}$ (with $3\sigma$ upper limit $\leqslant 1 \times 10^{37}$ erg s$^{-1}$) to $\sim 2 \times 10^{40}$ erg s$^{-1}$. They have X-ray colors consistent with those of Galactic LMXBs. Of these sources, 103 are found to vary long term by various amounts, including 13 transient candidates (>5 ratio variability).

can correlate the properties of these sources with those of the parent stellar population. By detecting an ever-growing sample of XRBs, we are also sensitive to extreme objects which may be missing in the Milky Way, such as the ultra-luminous X-ray sources (ULXs) that emit in excess of the Eddington luminosity for an NS or $\sim 10\ M_{\odot}$ stellar BH system ($>10^{39}$ erg s$^{-1}$). ULXs have been suggested to be intermediate-mass BHs, bridging the gap between the supermassive BHs of AGNs and the stellar BHs found in some XRBs, although now most systems are

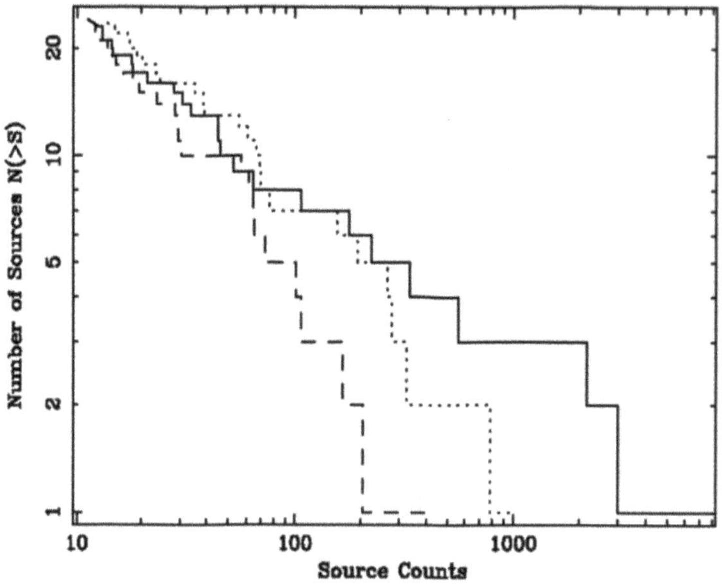


**Figure 7.2.** Integral XLFs of sources in M81: solid—arms; dotted—moving towards the disk; dashed—disk (reproduced from Swartz et al. 2003 ).

understood to be young, luminous HMXBs (see Fabbiano 1989, 2006; see also Section 7.2.1.2).

### 7.2.1 The XRB X-Ray Luminosity Functions and Scaling Laws in the Near Universe

The XRB populations of galaxies can be characterized by means of their X-ray luminosity function (XLF). In differential form, the XLF gives the number of XRBs detected at a given X-ray luminosity, in integral form, the number of XRBs detected above a given X-ray luminosity (Figure 7.2). The XLFs can be parameterized in terms of their normalization (i.e., the number of X-ray sources detected in a given galaxy) and shape (i.e., the distribution of these sources as a function of X-ray luminosity; functional slopes and break $L_X$). Both parameters reflect the composition of the XRB populations (see the review of Fabbiano 2006 and references therein).

The XLF normalization measures the total number of XRBs in a given galaxy, and thus the total X-ray luminosity of the galaxy, at least at energies >2 keV, where XRBs dominate the X-ray emission of normal galaxies. Beginning from the earlier studies of normal (nonactive) galaxies in X-rays with the *Einstein Observatory*, it has been evident that both total stellar mass and star formation rate are important factors in determining the total number of XRBs. Total stellar mass is the prevalent scaling factor for old stellar populations, i.e., elliptical galaxies and bulges, while star formation rate drives XRB production in the young stellar populations of spiral disks and irregular galaxies (see review by Fabbiano 1989, 2006 and references therein).

The shape of the XLF measures the relative number of low- and high-luminosity XRBs in a given population, and therefore provides the means to constrain the luminosity evolution of these sources (see, e.g., Fragos 2013b). Several monitoring studies of individual galaxies show that, although individual XRBs vary strongly in their X-ray output, this variability does not affect the general shape of the XLF (Zezas et al. 2007; Fridriksson et al. 2008; Mineo et al. 2014c).

*7.2.1.1 Old Stellar Populations*

In old stellar systems, such as elliptical galaxies, the XRB population is composed of low-mass stars accreting onto a compact remnant (LMXBs). As mentioned earlier in Section 7.2, two evolution channels are possible: (1) the evolution of a native X-ray binary composed of a high-mass star (now evolved into a compact remnant) and a long-lived low-mass stellar companion, or (2) dynamical binary formation in GCs. Because of its efficiency, the latter has been argued to be the main formation mechanism for LMXBs (Clark 1975). *Chandra* observations of nearby elliptical galaxies suggest that both formation channels are important. In particular, the normalization of the XLFs of these XRB populations is largely dependent on the integrated stellar mass of the parent galaxy and to a minor, but significant, extent, on the GC specific frequency (number of GCs per unit stellar mass; see review Fabbiano 2006; Kim et al. 2009; Boroson et al. 2011). These dependencies suggest that both LMXBs from the general stellar population and from GCs are involved.

The shape of the LMXB XLF also reflects the nature and evolution of the LMXBs. Based on joint *Chandra* and *Hubble* surveys of nearby elliptical galaxies, it has been possible to disentangle the XLFs of the LMXBs detected in the galaxy stellar field and those associated with GCs. While earlier reports suggested that the XLFs of the GC-LMXB and field-LMXB did not differ (see Fabbiano 2006), later works with deeper data have demonstrated that the GC-LMXB XLF is flatter than that of the field-LMXB. This difference was first suggested only for the lower luminosity portion of the XLF, $L_X < 5 \times 10^{37}$ erg s$^{-1}$ (Kim et al. 2006). More recent work that combines the LMXB populations of several galaxies has concluded that the GC-LMXB XLF is flatter overall (Lehmer et al. 2014; Peacock & Zepf 2016).

Both the metallicity and the age of the stellar populations can affect the XLF. Metal-rich, red GCs host on average ~3 times more LMXBs than metal-poor, blue GCs, affecting the normalization of the relative XLFs. However, no significant difference in XLF shape exists between LMXBs associated with metal-rich or metal-poor clusters (Kim et al. 2013). Instead, old stellar populations of relatively younger stellar ages have been associated with a relative excess of luminous LMXBs, resulting in flatter XLF slopes (Kim & Fabbiano 2010; Zhang et al. 2012). Further work that disentangled GC-XLF and field-XLF has clearly demonstrated that this increased number of luminous sources is a feature of the field population, which is also predicted by LMXB evolution models (Lehmer et al. 2014; Fragos 2013b).

#### *7.2.1.2 Young and Mixed Stellar Populations*

In the young stellar populations of actively star-forming galaxies, such as late-type spirals and irregulars in the Hubble sequence, the XRB population is dominated by young, massive, binary systems with a compact stellar remnant (NS or BH) and a high-mass stellar donor companion. These HMXBs are believed to be the product of the evolution of native binary systems and have relatively short lifetimes ($\sim 10^7$ yr), due to the rapid evolution of the massive donor star. This is the less massive of the original pair, the more massive having already evolved into the compact remnant.

The number of XRBs in star-forming galaxies (XLF normalization) is proportional to the star formation rate (SFR) of the stellar population of the galaxy. The shape of the XLF is a flat power-law extending to high X-ray luminosities. In galaxies with very high SFR, such as colliding/merging galaxies (e.g., the Antennae galaxy, Zezas et al. 2007; NGC 2207/IC 2163, Mineo et al. 2014c), the XLF extends up to luminosities of $\sim 10^{40}$ erg s$^{-1}$, including significant numbers of ULXs (sources detected at luminosities greater than $10^{39}$ erg s$^{-1}$). While it has been suggested that ULXs may host intermediate-mass BHs (see Fabbiano 2006), the continuity of these XLFs is part of the evidence for ULXs belonging to the HMXB population and being powered by super-Eddington accretion onto stellar-mass BHs (King et al. 2001; see also Section 7.2.3). Even an NS can power a ULX, as shown by the M82 pulsar ULX (Bachetti et al. 2014; Fragos et al. 2015).

Studies of individual galaxies with complex stellar populations, such as spiral and late-type galaxies, demonstrate that the characteristics of the XLF closely correlate with the local stellar population (see review by Fabbiano 2006). For example, in the interacting starburst galaxy NGC 2207/IC 2163, the number of ULXs (28 in all) correlates with the local SFR (Mineo et al. 2013, 2014c), again linking these sources to the normal stellar binary population. A new study of M51 exemplifies the link between the properties of the stellar population (SFR and mass-weighted stellar age) and the XLF (Lehmer et al. 2017). Differences are also reported in the XLF and source properties of the bulge and ring XRB populations in the ring galaxy NGC 1291 (Luo et al. 2012), reflecting the different ages of these stellar populations.

A mix of stellar populations is normal in star-forming galaxies. Therefore, their XRB populations should include both HMXBs and LMXBs, as is clearly the case in the Milky Way. Under this assumption, Lehmer et al. (2010) parameterized the total XRB X-ray luminosity of a sample of nearby luminous IR galaxies as a function of both stellar mass (for the LMXB population) and SFR (for the HMXB population), proposing the relation $L_X^{gal} = \alpha M_\star + \beta SFR$, where $\alpha = (9.05 \pm 0.37) \times 10^{28}$ erg s$^{-1}$ $M_\odot^{-1}$ and $\beta = (1.62 \pm 0.22) \times 10^{39}$ erg s$^{-1}$ $(M_\odot \text{ yr}^{-1})^{-1}$. These authors point out that HMXBs dominate the galaxy-wide X-ray emission for galaxies with specific star formation rate SFR/$M_\star \geqslant 5.9 \times 10^{-11}$ yr$^{-1}$. This baseline near-universe-based relationship can be used to investigate the evolution of XRB populations at different redshifts.

### 7.2.2 The Redshift Evolution of the XRB Emission

With the advent of *Chandra* surveys (e.g., the ever-deepening Deep Fields (Brandt et al. 2001) and the ever-widening wide-area, medium-depth COSMOS survey (Elvis et al. 2009; Civano et al. 2012)), observations of normal galaxies have been extended from the near universe out to $z \sim 5$, allowing studies of the redshift evolution of the XRB population. Multiwavelength coverage of these surveys has provided measurements of both the total stellar mass ($M_*$) and the SFR of the observed galaxies, leading to the parameterization of the integrated XRB emission of galaxies as a function of these parameters and of redshift.

Both Lehmer et al. (2008), from the stacking of galaxies covered by the *Chandra* Deep Fields (up to 2 Ms *Chandra* exposure), and Mineo et al. (2014b), using a sample composed of near-universe, high-luminosity IR galaxies and galaxies detected in the *Chandra* Deep Fields, find a linear relation between the total $L_X$ and SFR, and conclude that the X-ray emission can be used as a robust indicator of star formation activity out to $z \approx 1.4$. Subsequently, Lehmer et al. (2016), using the deeper 6 Ms *Chandra* Deep Field South (CDF-S), found that simple SFR scaling is insufficient for characterizing the average X-ray emission at all redshifts, and establish a scaling relation involving both SFR and stellar mass. Aird et al. (2017), using several deep surveys, proposed in addition a power-law dependence of the SFR. More recently, Fornasini et al. (2018), using the stacked luminosities of ~75,000 star-forming galaxies in the COSMOS survey with $0.1 < z < 5$, confirmed the functional dependence of the Aird et al. relationship: $L_X = \alpha(1 + z)^{\Upsilon} M_* + \beta (1 + z)^{\delta} \mathrm{SFR}^{\theta}$, and found best-fit values of the parameters: $\log(\alpha) = 29.98 \pm 0.12$, $\gamma = 0.62 \pm 0.64$, $\log(\beta) = 39.78 \pm 0.12$, $\delta < 0.2$, $\theta = 0.84 \pm 0.08$.

It is likely that future studies will provide more complex relationships for the $z$-dependence of the X-ray emission of XRB populations. For example, simulations of XRB populations and their XLFs have pointed out that metallicity and the initial stellar-mass function (IMF) slope are also important factors in XRB evolution and in the X-ray output (e.g., Fragos et al. 2013a, 2013b; Tremmel et al. 2013). These conclusions await a thorough observational verification.

Careful characterization of the redshift evolution of the XRB population is important not only for understanding the history of XRBs, but also for studying the other components of the cumulative X-ray emission of galaxies, in particular the occurrence and properties of low-brightness hot halos (Section 7.3) and nuclear emission (Section 7.4). In principle, once the parameters of the XLFs and their evolution are well constrained observationally and well reproduced by models, X-ray observations may provide a way to constrain galaxy evolution.

XRB energy input may be important both for the early universe, in particular for reionization (e.g., Venkasetan et al. 2001; Fragos et al. 2013b; Lehmer et al. 2016), and perhaps in later galaxy feedback.

### 7.2.3 The Spatial Distributions of the XRBs

Since the earliest detections of normal galaxies in X-rays, it has been known that the spatial distribution of the hard (>2 keV) XRB-dominated X-ray emission tends to follow that of the integrated stellar light, i.e., their parent stellar population (see Fabbiano 1989 and references therein). *Chandra* observations have clearly demonstrated this general association with the detection of individual XRBs in galaxies (see Fabbiano 2006 and references therein). As discussed in Section 7.2.1.2, in star-forming galaxies with complex stellar populations, *Chandra* imaging has allowed the spatially resolved investigation of the connection of XRB populations with the local underlying stellar population. These spatial associations have provided interesting constraints on the nature and evolution of XRBs in both young and old stellar populations.

The definitive associations of very luminous ULXs with intensely star-forming regions, (e.g., in the Antennae, Fabbiano et al. 2001; the Cartwheel galaxy, Wolter & Trinchieri 2004; and many other cases) support the conclusion that ULXs belong to these young stellar populations (see also Section 7.2.1.2). Given the typical stellar masses of these populations, ULXs are likely to be young binary systems with super-Eddington accretion (King et al. 2001; Soria et al. 2009). A comparison of the positions of ULXs with the Sloan survey database (Swartz et al. 2009) concludes that ULXs tend to be associated with OB associations, rather than with super star clusters, consistent with the results on the Antennae galaxies (Zezas et al. 2002). Interestingly, the colors of the stellar regions associated with ULXs tend to be redder than those associated with H II regions. If this reddening is not due to localized absorption, the redder colors would be consistent with an aging of the stellar population commensurate with the evolution time of a massive X-ray binary.

Observations of stellar populations too young for HMXB formation validate the models of massive binary systems evolving into HMXBs. This evolution requires a few $\sim 10^6$–$10^7$ yr. Therefore, although the normalization of the XLF is proportional to the SFR, it is more proper to say that the number of HMXBs represents the star formation that took place in a galaxy 5–60 Myr ago. Shtykovskiy & Gilfanov (2007) compared the distributions of XRBs and H II regions in the spiral arms of M51 and suggest that the distribution of HMXBs is not as peaked as that of the H II regions, reflecting earlier occurrences of star formation that have been left behind by the spiral compression wave. In the Large Magellanic Cloud (LMC), our nearest star-forming galaxy, Antoniou & Zezas (2016) found that the HMXBs are present in regions with star formation bursts of ~6-25 Myr age; in the Small Magellanic Cloud (SMC), on the other hand, the HMXB population peaks at later ages (~25-60 Myr; see also Antoniou et al. 2010). Antoniou & Zezas (2016) also found that the formation efficiency of HMXBs in the LMC is ~17 times lower than that in the SMC and attribute this difference primarily to the different ages (younger in the LMC) and possibly metallicities (higher in LMC) of the HMXB populations in the two galaxies. The SMC formation age is consistent with the findings of Williams et al. (2013) of the 40–55 Myr formation age for the HMXBs in NGC 300 and NGC 2403.

In elliptical galaxies, several studies of the radial distributions of detected X-ray sources have been pursued in attempts to associate the LMXBs with either the stellar field or with the GC systems. These studies led to inconclusive results, in part due to the relatively poor statistics of the LMXB populations and to incomplete mapping of the GC systems (see Fabbiano 2006). More recently, Zhang et al. (2013) investigated the cumulative radial distribution of LMXB in 20 early-type galaxies observed with *Chandra* and reported that sources more luminous than $\sim 5 \times 10^{38}$ erg $s^{-1}$ follow the distribution of the stellar light, while observing an overdensity of fainter sources, out to at least $\sim 10 r_e$ ($r_e$ is the effective radius). They proposed that the extended LMXB halos may comprise both LMXBs located in GCs, which are known to have a wider distribution than the stellar light, and NS LMXBs kicked out of the main body of the parent galaxy by supernova explosions.

The complete spatial study of the rich LMXB and GC populations of NGC 4649, a giant E in the Virgo cluster, was made possible by the coordinated, complete, deep coverage of these populations with *Chandra* and *Hubble* surveys. These observations resulted in complete samples of both LMXBs and GCs, and in reliable identifications of X-ray sources with GC counterparts (Strader et al. 2012; Luo et al. 2013). Using these rich data sets, Mineo et al. (2014a) showed that GC-LMXBs have the same radial distribution as the parent red and blue GCs. The radial profile of field LMXBs follows the *V*-band profile within the $D_{25}$[1] of NGC 4649, consistent with an origin of these sources from the evolution of native binaries in the stellar field. Mineo et al. (2014a) also suggest a possible excess of LMXBs in the field population at large radii ($\sim 400''$ or 6 effective radii) that may be consistent with the report of Zhang et al. (2013).

A different approach to the study of the spatial distribution of LMXBs and GCs is provided by the analysis of the two-dimensional distributions of LMXBs and GCs on the plane of the sky, and on the detection and characterization of significant localized discrepancies from the azimuthally smooth distributions conforming to the radial profiles of these objects (Bonfini et al. 2012; D'Abrusco et al. 2013). This technique has revealed significant anisotropies, suggesting streamers from disrupted and accreted dwarf companions in virtually all of the galaxies thus analyzed (D'Abrusco et al. 2013, 2014a, 2014b, 2015).

In both NGC 4649 (D'Abrusco et al. 2014a—Figure 7.3) and NGC 4278 (D'Abrusco et al. 2014b), arc-like distributions of GCs are associated with similar overdensities of X-ray sources. In NGC 4649 the GC-LMXBs follow the anisotropy of red GCs, where most of them reside. However, a significant overdensity of (high-luminosity) field LMXBs is also present to the south of the GC arc, suggesting that these LMXBs may be the remnants of star formation connected with a merger event. Alternatively, they may have been ejected from the parent red GCs, if the bulk motion of these clusters is significantly affected by dynamical friction. These sources occur at relatively large galactocentric radii and certainly contribute to the excess of field LMXBs reported by Mineo et al. (2014a).

[1] $D_{25}$ is the apparent major isophotal diameter, measured at or reduced to the surface brightness level 25.0 *B* mag per square arcsecond, from de Vaucouleurs et al. (1991).

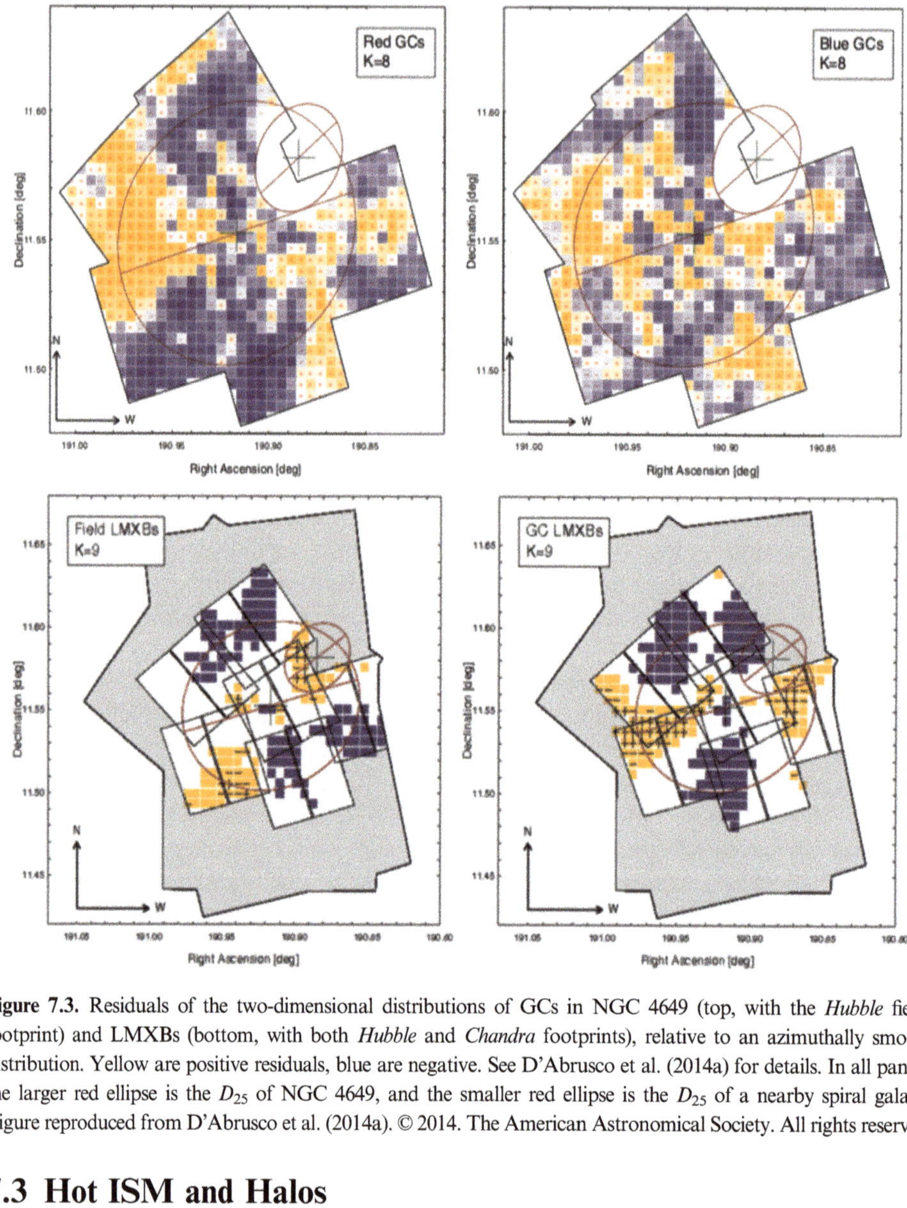


**Figure 7.3.** Residuals of the two-dimensional distributions of GCs in NGC 4649 (top, with the *Hubble* fields footprint) and LMXBs (bottom, with both *Hubble* and *Chandra* footprints), relative to an azimuthally smooth distribution. Yellow are positive residuals, blue are negative. See D'Abrusco et al. (2014a) for details. In all panels, the larger red ellipse is the $D_{25}$ of NGC 4649, and the smaller red ellipse is the $D_{25}$ of a nearby spiral galaxy. Figure reproduced from D'Abrusco et al. (2014a). 

## 7.3 Hot ISM and Halos

The first attempts to study the hot ISM systematically in normal galaxies were made with the *Einstein Observatory*: hot outflows and winds were detected in nearby starburst galaxies (e.g., NGC 253 and M82); extended hot gaseous components were found in some giant elliptical galaxies in the Virgo cluster. But the lack of good angular resolution in *Einstein* ($\sim 1'$) and all other pre-*Chandra* X-ray observatories left room for speculation, because the gaseous emission could not be cleanly separated from other sources, primarily the XRB populations discussed in Section 7.2. For example, the widespread presence of hot gaseous emission in elliptical galaxies was strongly debated (see Fabbiano 1989). Even when the spectral separation of these components was later attempted with the advent of X-ray CCDs in the Japanese X-ray satellite *ASCA*, some results were markedly strange, such as the very low metal abundances found in these gaseous components, in both star-forming and elliptical galaxies. These low abundances were in marked contrast with the expectations of chemical evolution (see, e.g., the review of the history of these studies for elliptical galaxies in Fabbiano 2012).

Because of its joint subarcsecond spatial resolution and spectral capabilities, the *Chandra* telescope with the ACIS CCD camera is well suited to the study of the extended hot ISM of galaxies. With the advent of *Chandra*, most issues and controversies from previous studies were resolved, and a new discovery space was opened to astronomers. Below we discuss some of these new developments for both the hot ISM of star-forming galaxies and that of early-type galaxies (ETGs; including both elliptical and S0s), where it is sometimes called a “hot halo.” The hot halos of ETGs are also discussed in Chapter 9, especially from the point of view of the physical processes governing these halos.

### 7.3.1 The Hot ISM of Star-forming Galaxies and Mergers

In star-forming galaxies, supernova explosions and winds from massive stars heat the ISM and enrich it with the elements shed by the stars. X-ray observations probe both the physical and the chemical properties of this hot component. In starburst regions, if the energy input into the ISM is enough to counterbalance or even surpass the pull of gravity, the hot ISM will expand above the disk of the galaxy, and in the most intense starbursts escape in the form of a hot wind (Chevalier & Clegg 1985; Heckman et al. 1990). These hot winds and outflows were first detected with the *Einstein Observatory* (e.g., in NGC 253 and M82; Fabbiano 1988), extending a few to ~10 kpc out of the plane of the galaxy. Outflows are frequently associated with optical line emission (Heckman et al. 1990) and may be common in star-forming galaxies at high redshift (Martin et al. 2012).

#### *7.3.1.1 Physical Evolution of the Hot ISM*

With *Chandra* observations, the connection between the heating of the ISM and star formation activity has been definitively established. Systematic *Chandra* studies of the soft, $kT \sim 0.2$–$0.7$ keV ($T \sim 2$–$8$ million K), diffuse emission of nearby galaxies show that the higher the star formation rate of a galaxy (and therefore the supernova rate), the higher is the luminosity of the hot ISM (Mineo et al. 2012). Observations of edge-on galaxies in the near universe also show that soft emission extending outside of the galaxy plane is common in intensely star-forming galaxies. The presence of extraplanar X-ray emission is always associated with extraplanar optical line emission of similar vertical extent (Strickland et al. 2000, 2004).

Joint *Chandra* and optical studies of the nuclear outflows of NGC 253 (Strickland et al. 2000) advanced the picture of the conical outflow of a hot (>2 keV) superwind, fueled by the starburst. The soft (<2 keV) X-rays and the optical Hα line emission would result from the interaction of this energetic wind with the swept-up and

**Figure 7.4.** Composite image of M82. Red: soft outflowing wind, 0.3–2.8 keV. Blue: nuclear starburst, 3–7 keV. Green: star light optical emission. The X-ray data are from *Chandra* ACIS, adaptively smoothed after point source subtraction and interpolation. The circle identifies the 500 pc radius region around the nucleus. Figure reproduced from Strickland & Heckman (2009).  The interactive figure shows separately the X-ray emission (color-coded as: 2.0–7.0 keV, blue; 1.2–2.0 keV, green; 0.5–1.2 keV, red), the optical image, and the composite image. Additional materials available online at http://iopscience.iop.org/book/978-0-7503-2163-1.

entrained gas from the denser ambient ISM. Modeling of *Chandra* and *XMM-Newton* observations of M82 (Strickland & Heckman 2009) suggests that the plasma within the starburst region (inside the circle in Figure 7.4) has a temperature $T \sim$ 30–80 million K ($kT \sim$ 2–7 keV), a mass-flow rate out of the starburst region of 1.4–3.6 $M_\odot$ yr$^{-1}$, and a terminal wind velocity, $v_\infty \sim$ 1410–2240 km s$^{-1}$. This velocity surpasses both the escape velocity from M82 ($v_{\rm esc} <$ 460 km s$^{-1}$) and the velocity of the Hα-emitting clumps and filaments entrained in M82's wind ($v_{\rm H\alpha} \sim$ 600 km s$^{-1}$).

Modeling of these data led Strickland & Heckman (2009) to estimate relatively high supernova and stellar wind thermalization efficiencies (30% $< \varepsilon <$ 100%). These high efficiencies are in contrast with the conclusions of Mineo et al. (2012) and Richings et al. (2010) that on average, only 5% of the mechanical energy of supernovae is converted into thermal energy of the ISM. The latter authors, however, only considered the thermal energy corresponding to the soft (<2 keV) ISM detected in their galaxy samples.

Interacting and merging galaxies provide a local laboratory where astronomers can easily observe phenomena that occur in the deeper universe. There, merging is common and may be an important step in the evolution of galaxies (e.g., Navarro et al. 1995). Major mergers of similar-mass spiral galaxies may give rise to an elliptical galaxy (Toomre & Toomre 1972). Galaxy interaction and merging triggers enhance star formation, which also means an increased injection of energy into the

ISM. This may cause galaxy-size outflows, and large-scale extended structures and hot halos (Figure 7.5).

As shown in Figure 7.5 (top left), a large-scale halo is detected in the interacting system NGC 4490/NGC 4485 (Richings et al. 2010), extending ~7.5 kpc outside the galaxy plane. This halo is not consistent with a freely expanding adiabatically

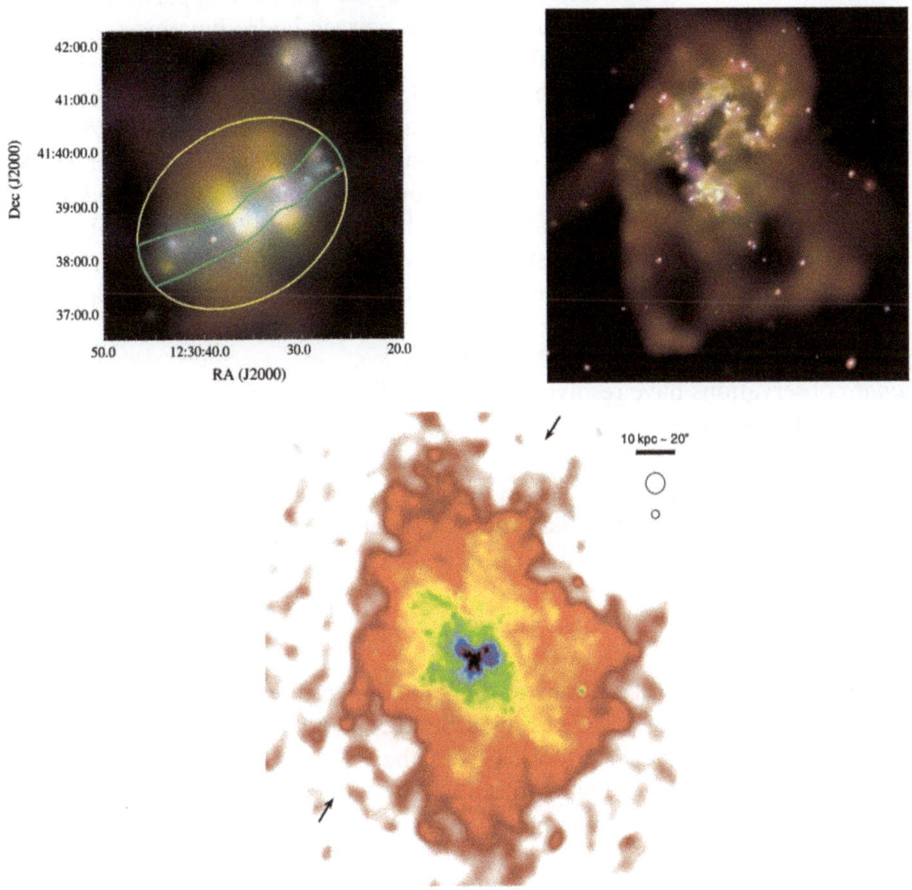


**Figure 7.5.** Three examples of *Chandra* images of large-scale structures in the hot gas associated with interacting and merging galaxies. North is up, east to the left in all these images. Top left: the hot ISM of the interacting galaxy NGC 4490 (reproduced from Richings et al. 2010 © 2010. The American Astronomical Society. All rights reserved), with the companion galaxy NGC 4485 identifiable with the clump of emission to the north. Point sources (XRBs) were subtracted from this image that shows four outflow regions from the galaxy plane and a fainter (red) halo extending ~7.5 kpc out of the plane; the green lines give the outline of the plane, and the yellow ellipse that of the halo. Top right: the image of the Antennae galaxies showing the two giant loops (10 kpc across) to the south (reproduced from Fabbiano et al. 2004 © 2004. The American Astronomical Society. All rights reserved; figure from the CXC web page.). Bottom: the image of NGC 6240 (blue higher intensity, red lower intensity, see Nardini et al. 2013 © 2013. The American Astronomical Society. All rights reserved) showing the loops and filaments in the hot ISM embedded in a giant hot halo (80 × 100 kpc).

cooling wind, as may occur in M82, so it could be at least partially gravitationally bound in the dark matter potential of the galaxy.

In the merging pair, the Antennae galaxies (NGC 4038/39; Figure 7.5, top right), aside from the widespread intense hot ISM emission throughout the stellar disks, two giant loops of hot gas (~10 kpc across, $kT\sim$ 0.3 keV) are seen extending to the south of the merging stellar disks. A cooler ~0.2 keV low surface brightness hot halo, extends out to ~18 kpc. These features may be related to superwinds from the starburst in the Antennae or result from the merger hydrodynamics. Their long cooling times (~1 Gyr) suggest that they may persist to form the hot X-ray halo of the emerging elliptical galaxy (Fabbiano et al. 2004).

NGC 6240 (Figure 7.5, bottom), a more advanced merger than the Antennae, displays several loops of hot X-ray-emitting gas, which are spatially coincident with H$\alpha$ filaments, and a very extended and luminous X-ray halo with projected physical size of ~$110 \times 80$ kpc (Nardini et al. 2013). This halo could have existed before the starburst. In a few cases of massive spiral galaxies, there have been reports of extended static hot coronae, trapped in the galaxy potential (Bogdán et al. 2013).

#### *7.3.1.2 Chemical Evolution of the Hot ISM*

*Chandra* observations have resolved a long-standing issue with the measurements of metal abundances in the hot ISM from spectral fitting of X-ray data. Pre-*Chandra* observations of star-forming galaxies suggested that the hot ISM had very low (subsolar) metal abundances. This is odd, given that the hot ISM should be enriched by the supernova ejecta and stellar winds in the star-forming regions (e.g., Weaver et al. 2000). In the Antennae, for example, *ASCA* spectra taken with CCDs with similar energy resolution to those in the *Chandra* ACIS suggested an overall extremely low abundance of heavy elements (~0.1 the solar value; Sansom et al. 1996).

The *Chandra* ACIS observations of the Antennae provided the opportunity for a direct comparison with the *ASCA* results. Baldi et al. (2006a) demonstrated that the puzzling subsolar abundances derived from the *ASCA* spectra were the result of mixing the signals from regions of the hot ISM with different spectral properties. They analyzed the entire *Chandra* emission from the Antennae as a single source, as it would have been seen by *ASCA* given its lower of angular resolution, and were able to reproduce the subsolar Sansom et al. (1996) results. Their detailed analysis of the *Chandra* data set of the Antennae, separating emission regions of different intensity and morphology, instead returned much higher—several times solar—abundance values, differing in different regions. *Chandra* studies of other starburst galaxies, analyzing separate emission regions within a given hot ISM complex, also returned higher metal abundances (e.g., Martin et al. 2002; Richings et al. 2010; Nardini et al. 2013). Mixing contributions of different temperature regions led to overlapping line emission that formed a pseudo continuum seen by ASCA.

In the Antennae, the ratios for several elemental abundances are consistent with those expected for SN II, the type of supernova resulting from the evolution of the massive stars found in starburst regions, but not SN Ia (Baldi et al. 2006b, Figure 7.6). A similar result was reported for NGC 4490 (Richings et al. 2010).

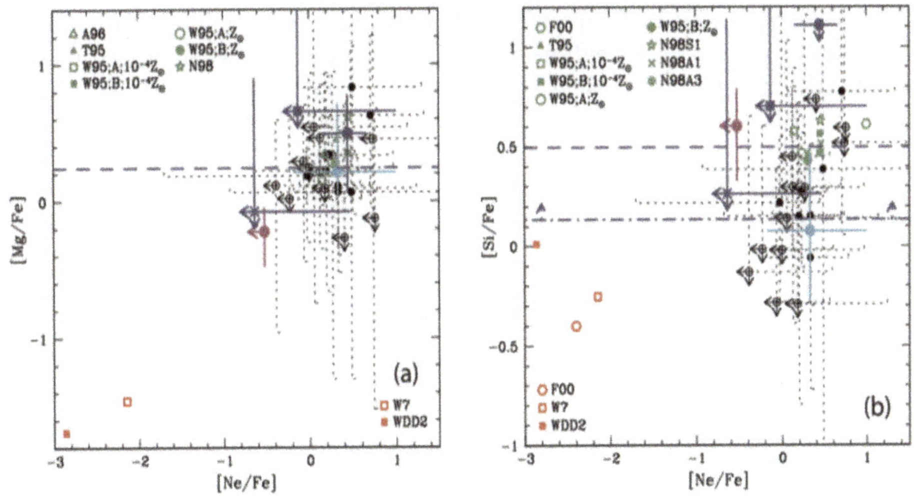


**Figure 7.6.** Elemental abundance ratios from distinct regions of the hot ISM of the Antennae (points with error bars), compared with similar ratios for SN II (green points) and SN Ia (red); SN II are the result of the evolution of massive young stars found in starburst regions, while SN Ia are associated with old stellar populations (reproduced from Baldi et al. 2006b © 2006. The American Astronomical Society. All rights reserved).

## 7.3.2 The Hot ISM of Early-type (Elliptical and S0) Galaxies

Early-type galaxies (ETGs) are characterized by an old stellar population. ETGs are believed to be the end product of galaxy evolution and to originate from major mergers of gas-rich disk galaxies, with subsequent accretion of smaller-mass companions (De Lucia et al. 2006; Oser et al. 2012). The hot ISM and halos of normal elliptical galaxies were discovered thanks to the imaging capabilities of the *Einstein Observatory* (Forman et al. 1985; Trinchieri & Fabbiano 1985; see reviews by Fabbiano 1989, 2012 and references therein). This discovery went against the accepted picture in astronomy at the time, that winds would occur in ETGs, dissipating the gaseous stellar ejecta outside the parent galaxy (e.g., Faber & Gallagher 1976; Mathews & Baker 1971). While many studies with *Einstein* (and the subsequent X-ray observatories *ROSAT* and *ASCA*) ensued, it is only with *Chandra* that these hot gaseous components can be studied in detail, with relatively uncontroversial results (see review by Fabbiano 2012).

With the subarcsecond imaging of *Chandra* and its spectral capabilities, the populations of LMXBs in ETGs can be detected and separated from the hot gaseous emission, both spatially and spectrally (Section 7.2). This capability has led to the resolution of previous hotly debated ambiguities regarding the amount of hot ISM and halos in different ETGs and has allowed the physical and chemical characterization of these hot halos. Studies of the interactions between hot ISM, gravity, and nuclear activity, based on these data, are setting stringent constraints on the amount of dark matter and nuclear feedback in ETGs.

Below we will review some of these topics, with emphasis on the observational picture of ETGs and its implications for their evolution. This discussion is complementary to that in Chapters 9 and 10, where a more in-depth discussion of the physics of the hot halos can be found (see also the reviews by Pellegrini 2012, Sarazin 2012, and Ciotti & Ostriker 2012).

### 7.3.3 Scaling Relations of ETGs

The first question following the discovery of hot halos in some elliptical galaxies in the Virgo cluster (Forman et al. 1979) was how widespread are these halos (Trinchieri & Fabbiano 1985), followed by the obvious corollary: how can we constrain the physical properties and the evolution of these halos, taking into account different scenarios for the heating of the gas (e.g., Forman et al. 1985; Canizares et al. 1987)? A basic tool for these studies was the comparison between the integrated X-ray luminosity ($L_X$) of the galaxies and their integrated stellar emission ($L_B$; earlier studies used *B*-band integrated stellar emission, while later studies used the *K*-band emission, which is more representative of the stellar population of these galaxies).

Because the emission of LMXB populations was also included in the total X-ray emission, it was highly controversial whether the pre-*Chandra* $L_X$–$L_B$ diagrams truly represented the behavior of the hot ISM as a function of the stellar mass of the galaxies. Moreover, early measurements of the temperatures of these halos were biased by the mixing in of the hard LMXB spectra, especially in galaxies with relatively small amounts of hot halo (those with lower $L_X/L_B$ ratios; Kim et al. 1992). While studies with *ROSAT* and *ASCA* have probed some of these issues, the angular resolution of *Chandra* was needed to change the observational paradigm (see Fabbiano 2012).

With *Chandra* ACIS, Boroson et al. (2011, hereafter BKF) derived scaling relations for the different components of the X-ray emission of a sample of 30 nearby ETGs, spanning a range of X-ray-to-optical ratios. The *K*-band near-IR integrated emission was used as a proxy for the total stellar mass of each galaxy. BKF detected samples of LMXBs in these galaxies and subtracted their contributions from the images. They subtracted the contribution of fainter LMXBs, below the detection threshold, following two approaches: (1) extrapolating the LMXB X-ray luminosity function to lower luminosities (see Section 7.2.1), and (2) spectral analysis, as the LMXB spectrum is harder than that of the hot ISM. They also considered the contribution of the coronally active binaries (ABs) and cataclysmic variables (CVs) in these galaxies, using *Chandra* observations of the bulge of M31 and of M32 to model this unresolved emission. Finally, the emission from sources detected at the galaxy nucleus, which could be X-ray-faint AGNs, was also subtracted.

The BKF $L_X$–$L_K$ diagram of the hot halo is shown in the left panel of Figure 7.7 (red points) together with the total ETG $L_X$ (black circles; effectively the pre-*Chandra* scaling relation) and the contributions of LMXBs (blue dots), ABs + CVs (black line), and nuclear sources (green triangles). There is no correlation between the nuclear luminosity (green triangles) and $L_K$. For the hot gas, the $L_X$–$L_K$ points (red) follow a much steeper relation than in the pre-*Chandra* studies (black circles), but still have considerable spread. The LMXB contribution is correlated with the stellar mass (represented by $L_K$), although a secondary correlation with $S_N$, the

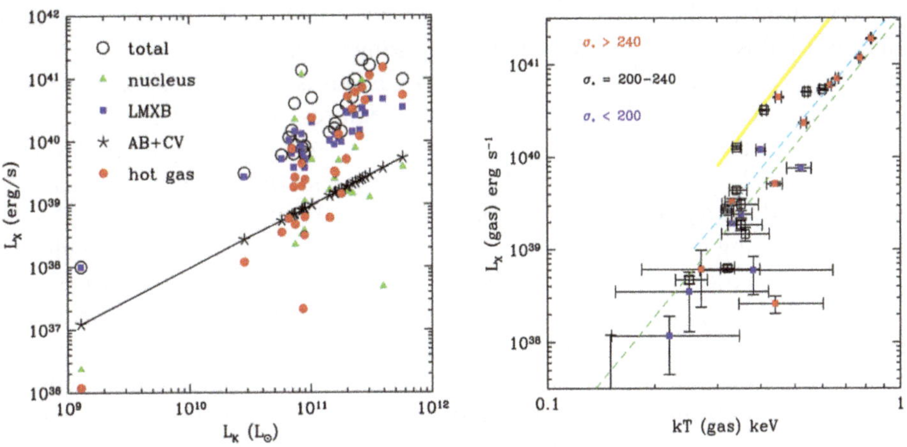


**Figure 7.7.** Left: $L_X$–$L_K$ diagram for the sample of 30 nearby ETGs observed with *Chandra*, including the total $L_X$ of each ETG (circles) and the contribution of each component as indicated. Right: $L_X$–$kT$ diagram for the hot gas. The dashed lines are best-fit relations for the entire sample (green) and galaxies with $L_X$ (gas) > $10^{39}$ erg s$^{-1}$ (cyan); the yellow line represents the relation for cD galaxies and groups. (Reproduced from Boroson et al. 2011. © 2011. The American Astronomical Society. All rights reserved.)

number of GCs per unit stellar mass in a galaxy (see Section 7.2), was also found by BKF. The AB + CV contribution is by construction correlated with $L_K$. The average LMXB contribution to the integrated $L_X$ is 10 times larger than that from the unresolved AB + CV emission:

$$L_X(\text{LMXB})/L_K = 7.6 \times 10^{28} \times S_N{}^{0.334} \text{erg s}^{-1} L_K{}^{-1} \quad \text{and}$$
$$L_X(\text{AB} + \text{CV})/L_K = 9.5 \times 10^{27} \text{erg s}^{-1} L_K{}^{-1}.$$

One of the most striking results of the BKF study is the strong positive correlation found between the luminosity and temperature of the hot gas component (right panel of Figure 7.7), which follows the best-fit relation $L_X$ (gas) ~ $T^{4.6 \pm 0.7}$ (green line in the figure; the cyan line is the similar best-fit relation obtained excluding ETGs with $L_X$ (gas) < $10^{39}$ erg s$^{-1}$, where the errors are larger). The yellow line represents the relation (similar, but shifted towards higher $L_X$) for the central galaxies of groups and clusters. A correlation between gas luminosity and temperature is expected for gravitationally confined hot halos (see also Section 7.3.4).

The $L_{X,GAS}$–$L_K$ and $L_{X,GAS}$–$T_{GAS}$ relations were further investigated by Kim & Fabbiano (2015) using a larger sample of ETGs, the 61 ATLAS3D E and S0 galaxies observed with *Chandra*, including *ROSAT* results for a few X-ray-bright galaxies with extended hot gas. This work uncovered a dependence in these relations, on the structural and dynamical properties of ETGs, suggesting that the correlations are carried by the "core" ETGs in the sample, ETGs with central surface brightness cores, slow stellar rotations, and uniformly old stellar populations. For these galaxies, the $L_{X,GAS} \sim T_{GAS}{}^{4.5\pm0.3}$ correlation extends down into the $L_{X,GAS} \sim 10^{38}$ erg s$^{-1}$ range, where simulations predict the gas to be in an outflow/wind state (Negri et al. 2014), with the resulting $L_X$ values much lower than observed. Instead, the observed correlation may suggest the presence of small, bound, hot halos even in this low-luminosity range.

The $L_{X,GAS} \sim T_{GAS}^{4.5\pm0.3}$ correlation of core ETGs is consistent with the presence of virialized hot halos. The virial theorem requires that, if the hot gas is in equilibrium in the gravitational potential, $M_{Total} \sim T_{GAS}^{3/2}$. Kim & Fabbiano (2013; see also Forbes et al. 2017) found a tight relation between the gas X-ray luminosity and the dynamical galaxy mass: $L_{X,GAS}/10^{40}$ erg s$^{-1}$ = $(M_{Total}/3.2 \times 10^{11}\ M_{\odot})^3$. Substituting $L_{X,GAS} \sim T_{GAS}^{4.5}$, we obtain the virial relation. Therefore, in these galaxies, dark matter is primarily responsible for retaining the hot gas.

Among the gas-poor galaxies, which also tend to be cuspy in their internal isophotes, the scatter in the scaling relations increases, suggesting that secondary factors (e.g., rotation, flattening, star formation history, cold gas, environment, etc.) may become important (Kim & Fabbiano 2015). In these galaxies, $L_{X,GAS}$ is $< 10^{40}$ erg s$^{-1}$ and is not correlated with $T_{GAS}$. The $L_{X,GAS}$–$T_{GAS}$ distribution is a scatter diagram similar to that reported for the hot interstellar medium (ISM) of spiral galaxies (Li & Wang 2013), suggesting that both the energy input from star formation and the effect of galactic rotation and flattening may disrupt the hot ISM.

Figure 7.8, from Kim & Fabbiano (2015), compares the observed distributions of ETGs in the $L_{X,GAS}$–$T_{GAS}$ plane with those of cD galaxies (the dominant galaxies

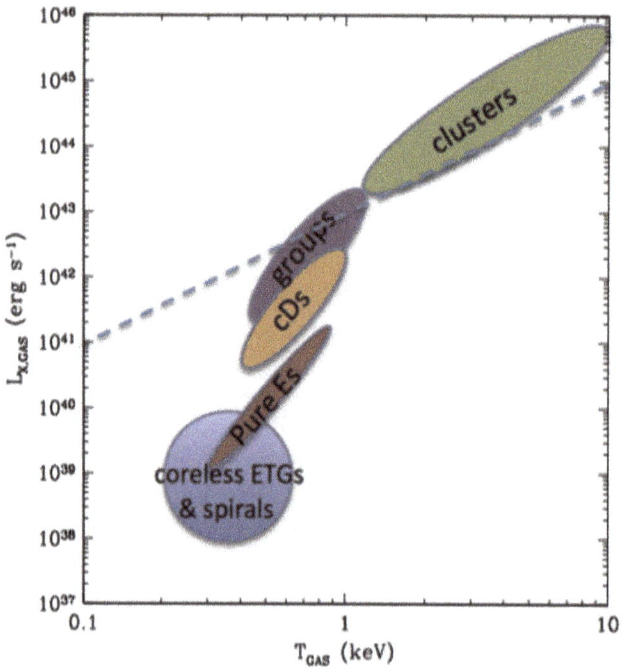


**Figure 7.8.** Comparison of the $L_{X,GAS}$–$T_{GAS}$ of different classes of objects, as indicated. No correlation is found in spiral galaxies and coreless (cuspy) ETGs. Normal core E galaxies follow a tight $L_{X,GAS} \sim T_{GAS}^{4.5}$ correlation. CD galaxies and groups follow a similar correlation, but shifted towards higher $L_{X,GAS}$. Cluster of galaxies follow a flatter $L_{X,GAS} \sim T_{GAS}^3$ relation. The dashed line shows the expectation for gravitational confinement alone ($L_{X,GAS} \sim T_{GAS}^2$). Reproduced from Kim & Fabbiano (2015). © 2015. The American Astronomical Society. All rights reserved.

of groups and clusters, sitting at the bottom of the potential well imposed by the group dark matter, e.g., M87 in the Virgo Cluster), and with groups and clusters of galaxies. The dashed line shows the expectation for gravitational confinement alone ($L_{X,GAS} \sim T_{GAS}^2$), which does not even represent the correlation for galaxy clusters $L_{X,GAS} \sim T_{GAS}^3$ (see Arnaud & Evrard 1999; Maughan et al. 2012). Nongravitational effects may be responsible for the increasingly steeper relations for less massive systems. The $L_{X,GAS}$–$T_{GAS}$ correlation of core elliptical galaxies ("Pure Es" in the figure) is similar to that found in samples of cD galaxies and groups, but shifted down toward relatively lower $L_{X,GAS}$ for a given $T_{GAS}$. Enhanced cooling in cD's, which have higher hot gas densities and lower entropies, could lower $T_{GAS}$ to the range observed in giant Es, a conclusion supported by the presence of extended cold gas in several cD's. In the smaller halos of ETGs, the effects of supernova heating and nuclear feedback would be the strongest, depleting the hot halos (i.e., lowering $L_{X,GAS}$).

### 7.3.4 Constraints on the Binding Mass of ETGs

As first applied to M87, the central galaxy of the Virgo cluster, the binding mass of hot halos in gravitational equilibrium can be measured using the equation (Fabricant et al. 1980):

$$M(r) = -kT_{GAS}/G\mu m_H(d \log \rho_{GAS}/d \log r + d \log T_{GAS}/d \log r)r,$$

where $M(r)$ is the binding mass within radius $r$, $T_{GAS}$, and $\rho_{GAS}$ are the gas temperature and density at the same radius, $k$ is the Boltzmann constant, $G$ is the gravitational constant, $\mu$ is the molecular weight, and $m_H$ is the mass of the hydrogen atom. Based on this equation, the mass enclosed within the outer detected halo radius is a function of four measurable quantities: temperature, density, and their radial gradients. The uncertainty on these mass measurements derives from the uncertainties on each of these four quantities.

The use of the hot halos of ETGs as a way to measure the gravitational mass of the galaxy has been intensely debated. Issues include the uncertainties in these measurements, the potentially large biases resulting from the contaminations of the halos with undetected LMXB populations, and the physical status of the halos themselves, which may not be solely in gravitational equilibrium. Large halos can be affected by nuclear feedback at small radii and by interaction with their environment at large radii, while small halos could be escaping as galactic winds (see, e.g., the case of the Fornax A galaxy, NGC 1316, Kim & Fabbiano 2003; and reviews by Fabbiano 2012; Statler 2012; Buote & Humphrey 2012).

While this discussion continues, recent *Chandra* work has sought to compare dynamical mass measurements (from stellar, planetary nebula, and GC kinematics) with X-ray mass measurements. The purpose of these comparisons is twofold: to provide a way to establish the validity of X-ray mass measurements and to constrain the physical state of the halos where departures of the X-ray mass estimate from the dynamical mass estimate are observed. For this work, the halos observed with *Chandra* are "cleaned" of LMXB contaminants (see Section 7.3.3). In some cases, lower angular resolution *XMM-Newton* data was also used at large radii, where the effect of the LMXB population is small.

In NGC 4649, a giant E galaxy in Virgo with an extensive hot halo, Paggi et al. (2014, 2017b) found a clear difference in the X-ray and dynamical mass profiles that can be related to the effect of nuclear feedback from the expanding radio lobes in the hot halo (Figure 7.9). This nonthermal pressure amounts to ~30% of the gravitational pressure, counterbalancing somewhat the effect of gravity in the inner radii and resulting in an apparently smaller X-ray mass than dynamical mass.

In NGC 5846, Paggi et al. (2017) found significant azimuthal asymmetries in the X-ray mass profiles. Comparison with optical mass profiles suggests significant departures from hydrostatic equilibrium, consistent with bulk gas compression and decompression, due to sloshing on ~15 kpc scales. Only in the NW direction, where the emission is smooth and extended, do they find consistent X-ray and optical mass profiles, suggesting that the hot halo is not affected by strong nongravitational forces. These authors also note how the results are dependent on the assumptions made for the metal abundance of the gas, because abundance and temperature are coupled in the spectral fit.

The above examples demonstrate how careful an analysis is needed when attempting to characterize the properties of the hot gas, and to measure the galaxy mass, in each individual case. Instead, as discussed in Section 7.3.3, the scaling relations of ETGs, in comparison with dynamical mass measurements, provide a more encouraging picture, at least for a class of ETGs: the core elliptical galaxies,

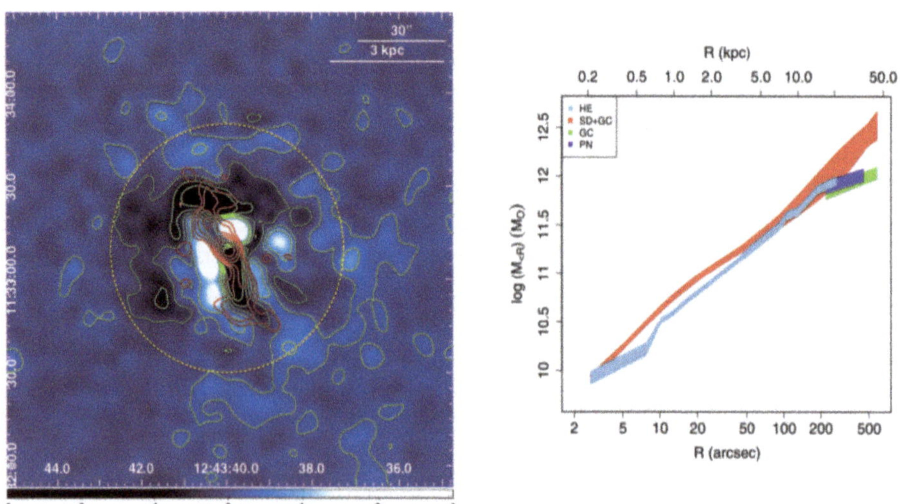


**Figure 7.9.** Left: ridges in the hot gas distribution of NGC 4649, resulting from the interaction with the nuclear radio source (the red contours). Right: comparison of the radial mass profile from the hot gas in the assumption of hydrostatic equilibrium (pale blue) with mass profiles derived from the kinematics of stars, planetary nebulae, and globular clusters. Note the smaller values of the X-ray-derived mass in the central ~3 kpc, where the gas is subject to the pressure of the expanding radio jet/lobes (see left panel). Reproduced from Paggi et al. (2014). 

where the scaling relations appear strongest (Kim & Fabbiano 2013, 2015; Forbes et al. 2017).

### 7.3.5 Metal Abundances of the Hot Halos of ETGs

The physical evolution of the hot halos is closely linked to their chemical evolution, as these halos are enriched in metals by stellar and supernova ejecta. In particular, large hot halos in gravitational equilibrium should have Fe content commensurate with the integrated output of SNe Ia over their lifetimes. In these halos, the Fe-to-alpha-element ratios should be solar or higher, unless inflow of intracluster gas, enriched by SN II-powered winds early in the galaxy's lifetime, alter these values (David et al. 1991; Ciotti et al. 1991; Renzini et al. 1993; Arimoto et al. 1997). The X-ray spectra should contain emission lines revealing the imprint of these metals.

As discussed in Section 7.3.1 for the metal abundances of the hot ISM and outflows of star-forming galaxies, most of the early measurements in ETGs returned puzzling subsolar abundances. More recent work, both based on a careful analysis of *ASCA* spectra (Matsushita et al. 2000), and on *Chandra* and *XMM-Newton* observations, has produced results more in keeping with the expectations (Kim & Fabbiano 2003, 2004). The problem with past results resided both in the contamination of the spectra by the LMXB population, and by radial temperature and abundance gradients in the hot halos. In NGC 507, in particular (Figure 7.10; Kim & Fabbiano 2004), the supersolar Fe abundances at small radii are consistent with chemical evolution models of the enrichment by SNe Ia throughout the lifetime of the galaxy, after the initial star formation when SNe II are responsible for the enrichment.

The ratio of Si to Fe is also consistent with this picture, because significantly larger ratios result from SN II yields. For an in-depth discussion of observational and

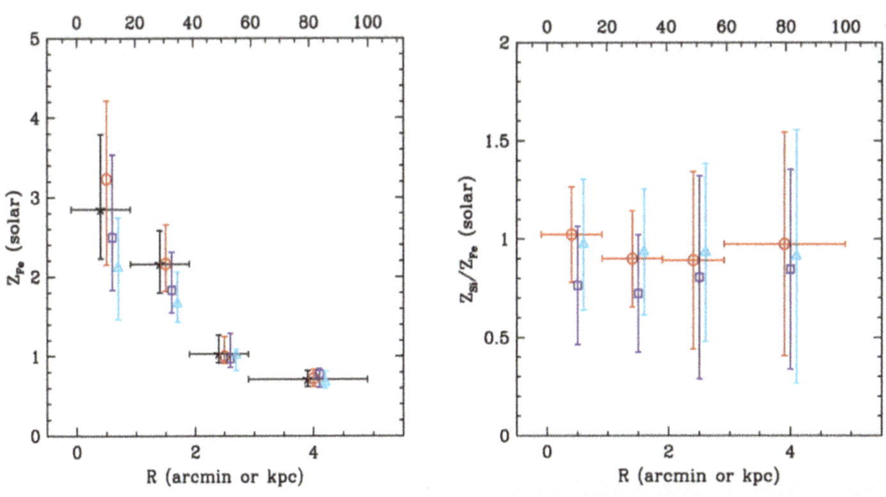


**Figure 7.10.** Reproduced from Kim & Fabbiano (2004). © 2004. The American Astronomical Society. All rights reserved. Radial distributions of (a) Fe abundance and (b) Si-to-Fe abundance ratios measured with *Chandra* data for NGC 507. Different symbols indicate different ways of tying the elemental abundances in the spectral analysis.

analysis issues, see Kim (2012). A discussion of the chemical evolution of the hot gas of ETG, enriched by stellar winds and supernovae throughout the lifetime of the galaxy, can be found in Pipino (2012).

### 7.3.6 ETGs at Higher Redshift

The studies of ETGs discussed so far were all based on nearby galaxies, mostly within 30 Mpc. To study the evolution of these systems, it is necessary to look at the higher redshifts explored with *Chandra* deep and medium-depth surveys. Studies including both ETGs detected in the *Chandra* Deep Fields and stacking analysis of undetected galaxies in these fields suggest no or mild evolution of the X-ray luminosity (relative to the optical luminosity) up to $z \sim 1.2$ (Tzanavaris & Georgantopoulos 2008; Lehmer et al. 2007; Danielson et al. 2012). The mild increase in luminosity with redshift for ETGs was confirmed by the stacking analysis of a sample of *K*-band selected galaxies in the COSMOS field surveyed with *Chandra* by Jones et al. (2014). These authors suggested mechanical heating from radio AGNs as the heating mechanism.

Civano et al. (2014) took a more direct approach, comparing the sample of ETGs detected with *Chandra* in the C-COSMOS field (Elvis et al. 2009) with the local-universe ETGs studied with *Chandra*. The C-COSMOS detections include 69 ETGs with $L_X \sim 10^{40}$–$10^{43.5}$ erg s$^{-1}$ and redshift $z \leqslant 1.5$. The optical spectra are consistent with a passive old stellar population, but a few "possible" AGNs (see Chapter 8) are included. As we know from detailed studies of nearby galaxies, the X-ray emission of ETGs is complex: it includes the integrated emission of the stellar LMXB population, the emission of hot gaseous ISM and halos, and possible AGN emission from the nuclear supermassive black hole (see Sections 7.2 and 7.3.3). The purpose of Civano et al. (2014) was to explore the $z$-evolution of hot halos and the occurrence of low-luminosity AGNs. To this end, the expected LMXB contribution was subtracted using the BFK scaling relation (Section 7.3.3), derived from the ETGs in the local universe, modified to take into account the expected $z$-evolution of these populations (Fragos et al. 2013a). The resulting counts were converted to luminosity assuming a typical range of hot halo temperatures and compared with the BKF $L_X$(gas)–$L_K$ diagram of local ETGs (Figure 7.11).

Civano et al. (2014) found that most galaxies with estimated $L_X < 10^{42}$ erg s$^{-1}$ and $z < 0.55$ follow the $L_{X,GAS}$–$L_K$ relation of the local-universe ETGs (the red Local Strip in Figure 7.11). The stacking experiment of Paggi et al. (2016; see Section 7.4.1) showed that the average X-ray spectra of Local Strip galaxies are soft, consistent with gaseous emission. All the X-ray ETGs with stellar age >5 Gyr follow the Local Strip reasonably well, suggesting that the hot halos are similar to those observed in the local universe (Figure 7.12, right). This result is consistent with the predictions of evolutionary gas-dynamical models including stellar mass losses, supernova heating, and AGN feedback (Pellegrini 2012; Pellegrini et al. 2012). For these galaxies, total masses may be derived using the virial relation of the local sample (Kim & Fabbiano 2013; see Section 7.3.3).

There are a few galaxies to the left of this strip (in the green "X-ray excess" locus in Figure 7.11). These have X-ray emission well in excess of the $L_{GAS}$ expected from

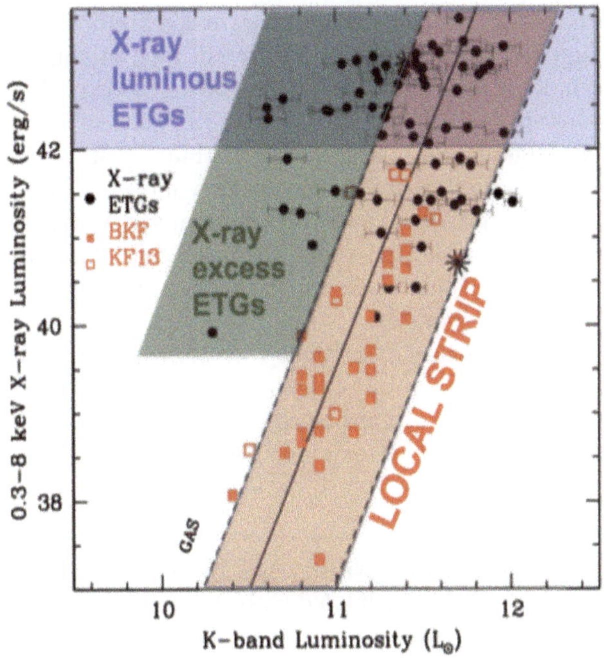


**Figure 7.11.** Nonstellar X-ray luminosity (see text) vs. *K*-band luminosity (a proxy of the total stellar mass of the ETG) for the COSMOS-detected ETGs (black dots), compared with the local sample gaseous halos (BKF: Boroson et al. 2011; KF13: Kim & Fabbiano 2013; red points). See text for the colored strips. Reproduced from Civano et al. (2014). © 2014. The American Astronomical Society. All rights reserved.

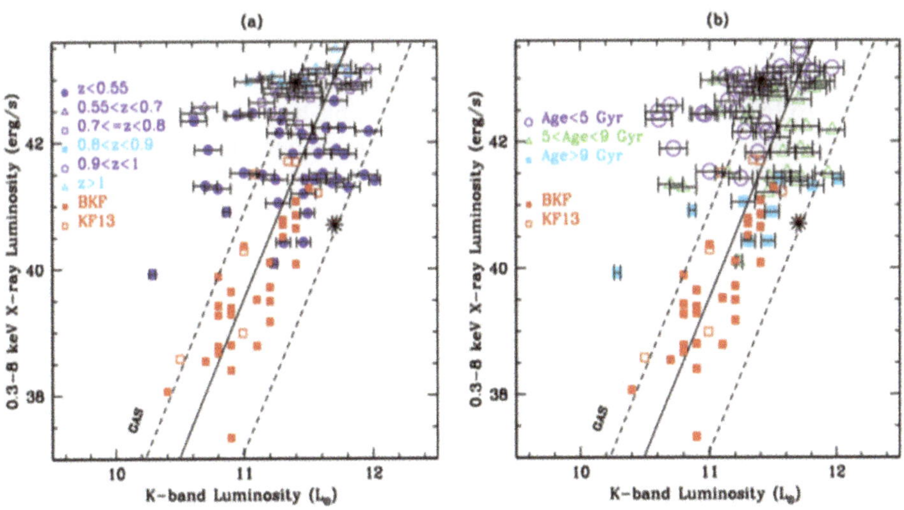


**Figure 7.12.** Same as Figure 7.11, except that the COSMOS ETGs are labeled according to their redshift (left panel) and average stellar age (right panel). The dashed lines identify the boundaries of the Local Strip of Figure 7.11. Reproduced from Civano et al. (2014). © 2014. The American Astronomical Society. All rights reserved.

the local sample ETGs given their stellar mass (or $K$-band luminosity). This excess X-ray emission suggests the presence of a different type of source, neither LMXB nor hot gaseous halo. Likely candidates are low-luminosity AGNs (see Section 7.4.1). The more luminous ($10^{42}$ erg s$^{-1}$ < $L_X$ < $10^{43.5}$ erg s$^{-1}$) and distant galaxies (higher $z$) present significantly larger scatter (blue strip in Figure 7.11; see Figure 7.12, left). These galaxies typically have younger stellar ages. They also tend to have $L_X > 10^{42}$ erg s$^{-1}$ and to be overluminous in X-rays for their $L_K$ when compared to the local sample and older stellar-age galaxies (Figure 7.12). The *Hubble* images of several of these galaxies show close companions, suggesting that galaxy merging may be responsible for the high X-ray luminosity. Merging may enhance the star formation rate (consistent with the younger stellar ages of these ETGs), thus producing a population of luminous X-ray binaries (see Section 7.2.1); enhance the X-ray luminosity of the halo (Section 7.3.1); and also induce nuclear accretion awakening in an AGN (Cox et al. 2006; see Section 7.4.2).

## 7.4 Nuclear BHs and AGNs

Supermassive black hole (SMBHs), with masses $10^8$ $M_\odot$ or larger, are ubiquitous in giant galaxies, and their mass is correlated with the stellar mass of the galaxy bulge (Magorrian et al. 1998; Gültekin et al. 2009). This relation suggests a "coevolution" scenario, with the growth of the SMBH linked to the growth of the host galaxy. In the picture of growth via merger and accretion, simulations show that the SMBH of one or both merging galaxies undergoes periods of accretion and activity, becoming an AGN (Hopkins et al. 2008; Volonteri et al. 2016). The AGN, in turn, interacts with the host galaxy, either stimulating or dampening star formation (see Heckman & Kauffmann 2011 and references therein).

Below, we discuss briefly two aspects of nuclear properties that are relevant for having a complete picture of normal galaxies: (1) the presence of low-luminosity "hidden" AGNs in normal galaxies, and (2) the link between galaxy merging and SMHB activity and growth. This discussion is complementary to Chapter 8, which addresses the X-ray properties of AGNs.

### 7.4.1 Hidden AGNs in Normal Galaxies

Two still open questions are: (1) what is the full range of nuclear BH masses, i.e., how small can a nuclear BH be? Do dwarf galaxies host smaller-mass nuclear BHs? And (2) what is the extent of nuclear activity in the universe, i.e., the cycle of accretion and starvation of the BH? Answering the first question is relevant to constrain theories of the formation and characteristics of SMBH seeds in the early universe (Volonteri 2012). As for the second question, because normal galaxies host nuclear massive BHs (Magorrian et al. 1998), the distinction between normal galaxies and AGNs is a matter of how much the nuclear BH is accreting.

Observationally, AGNs have only been considered in those sources detected in X-ray surveys with $L_X > 10^{42}$ erg s$^{-1}$, because below this threshold, the X-ray luminosity can be explained with stellar and hot ISM emission (Fabbiano 1989). However, with *Chandra*, one can explore more fully the range of nuclear emission.

In nearby galaxies, for which the emission components can be resolved, low-luminosity nuclear sources are frequently detected, associated with SMBH masses ranging from the $(4.1 \pm 0.6) \times 10^6\ M_\odot$ SMBH at the center of the Milky Way (Ghez et al. 2008) to the $\sim 10^9$–$10^{10}\ M_\odot$ SMBHs in giant ETGs (e.g., Paggi et al. 2014). For more distant galaxies, even if the emission components cannot be resolved via imaging, *Chandra* studies of the properties and scaling laws of the XRB populations and hot halos (see Sections 7.2 and 7.3) provide the tools for estimating the AGN contribution to the X-ray luminosity. Low-luminosity nuclei may be too faint or too obscured to leave an AGN signature in the optical spectra, and therefore may be missed by multiwavelength surveys, such as COSMOS, but can be revealed by X-ray studies.

Paggi et al. (2016; see Figure 7.13) performed a stacking experiment on the 6388 ETGs undetected by *Chandra* in the C-COSMOS sample by adding up the *Chandra* data of galaxies in bins of optical luminosity ($L_K$) and redshift ($z$). The expected LMXB contribution to the stack was subtracted, following the procedure of Civano et al. (2014; see Section 7.3.6). Similar stacking experiments, using galaxies in the COSMOS survey, were performed by Mezcua et al. (2016) for the ~50,000

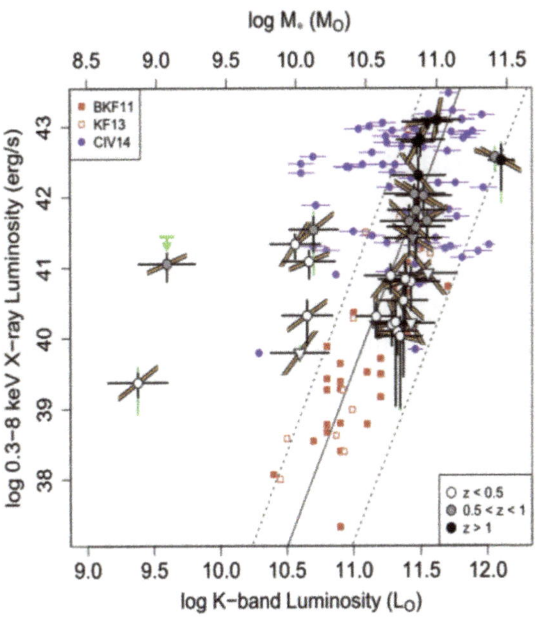


**Figure 7.13.** Average nonstellar (LMXB contribution subtracted) X-ray luminosity of $L_K$ stacks (circles, shaded according to the average redshift of the stack), for non-X-ray-detected COSMOS ETGs. The blue points are the COSMOS detections (Civano et al. 2014, Section 7.3.2.4), and the red points are the near-universe gaseous halos. The yellow diagonal bars across the points represent the average power-law spectrum of the stack. Hard spectra (upward-pointing power laws) are found both in the highest stellar-mass ($L_K$)/ highest $L_X$ stacks, and in the X-ray excess dwarf ETG stacks. The stacks that follow the local strip of Figure 7.11 (between the dashed lines) tend to have soft X-ray spectra (downward-pointing power law), consistent with the emission of hot halos. Reproduced from Paggi et al. (2016). © 2016. The American Astronomical Society. All rights reserved.

undetected star-forming dwarf galaxies with redshift up to $z = 1.5$, and Fornasini et al. (2018) for 75,000 giant star-forming galaxies between redshifts 0.1 and 5.

In dwarf ETGs ($\log(L_K) < 10.5$ in solar units), Paggi et al. (2016) found significant emission above that expected from the LMXB populations. Moreover, this emission has a hard average X-ray spectrum, consistent with AGN emission. The luminosities (a few $10^{39}$–$10^{41}$ erg s$^{-1}$) are consistent with those produced by inefficient accretion ($10^{-5}$–$10^{-4}$ of the Eddington rate) onto BHs of mass $\sim10^6$–$10^8$ $M_{\odot}$. In star-forming dwarfs, with stellar mass $<10^{9.5}$ $M_{\odot}$, Mezcua et al. (2016) reported an X-ray excess that suggests nuclear accreting BHs, with average nuclear X-ray luminosities in the range $10^{39}$–$10^{40}$ erg s$^{-1}$ and inferred masses $\sim1$–$9 \times 10^5$ $M_{\odot}$. These nuclear BH masses derived from the stacking of survey data are in the range of the intermediate-mass BH AGNs studied by Greene & Ho (2007), and of those inferred from *Hubble–Chandra* studies of nearby single dwarf galaxies (e.g., Baldassare et al. 2017) and of low-luminosity AGNs in small-bulge galaxies (Chilingarian et al. 2018).

Highly obscured hidden AGNs in giant galaxies are reported by Paggi et al. (2016) in the high-redshift ETG COSMOS sample (Figure 7.13), in particular those with X-ray luminosity $>10^{42}$ erg s$^{-1}$. Similarly, hidden highly obscured AGNs at $z >$ 1.5 are also found in the stacking analysis of star-forming giant galaxies by Fornasini et al. (2018). In both cases, these AGNs are not visible in the COSMOS optical–IR colors and spectra.

### 7.4.2 AGNs in Merging Galaxies

Simulations (Hopkins et al. 2008) suggest that galaxy merging and the resulting AGN activity are key steps of galaxy evolution. During this process, the SMBH may be "buried" by thick molecular gas, which feeds the SMBH at high rates, causing the birth of an obscured Compton-thick (CT; Risaliti et al. 1999) AGN. Uncontroversial statistics are missing on the occurrence of dual AGNs that may then merge into a single SMBH after releasing gravitational waves (Shannon et al. 2013). However, some evidence of merger-related AGN activity has been provided by *Chandra* observations.

The first dual CT AGNs were discovered with *Chandra* in the merger NGC 6240 (Komossa et al. 2003). Both nuclei are detected in the hard continuum and have the strong 6.4 keV Fe K$\alpha$ emission lines characteristic of CT AGNs. Extended, hard X-ray continuum and Fe XXV line emission in this galaxy has been related either to shocks caused by the intense star formation activity near the nuclei or to AGN winds (Wang et al. 2014).

Deep *Chandra* observations of the merger Arp 220 (Paggi et al. 2017a) clearly show that the nuclear regions are dominated by Fe XXV emission in X-rays, suggesting either highly shrouded nuclei or the presence of intense star formation. One active AGN contributes to the X-ray luminosity of the merger Arp 299 (Anastasopoulou et al. 2016). Ultraluminous infrared galaxies (ULIRGs) could be the result of recent mergers. A *Chandra* survey of these galaxies found widespread Fe XXV emission (Iwasawa et al. 2009).

While most ULXs have been convincingly related to the X-ray binary population of the host galaxy, some very luminous ones may indicate a merging remnant. A compelling example of an accreting black hole with a mass of $\geqslant 10^4\ M_\odot$, which may be the stripped remnant of a merging dwarf galaxy, is the very luminous off-nuclear X-ray source discovered in the Seyfert galaxy NGC 5252 (Kim et al. 2015). With an X-ray luminosity of $1.5 \times 10^{40}$ erg $s^{-1}$, this source is associated with radio emission and optical emission lines (at the redshift of NGC 5252). The flux of [O III] appears to be correlated with both X-ray and radio luminosity in the same manner as ordinary AGNs.

## 7.5 AGN–Galaxy Interaction in Nearby Spiral Galaxies

As discussed in Section 7.4, nuclear SMBHs are ubiquitous in galaxies. When accreting, they give rise to AGNs, which produce a significant fraction of (or exceed) the total energy output of the galaxy from other sources. The physical processes related to AGN activity are discussed in Chapter 8. However, both AGN radiation and the mechanical energy associated with jets and outflows have an impact on the host galaxy. AGN feedback is believed to be one of the important factors in galaxy evolution (see Section 7.4). Deep *Chandra* observations of nearby galaxies provide a new perspective on this interaction. Nuclear radio sources (jet and lobes) in elliptical galaxies carve cavities in the hot ISM (see Section 7.3.4 and Figure 7.9) and effectively stop cooling flows. These interactions are extensively discussed in Chapters 9 and 10. Here, instead, we address the interaction of the AGN with the colder ISM of spiral galaxies.

The X-ray emission of AGNs in the ~0.3–8.0 keV *Chandra* energy range is characterized by a spatially extended soft component at energies <2.5 keV, which is dominated by line emission, and a harder continuum component at higher energies. In some heavily obscured AGNs, prominent Fe Kα 6.4 keV neutral lines are also seen. In some cases, Fe XXV at 6.7 keV has also been detected. Figure 7.14 shows the *Chandra* X-ray spectrum of one such AGN, observed with *Chandra* ACIS (Fabbiano et al. 2017): ESO 428-G014, which has a heavily absorbed ($N_H > 10^{25}$ $cm^{-2}$), CT Seyfert type 2 nucleus (Risaliti et al. 1999). In the "standard model" applied to a CT AGN (e.g., Urry & Padovani 1995), the soft component would by emitted by clouds in the galaxy ISM, photoionized by the AGN (Bianchi et al. 2006), while the hard continuum and the Fe Kα lines are believed to arise from the interaction of hard photons from near the nuclear SMBH with a small-scale ~1 pc or less circumnuclear obscuring torus (e.g., Matt et al. 1996; Gandhi et al. 2015).

Below we discuss how the "standard model" paradigm is being explored, and in some cases challenged, by *Chandra* images of the different spectral components of AGNs. This work makes use of the ultimate (~0.25″) resolution of the *Chandra* telescope, and of multiwavelength comparison with similar angular resolution radio (*VLA*), optical (*Hubble*), and millimeter (*ALMA*) images. We first discuss the related observational issues and methods (Section 7.5.1). We then summarize what these observations imply for the nature of the soft diffuse emission (energies <2.5 keV; Section 7.5.2) and of the hard continuum and Fe Kα emission (Section 7.5.3).

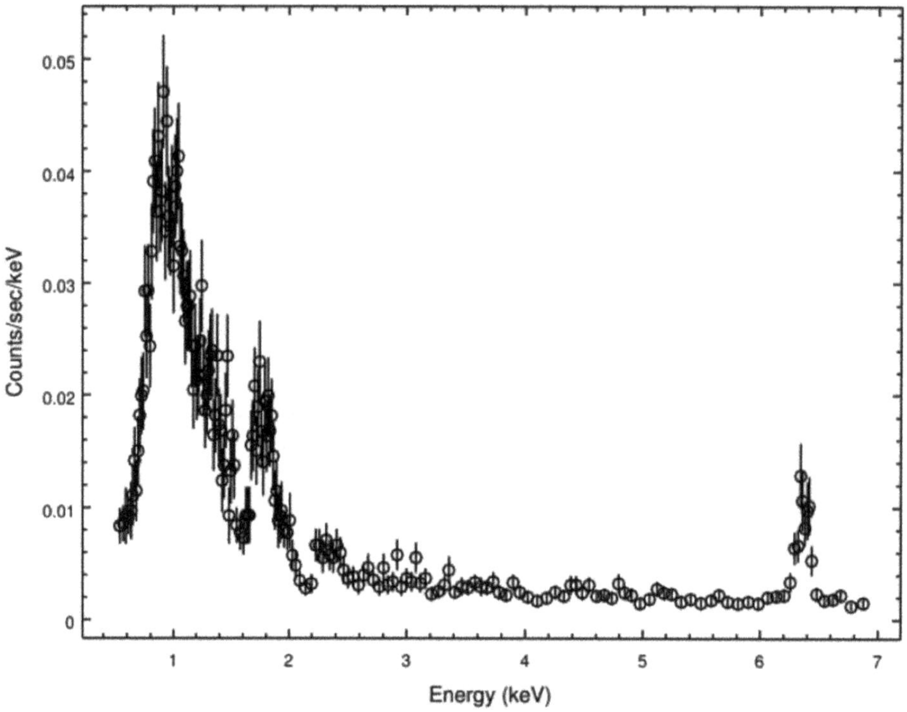


**Figure 7.14.** *Chandra* ACIS-observed spectrum of ESO 428-G014 (reproduced from Fabbiano et al. 2018a © 2018. The American Astronomical Society. All rights reserved). It includes all the photons detected within a circle of 8″ radius centered on the peak of the hard (>3 keV) nuclear source.

Finally, we discuss the resulting constraints on the nature of the obscuring torus of the standard model (Sections 7.5.4 and 7.5.5).

### 7.5.1 Methods

Observationally, the subarcsecond angular resolution of the *Chandra* telescope allows the resolution of physical regions on a scale of ~50–100 pc in galaxies at distances of ~10–20 Mpc. However, there are some obstacles to overcome, both of them linked to the instrument of choice for most of these studies, the ACIS. This CCD detector can measure both the energy and position of the incoming photons (as well as the time of arrival, which is not relevant for the study of extended features). However, the CCD pixel is a 0.49″ sided square and therefore undersamples the *Chandra* PSF. Moreover, the relatively slow CCD readout time (3.2 s if reading out the full chip) results in "pileup" of multiple photons in a single pixel from an intense point source (such as a bright AGN), causing both spectral and spatial unwelcome effects (see the Chandra Proposers' Observatory Guide, POG[2]). The other *Chandra* imager, the HRC (see POG), has instrumental pixels smaller than the mirror PSF,

[2] http://cxc.cfa.harvard.edu/proposer/POG/.

but a lower effective area than the ACIS (so that much longer exposure times are needed), and virtually no energy resolution.

These problems can be overcome with:

(1) A careful selection of the AGNs suited for these studies (e.g., Fabbiano et al. 2018a). To minimize the effects of ACIS pileup, obscured and highly obscured AGNs are the best, so that the nuclear photons are not directly visible at low energies (<2 keV). At these low energies, the extended emission is prevalent (Section 7.5.2). The obscuring clouds form a natural coronagraph.

(2) Complementing ACIS observations of less-obscured AGNs with HRC observations to image the innermost circumnuclear regions. This method was used by Wang et al. (2009) in their study of NGC 4151 and by Wang et al. (2012) in their study of the innermost regions of NGC 1068. Neither AGN could be studied with ACIS in the innermost regions, because of pileup.

(3) Using subpixel binning to produce images from ACIS data. This method is made possible by the fact that the *Chandra* telescope scans over the source position in a well-known Lissajous figure pattern (see POG). The method is the continuous limit of the "multidrizzle" technique employed on *Hubble*. It was applied to the ACIS data of NGC 4151 by Wang et al. (2011a), who demonstrated that it recovered the spatial information of the HRC observations of the same region (Wang et al. 2009). To illustrate the advantage of subpixel binning, Figure 7.15 compares the ACIS native image of the nuclear region of ESO 428– G014 (top-left panel) with that obtained by binning the same photons in bins 1/16 of the instrument pixel (top-right panel).

(4) Image processing with adaptive smoothing and/or image reconstruction (e.g. EMC2, Esch et al. 2004) can then be used to enhance visually detected features. The bottom panels of Figure 7.15 show the 1/16 bin data processed with adaptive smoothing (left) and EMC2 image restoration plus 2 pixel Gaussian smoothing (right). The contours are from the *Hubble* H$\alpha$ image (Falcke et al. 1996). The correspondence between H$\alpha$ features and subpixel images is remarkable.

(5) There are some issues with the *Chandra* PSF one has to be careful about. The first is the well-documented "hook" feature that requires some checking of the images to exclude spurious results.[3] The second is a marked splitting of the image at the core of the PSF at energies >7 keV (Fabbiano et al. 2019).

[3] http://cxc.harvard.edu/ciao/caveats/psf_artifact.html.

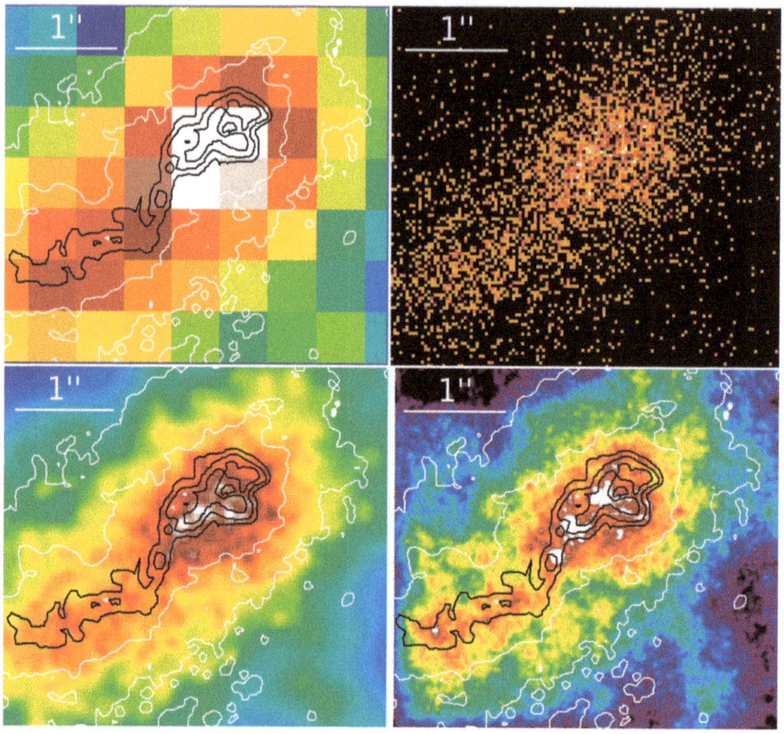


**Figure 7.15.** Data from Fabbiano et al. (2018b). © 2018. The American Astronomical Society. All rights reserved. (Top left) Raw 0.3–3 keV ACIS image of the circumnuclear region of ESO 428+G014 (1″ = 170 pc), (top right) ACIS image of the same region with subpixel binning (1/16 native pixel), (bottom left) 1/16 bin image processed with adaptive smoothing, and (bottom right) EMC2 image restoration plus 2 pixel Gaussian smoothing. Hα contours from Falcke et al. (1996), superimposed.

### 7.5.2 The Soft ($E$ < 2.5 keV) X-Ray Emission of AGN Photoionization Cones and Soft X-Ray Constraints on AGN Feedback

The soft X-ray emission (energies $\lesssim$2.5 keV; see Figure 7.14) is typically extended and has been associated with the emission of galaxy ISM clouds photoionized by nuclear photons at a distance from the AGN, or with a wind. This is the same region that also gives rise to the optical narrow-line emission of the ionization cones (e.g., Bianchi et al. 2006; Levenson et al. 2006). Figure 7.16 shows the striking ionization cone of Mkn 573 (Paggi et al. 2012), as seen with *Chandra* (blue), compared with the [O III] (*Hubble*, orange), and radio 6 cm (VLA, magenta) images.

The ACIS spectra of the "ionization cones" are generally consistent with photoionization, except for the regions where the radio lobes terminate. Here, additional thermal components are required, suggesting shocks driven by the impact of the radio jet on the ISM clouds (see Figure 7.17, bottom).

This feature was first observed by Wang et al. (2011b) in the *Chandra* Ne IX/O VII line ratio map of NGC 4151, where it appears as localized excesses of the line ratio at a distance ~1″ (65 pc) from the nucleus (Figure 7.17, top). A similar effect was also observed in Mkn 573 (Paggi et al. 2012).

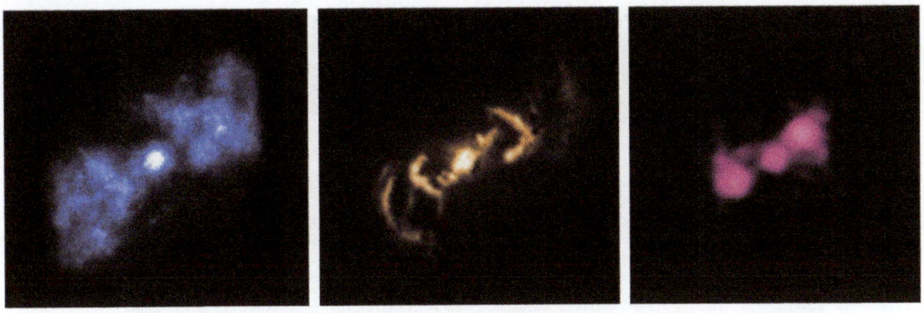

**Figure 7.16.** Left, *Chandra* image of the ionization cone of Mkn 573; center, [O iii]; right, radio, 6 cm. Each image is 10″ across, corresponding to ~3.3 kpc at the distance of this galaxy (reproduced from Paggi et al. 2012 ). Images from the CXC. The interactive figure shows separately the X-ray emission (blue), optical [OIII] emission (yellow), radio emission (magenta), and the composite image. Additional materials available online at http://iopscience.iop.org/book/978-0-7503-2163-1.

Measuring individual regions of the ionization cone in both *Hubble* and *Chandra* images, Wang et al. (2011c) found that the ratios of [O iii]/soft X-ray flux are approximately constant from ~30 pc from the nucleus to the 1.5 kpc radius spanned by their measurements. These ratios are consistent with the average values measured from the soft components of several Seyfert galaxies with *XMM/Newton* (Bianchi et al. 2006). If the [O iii] and X-ray emissions both arise from a single photoionized medium, this constant ratio implies a constant ionization parameter (flux/density ratio) and thus an ISM density decreasing as $r^{-2}$. This naturally suggests an outflow with a wind-like density profile. The regions corresponding to the termination of the radio jets/lobes have smaller [O iii]/soft X-ray ratios, consistent with the presence of the additional thermal X-ray components that we have discussed above (see Figure 7.17).

Using spatially resolved X-ray features, Wang et al. (2011c) estimated that the mass outflow rate in NGC 4151 is ~2 $M_\odot$ yr$^{-1}$ at 130 pc from the nucleus and that the kinematic power of the ionized outflow is $1.7 \times 10^{41}$ erg s$^{-1}$, approximately 0.3% of the bolometric luminosity of the active nucleus in NGC 4151. The contribution from the radio-termination shocks is similarly small, suggesting that ~0.1% of the jet power is deposited in the ISM (Wang et al. 2011b). These values are significantly lower than the expected efficiency in the majority of quasar feedback models, but comparable to the two-stage model described in Hopkins & Elvis (2010).

In NGC 4151, the AGN–galaxy interaction extends on spatial scales larger than the ionization cones. Wang et al. (2010) discovered soft diffuse X-ray emission in NGC 4151, with a luminosity $L_{0.5\text{–}2\,\mathrm{keV}} \sim 10^{39}$ erg s$^{-1}$, extending out to ~2 kpc from the active nucleus and filling in a cavity surrounded by cold H i material (Figure 7.18).

The best fit to the spectrum of the X-ray emission of the "cavity" of NGC 4151 requires either a $kT \sim 0.25$ keV thermal plasma, or a photoionized component from an Eddington-limited nuclear outburst. For both scenarios, the AGN–host interaction must have occurred relatively recently (some $10^4$ yr ago). This very short timescale to the last episode of high-activity phase may imply such outbursts occupy ~1% of an AGN's lifetime

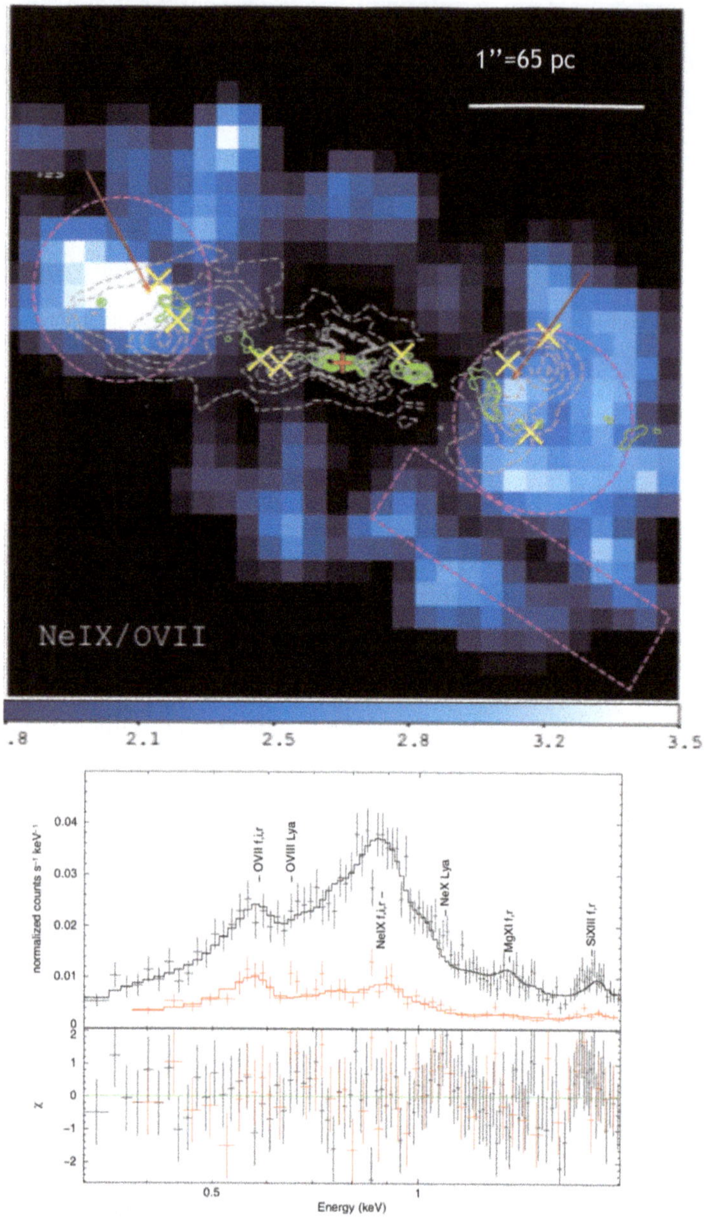


**Figure 7.17.** Top, Ne IX /O VII ratio map from the ACIS observations of the nucleus of NGC 4151 (1/8 subpixel binning); green contours represent the radio jet (red arrows indicate the termination points), and dashed contours the near-IR [Fe II] emission; yellow crosses mark the locations of the *Hubble* clouds with high velocity dispersion. The magenta circles are the extraction regions for the spectrum shown in black in the right panel, while the spectrum from the rectangular area is shown in magenta. The latter can be fitted with photo-ionization models, while the spectrum from the Ne IX excess areas requires an additional thermal component; the best-fit residuals are shown in the bottom of the bottom panel (reproduced from Wang et al. 2011b, © 2011. The American Astronomical Society. All rights reserved, and references therein).

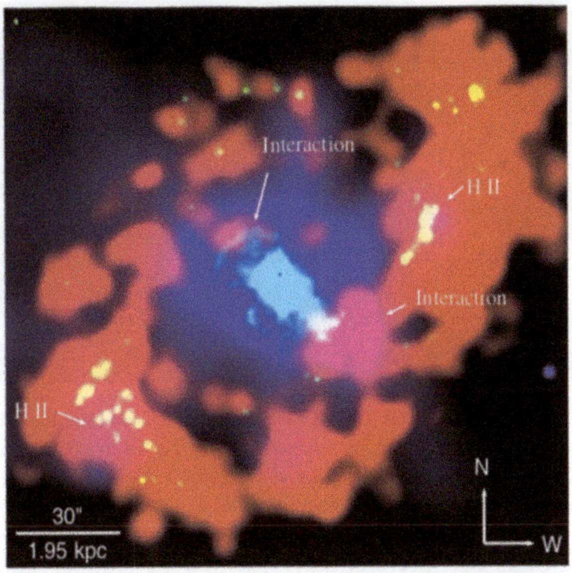


**Figure 7.18.** The hot, circumnuclear, 2 kpc radius X-ray bubble of NGC 4151 (blue), within the ring of H I emission, in red. Star-forming H II regions are shown in yellow (reproduced from Wang et al. 2010 © 2010. The American Astronomical Society. All rights reserved). The interactive figure shows separately the X-ray emission (blue), optical HII emission (yellow), radio HI emission (red), and the composite image. Additional materials available online at http://iopscience.iop.org/book/978-0-7503-2163-1.

### 7.5.3 A *Chandra* Surprise: The Extended Hard and Fe Kα Emission of AGNs

As mentioned at the beginning of Section 7.5, it is commonly assumed that the hard continuum emission (>2.5 keV; see Figure 7.14) and the Fe Kα 6.4 keV line of CT AGNs both originate from the interaction of hard nuclear photons with the dense CT clouds that constitute a 0.1 to ~ a few pc radius "obscuring torus" surrounding the nucleus.

Recent *Chandra* observations are beginning to challenge this paradigm, showing that the hard continuum and Fe Kα emission are not confined to a circumnuclear AGN torus. In particular, in the Circinus galaxy (Arevalo et al. 2014) and in NGC 1068 (Bauer et al. 2015), both of these emissions also have extended components in the direction of the ionization cone (~600 pc in Circinus, >140 pc in NGC 1068). Kiloparsec-scale hard continuum and Fe Kα emission, extended along the ionization cone, have also been discovered in the CT AGN ESO 428-G014 (Fabbiano et al. 2017; Figure 7.19). These results suggest that hard photons from the AGN, escaping along the same torus axis as the soft photons that give rise to the ionization cone, are then scattered/reflected by dense molecular clouds in the galactic disk of ESO 428-G014.

At energies >3 keV, the extended emission in the central 1.5″ (170 pc) radius circumnuclear region amounts to ~30%–70% of the contribution of a point source in that area (or ~25%–40% of the total counts in the region). Within a 5″ radius, the contribution from the extended emission is greater than that from a nuclear point source in the 3–4 keV band. The size of the extended region of ESO 428-G014 decreases with increasing energy in the *Chandra* range, suggesting that the optically

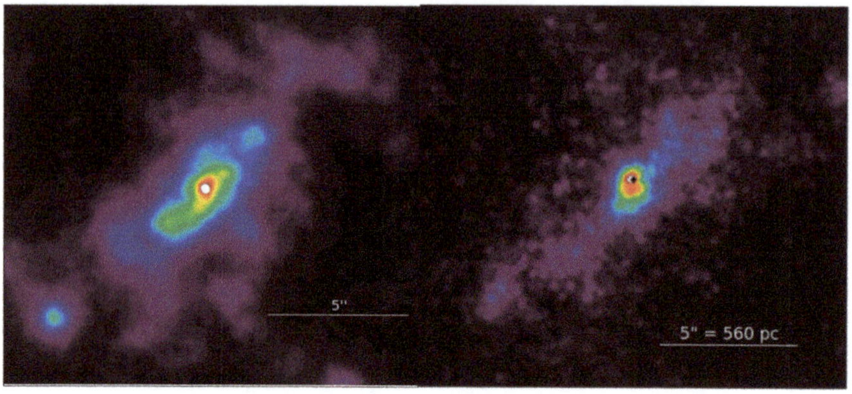


**Figure 7.19.** Left, 3–6 keV continuum image of ESO 428-G014 (reproduced from Fabbiano et al. 2017 © 2017. The American Astronomical Society. All rights reserved). Right, Fe Kα image (reproduced from Fabbiano et al. 2018a © 2018. The American Astronomical Society. All rights reserved). The color intensity scale goes from purple (lowest intensity) to white (highest intensity). This is the same object used in Figure 7.15.

thick scattering clouds are preferentially located at smaller galactocentric radii (Fabbiano et al. 2018a, 2018b), as is the case for the Milky Way molecular clouds (e.g., Nakanishi & Sofue 2006).

The presence of extended hard continuum and Fe Kα emission does not in itself challenge the "standard model," as this model accounts for AGN photons escaping along the axis of the torus. However, it illustrates how different properties of the ISM in the host galaxies may cause different phenomena. In particular, AGNs in galaxies with thick molecular disks should be more prone to present extended hard components, because they have the molecular clouds needed to interact with the hard photons.

Extended hard components, however, may adversely bias the torus modeling of spectra from X-ray telescopes with inferior angular resolution to *Chandra*, such as *NuSTAR* and *XMM-Newton*, because these spectra will include a substantial extended fraction of reprocessed photons not originating in the torus. In ESO 428-G014 the Fe Kα emission within a circle of 8″ (~900 pc) radius centered on the nuclear peak is commensurate with that of a nuclear point source centered on that peak (Fabbiano et al. 2018a, 2018b; Fabbiano et al. 2019). High-resolution X-ray images are needed to complement these spectral studies, demonstrating that the total emission is not dominated by a nuclear point-like source.

### 7.5.4 Imaging the Obscuring Torus

High-resolution *Chandra* imaging of the inner circumnuclear regions of CT AGNs, in the spectral bands representative of the hard continuum (2–6 keV) and Fe Kα (6–7 keV) emission (see the spectrum shown in Figure 7.14), has provided some direct observational constraints on the central obscuring structures.

The *Chandra* ACIS observations of NGC 4945 (Marinucci et al. 2012) provided the first opportunity to image circumnuclear regions commensurate with the scale of the obscuring torus. NGC 4945 hosts a highly obscured CT AGN ($N_H \sim 4 \times 10^{24}$ $cm^{-2}$) and is nearby. At the distance of this galaxy (~3.7 Mpc), 1″ corresponds to 18

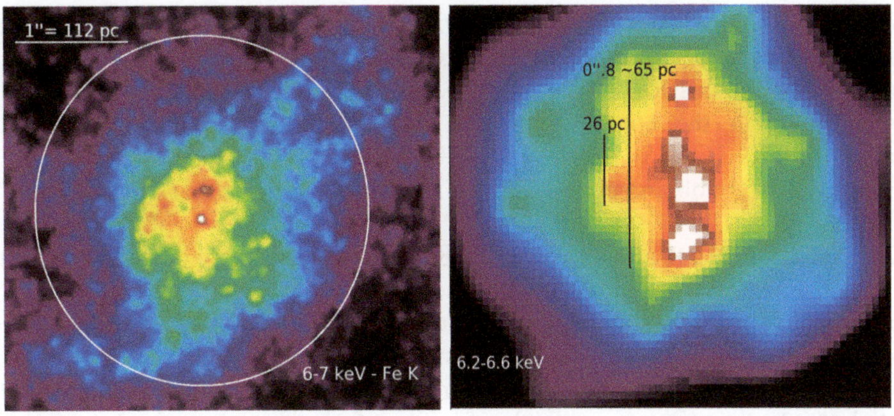


**Figure 7.20.** Left, EMC2 reconstruction of the central region of ESO 428-G014 in the spectral band dominated by the Fe Kα emission observed with *Chandra*, showing both extended and clumpy emission (reproduced from Fabbiano et al. 2019 © 2019. The American Astronomical Society. All rights reserved). Right, adaptive smoothing of the central north–south Fe Kα structure discovered with *Chandra* in NGC 5643 (reproduced from Fabbiano et al. 2018c, © 2018. The American Astronomical Society. All rights reserved); this structure extends in the same direction as the CO (2–1) line emission and the central rotating CO (2–1) 26 pc disk discovered with ALMA (Alonso-Herrero et al. 2018).

pc. In addition to the ~kiloparsec extended soft emission component consistent with the ionization cone, the *Chandra* images showed a flattened feature of hard continuum and Fe Kα circumnuclear emission with approximately 150 pc diameter, extending in the cross-cone direction. Although larger than would be expected for the torus in this AGN, this feature could be part of a structure surrounding the torus. More recent deep observations of this structure (Marinucci et al. 2017) revealed clumpy emission in both the neutral and ionized Fe Kα lines.

Recent high-resolution *Chandra* imaging of the Fe Kα emission of ESO 428-G014 (Figure 7.20, left) provides another example of clumpy emission in the inner ~30 pc, which could be associated with the obscuring torus (Fabbiano et al. 2019). In the CT AGN NGC 5643, a central flattened structure is visible with *Chandra* in the Fe Kα line (Fabbiano et al. 2018c), in the direction perpendicular to that of the ionization cone (Figure 7.20, right). This structure is spatially coincident with the circumnuclear rotating molecular disk discovered with ALMA in this galaxy (Alonso-Herrero et al. 2018). The availability of both ALMA and *Chandra* observations gives us the first example in which both the obscuring structure ("torus") and the result of the nuclear X-rays interacting with this structure are directly imaged.

### 7.5.5 Is the Torus Porous?

In the standard model, AGN photons escape along the axis of the torus and are collimated by the torus. However, at least in the case of NGC 4151, the presence of peculiar cloud velocities may suggest outflows in the cross-cone direction (Das et al. 2005). Moreover, the spectral variability of NGC 1365 (Risaliti et al. 2005) and NGC 4945 (Marinucci et al. 2012) shows that the obscuring structure, at least in these two CT AGNs, is composed of discrete clouds that at times move out of the

AGN line of sight. However, the time variability implies cloud velocities consistent with the broad-line region, too fast for these clouds to be in the torus. With *Chandra* imaging, X-ray emission has been detected in several AGNs in the direction perpendicular to the cone, where the torus should block their propagation. Examples are NGC 4151 (Wang et al. 2011c), Mkn 573 (Paggi et al. 2012), and ESO 428-G014 (Fabbiano et al. 2018a). These results suggest that the torus may be somewhat porous.

In ESO 428-G014 (Fabbiano et al. 2018a), the opening angle of the cone and the ratio of the X-ray emission detected in the cross-cone and cone regions imply that the cross-cone transmission of the obscuring torus is ~10% of that in the cone direction. In this CT AGN, the lack of energy dependence of the ratio of photons escaping from the cross-cone and cone regions would be in agreement with the partial obscuration picture, in which a fraction of photons escape in the cross-cone region with similar spectral distribution to those escaping in the cone region.

However, a different phenomenon may also contribute to the cross-cone X-ray emission. The interaction of radio jets with dense molecular disks may also produce cross-cone emission. Fabbiano et al. (2018b) noted that in ESO 428-G014, the overall similarity of the morphology of the radio jet and the warm and the hot ISM are consistent with the predictions of recent 3D relativistic hydrodynamic simulations by Mukherjee et al. (2018) of the interaction of radio jets with a dense molecular disk. These simulations also predict a hot cocoon enveloping the interaction region, offering a possible alternative explanation for the presence of X-ray emission perpendicular to the bicone. All CT AGNs where cross-cone extent has been reported so far host small radio jets.

Even while the interpretation of the *Chandra* results is still evolving, it is clear that these high-angular-resolution observations of AGNs are producing important new constraints on the AGN–galaxy interaction and will also have repercussions for AGN modeling.

## 7.6 Looking Forward

The subarcsecond angular resolution and simultaneous spectral resolution available with *Chandra* ACIS have been proven a winning combination for the study of galaxies in X-rays. As discussed in this chapter, these unique *Chandra* capabilities have allowed the separation of point-like and extended emission in galaxies up to a few 10 Mpc away, resulting in the “clean” observational characterization of these components and resolving long-standing ambiguities from studies with inferior angular resolution. Comparison with similar angular resolution multiwavelength data (*Hubble*, *Spitzer*, JVLA, and ALMA) has further advanced our knowledge of the properties and evolution of galaxies and their active nuclei.

These near-universe observations have provided the baseline for studies of the evolution of galaxies and their X-ray emission components with redshift. Unique *Chandra* observational achievements include (1) the study of populations of XRBs in a variety of stellar environments and the realization that XRBs may be an important source of feedback in the early universe; (2) the physical and chemical characterization of the hot ISM, hot wind, and halos, and their connection with star formation activity and galaxy mass; (3) the investigation of the full gamut of AGN activity, including the discovery of low-luminosity AGNs in dwarf galaxies both nearby and at higher redshift, which is consistent with the widespread presence of lower-mass nuclear BHs in these systems; and (4) the study of AGN–ISM interaction and feedback in both radio-loud and radio-quiet AGNs.

While these results are impressive, looking forward it is also important to point out their limitations, resulting chiefly from the small collecting area of the *Chandra* telescope, which requires extremely long observations to achieve significant results. While a larger area X-ray telescope (*Athena*) is being developed in Europe, its angular resolution is several times worse than *Chandra*'s, making it unsuitable for future resolved studies of galaxies.

The *Lynx* mission (see Chapter 11) instead, under study by NASA, is the natural telescope for progressing in this field. *Lynx*, with a collecting area similar to that projected for *Athena* (30–50 times larger than *Chandra*'s), seeks to maintain *Chandra*'s angular resolution, while providing substantially better spectral imagers in both soft and harder energy ranges, a winning combination for the study of galaxies and their evolution.

## References

Aird, J., Coil, A. L., & Georgakakis, A. 2017, MNRAS, 465, 3390
Alonso-Herrero, A., Pereira-Santaella, M., García-Burillo, S., et al. 2018, ApJ, 859, 144
Anastasopoulou, K., Zezas, A., Ballo, L., & Della Ceca, R. 2016, MNRAS, 460, 3570
Antoniou, V., & Zezas, A. 2016, MNRAS, 459, 528
Antoniou, V., Zezas, A., Hatzidimitriou, D., & Kalogera, V. 2010, ApJL, 716, 140
Arevalo, P., Bauer, F. E., Puccetti, S., et al. 2014, ApJ, 791, 81
Arimoto, N., Matsushita, K., Ishimaru, Y., Ohashi, T., & Renzini, A. 1997, ApJ, 477, 128
Arnaud, M., & Evrard, A. E. 1999, MNRAS, 305, 631
Bachetti, M., Harrison, F. A., Walton, D. J., et al. 2014, Natur, 514, 202
Baldassare, V. F., Reines, A. E., Gallo, E., & Greene, J. E. 2017, ApJ, 836, 20
Baldi, A., Raymond, J. C., Fabbiano, G., et al. 2006a, ApJS, 162, 113
Baldi, A., Raymond, J. C., Fabbiano, G., et al. 2006b, ApJ, 636, 158
Bauer, F. E., Arevalo, P., Walton, D. J., et al. 2015, ApJ, 812, 116
Bianchi, S., Guainazzi, M., & Chiaberge, M. 2006, A&A, 448, 499
Bildsten, L., & Deloye, C. J. 2004, ApJ, 607, L119
Bogdán, Á., Forman, W. R., Kraft, R. P., & Jones, C. 2013, ApJ, 772, 98
Bonfini, P., Zezas, A., Birkinshaw, M., et al. 2012, MNRAS, 421, 2872
Boroson, B., Kim, D.-W., & Fabbiano, G. 2011, ApJ, 729, 12
Brandt, W. N., Hornschemeier, A. E., Alexander, D. M., et al. 2001, AJ, 122, 1
Brassington, N. J., Fabbiano, G., Kim, D.-W., et al. 2008, ApJS, 179, 142
Brassington, N. J., Fabbiano, G., Kim, D.-W., et al. 2009, ApJS, 181, 605
Brassington, N. J., Fabbiano, G., Blake, S., et al. 2010, ApJ, 725, 1805
Buote, D. A., & Humphrey, P. J. 2012, Hot Interstellar Matter in Elliptical Galaxies, Astrophysics and Space Science Library, Vol. 378, ed. D.-W. Kim, & S. Pellegrini (Berlin: Springer), 235

Canizares, C. R., Fabbiano, G., & Trinchieri, G. 1987, ApJ, 312, 503
Chevalier, R. A., & Clegg, A. W. 1985, Nature, 317, 44
Chilingarian, I. V., Katkov, I., Yu., Zolotukhin, I., Yu., et al. 2018, ApJ, 863, 1
Ciotti, L., D'Ercole, A., Pellegrini, S., & Renzini, A. 1991, ApJ, 376, 380
Ciotti, L., & Ostriker, J. P. 2012, Hot Interstellar Matter in Elliptical Galaxies, Astrophysics and Space Science Library, Vol. 378, ed. D.-W. Kim, & S. Pellegrini (Berlin: Springer), 83
Civano, F., Elvis, M., Brusa, M., et al. 2012, ApJS, 201, 30
Civano, F., Fabbiano, G., Pellegrini, S., et al. 2014, ApJ, 790, 16
Clark, G. W. 1975, ApJ, 199, L143
Cox, T. J., Di Matteo, T., Hernquist, L., et al. 2006, ApJ, 643, 692
D'Abrusco, R., Fabbiano, G., & Brassington, N. J. 2014b, ApJ, 783, 19
D'Abrusco, R., Fabbiano, G., Mineo, S., et al. 2014a, ApJ, 783, 18
D'Abrusco, R., Fabbiano, G., Strader, J., et al. 2013, ApJ, 773, 87
D'Abrusco, R., Fabbiano, G., & Zezas, A. 2015, ApJ, 805, 26
Das, V., Crenshaw, D. M., Hutchings, J. B., et al. 2005, AJ, 130, 945
Danielson, A. L. R., Lehmer, B. D., Alexander, D. M., et al. 2012, MNRAS, 422, 494
David, L. P., Forman, W. R., & Jones, C. 1991, ApJ, 369, 121
De Lucia, G., Springel, V., White, S. D. M., et al. 2006, MNRAS, 366, 499
de Vaucouleurs, et al. 1991, Third Reference Catalogue of Bright Galaxies (New York: Springer)
Ellis, R. S., McLure, R. J., Dunlop, J. S., et al. 2013, ApJL, 763, 7
Elvis, M., Civano, F., Vignali, C., et al. 2009, ApJS, 184, 158
Esch, D. N., Connors, A., Karovska, M., & van Dyk, D. A. 2004, ApJ, 610, 1213
Fabbiano, G. 1988, ApJ, 330, 672
Fabbiano, G. 1989, ARAA, 27, 87
Fabbiano, G. 2006, ARAA, 44, 323
Fabbiano, G. 2012, Hot Interstellar Matter in Elliptical Galaxies, Astrophysics and Space Science Library, Vol. 378, ed. D.-W. Kim, & S. Pellegrini (Berlin: Springer), 1
Fabbiano, G., Baldi, A., King, A. R., et al. 2004, ApJ, 605, L21
Fabbiano, G., Brassington, N. J., Lentati, L., et al. 2010, ApJ, 725, 1824
Fabbiano, G., Elvis, M., Paggi, A., et al. 2017, ApJL, 842L, 4
Fabbiano, G., Paggi, A., Karovska, M., et al. 2018a, ApJ, 855, 131
Fabbiano, G, Paggi, A., Karovska, M., et al. 2018b, ApJ, 865, 83
Fabbiano, G., Paggi, A., Siemiginowska, A., & Elvis, M. 2018c, ApJ, 859, L36
Fabbiano, G., Siemiginowska, A., Paggi, A., et al. 2019, ApJ, 870, 69
Fabbiano, G., Zezas, A., & Murray, S. S. 2001, ApJ, 554, 1035
Faber, S. M., & Gallagher, J. S. 1976, ApJ, 204, 365
Fabricant, D., Lecar, M., & Gorenstein, P. 1980, ApJ, 241, 552
Falcke, H., Wilson, A. S., Simpson, C., & Bower, G. A. 1996, ApJ, 470, L31
Forbes, D. A., Alabi, A., Romanowsky, A. J., et al. 2017, MNRAS, 464L, 26
Forman, W., Jones, C., & Tucker, W. 1985, ApJ, 293, 102
Forman, W., Schwarz, J., Jones, C., Liller, W., & Fabian, A. C. 1979, ApJ, 234L, 27
Fornasini, F. M., Civano, F., Fabbiano, G., et al. 2018,865, 43
Fragos, T., Linden, T., Kalogera, V., & Sklias, P. 2015, ApJ, 802, L5
Fragos, T., Lehmer, B. D., Naoz, S., Zezas, A., & Basu-Zych, A. 2013b, ApJ, 776, L31
Fragos, T., Lehmer, B., Tremmel, M., et al. 2013a, ApJ, 764, 41
Fridriksson, J. K., Homan, J., Lewin, W. H. G., Kong, A. K. H., & Pooley, D. 2008, ApJS, 177, 465

Gandhi, P., Hönig, S. F., & Kishimoto, M. 2015, ApJ, 812, 113
Ghez, A. M., Salim, S., Weinberg, N. N., et al. 2008, ApJ, 689, 1044
Giacconi, R., Gursky, H., Paolini, F. R., & Rossi, B. B. 1962, PhRvL, 9, 439
Greene, J. E., & Ho, L. C. 2007, ApJ, 656, 84
Grindlay, J. E. 1984, AdSpR, 3, 19
Gültekin, K., Richstone, D. O., Gebhardt, K., et al. 2009, ApJ, 698, 198
Heckman, T. M., Armus, L., & Miley, G. K. 1990, ApJS, 74, 833
Heckman, T. M., & Kauffmann, G. 2011, Sci., 333, 182
Hopkins, P. F., & Elvis, M. 2010, MNRAS, 401, 7
Hopkins, P. F., Hernquist, L., Cox, T. J., & Kereš, D. 2008, ApJS, 175, 356
Iwasawa, K., Sanders, D. B., Evans, A. S., et al. 2009, ApJ, 695, L103
Jones, T. M., Kriek, M., van Dokkum, P. G., et al. 2014, ApJ, 783, 25
Kim, D.-W. 2012, Hot Interstellar Matter in Elliptical Galaxies, Astrophysics and Space Science Library, Vol. 378, ed. D.-W. Kim, & S. Pellegrini (Berlin: Springer), 121
Kim, D.-W., & Fabbiano, G. 2003, ApJ, 586, 826
Kim, D.-W., & Fabbiano, G. 2004, ApJ, 613, 933
Kim, D.-W., & Fabbiano, G. 2010, ApJ, 721, 1523
Kim, D.-W., & Fabbiano, G. 2013, ApJ, 776, 116
Kim, D.-W., & Fabbiano, G. 2015, ApJ, 812, 127
Kim, D.-W., Fabbiano, G., Brassington, N. J., et al. 2009, ApJ, 703, 829
Kim, D.-W., Fabbiano, G., Ivanova, N., et al. 2013, ApJ, 764, 98
Kim, D.-W., Fabbiano, G., Kalogera, V., et al. 2006, ApJ, 652, 1090
Kim, D.-W., Fabbiano, G., & Trinchieri, G. 1992, ApJ, 393, 134
Kim, M., Ho, L. C., Wang, J., et al. 2015, ApJ, 814, 8
King, A. R., Davies, M. B., Ward, M. J., Fabbiano, G., & Elvis, M. 2001, ApJ, 552, L109
Komossa, S., Burwitz, V., Hasinger, G., et al. 2003, ApJ, 582L, 15
Lehmer, B. D., Alexander, D. M., Bauer, F. E., et al. 2010, ApJ, 724, 559
Lehmer, B. D., Brandt, W. N., Alexander, D. M., et al. 2007, ApJ, 657, 681
Lehmer, B. D., Brandt, W. N., Alexander, D. M., et al. 2008, ApJ, 681, 1163
Lehmer, B. D., Basu-Zych, A. R., & Mineo, S. 2016, ApJ, 825, 7
Lehmer, B. D., Berkeley, M., Zezas, A., et al. 2014, ApJ, 789, 52
Lehmer, B. D., Eufrasio, R. T., Markwardt, L., et al. 2017, ApJ, 851, 11
Levenson, N. A., Heckman, T. M., Krolik, J. H., Weaver, K. A., & Życki, P. T. 2006, ApJ, 648, 111
Lewin, W. H. G., & van der Klis, M. 2006, Compact Stellar X-ray Sources (Cambridge: Cambridge Univ. Press)
Lewin, W. H. G., van Paradijs, I., & van den Heuvel, E. P. J. 1995, X-ray Binaries (Cambridge: Cambridge Univ. Press)
Li, J.-T., & Wang, Q. D. 2013, MNRAS, 428, 2085
Luo, B., Fabbiano, G., Fragos, T., et al. 2012, ApJ, 749, 130
Luo, B., Fabbiano, G., Strader, J., et al. 2013, ApJS, 204, 14
Magorrian, J., Tremaine, S., Richstone, D., et al. 1998, AJ, 115, 2285
Marinucci, A., Bianchi, S., Fabbiano, G., et al. 2017, MNRAS, 470, 4039
Marinucci, A., Risaliti, G., Wang, J., et al. 2012, MNRAS, 423, L6
Martin, C. L., Kobulnicky, H. A., & Heckman, T. M. 2002, ApJ, 574, 663
Martin, C. L., Shapley, A. E., Coil, A. L., et al. 2012, ApJ, 760, 127

Mathews, W. G., & Baker, J. C. 1971, ApJ, 170, 241
Matsushita, K., Ohashi, T., & Makishima, K. 2000, AdSpR, 25, 583
Matt, G., Brandt, W. N., & Fabian, A. C. 1996, MNRAS, 280, 823
Maughan, B. J., Giles, P. A., Randall, S. W., Jones, C., & Forman, W. R. 2012, MNRAS, 421, 1583
Mezcua, M., Civano, F., Fabbiano, G., Miyaji, T., & Marchesi, S. 2016, ApJ, 817, 20
Mineo, S., Fabbiano, G., D'Abrusco, R., et al. 2014a, ApJ, 780, 132
Mineo, S., Gilfanov, M., Lehmer, B. D., Morrison, G. E., & Sunyaev, R. 2014b, MNRAS, 437, 1698
Mineo, S., Gilfanov, M., & Sunyaev, R. 2012, MNRAS, 426, 1870
Mineo, S., Rappaport, S., Levine, A., et al. 2014c, ApJ, 797, 91
Mineo, S., Rappaport, S., Steinhorn, B., et al. 2013, ApJ, 771, 133
Mukherjee, D., Wagner, A. Y., Bicknell, G. V., et al. 2018, MNRAS, 476, 80
Nakanishi, H., & Sofue, Y. 2006, PASJ, 58, 847
Nardini, E., Wang, J., Fabbiano, G., et al. 2013, ApJ, 765, 141
Navarro, J. F., Frenk, C. S., & White, S. D. M. 1995, MNRAS, 275, 56
Negri, A., Posacki, S., Pellegrini, S., & Ciotti, L. 2014, MNRAS, 445, 1351
Oser, L., Naab, T., Ostriker, J. P., & Johansson, P. H. 2012, ApJ, 744, 63
Paggi, A., Fabbiano, G., Civano, F., et al. 2016, ApJ, 823, 112
Paggi, A., Fabbiano, G., Kim, D.-W., et al. 2014, ApJ, 787, 134
Paggi, A., Fabbiano, G., Risaliti, G., et al. 2017, ApJ, 841, 44
Paggi, A., Fabbiano, G., Risaliti, G., et al. 2017a, ApJ, 841, 44
Paggi, A., Kim, D.-W., Anderson, C., et al. 2017b, ApJ, 844, 5
Paggi, A., Wang, J., Fabbiano, G., Elvis, M., & Karovska, M. 2012, ApJ, 756, 39
Peacock, M. B., & Zepf, S. E. 2016, ApJ, 818, 33
Pellegrini, S. 2012, Hot Interstellar Matter in Elliptical Galaxies, Astrophysics and Space Science Library, Vol. 378, ed. D.-W. Kim, & S. Pellegrini (Berlin: Springer), 21
Pellegrini, S., Ciotti, L., & Ostriker, J. P. 2012, ApJ, 744, 21
Pipino, A. 2012, Hot Interstellar Matter in Elliptical Galaxies, Astrophysics and Space Science Library, Vol. 378, ed. D.-W. Kim, & S. Pellegrini (Berlin: Springer), 163
Remillard, R. A., & McCLintock, J. E. 2006, ARAA, 44, 49
Renzini, A., Ciotti, L., D'Ercole, A., & Pellegrini, S. 1993, ApJ, 419, 52
Richings, A. J., Fabbiano, G., Wang, J., & Roberts, T. P. 2010, ApJ, 723, 1375
Risaliti, G., Elvis, M., Fabbiano, G., Baldi, A., & Zezas, A. 2005, ApJ, 623, L93
Risaliti, G., Maiolino, R., & Salvati, M. 1999, ApJ, 522, 157
Sansom, A. E., Dotani, T., Okada, K., Yamashita, A., & Fabbiano, G. 1996, MNRAS, 281, 48
Sarazin, C. L. 2012, Hot Interstellar Matter in Elliptical Galaxies, Astrophysics and Space Science Library, Vol. 378, ed. D.-W. Kim, & S. Pellegrini (Berlin: Springer), 55
Shannon, R. M., et al. 2013, Sci., 342, 334
Shtykovskiy, P. E., & Gilfanov, M. R. 2007, AstL, 33, 299
Soria, R., Risaliti, G., Elvis, M., et al. 2009, ApJ, 695, 1614
Statler, T. S. 2012, Hot Interstellar Matter in Elliptical Galaxies, Astrophysics and Space Science Library, Vol. 378, ed. D.-W. Kim, & S. Pellegrini (Berlin: Springer), 207
Strader, J., Fabbiano, G., Luo, B., et al. 2012, ApJ, 760, 87
Strickland, D. K., & Heckman, T. M. 2009, ApJ, 697, 2030

Strickland, D. K., Heckman, T. M., Colbert, E. J. M., Hoopes, C. G., & Weaver, K.A. 2004, ApJS, 151, 193

Strickland, D. K., Heckman, T. M., Weaver, K. A., & Dahlem, M. 2000, AJ, 120, 2965

Swartz, D. A., Ghosh, K. K., McCollough, M. L., et al. 2003, ApJS, 144, 213

Swartz, D. A., Tennant, A. F., & Soria, R. 2009, ApJ, 703, 159

Toomre, A., & Toomre, J 1972, ApJ, 178, 623

Tremmel, M., Fragos, T., Lehmer, B. D., et al. 2013, ApJ, 766, 19

Trinchieri, G., & Fabbiano, G. 1985, ApJ, 296, 447

Tzanavaris, P., & Georgantopoulos, I. 2008, A&A, 480, 663

Urry, C. M., & Padovani, P. 1995, PASP, 107, 803

Venkasetan, A., Giroux, M. L., & Shull, M. J. 2001, ApJ, 563, 1

Verbunt, F., & van den Heuvel, E. P. J. 1995, X-ray Binaries, ed. W. H. G. Lewin, I. van Paradijs, & E. P. J. van den Heuvel (Cambridge: Cambridge Univ. Press), 457

Volonteri, M. 2012, Sci., 337, 544

Volonteri, M., Dubois, Y., Pichon, C., & Devriendt, J. 2016, MNRAS, 460, 2979

Wang, J., Fabbiano, G., Elvis, M., et al. 2011c, ApJ, 742, 23

Wang, J., Fabbiano, G., Karovska, M., Elvis, M., & Risaliti, G. 2012, ApJ, 756, 180

Wang, J., Fabbiano, G., Karovska, M., et al. 2009, ApJ, 704, 1195

Wang, J., Fabbiano, G., Risaliti, G., et al. 2010, ApJ, 719, L208

Wang, J., Fabbiano, G., Risaliti, G., et al. 2011a, ApJ, 729, 75

Wang, J., Fabbiano, G., Risaliti, G., et al. 2011b, ApJ, 736, 62

Wang, J., Nardini, E., Fabbiano, G., et al. 2014, ApJ, 781, 55

Weaver, K. A., Heckman, T. M., & Dahlem, M. 2000, ApJ, 534, 684

Weisskopf, M. C., Tananbaum, H. D., Van Speybroeck, L. P., & O'Dell, S. L. 2000, Proc. SPIE, 4012, 2

Williams, B. F., Binder, B. A., Dalcanton, J. J., Eracleous, M., & Dolphin, A. 2013, ApJ, 772, 12

Wolter, A., & Trinchieri, G. 2004, A&A, 426, 787

Zezas, A., Fabbiano, G., Baldi, A., et al. 2007, ApJ, 661, 135

Zezas, A., Fabbiano, G., Rots, A. H., & Murray, S. S. 2002, ApJ, 577, 710

Zhang, Z., Gilfanov, M., & Bogdán, Á. 2012, A&A, 546, 36

Zhang, Z., Gilfanov, M., & Bogdán, Á. 2013, A&A, 556, 9

# Chapter 8

# Supermassive Black Holes (Active Galactic Nuclei)

**Aneta Siemiginowska and Francesca Civano**

The first image of a supermassive black hole (SMBH) was recently made at 1.3 mm wavelength by the Event Horizon Telescope (EHT; see Figure 8.1, Event Horizon Telescope Collaboration et al. 2019a, 2019b). It provides the best evidence to date for the presence of a black hole in the center of a giant elliptical galaxy, M87. This is the only galaxy with a visible black hole "shadow" that has been imaged to date. Long before this observation, there were many indications that most galaxies harbor an SMBH (e.g., Ferrarese & Ford 2005, Volonteri 2010). The high luminosity of active galactic nuclei (AGN), including luminous quasars, originates from a compact, unresolved, parsec-scale center, and often exceeds the luminosity of its host galaxy. The compactness, high luminosity, variability, and stellar dynamics point to an accreting black hole as the only possible energy source required by the observations. The EHT data show remarkable agreement with theoretical predictions of SMBH radiation at the event horizon (Mościbrodzka et al. 2016), adding direct evidence for the presence of SMBHs in galaxies.

During the last two decades, *Chandra* has brought many exciting discoveries highlighting the SMBH activity, e.g., high X-ray luminosity, variability, relativistic outflows, hidden nuclei, recoil of a black hole during merger, shocks in the interstellar medium, "sound waves" in clusters, etc. In this chapter, we focus on the *Chandra* contribution to studies of SMBH phenomenon. We discuss *Chandra* discoveries and progress in our understanding of physics and SMBH phenomena made possible by the highest angular resolution X-ray observations to date.

## 8.1 Observing SMBHs

Black holes, unlike stars, do not emit observable electromagnetic light, but we can study them using indirect observational methods. The gravity of an SMBH generates an extreme amount of energy sufficient to power luminous quasars and

doi:10.1088/2514-3433/ab43dcch8  

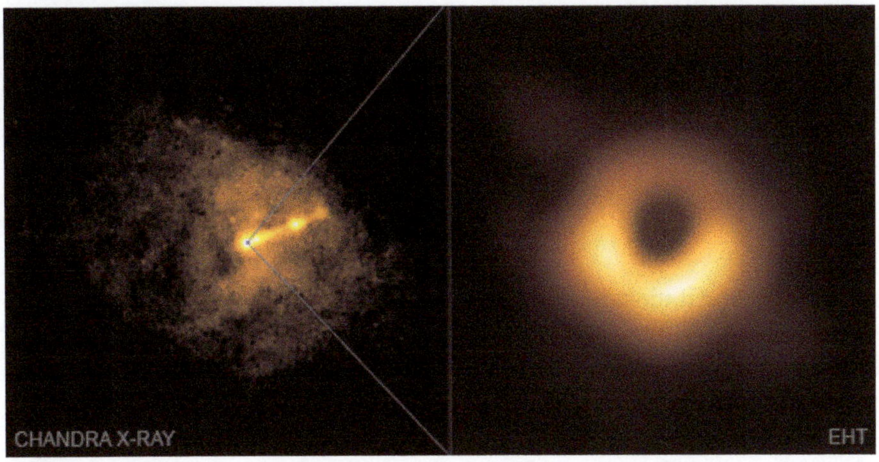


**Figure 8.1.** Left: *Chandra* X-ray image of the M87 galaxy with the X-ray jet extending to the right. Right: the zoom toward the central $10r_g$ (where $r_g$ is the gravitational radius of a black hole) obtained by the Event Horizon Telescope. The ring of the radiation at 1.3 mm wavelength originates from an optically thin accretion flow, and the gravity of the SMBH causes some of the photon's trajectories to cross the event horizon and the observed "black hole shadow." The asymmetry of the ring is due to the rotation of the radiating flow combined with Doppler effects. The matter in the bottom of the ring is moving toward the observer in the clockwise direction as seen from Earth. (From http://chandra.si.edu/blog/node/719, X-ray image courtesy of NASA/CXC/Villanova University/J. Neilsen, Radio image courtesy of EHT.)

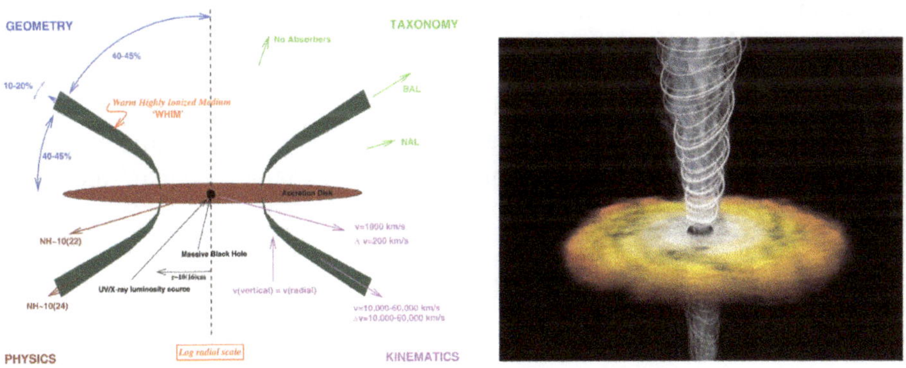


**Figure 8.2.** Left: a schematic view of the structure components in a model of radio-quiet quasars showing geometry, physics, taxonomy and kinematics (reproduced from Elvis 2000, © 2000. The American Astronomical Society. All rights reserved). Right: an artist's view of an additional component in a radio-loud quasar model, a jet perpendicular to an accretion disk and torus plane.

relativistic outflows, and impacts the interstellar and intergalactic environments. This energy is released in a turbulent accretion flow formed around the SMBH (see Figure 8.2), and its amount depends on three physical parameters: black hole mass, spin, and accretion rate. With these parameters, we can study the details of the black hole physics, their evolution, and their part in the formation of structures in the universe. However, these parameters are degenerate, and with additional physical processes contributing to the observed radiation, their measurements are quite challenging (see the radiation components shown in Figure 8.2).

There are several observables that characterize the properties of an accreting black hole. The Eddington luminosity, $L_{\rm Edd}$, defines the maximum luminosity at the equilibrium between the radiation pressure and the gravitational force:

$$L_{\rm Edd} = \frac{4\pi G M_{\rm bh} m_{\rm p} c}{\sigma_{\rm T}}, \tag{8.1}$$

where $G$ is the gravitational constant, $M_{\rm bh}$ is the black hole mass, $m_{\rm p}$ is the proton mass, $c$ is the speed of light, and $\sigma_T$ is the Thompson cross section (see also Chapter 4).

The total luminosity relates to the accretion rate $\dot{M}$ as

$$L = \eta \dot{M} c^2 \tag{8.2}$$

and depends on the efficiency, $\eta$, of converting the gravitational energy into radiation. Typically, $\eta$ is assumed to be between 0.06 for a nonrotating black hole and 0.4 for a maximally rotating black hole. The Eddington accretion rate describes the inflowing supply of matter at the rate corresponding to the Eddington luminosity, so

$$\dot{M}_{\rm Edd} = \frac{L_{\rm Edd}}{\eta c^2}. \tag{8.3}$$

The above equations imply the luminosity dependence on the black hole mass and accretion rate, and hide the dependence on the spin in the $\eta$ parameter. The spin plays an important role in the generation of relativistic jets via the Blandford–Znajek process (Blandford & Znajek 1977; Tchekhovskoy 2015) and defines the total power of a jet.

The gravitational radius of a black hole, $r_{\rm g} = GM_{\rm bh}/c^2$, depends on the black hole mass $M_{\rm bh}$. The spin of a black hole, $a = J_{\rm bh}/J_{\rm max}$, is defined by the ratio of the black hole angular momentum to the maximum achievable angular momentum $J_{\rm max} = M_{\rm bh} r_{\rm g} c$.

## 8.2 Accretion Flow onto SMBH

### 8.2.1 Modes of Accretion

The geometry and physical properties of the accretion flow depend on the rate and angular momentum of the matter flowing into the gravitational potential of a black hole (Shakura & Sunyaev 1973; Paczyńsky & Wiita 1980; Abramowicz & Zurek 1981). Theory defines three main regimes of accretion flow (Begelman 1985; Abramowicz et al. 1988; Abramowicz 2005):

(1) The standard geometrically thin and optically thick accretion disk is formed at intermediate accretion rates with respect to the Eddington accretion rate, approximately within $0.001 < \dot{M}_{\rm Edd} < 0.1$.

(2) At the lowest accretion rates, $\dot{M}_{\rm Edd} \ll 0.001$, the disk becomes geometrically thick, and the heat released via viscosity cannot be radiated away locally, and it is advected toward the black hole (e.g., inefficient accretion;

advection-dominated accretion flow, ADAF; convection-dominated accretion flow, CDAF).

(3) Similarly at the highest accretion rates, close to or higher than the Eddington rate, the disk becomes geometrically thick, hot, and dominated by the radiation and (or) magnetic pressure (e.g., magnetically arrested disks, MAD) possibly leading to the formation of a relativistic jet (e.g., Janiuk et al. 2002; Tchekhovskoy et al. 2011).

These accretion modes lead to different manifestations of the SMBH activity. The accretion flow and SMBH are contained within the nuclear region of a host galaxy, much smaller than the central parsec. They remain unresolved in X-rays. However, we can model the observed spectra and estimate the physical parameters of the accreting SMBH. The emission from the standard accretion disk is simply modeled as a sum of the blackbody emission assuming the disk's radial temperature dependence $T(r) \sim r^{-3/4}$, with the peak temperature at $10^4$–$10^5$ K. For AGN, this emission is identified as a "big blue bump" and observed in the optical-UV band (Czerny & Elvis 1987). X-rays originate in a hot corona, which upscatters the disk photons (Haardt & Maraschi 1991, 1993; Różańska & Czerny 2000; Czerny et al. 2003). The geometry of the corona has yet to be established, but *Chandra* observations of resolved lensed quasar images (Chartas et al. 2012) provide constraints on the size of the X-ray-emitting region and place it within about $\sim 10 r_g$ from the SMBH (see Section 8.4).

Inefficient accretion flows (i.e., ADAF, CDAF) are optically thin and dominated by nonthermal radiation (synchrotron and Compton scattering processes) from two-temperature plasma. These flows generate low-luminosity radiation across the whole electromagnetic spectrum (often called "quiescent" SMBHs). Depending on the SMBH mass, the X-ray luminosity could be $<10^{42}$ erg s$^{-1}$, a threshold typically assumed for AGN in X-ray surveys. The *Chandra* angular resolution has been critical in identifying quiescent SMBHs in local galaxies. The high-resolution *Chandra* images can be used to decompose and model contributions from X-ray binaries, hot gas, and accretion flow to the observed X-ray spectrum of the nuclear region (see Chapter 7; Section 7.4). For a few local galaxies with known SMBH mass, the estimated accretion rates are extremely low, reaching $\sim 10^{-8} \dot{M}_{\rm Edd}$ (e.g., M87, Di Matteo et al. 2003; NGC821, Pellegrini et al. 2007; Sgr A*, Baganoff et al. 2001). The importance of these detections has been underscored by studies of feedback, black hole growth, and black hole occupation fraction, and each of these has impacted our understanding of the evolution of galaxies across the universe (see Sections 8.2.2 and 8.6).

At very high accretion rates, $\sim \dot{M}_{\rm Edd}$, the flow is dominated by radiation and potentially unstable (Pringle et al. 1973; Lightman & Eardley 1974), can form a "slim" disk (Abramowicz et al. 1988), generate outflows (Janiuk et al. 2002), or can be supported by magnetic pressure (Dexter & Begelman 2019). The super-Eddington accretion onto a rotating SMBH with strong magnetic fields provides conditions for launching powerful collimated relativistic jets (Tchekhovskoy 2015). The jets transfer the SMBH energy to distances of hundreds of kiloparsecs from the origin and along the way deposit the energy into the environment. Many detections of X-ray emission from relativistic jets associated with quasars have been made by *Chandra* thanks to its angular resolution and high dynamic range observations (see Harris & Krawczynski 2006; Worrall 2009 for a review). The most distant X-ray jets are faint, but *Chandra* pointed observations provided the first resolved X-ray images of jets at high redshift $z > 4$ (Siemiginowska et al. 2003a; Cheung et al. 2012) associated with luminous quasars (so at high accretion rates onto the SMBH, $> 0.1\ \dot{M}_{\rm Edd}$). X-ray jets have also been observed in low-power radio galaxies associated with inefficient accretion flows (see Section 8.5). In X-ray observations of galaxy clusters, the morphology of the hot gas exposed in the *Chandra* images indicates multiple epochs of past, powerful radio outbursts of the SMBH (McNamara & Nulsen 2007), implying the intermittent jet activity of the SMBH in the dominant central cluster galaxy.

Quasars exhibit a high luminosity across the entire electromagnetic spectrum, and they have been detected in X-ray surveys. *Chandra* low background has been important for the surveys and detections of the faintest point sources at high redshifts. The *Chandra* Deep Field South survey with a 7 Ms exposure has the highest sensitivity of any X-ray image to date (Liu et al. 2017; Vito et al. 2018), allowing for the study of the number density of AGN in the early universe. Such sensitive data are required for the study of quasar evolution and black hole mass function across the universe (see Section 8.6).

### 8.2.2 Bondi Radius and Inflow Rates

*Chandra* observations delivered the first measurements of gas inflow rates at the Bondi radius, defined as the radius at which the gravitational potential of the SMBH is equal to the thermal energy of the gas ($r_{\rm Bondi} = 2GM_{\rm bh}/c_{\rm s}^2$, where $c_{\rm s}$ is the sound speed in the gas; Bondi 1952). This radius marks an important physical scale in studies of SMBH accretion. The classic Bondi accretion flow is assumed to be spherical onto a point mass with negligible angular momentum. It is an ideal model, but important in analytical estimates of inflow rates at the SMBH sphere of influence. Given the mass of the SMBH, gas temperature, and density, the Bondi radius and the accretion rate can be calculated. X-ray data provide the key measurements, but require careful analysis of the X-ray surface brightness in the close vicinity of a strong AGN point source, and other diffuse emission (e.g., from unresolved X-ray binaries or an X-ray cluster). Subarcsecond resolution and high dynamic range capabilities are the key requirements for these studies. Currently, only the *Chandra* X-ray Observatory has the capacity to resolve X-ray radiation on these scales, allowing for detailed studies of the gas properties close to the nucleus.

Deep *Chandra* images (see Figure 8.3) show a multiphase gas structure across the Bondi radius of an SMBH located at the center of a giant elliptical galaxy, M87 (Di Matteo et al. 2003; Russell et al. 2018) at 16.1 Mpc distance (Tonry et al. 2001). The SMBH has an estimated mass of $M_{\rm bh} = (3 - 6) \times 10^9\ M_\odot$ (Gebhardt et al. 2011; Walsh et al. 2013) with the corresponding Bondi radius of $r_{\rm Bondi} = 0.12$–$0.22$ kpc (1.5″–2.8″). The recent EHT mass measurement of $M_{\rm bh} = 6.5 \pm 0.7 \times 10^9\ M_\odot$ (Event Horizon Telescope Collaboration et al. 2019a) is consistent with the larger mass. Russell et al. (2018) detected multitemperature gas in the central 1 kpc region with temperatures of 0.2 keV and 0.8–1 keV. The density structure is nonspherical

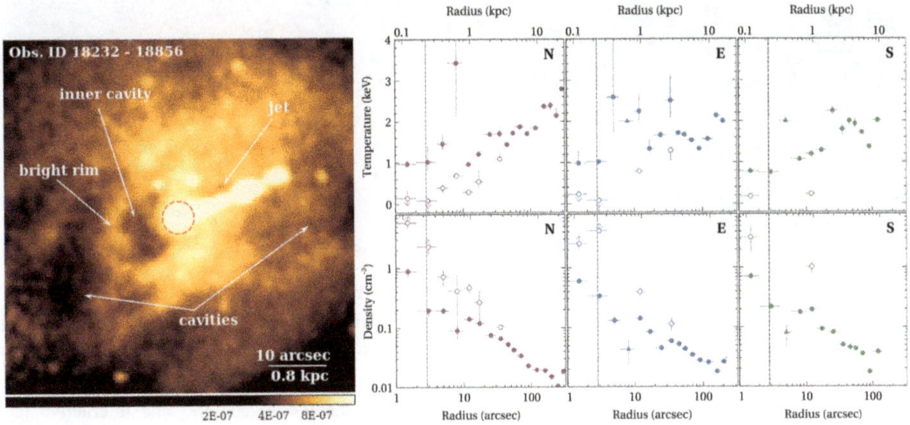


**Figure 8.3.** Left: exposure-corrected *Chandra* image in the 0.5–7 keV energy range showing X-rays from the hot gas surrounding the $6.5 \pm 0.7 \times 10^9$ $M_\odot$ SMBH in the center of M87 galaxy. The Bondi radius, 0.22 kpc (2.8″), is shown by the red dashed circle. The cavities, bright rim, and the jet are marked. The image is in logarithmic color scale with color bar units in (counts $\mathrm{cm^{-2}\,s^{-1}\,pixel^{-1}}$). Right: deprojected temperature and electron density profiles in the north (N), east (E), and south (S) sectors (excluding the jet in the west) from the left image. Two spectral components are shown in solid and open points indicating at two-temperature plasma models fit to the spectra. Regions affected by the cavity where the temperature was fixed are shown by the triangles. The classical Bondi radius is marked by vertical dashed lines. Figures adapted from Russell et al. (2018), by permission of Oxford University Press on behalf of the Royal Astronomical Society.

and consistent with outflow along the jet axis and inflow perpendicular to the jet axis. The overall density of the gas is increasing toward the nucleus with no indication of temperature increase (see Figure 8.3 right). The radiative cooling time of the cooler gas within the Bondi radius is relatively short ($\sim 10^5$ yr), so the most rapidly cooling gas in this galaxy is located within the innermost 100 pc nuclear region. The estimated limit on the spherical accretion rate at the Bondi radius of $<0.01$ $M_\odot$ yr$^{-1}$ is higher than the rate of $2.7 \times 10^{-3}$ $M_\odot$ yr$^{-1}$ ($\sim 10^{-5} \dot{M}_{\rm Edd}$) measured by the EHT at $5r_g$ (Event Horizon Telescope Collaboration et al. 2019c) and is more than enough to power the SMBH activity (i.e., jets and observed radiation).

### 8.2.3 SMBH in the Galactic Center

X-ray images of the Galactic Center (see Figure 8.4 and 8.5) show a large number of X-ray sources embedded in a relatively smooth diffuse emission from hot gas and unresolved sources (Baganoff et al. 2001; Muno et al. 2003, 2004, 2009). The dynamical center of the Milky Way has been identified with a strong radio source, Sagitarius A* (Sgr A* Reid et al. 1999). An optical monitoring of stellar orbits in the Galactic Center confirmed this location and showed that the center harbors an SMBH with a mass of $(4.3 \pm 0.4) \times 10^6$ $M_\odot$ (Ghez et al. 2008; Gillessen et al. 2009). *Chandra* discovered faint X-ray radiation from Sgr A* associated with the radiation from the accreting SMBH (Baganoff et al. 2001, 2003). Surprisingly, the X-ray luminosity of $L_{2-10\,\rm keV} = 2.4 \times 10^{33}$ erg s$^{-1}$ was lower than expected, and the estimated bolometric luminosity of $\sim 10^{-9} L_{\rm Edd}$ implied a "quiescent" SMBH. The *Chandra* measurements of

**Figure 8.4.** *Chandra* panoramic view of the Galactic Center covering a 900-by-400 lt-yr swath based on 88 *Chandra* pointings. Sagittarius A* is marked in the center of the image. A few bright X-ray sources have been labeled; however, many more sources contribute to the rich X-ray radiation shown in the image. Color represents different energy ranges: red (1–3 keV), green (3–5 keV), blue (5–8 keV). (Courtesy of NASA/CXC/UMass/D. Wang et al.) See animation at http://chandra.harvard.edu/edu/gcenter/. The interactive figure provides interaction with this figure, including pan and zoom. Additional materials available online at http://iopscience.iop.org/book/978-0-7503-2163-1.

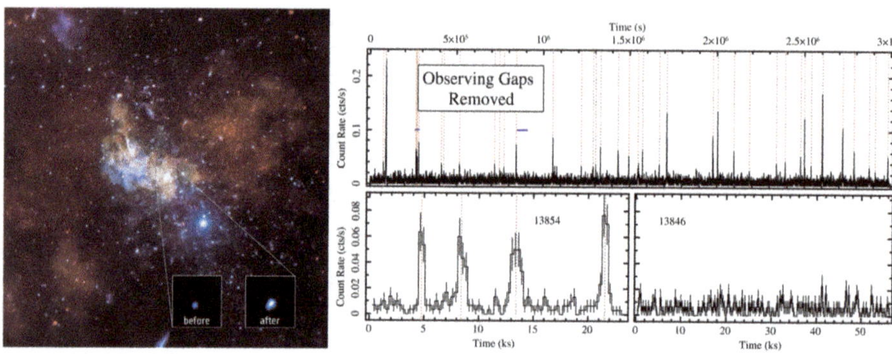


**Figure 8.5.** Left panel: *Chandra* image of the Sgr A* region. Courtesy of NASA/CXC/Amherst College/D. Haggard et al. Two inserts show X-ray emission from Sgr A* in quiet state and during the largest flare. See animation of the largest flare: http://chandra.harvard.edu/photo/2015/sgra/. The time-lapse video shows a movie of the Sgr A* flare, captured by *Chandra* on 2013 September 14, as shown in the left hand panel. Right panel: Sgr A* variability. Top: combined zeroth- and first-order 2–8 keV *Chandra* X-ray light curve of Sgr A* in 300 s bins for the entire 2012 XVP campaign with observation gaps removed. A number of flares of varying intensity are apparent and are indicated by dotted red lines. Short blue horizontal lines indicate sample observations shown in the bottom panel. Bottom: sample light curves of an observation with (left) and without (right) detected flares. ObsID 13854 (left) shows four moderately bright flares within 20 ks. Reproduced from Neilsen et al. (2013). © 2013. The American Astronomical Society. All rights reserved. Additional materials available online at http://iopscience.iop.org/book/978-0-7503-2163-1.

the density and temperature of hot gas at the Bondi radius gave an estimate of the accretion rate that was higher than the predictions based on radio measurements, and therefore inconsistent with the standard ADAF flow models developed in the 1990s.

Many new theoretical models of low-luminosity accretion flows or radiation from a base of a jet have been applied to the Sgr A* spectral energy distribution (see Melia & Falcke 2001 for a review), and the *Chandra* observations have provided the critical constraints for the model parameters. However, there is still no clear resolution to the origin of the observed X-ray radiation (e.g., accretion flow, corona, a base of a jet).

Sophisticated GRMHD simulations of the expected radio-band images from a low-radiation accretion flow or a jet (Mościbrodzka et al. 2014) in Sgr A* can be tested with high-resolution radio images. The X-ray spectral shape and the X-ray luminosity can restrict model parameters, but the high-angular-resolution X-ray data must be paired with simultaneous radio observations.

*Chandra* detected strong X-ray variability of Sgr A* (see right panel in Figure 8.5), characterized by short-duration (<5 ks) rapid flaring events with an average factor of ~50 change in the X-ray luminosity (Baganoff et al. 2001; Nowak et al. 2012; Neilsen et al. 2013, 2015). A low-luminosity quiescent (nonflaring) X-ray radiation with luminosity $\sim 10^{33}$ erg $s^{-1}$ is identified with radiation from accretion flow, while the process responsible for the observed flaring activity (with the flare luminosity reaching up to $\sim 10^{35}$ erg $s^{-1}$) is still not fully understood (see Wang et al. 2013 for a review). A plethora of theoretical models have been proposed; however, there is still no conclusion on the physical process responsible for the flares (e.g., Neilsen et al. 2015).

### 8.2.4 Quasars and the Eddington limit

Quasars are the most luminous class of AGN. They host the most-massive black holes ($M_{\rm bh} \sim 10^9\, M_\odot$ at $z > 6$), which accrete at rates corresponding to the Eddington limit or even higher. This high luminosity means that quasars can be detected at very high redshifts and thus be used to probe conditions in the early universe and to test cosmological models (Risaliti & Lusso 2015). The fact that SMBHs with such high masses have been formed already at $z > 6$ gives constraints on black hole formation and growth. This is an active area of theoretical research with many unsolved questions (see, e.g., Volonteri 2010).

At the high accretion rates, the temperature of the standard accretion disk surrounding an SMBH corresponds to the blackbody emission peaking in the optical/UV band. The X-ray radiation is generated in a hot medium (e.g., optically thin corona) that Compton-upscatters the disk photons (see, e.g., Czerny et al. 2003; Kubota & Done 2018). The X-ray spectra from such a process are relatively simple and can be described by a power law with a slope dependent on the properties of the corona. Sobolewska et al. (2004a, 2004b) applied a disk-corona model to a small sample of $z > 4$ radio-quiet quasars observed with *Chandra* and showed that a high accretion rate $> 0.2\dot{M}_{\rm Edd}$ onto a high-mass SMBH, $M_{\rm bh} > 5 \times 10^9\, M_\odot$, is required to explain these spectra. Models with a relatively low optical depth ($\tau \sim 0.015 - 0.5$) and high electron temperatures, $kT_{\rm e} \sim (100 - 500)$ keV, were valid in both types of the corona geometry (i.e., plane parallel and hot inner sphere). The geometry of the corona is still uncertain (see, e.g., Reis & Miller 2013; Wilkins & Fabian 2013; Czerny & Naddaf 2018). In general, any geometry of the hot compact medium with suitable conditions for inverse-Compton radiation would be valid. Time-dependent broadband observations can be used to test details of the possible geometries (see Wilkins & Fabian 2012; Edelson et al. 2015, 2017; Uttley et al. 2014; Kara et al. 2016).

An X-ray loudness parameter, $\alpha_{\rm ox}$ (Tananbaum et al. 1979), defined as the ratio of the intrinsic monochromatic luminosity measured at 2500 Å and at 2 keV, indicates the relative amount of the energy radiated by the accretion disk and by the corona.

Studies of radio-quiet quasars with a range of UV luminosities found that the X-ray loudness correlates with the UV luminosity in such a way that the importance of the corona decreases with increasing UV luminosity, i.e., radio-quiet quasars become more X-ray quiet with increasing accretion rates, and the importance of the corona declines (Bechtold et al. 2003; Kelly et al. 2008; Nanni et al. 2017). Other correlations between the X-ray loudness, the UV spectral slope, and X-ray photon index can provide further constraints on the accretion process, but they are currently not conclusive.

Correlation studies of spectral parameters with redshift probe the evolution of accretion, SMBH growth, and feedback across the universe. Newly discovered very high-redshift ($z > 6$) quasars provide critical data points for such studies; however, these samples are still very small and biased to the most-luminous and highly accreting sources. Many observations of high-redshift quasars have been made by *Chandra*, often providing the first X-ray data for these sources[1] (e.g., Bechtold et al. 2003; Bassett et al. 2004; Vignali et al. 2005; Shemmer et al. 2006; see Section 8.6). Recently, *Chandra* performed observations of the most distant quasar observed in X-rays to date, ULAS J1342+0928 at $z = 7.54$ (Bañados et al. 2018). The measured X-ray luminosity of $L_{2-10\mathrm{keV}} = 1.16^{+0.43}_{-0.35} \times 10^{45}$ erg s$^{-1}$ and X-ray loudness parameter, $\alpha_{\mathrm{ox}} = 1.67^{+0.10}_{-0.16}$ are consistent with those of other luminous quasars at $z > 5.5$ (e.g., Nanni et al. 2017). This places even more stringent constraints on the models of black hole formation in the early universe. The SMBH and accretion process must have been in place already when the universe was only at 5% of its current age.

*Radio-loud Quasars*

About ~10% of all quasars have relatively strong radio emission with high radio loudness (ratio of the observed radio to optical flux) and are classified as radio-loud. These quasars have relativistic jets, with the jet emission contributing to the observed radiation at levels related to the overall source properties, such as jet power or viewing angle. Many large-scale X-ray jets have been resolved by *Chandra* (see Section 8.5.1). Compact, parsec- to a few-kiloparsec-scale jets in the nuclear regions cannot be spatially resolved, and their emission contributes to the total observed X-ray emission of the quasar core.

Spectral studies of radio-quiet and radio-loud quasars have focused on the differences between the two populations. The radio-loud quasars are more X-ray loud than the radio-quiet ones (Wilkes & Elvis 1987; Elvis et al. 1994; Bechtold et al. 1994; Miller et al. 2011), with the X-ray photon index indicating a harder spectrum due to additional jet emission. Dedicated *Chandra* observations of the most radio-loud quasars at $z > 4$ suggest an enhanced X-ray emission in these sources in comparison to radio-loud quasars at low redshift (Wu et al. 2013; Zhu et al. 2019). The jet radiation processes and overall jet properties might be different at high-$z$ (see also McKeough et al. 2016). X-ray observations of resolved jets at high-$z$ and of

[1] *Chandra* has a very low background for point-source detection.

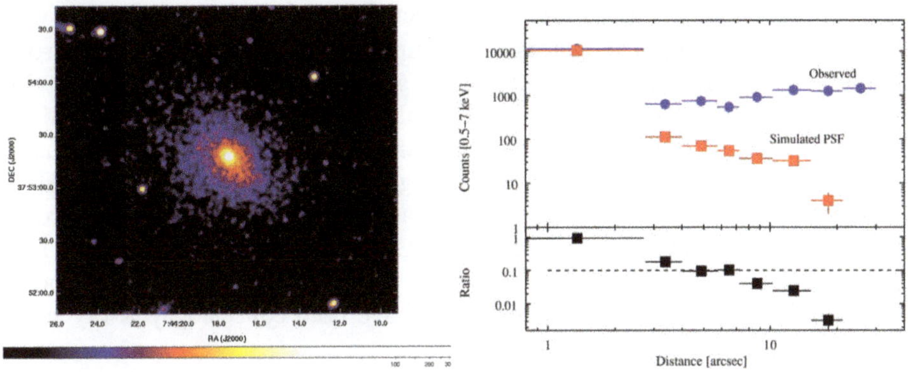


**Figure 8.6.** Left: *Chandra* ACIS-S X-ray image of 3C 186, a radio-loud quasar at $z = 1.16$ embedded in a galaxy cluster seen as a diffuse structure surrounding a strong point-source emission. The *Chandra* image was smoothed using adaptive methods to highlight the emission on different scales. The size of the point-spread function (PSF) can be seen in the emission of a few bright point sources outside the diffuse cluster emission. Color bar shows the intensity scale. Right, upper panel: a number of counts in the spectral extraction regions for the observed (quasar + cluster) X-ray emission (blue circles) and the simulated *Chandra* PSF—red squares. The first data point shows the PSF normalized to the observed counts in the circular region with $r = 2.7''$ centered on the quasar. The strong X-ray cluster emission is indicated by the excess in the observed counts. Right, lower panel: the ratio of the simulated to the observed counts in each spectral region. The dashed line marks the 0.1 value of the ratio. Figures adapted from Siemiginowska et al. (2005, 2010). 

larger samples of high-$z$ radio-loud quasars are needed to better understand the physics of these quasars.

Radio-loud quasars host the most-massive SMBHs and reside in rich galaxy environments. However, the central galaxy in most nearby galaxy clusters has a low Eddington luminosity ratio. Prior to *Chandra*, there were no detections of X-ray clusters associated with radio-loud luminous quasars. The bright quasar radiation would dominate over the low surface brightness diffuse cluster emission, and a sharp PSF (point-spread function) was needed to clearly separate these two radiation components. *Chandra* provided the first significant detections of X-ray clusters associated with radio-loud quasars (e.g. Figure 8.6; Siemiginowska et al. 2005, 2010; Russell et al. 2012). X-ray clusters associated with bright radio-quiet quasars (Russell et al. 2010) and young radio sources (Kunert-Bajraszewska et al. 2013; Hlavacek-Larrondo et al. 2017) were also found in the *Chandra* images. Such detections of X-ray clusters require relatively long exposures ( >100 ks), and only a small number of quasars could have been observed with *Chandra*. Any systematic studies of the quasar X-ray environments have to wait until the new generation of high-angular-resolution X-ray missions, such as *Lynx* (see Chapter 11, Section 11.2.1), become available in the future.

## 8.3 SMBH Formation and Growth

Observations of distant luminous quasars indicate that massive SMBHs were already present in the early universe. The most distant quasar at $z = 7.54$ (at the time when the universe was <1 Gyr old) detected to date harbors an SMBH with a mass of at least $10^9\ M_\odot$. This places constraints on SMBH formation and growth. There are two main scenarios considered for growing an SMBH: (1) a hierarchical growth by consequent mergers of the initial small black hole "seeds" left over by Pop III stars (see, e.g., Volonteri & Rees 2005) and (2) a direct collapse of a massive object formed from primordial clouds (see, e.g., Begelman et al. 2006). The SMBH mass function (the number of black holes in a given volume and redshift) provides constraints on different formation scenarios. Observations of both local and high-$z$ AGN provide constraints on the evolution and growth of SMBHs. Important statistical samples of AGN for these studies have been collected by X-ray surveys (see Section 8.6).

The growth of SMBHs in the early universe, from light $\sim 100\ M_\odot$ or heavy $\sim 10^4 M_\odot$ black hole seeds, could proceed via super-Eddington accretion episodes which require a large reservoir of gas. The duration of these episodes depends on the gas angular momentum and also on the disk outflows and jets reducing the efficiency of the black hole growth (e.g., Pezzulli et al. 2017; Regan et al. 2019). Theoretical simulations indicate that efficiently accreting SMBHs may be buried by a large amount of obscuring matter (Trebitsch et al. 2019), impacting their optical detectability and biasing the observed duty cycles. While these accreting black holes could be detectable in X-rays, at high-$z$ they would be relatively faint, and finding such systems with current X-ray instruments is challenging. Interestingly, a heavily obscured Compton-thick quasar candidate at $z > 6$ was recently detected in the ~60 ks *Chandra* observation by Vito et al. (2019). If confirmed, this would be the first detection of an obscured SMBH system in the early universe.

In order for an SMBH to reach a mass of $\sim 10^9 M_\odot$ at early times, additional processes, such as mergers, have to take place. Dual SMBHs (with separation <10 kpc) are at the last stage of the SMBH pair merger and typically last long enough (~100 Myrs) so they can be easily detected. Once the SMBHs are close enough in a subparsec binary, they will eventually coalesce, generating a gravitational wave signal. The newly merged BH will likely have a velocity with respect to the host galaxy depending on the SMBH mass and spin (see, e.g., Blecha et al. 2019). Asymmetric gravitational wave radiation during the merger may lead to a high-velocity kick of the remnant SMBH, which can escape the system (i.e., recoil). The merging process funnels gas toward the SMBHs and results in X-ray radiation; however, detecting recoiling SMBHs depends on their separation. Dual AGN and binary BHs have been detected by *Chandra* in nearby galaxies (see Chapter 7) and also low-redshift galaxies (see, e.g., Figure 8.7 from Comerford et al. 2015; Liu et al. 2013), providing the evidence for SMBH mergers. *Chandra* also found a few strong candidates for recoiling SMBHs using the data from surveys (CID-42, Figure 8.7; Civano et al. 2010, 2012; Blecha et al. 2013), as well as the *Chandra* Source Catalog (see, e.g., Kim et al. 2017).

## 8.4 AGN Structure

The standard AGN structure model consists of an accretion disk, a corona, line-emitting clouds, outflows in the form of winds or jets, hot ionization cones possibly

**Figure 8.7.** Left: composite image of the galaxy SDSS J1126+2944 hosting dual SMBHs both emitting in the X-ray band from Comerford et al. (2015). X-ray courtesy of NASA/CXC/Univ of Colorado/J.Comerford et al; Optical courtesy of NASA/STScI. Right: the *Chandra* and *HST* composite image of the candidate recoiling SMBH, CID-42, from Civano et al. (2012). X-ray courtesy of: NASA/CXC/SAO/F.Civano et al; Optical courtesy of NASA/STScI; Optical (wide field) courtesy of CFHT, NASA/STScI. The SDSS J1126+2944 interactive figure shows separately the optical emission in three bands (HST/WFC3 filters: F160W (H band, red), F814W (I band, green), and F438W (B band, blue)), the X-ray emission (0.3–8 keV, pink), and the composite image of SDSS J1126+2944, as shown in the left hand panel. The CID42 interactive figure shows separately the optical emission (yellow), X-ray emission (blue), and the composite image of CID42, shown as insets, and a wide-field composite image, shown in the right hand panel. Additional materials available online at http://iopscience.iop.org/book/978-0-7503-2163-1.

associated with the outflows, and obscuring medium in the form of a torus or outflowing wind (see Figure 8.2; e.g., Elvis 2000, 2017; Czerny & Naddaf 2018). The X-ray emission originates in compact regions within the immediate vicinity of a SMBH, due to radiation from a corona, a hot accretion flow, or a beamed jet.

The compact AGN (<1 pc) is located in the center of the host galaxy, and it is spatially unresolved. Multiwavelength observations provide a general idea about the emission and absorption components contributing to the spectrum, but the exact details of their location, size, and dependence on the accretion mode are still under active investigation. While the two main AGN classifications into type I (unobscured) and type II (obscured) have been based on the viewing angle, the SMBH accretion modes provide an additional complication into the zoo of AGN types (see Netzer 2015; Padovani et al. 2017 for a review). The X-ray observations provide a measure of the total obscuration (see Hickox & Alexander 2018 for a review). Complex absorption with several absorption components has been seen in *Chandra* spectra (Kaspi et al. 2000; Krongold et al. 2003); however, the *Chandra* sensitivity limits these studies to Compton-thin and warm absorbers (i.e., equivalent column density of hydrogen $N_{\rm H} < 10^{24}$ cm$^{-2}$). The obscured X-ray-emitting region has to be located well within the unresolved central parsec region (see Section 8.6.2.4). In addition, illumination and reflection of the accretion disk, or scattering off a relatively cold material surrounding the primary emission, imprints lines and continuum features on the X-ray spectra.

Several observational methods can be used with *Chandra* to resolve structures in the X-ray nuclear regions. Below, we discuss results based on spectral variability, direct subarcsec-resolution imaging, microlensing events in lensed quasars, and high-resolution spectroscopy.

### 8.4.1 Variability and X-Ray Reverberation

Variability studies can identify radiation components and the geometry of the unresolved AGN emission (e.g., see Edelson et al. 2015 for a multiband analysis of NGC 5548 variability). Intrinsic variations give the characteristic size scales, while the variable obscuration allows for measurements of the relative locations of the obscuring medium and the emission source or gives the total size of the emitting regions. If the velocity-broadened emission lines are observed, then their profile (and profile variations) can be used to determine the size and scale of the line-emitting regions.

Reverberation-mapping techniques (see Uttley et al. 2014 for a review) have been used to study correlations between X-ray emission components. They show compact $< 10r_g$ X-ray emission regions in nearby galaxies (see Kara et al. 2016). However, time-dependent data are challenging as the AGN spectra have a low number of counts per time resolution bin with contributions from many variable emission components, so only a few AGN have been studied to date. A broadband X-ray coverage is required for successful X-ray reverberation experiments. *Chandra*, *XMM-Newton*, *Swift*, and *NuSTAR* can be combined to cover a broad, 0.5–70 keV energy range.

Accurate study of broad X-ray emission requires the monitoring data to be taken simultaneously by all involved observatories and collected over a long period of time. Such observational campaigns of a few targets are being performed. The first results from *Swift* mapping of accretion disks in a few local AGN (Edelson et al. 2019) show variations consistent with the prediction of a thin accretion disk model and indicate challenges for the reprocessing model with the central X-ray corona illuminating the disk. Long-term X-ray monitoring of AGN is necessary to disentangle individual radiation components and to understand their physical origin.

### 8.4.2 Microlensing

Gravitationally lensed quasars with microlensing events can be used to probe the geometry of the accretion flow and radiation in the close vicinity of an SMBH. Light curves from the resolved lensed images (see Figure 8.8) have patterns related to the known time delay between each image, but microlensing imprints uncorrelated variations on the images. Several quasars have been monitored with *Chandra*, placing the most stringent constraints on the radiation regions to date (Morgan et al. 2008; Pooley et al. 2007; Chartas et al. 2009; Dai et al. 2010; Chartas et al. 2012). In all monitored quasars, the X-ray emission region is more compact than the optical-UV emission region and located within $< 10r_g$ from the SMBH.

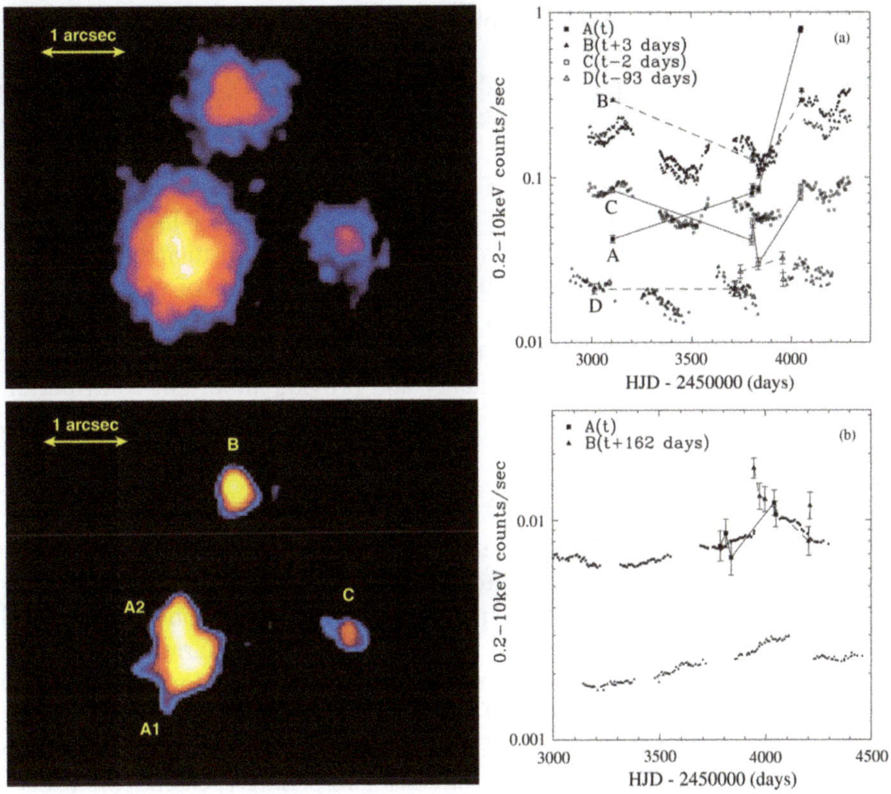


**Figure 8.8.** Left: *Chandra* X-ray images of the lensed radio-loud quasar, MG J0141+0534 ($z$ = 2.64). The smoothed *Chandra* image (top) and the deconvolved image (bottom) showing the four resolved components of the lensed quasar (reproduced from Chartas et al. 2002 © 2002. The American Astronomical Society. All rights reserved). Right: *Chandra* light curves from the individual images of two lensed quasars, RX J1131−1231, $z$ = 0.658 (top), and HE 1104−1805, $z$ = 2.32 (bottom). The light curves show noncorrelated X-ray variability between four lensed quasar images (A, B, C, D), indicating microlensing events (reproduced from Chartas et al. 2009; © 2009. The American Astronomical Society. All rights reserved; Dai et al. 2010). The light curves have been adjusted for the time delays with respect to the A image.

Significant variations due to microlensing were detected in deep *Chandra* observations of RX J1131−1231 by Chartas et al. (2012). The differences between the soft and hard X-ray variations could be explained by soft emission from a compact hot corona and a slightly more extended reflected component generating the hard X-rays. In addition, the profile changes of the Fe–K$\alpha$ line indicated a compact reflection region. Interestingly, blueshifted and redshifted Fe–K$\alpha$ were detected in individual images of three quasars, placing a limit of $\sim 20 r_g$ on the location of the emission region if they are the result of microlensing (Chartas et al. 2017). Detailed modeling of the line profiles and the continuum variability due to microlensing can provide measurements of inclination angles and SMBH spin, and an important test of the strong gravity at the innermost stable orbit of the SMBH.

These studies require high-quality observations and long-term monitoring (see Chartas et al. 2019).

Chartas et al. (2002) reported the discovery of the first X-ray broad absorption lines (XBALs) at 8.05 keV and 9.79 keV (rest frame) in the gravitationally lensed quasar APM 08279+5255 ($z = 1.51$). The lines are due to two distinct absorption systems with velocities of $0.2c$ and $0.4c$, respectively. The estimated outflow launching radius was relatively small, with the X-ray absorber located within the UV outflow, possibly representing a shielding gas that is postulated in theoretical AGN wind models (e.g., see Proga 2003). Later observations of this source and a larger sample of radio-quiet quasars confirmed the existence of ultrafast winds originating near the SMBH (see, for example, Tombesi et al. 2010, 2013). These outflows carry out large amounts of energy and provide important contribution to the feedback.

### 8.4.3 Imaging of the Torus and X-Ray-scattering Region

A dusty "torus" is prominently present in the AGN models to explain hot dust emission observed in the near-IR and to obscure the primary radiation in Compton-thick sources (i.e., $N_H > 10^{24}$ cm$^{-2}$). The X-ray spectra and the observed variability suggest a clumpy structure (Risaliti et al. 2002, 2005, 2010; Marinucci et al. 2013), but in general the nature of the torus is not well understood (e.g., rotating multiphase structure, outer regions of the accretion flow, outflows). Recent high-angular-resolution *Chandra* images of a few nearby Compton-thick AGN show a spatially extended soft X-ray emission, and also, surprisingly, extended hard X-ray (3–6 keV) and Fe–K$\alpha$ line emission (Figure 8.9 Fabbiano et al. 2018, 2019; see also Chapter 7). ALMA observations of these AGN provided high-resolution maps of the molecular gas showing linear structures on similar scales (~10–100 pc) to the *Chandra* features in the nuclear region (Alonso-Herrero et al. 2018; Feruglio et al. 2019). The correspondence between the hard X-ray features and the distribution of molecular gas points to the torus for the origin of this emission (Fabbiano et al. 2018; Feruglio et al. 2019). Future high-resolution imaging with *Chandra* and ALMA are necessary to decide on the structure and nature of the torus.

### 8.4.4 Resolving Layers of Ionized Gas with High-resolution X-Ray Spectra

"Warm absorbers" specify a partially ionized medium with variable ionization structures (see Turner & Miller 2009; King & Pounds 2015). They were initially proposed by Halpern (1984) to explain a large change in the X-ray absorption column detected between two *Einstein* observations, separated by about a year, in a quasar MR2251 − 178. The observed variations would be due to ionization changes of such a "warm absorber" ($T \sim 10^5$ K). During the last two decades *Chandra* and *XMM-Newton* high-resolution grating observations have shown that "warm absorbers" have a very complex ionization structure, dominated by K-shell ions of the lighter metals (C, N, O) and Fe–L shell, and have outflow velocities of several hundred km s$^{-1}$ (McKernan et al. 2007).

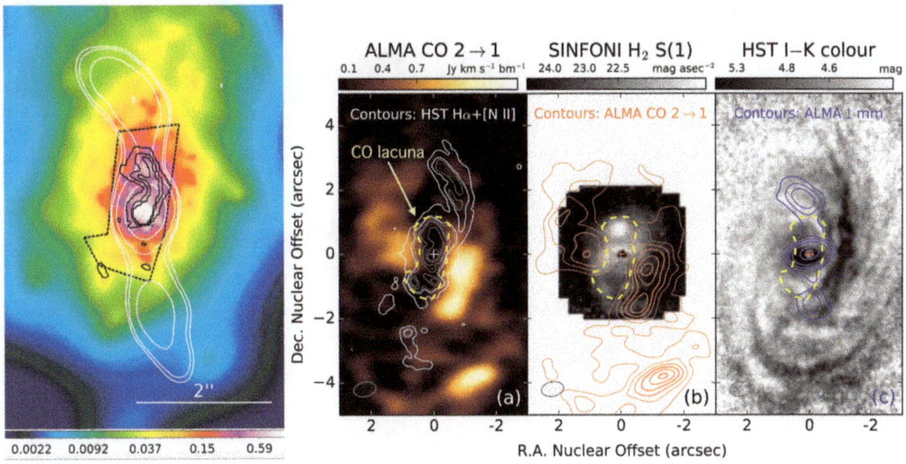


**Figure 8.9.** Left: *Chandra* ACIS-S adaptively smoothed image of the central high surface brightness region in NGC 2110, in the 0.1–1 keV band. Superimposed are the *HST* H$\alpha$ + [NII] contours (black), a sketch of the CO 2–1 "lacuna," and the 5 GHz (VLA) contours from Evans et al. (2006), in white. Reproduced from Fabbiano et al. (2019). © 2019. The American Astronomical Society. All rights reserved. (a) ALMA CO 2–1 line map. Contours are from the *HST* map of the H$\alpha$ + [NII] emission, smoothed to match the angular resolution of the ALMA data. (b) A map of the $H_2$ 1–0 S(1) line at 2.12$\mu$m from VLT/SINFONI. The contours of the ALMA CO 2–1 emission from (a). (c) A map of the ratio of the *HST* F791W (optical) and *HST* F200N (near-infrared) images, which emphasizes dust absorption as dark features. The blue contours show the shape of the ALMA 1 mm continuum, which traces the bipolar radio jet. Reproduced from Rosario et al. (2019). © 2019. The American Astronomical Society. All rights reserved.

The first high-resolution X-ray spectrum of an AGN obtained by the *Chandra* LETG showed several strong and highly ionized narrow absorption lines (NGC 5548, Kaastra et al. 2000, 2002; Steenbrugge et al. 2005). These lines were blueshifted by a few hundred km $s^{-1}$, indicating an outflow. There were also emission lines, the O VII triplet and OVIII Ly-*$\alpha$, from a warm ionized medium present in this spectrum. Only a few AGN have been observed using gratings, as they are quite faint in X-rays and require very long exposures (see for example Kaspi et al. 2000; Lee et al. 2001; Krongold et al. 2003, 2005; Reeves et al. 2010; King et al. 2012; Lee et al. 2013; Tombesi et al. 2016). However, these spectra provide a rich amount of information (see, for example, *Chandra* grating spectra of NGC 3783 in Figure 8.10) showing complex absorption and emission features, stratified ionization and densities, and a range of outflow velocities originating at a range of distances from the SMBH. In particular, ultrafast velocity outflows (UFOs) have also been detected, with blueshifted Fe–K lines indicating velocities exceeding $>0.033c$ (Tombesi et al. 2010). Deep *Chandra* grating observations of PG 1211+143, by Danehkar et al. (2018), confirmed the presence of the UFO discovered in the *XMM-Newton* data (Pounds et al. 2003, 2016) and showed that the outflow is highly variable, with a range of velocities.

Powerful ionized winds observed in X-rays are common, span a range of velocities with a layered ionization structure (see Tombesi et al. 2013 for details).

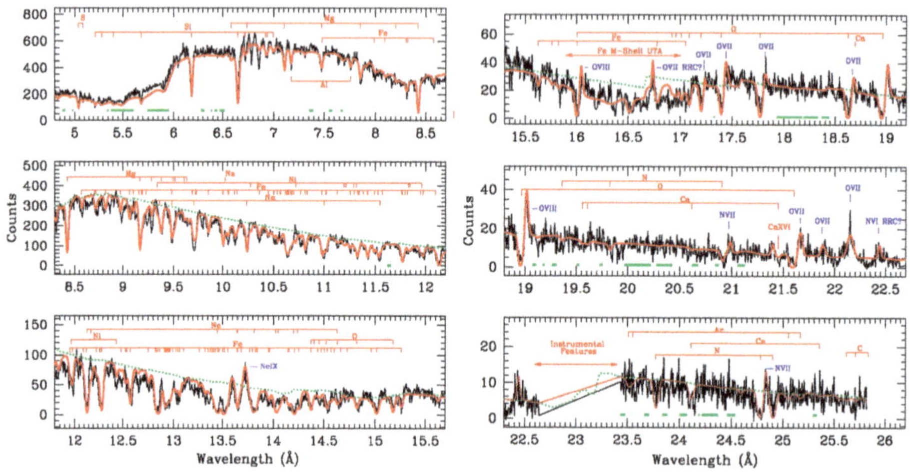


**Figure 8.10.** Two-phase absorber model plotted against the first-order *Chandra* MEG spectrum of NGC 3783. Model-predicted absorption lines are marked in red, and single emission lines are labeled in blue. The line-free zones are indicated at the bottom of each panel (green). The continuum level (including edge continuum absorption) is overplotted for comparison (dotted green line). The spectrum is presented in the rest-frame system of the absorbing gas. Reproduced from Krongold et al. (2003). 

The recent multiband observations of AGN outflows show an important link between the X-ray-ionized winds, UFOs, and molecular outflows seen in ALMA (see Mkn 231; Feruglio et al. 2015), indicating that the outflows driven by SMBHs have relatively large mass-loss rates. These outflows contribute to AGN feedback and provide enough energy to evacuate large amounts of gas and stop star formation (Hopkins & Elvis 2010; King & Pounds 2015).

## 8.5 Jets and Extended Radio Structures

Relativistic jets have been observed and studied in the radio band since the 1950s. The first X-ray detection of a resolved jet and the first discovery of an X-ray jet were made by the *Einstein* High Resolution Imager (HRI) in the early 1980s (M87; Feigelson 1980; Schreier et al. 1982; and Centaurus A; Schreier et al. 1979). A few X-ray detections of radio jets were later made by *ROSAT*, but detailed X-ray studies of resolved large-scale jets were only able to be performed at the start of this millennium with *Chandra*.

During the past two decades, *Chandra* has obtained X-ray data for many resolved jets associated with radio galaxies, AGN, and quasars (see examples in Figure 8.11 and Figure 8.12). These observations show that the jet X-ray radiation is ubiquitous and that multiband studies can lead to a better understanding of jet physics. They also underscored the significance of jets in the evolution of galaxies and clusters of galaxies. However, there are still many unanswered questions about the nature of relativistic jets, particle content, particle acceleration, or jet emission processes (see Perlman et al. 2019). *Chandra* archives have sufficiently deep observations for only a few jets and

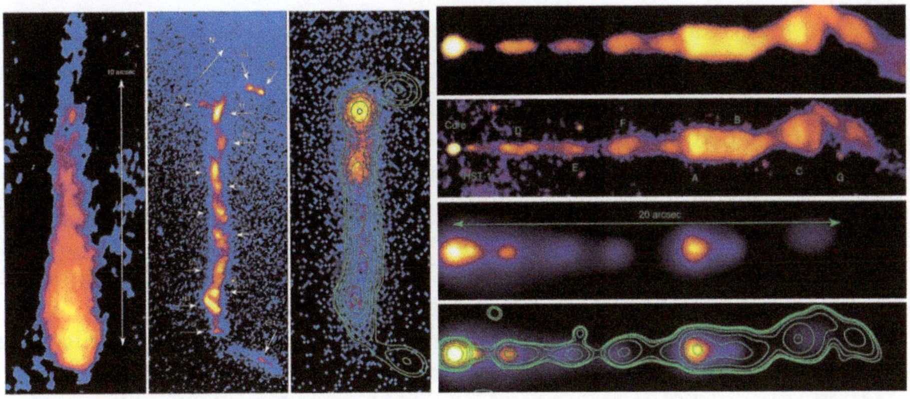

**Figure 8.11.** Left: 3C 273 jet in: radio 1.647 GHz (left), optical *HST* (F622W) centered at 6170 Å (middle), and X-rays from *Chandra* ACIS-S (0.1″ pix$^{-1}$) overlaid with the optical contours (right). Colors indicate the intensity. The overall shape of the jet is remarkably similar in length and curvature, and the X-ray emission fades to the end of the jet. (Reproduced from Marshall et al. 2001 © 2001. The American Astronomical Society. All rights reserved.) Right: images of the M87 jet in radio, 14.435 GHz (top, ~0.2″ resolution), the optical *HST* (F814W) image (second panel), adaptively smoothed *Chandra* image (0.2″ pix$^{-1}$; third panel). *Chandra* image overlaid with the optical contours. The *HST* and VLA images are displayed using a logarithmic stretch to bring out faint features, while the X-ray image scaling is linear. (Perlman et al. 2001; Marshall et al. 2002) © 2002. The American Astronomical Society. All rights reserved.

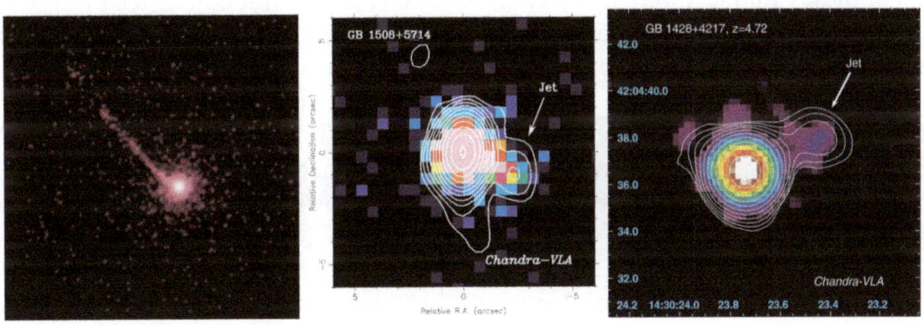


**Figure 8.12.** X-ray jets in quasars. Left: *Chandra* smoothed image of ~30″ X-ray jet discovered in PKS 1127 +145, $z$ = 1.18 (Siemiginowska et al. 2002). Courtesy of NASA/CXC/A.Siemiginowska(CfA)/J.Bechtold(U. Arizona). Center: *Chandra* color image of GB 1508+5714 and the ~3″ jet overlaid with the radio contours ($z$ = 4.3; Siemiginowska et al. 2003a; Cheung 2004 © 2004. The American Astronomical Society. All rights reserved). Right: the most distant X-ray jet detected by *Chandra* to date is associated with GB 1428+4217 ($z$ = 4.72) displayed in a color image with the radio contours (Cheung et al. 2012 © 2012. The American Astronomical Society. All rights reserved).

unfortunately do not contain a large sample of jets suitable for statistical studies. In this section, we highlight important results from *Chandra* studies of extragalactic jets. Reviews of many aspects of X-ray jets phenomena can be found in several publications, e.g., see Harris & Krawczynski (2006), Worrall (2009), Pudritz et al. (2012), and Hardcastle (2015). Specific *Chandra* results have been summarized in two *Chandra* Newsletter articles by Schwartz & Harris (2006) and Siemiginowska (2014).

### 8.5.1 Resolving X-Ray Jets—Knots and Hotspots

The discovery of a resolved ∼11″-long jet of PKS 0637−752 (see figure in Chapter 2) in the first *Chandra* observation designed to test the focus of the mirrors on orbit (Schwartz et al. 2000) marks the beginning of jet studies in X-rays. Indeed, the first few years of the *Chandra* mission brought discoveries of new X-ray jets (Siemiginowska et al. 2002, 2003a, 2003b, 2007; Cheung 2004; see Figure 8.12), followed by several systematic surveys to find X-ray counterparts to known radio jets (Marshall et al. 2002; Sambruna et al. 2004; Marshall et al. 2010; Hogan et al. 2011). A few long jets were studied in detail.

X-ray jets form diffuse linear and bending structures with enhancements due to knots or hotspots. An X-ray jet is often located close to a strong AGN core emission or embedded in the diffuse emission of a host galaxy with a total length very rarely exceeding ∼30″. The X-ray jets are rather faint, and their emission is only a small fraction (<3%) of a strong core. This was one of the reasons for finding only a few jets with the earlier X-ray missions. For the *Chandra* images with Poisson nature and low background counts, the detection of faint diffuse structures, such as jets, has been challenging (Stein et al. 2015; McKeough et al. 2016).

A few surveys of X-ray jets have been completed to date. They differ in their selection criteria, but all originated from samples of known radio jets. Sambruna et al. (2004) and Marshall et al. (2011, 2018) targeted jets associated with radio galaxies and quasars. *Chandra* observations were relatively short (<10 ks), but about 60% of these radio jets were detected in X-rays. A higher fraction of resolved X-ray jets, 78%, was found by Hogan et al. (2011) in a sample of blazars selected from the flux-limited MOJAVE sample of relativistic jets associated with quasars and FR II radio galaxies. Massaro et al. (2011) studied the properties of knots, hotspots, and diffuse jet emission using all of the X-ray jets observed during the first decade of the *Chandra*[2] mission. This archival sample is heterogeneous, consists of 106 sources (FR I and FR II radio galaxies and quasars) and shows a total of 236 detected jet-related features.

Most of the jet surveys focused on detecting jets and understanding the X-ray emission process. X-rays can be due to synchrotron emission from highly relativistic particles in the jet or due to inverse-Compton emission where the jet relativistic particles transfer their energy to low-energy photons. The photon field could be internal to the jet (synchrotron self-Compton, SSC) or external to the jet. The cosmic microwave background (CMB) photons can illuminate highly relativistic kiloparsec-scale jets resulting in the X-ray emission (IC-CMB; Celotti et al. 2001; Harris & Krawczynski 2002).

### 8.5.2 Jet X-Ray Radiation Processes

Current observations indicate that the X-ray synchrotron emission is primarily seen in jets associated with low-power radio galaxies (FR I). For example, the jets of nearby FR I galaxies, M87 and Centaurus A, were studied in great detail. Both jets

[2] http://hea-www.harvard.edu/XJET/.

exhibit a correlated morphology with an almost constant spectral index across the broad frequency range from radio through optical to X-rays (see M87 jet in Figure 8.11) suggesting the synchrotron process as the primary X-ray emission. However, the simple "one-zone" model, where the radio and X-ray emission are produced by the same population of relativistic particles, cannot fully explain the observations, and some particle evolution along the jet is required. Studies of FR I jets by Harwood & Hardcastle (2012) show no statistically significant correlations between the properties of the host galaxy and the jet, but do show that the luminosity of X-ray jets scales linearly with the jet radio luminosity and that the spectral slope is related to both X-ray and radio luminosity of the jet. Because the X-ray emission in these jets is due to the synchrotron process, the particle acceleration mechanisms could be tested with a larger sample of FR I jets in the future.

Studies by Massaro et al. (2011) indicated differences in the radio to X-ray flux ratios between the hotspots and knots in powerful FR II radio galaxies, and no significant difference in these ratios for the knots in FR I and quasar jets. This second result suggests that either the knots in both types of jets are due to the synchrotron process or that the inverse-Compton process in quasar knots has very specific conditions resulting in the same flux ratios. A large sample of quasar jets, especially at high $z$ is needed to address these issues.

Results of X-ray and $\gamma$-ray (*Fermi*-LAT) spectral studies show inconsistencies with the IC-CMB emission model for the knots in a few quasar jets (Meyer & Georganopoulos 2014; Meyer et al. 2015, 2017; Breiding et al. 2017). However, the IC-CMB emission should contribute to the observed X-ray radiation of high-$z$ jets (Siemiginowska et al. 2003a; Cheung et al. 2012; Simionescu et al. 2016). Recently, Marshall et al. (2018) summarized the critical observations that do not agree with the IC-CMB model predictions, e.g., differences in radio and X-ray morphology, disagreement between radio and X-ray spectral indices, low-jet bulk Lorentz factor, and inconsistencies between X-ray and $\gamma$-ray emission. On the other hand, the X-ray emission due to the synchrotron process requires *in situ* particle acceleration to take place at a distance of hundreds of kiloparsec from the SMBH. This poses interesting theoretical questions. The quasar jets and their high-energy radiations are still not understood, requiring both high-quality multiband data and theoretical development of the field to answer such questions.

### 8.5.3 Variability and Proper Motions of Resolved X-Ray Jets

Relativistic parsec-scale blazar jets, unresolved in X-rays, are variable with large-amplitude flares observed across the entire electromagnetic spectrum on timescales from days to years. However, the X-ray variability of large-scale resolved jets was not known before *Chandra*. During the last two decades, only a few jets have had multiple *Chandra* observations suitable for studies of intensity variations and motions of the resolved jet features.

The best monitoring X-ray data have been collected for the M87 jet, which was observed with *Chandra* more than 60 times between the years 2002–2008 (Harris

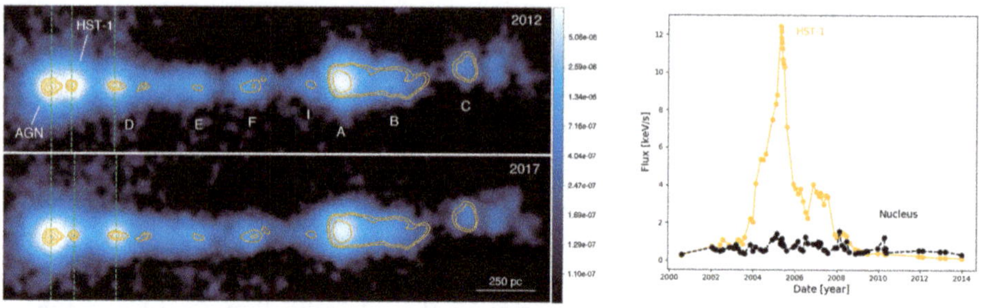


**Figure 8.13.** Left: M87 proper motions. *Chandra* HRC-I images of M87 in the 0.08–10.0 keV energy band from two epochs, 2012 and 2017, overlaid with *HST* WFC3/UVIS F275W contours observed during the same years. The X-ray and optical emission regions are spatially consistent in each epoch. Vertical lines (green, dashed) correspond to the X-ray centroid position of the AGN, *HST*-1 knot, and Knot D in each epoch. A small shift in the locations of *HST*-1 and Knot D is visible between the two epochs and indicates the observed proper motion of the features. Figure adapted from Snios et al. (2019a) © 2019. The American Astronomical Society. All rights reserved. Right: *Chandra* X-ray light curve of M87 nucleus (black) and *HST*-1 knot (orange) collected over 13 years with the monitoring program led by D. A. Harris (Harris et al. 2009). A strong and long-lasting outburst of the *HST*-1 knot dominates the light curve. The nucleus shows short-term variability during the same time. Figure adapted from Siemiginowska (2014).

et al. 2009) and shows the most dramatic flux variability of the *HST*-1 knot with an increase in the X-ray flux by a factor of ~50 (see Figure 8.13, right). Multiband studies of the rise and decay times provided constraints on the emission size and the energy losses of the relativistic electrons. The data were consistent with the energy losses dominated by synchrotron cooling with an average magnetic field of 0.6 mG, consistent with the earlier dynamical constraints by Stawarz et al. (2005, 2006). Most surprising was the discovery of the flux oscillation on a 0.5–0.8 yr timescale during the flare. The origin of these oscillations remained unidentified, but Harris et al. (2009) suggested that it could be related to quasi-periodic variations in the conversion of the jet kinetic power to the internal energy of the radiating plasma. The *HST*-1 X-ray flare remains the strongest one observed for any jet knot at a large distance ( >50 pc) from the SMBH.

The other nearby, well-studied jet is that of Centaurus A, resolved into more than 40 knots in *Chandra* images (Kraft et al. 2000; Hardcastle et al. 2003, 2007; Goodger et al. 2010). Some of these knots have corresponding radio emission, but there are also knots with no radio counterpart. Interestingly, there was no strong X-ray flux variability in any of the knots, with only a small, 25% level decrease in flux detected in a few knots. However, some interesting differences in morphology and brightness were detected between images taken 15 years apart (Snios et al. 2019c see Figure 8.14).

Multiple *Chandra* observations of both jets, M87 and Centaurus A, over the past decade also offered the possibility to study, for the first time, the movement (proper motions) of the X-ray jet features (see Figure 8.13; Snios et al. 2019b, 2019c). The detection of proper motions confirms the jet's relativistic nature with the apparent superluminal speeds for some innermost knots. The apparent deceleration is seen in both jets with decreasing apparent speed of the knots along the jets. The X-ray data

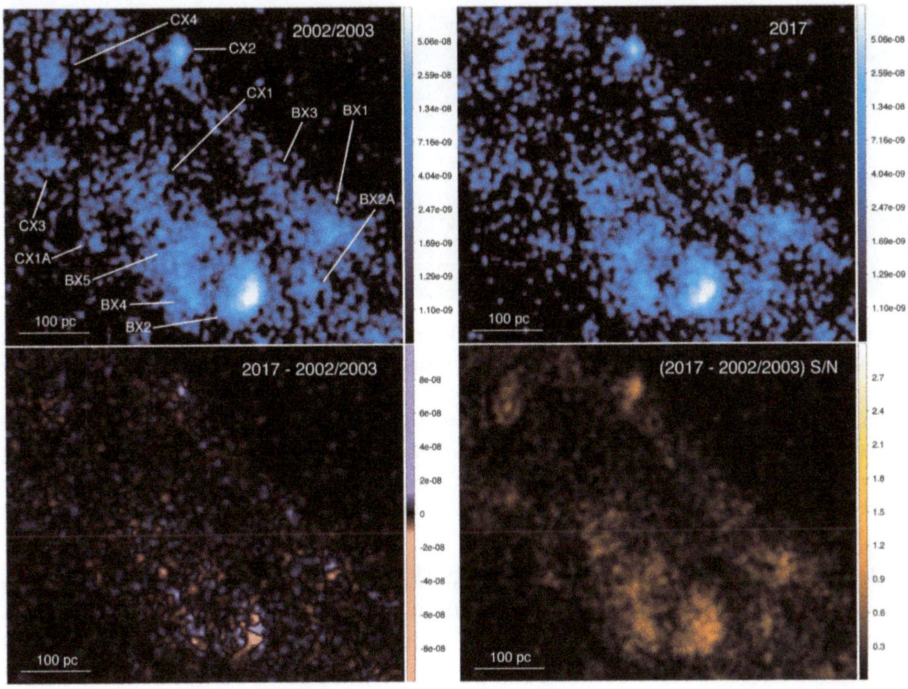


**Figure 8.14.** Centaurus A proper motions. *Chandra* ACIS-S images at 0.9–2.0 keV (upper panels), difference map (lower-left panel), and S/N map (lower-right panel) are shown of the Knot A complex in the Centaurus A jet. The images are binned on a scale of $0.123''$ pix$^{-1}$ and smoothed with a 3 pixel Gaussian. The exposure-corrected images are in units of (photon cm$^{-2}$ s$^{-1}$). In the difference map, the red regions are areas that are brighter in the 2002/2003 data set, while the blue regions are brighter in the 2017 data. Reproduced from Snios et al. (2019c) 

provide strong constraints on the equipartition magnetic fields and acceleration processes in these two jets (see discussion in Snios et al. 2019b, 2019c).

Only a small number of quasar jets was observed more than once with *Chandra*, and variability of the X-ray knot located at a large distance from the nucleus was detected only in the Pictor A jet shown in Figure 8.15 (Marshall et al. 2010; Hardcastle et al. 2016), where the X-ray knot disappeared between 2000 and 2002. Pictor A is an FR II radio galaxy with a projected jet length of about 150 kpc. An X-ray flare of the knot located at $\sim 35''$ distance from the core is a surprise. Given the timescales of the synchrotron losses, the size of the flaring region must be significantly smaller than the knot and characterized by a much larger magnetic field than the average over the entire jet.

### 8.5.4 The Impact of Jets on the ISM

SMBHs are driving the evolution of large radio structures. Longstanding questions about the origin and nature of large radio sources, the conditions required for jet formation, and the nature of the feedback process could be addressed by studies of compact radio sources with their radio structures (i.e., core, jets, lobes) contained

**Figure 8.15.** A long X-ray jet of Pictor A, an FR II radio galaxy at $z = 0.035$. Left: the composite VLA radio (red) and *Chandra* X-ray images (blue). Center: X-ray image of a long jet and hotspots, which diffuse X-ray gas. Right: radio structure of Pictor A with a strong core, lobes, and hotspots (Hardcastle et al. 2016; http://chandra.harvard.edu/photo/2016/pictora/). X-ray courtesy of NASA/CXC/Univ of Hertfordshire/M. Hardcastle et al.; Radio courtesy of CSIRO/ATNF/ATCA. The interactive figure shows separately the X-ray emission (blue), radio emission (red), and the composite image of Pictor A, as shown in the figure. Additional materials available online at http://iopscience.iop.org/book/978-0-7503-2163-1.

within the central (<1 kpc) regions of the host galaxy (see O'Dea 1998 for a review). At the early evolution phase, the radio source delivers the energy to the gas directly responsible for black hole feeding and influences the supply of matter to the accretion flow (Wagner et al. 2012, 2016). The nature of compact radio sources has been greatly discussed in the context of the "frustrated" (confined by a dense medium) and "evolutionary" (young) models. The current data suggest that they are both frustrated, and young (and perhaps "intermittent", Czerny et al. 2009). We can probe the physical conditions during this initial phase of the radio source evolution with X-ray observations (Heinz et al. 1998). However, the compact radio sources are faint, and only a few sources were detected in X-rays before *Chandra*.

During the past two decades, *Chandra* performed the first X-ray observations, forming the first X-ray sample of young radio sources (see Siemiginowska 2009 for a review). *Chandra* results show that these sources are quite heterogeneous in their X-ray properties and also indicate that they reside in diverse environments, with the X-ray spectra showing strong absorption ($N_H > 10^{23}\,cm^{-2}$) present in the most compact sources (i.e., compact symmetric objects with the radio source linear size <50 pc; see Figure 8.16). This result suggests that dense medium may be able to confine radio jets, or that dense medium is linked to the formation of more powerful jets (Siemiginowska et al. 2016; Sobolewska et al. 2019). Larger samples of compact radio sources observed in X-rays are necessary for statistically robust studies.

Signatures of the jet interaction with the environment on galaxy scales (so larger than >1 kpc) could be traced in several deep *Chandra* observations of nearby radio galaxies (see also Chapter 7). Figure 8.17 displays an example of complex morphology in a nearby radio galaxy, 4C+29.30 at $z = 0.0647$ (Siemiginowska et al. 2012) with a prominent one-sided radio jet and lobes located ~30 kpc from the SMBH. The *Chandra* X-ray image shows an X-shaped structure with the filamentary X-ray emission extending beyond the radio source, a pronounced center, and two hotspots at the jet termination regions.

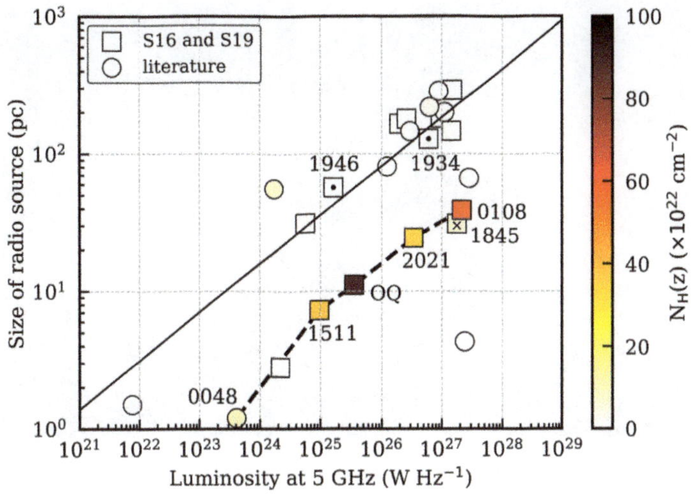


**Figure 8.16.** Luminosity at 5 GHz versus radio source size for a sample of compact radio sources observed with *Chandra* and/or *XMM-Newton*. Squares mark radio sources with kinematic age measurements and X-rays (Siemiginowska et al. 2016; Sobolewska et al. 2019, hereafter S16 and S19); circles mark other young sources from literature. Color coding represents the intrinsic equivalent hydrogen column density measured from the X-ray spectra. Solid line marks the relation from Tremblay et al. (2016). Dashed line connects sources with intrinsic $N_H > 10^{23}$ cm$^{-2}$. Figure adapted from Sobolewska et al. (2019) 

The spectral analysis of the brightest X-ray features indicates a mixture of thermal and nonthermal emission components, characterized by a variety of temperatures and spectral slopes, and a possible difference in metal abundances across the Galaxy. A significant fraction (~10%) of the jet energy ($E_{jet} \sim 10^{56}$ erg) goes into heating the surrounding gas ($kT \sim$0.1 keV) via weak shocks (Mach number ~1.6). Given the total kinetic energy of the line-emitting clouds ($E_{kin} \sim 10^{54}$ erg), only a small amount of jet power is needed to accelerate clouds. The nucleus is luminous in X-rays (~$10^{44}$ erg s$^{-1}$) and heavily obscured by infalling matter (Sobolewska et al. 2012). This infall may be related to feeding the nucleus and triggering the jet activity. The *Chandra* results for this galaxy support the idea of feedback between the central SMBH and the environment, which is critical to the evolution of galaxies. This radio galaxy is also an example of intermittent jet activity detected in radio structures (Jamrozy et al. 2007), but X-rays from the relic radio source were not detected by *Chandra*.

Shocks can form sharp narrow features in the X-ray surface brightness as the physical properties of the emitting gas drastically change across the discontinuity. They have been detected by *Chandra* in nearby galaxies. Deep *Chandra* images of the Centaurus A radio galaxy indicated that particle acceleration is taking place along the jet and at the sites of the shocks (Hardcastle et al. 2007; Croston et al. 2009). Signatures of direct shock heating have also been observed in other nearby radio galaxies (e.g., Croston et al. 2007; Massaro et al. 2009; Hardcastle et al. 2010,

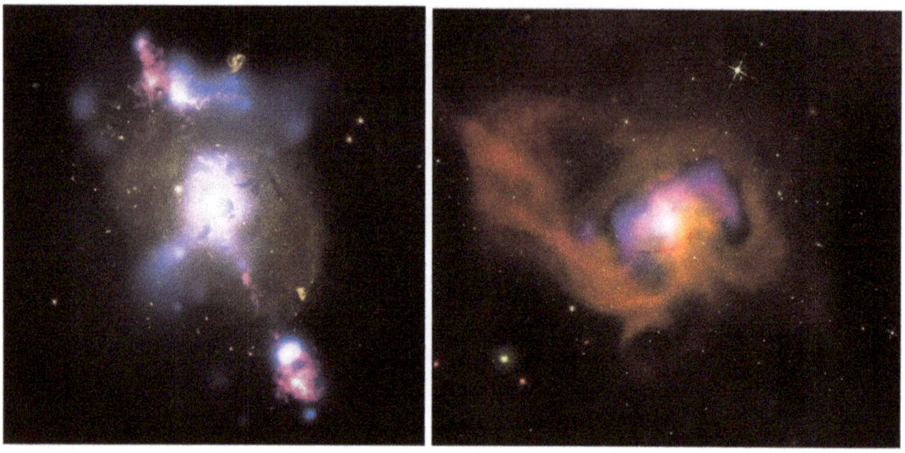

**Figure 8.17.** Left: the multiwavelength view of a radio galaxy 4C+29.30. The radio emission (pink) comes from two jets extending beyond the optical image of the galaxy (yellow), due to stars. The X-rays (blue) trace the location of hot gas: surrounding the central SMBH, hotspots at the end of the jets, and diffuse emission associated with the extended emission-line regions (Siemiginowska et al. 2012; Sobolewska et al. 2012; http://chandra.si.edu/photo/2013/4c2930/). X-ray courtesy of NASA/CXC/SAO/A.Siemiginowska et al; Optical courtesy of NASA/STScI; Radio courtesy of NSF/NRAO/VLA. Right: a composite image of the Centaurus galaxy cluster dominated by the central elliptical galaxy NGC 4696. *Chandra* X-rays (red) reveal the hot gas in the cluster, and JVLA radio (blue) shows high-energy particles produced by the black-hole-powered jets filling up the cavity in X-ray-emitting gas; optical *HST* image (green) shows galaxies in the cluster as well as galaxies and stars outside the cluster (Sanders et al. 2016; http://chandra.si.edu/photo/2017/ngc4696/). X-ray courtesy of NASA/CXC/MPE/J.Sanders et al.; Optical courtesy of NASA/STScI; Radio courtesy of NSF/NRAO/VLA. The 4C+29.30 interactive figure shows separately the X-ray emission (blue), optical emission (yellow), radio emission (pink), and the composite image of 4C+29.30, as shown in the left hand panel. The NGC4696 interactive figure shows separately the X-ray emission (red), optical emission (green), radio emission (blue), and the composite image of NGC4696, as shown in the right hand panel. Additional materials available online at http://iopscience.iop.org/book/978-0-7503-2163-1.

2012; Maselli et al. 2018). Such energy transfer between the jet and ISM is an important component of the feedback between the SMBH and the host galaxy.

The history of interactions between the jet and the large-scale environment in clusters of galaxies has been also imprinted onto the morphology of X-ray clusters (see McNamara & Nulsen 2007 for a review). The multiple cavities and presence of ripples in deep *Chandra* images of, for example, nearby Virgo (Forman et al. 2005, 2007) or Perseus (Fabian et al. 2000, 2003) clusters suggest many episodes of jet activity. Thus, the intermittent behavior of an SMBH can be traced with the high-resolution images of X-ray clusters. In addition, the total jet power in each episode can be measured using the X-ray cavities. The nature and triggering of the intermittent jet behavior have not been fully determined and may be related to the accretion flow, larger scale feedback, and/or mergers.

Cygnus A ($z$ = 0.0562), the powerful nearby FR II radio galaxy located in a galaxy cluster (Owen et al. 1997), displayed in Figure 8.18, has been extensively studied at all wavebands (see Carilli & Barthel 1996 for a review). It provided a strong case for theoretical models of radio source evolution (Scheuer 1974;

**Figure 8.18.** A composite image of Cygnus A, a powerful FR II radio galaxy in a galaxy cluster: X-rays from *Chandra* combined with an optical view from the *Hubble Space Telescope*. The main figure marks the location of the SMBH, the jets, and hotspots: E—jet "reflected" off the ISM; D—secondary. The image in the insert displays the fine X-ray structures of the jet–ISM interaction at the hotspots (http://chandra.harvard.edu/photo/2019/cyga/) X-ray courtesy NASA/CXC/Columbia Univ./A. Johnson et al.; Optical courtesy of NASA/STScI.

Blandford & Rees 1974; Begelman 1985; Begelman & Cioffi 1989; Reynolds et al. 2001; Clarke et al. 1997). The impact of the radio jet and lobes of Cygnus A on the X-ray cluster gas, in the form of cavities, enhanced X-ray emission, at the edges of the radio lobes, and hotspots, have been studied using *ROSAT* High Resolution Imager X-ray observations (Carilli et al. 1994). *Chandra* high-resolution X-ray images of Cygnus A revealed an X-ray jet, double hotspots, X-ray cavities, and overall complex X-ray morphology of the cluster gas, indicating significant impact of the jet on the intercluster medium (Wilson et al. 2000; Smith et al. 2002; Wilson et al. 2006; Duffy et al. 2018). Twisted filaments, jets, hotspots, and "bubbles" are visible in exquisite detail in the deep *Chandra* image displayed in Figure 8.18. The cocoon shock[3] with Mach numbers between 1.18–1.66 marks the volume impacted by the radio source (Snios et al. 2018). In particular, the double hotspots at opposite ends of the jet (Harris et al. 1994; Carilli et al. 1994) coincident with the radio hotspots have been resolved by *Chandra* (Figure 8.18 insert). There have been

[3] The edge of the X-ray emission.

several theoretical explanations for the origin of multiple hotspots in radio galaxies (Scheuer 1982; Williams & Gull 1985; Carilli & Barthel 1996; Steenbrugge et al. 2008; Pyrzas et al. 2015). The deep *Chandra* observations of Cygnus A may have provided a possible answer: a jet is being deflected at a primary hotspot, then travels onward to deposit the energy at the secondary hotspot (Nulsen et al. 2019; B. Snios et al. 2019, in preparation). This is supported by several arguments: the secondary hotspots of Cygnus A are more luminous (Stawarz et al. 2007; Pyrzas et al. 2015), the light travel time and particle lifetimes do not agree with the recurrent jet models, and an X-ray "hole" in the vicinity of hotspot E has a natural explanation in the "deflected" jet model. The multiple shocks present at the primary hotspot resulting in jet deflection instead of jet termination can potentially change our understanding of both the radio source evolution and its impact on the environment.

## 8.6 Finding Supermassive Black Holes in X-Ray Surveys

X-ray surveys enable the detailed study of the extragalactic X-ray source population (see Brandt & Hasinger 2005 for a review). Such surveys provide the most efficient selection of AGN, relatively unbiased by moderate-to-high obscuration, and not easily contaminated by non-nuclear emission, mainly due to star formation processes, which is far less significant than in optical and infrared surveys.

### 8.6.1 X-Ray Surveys Overview

Over the last 20 years, the *Chandra* and *XMM-Newton* satellites observed both deep and wide fields of the extragalactic sky in the 0.5–8 keV energy range (see Brandt & Alexander 2015 for a review). These surveys follow a "wedding cake" strategy, tiered by decreasing area and increasing depth (see Figure 8.19), such that the total exposure time dedicated to each field is comparable. The bottom tiers are wide and shallow (short exposures) surveys that are designed to cover a large volume of the universe and thus find rare sources, i.e., high-luminosity and/or high-redshift AGN. At the opposite end are narrow/ultra-deep (very long exposures) surveys that cover a small area to detect very faint sources. These surveys have detected AGN at $z \sim 5$ to very faint limits and nonactive galaxies (with X-ray emission dominated by star-forming processes or hot gas) up to $z \sim 2$, but have statistically small samples of sources at any redshift because of the small volumes covered. To trace the full population of AGN at all redshifts and luminosities, it is necessary to sample the full parameter space between the ultra-deep and ultra-wide surveys.

Thanks to its higher spatial resolution and lower instrumental background, *Chandra* surveys can reach the faintest sensitivity limits and thus fill in the lower-left part of the surveys' sensitivity plot (see Figure 8.19). So far, *Chandra* has performed extragalactic (contiguous and noncontiguous) surveys totaling ~30 Ms of exposure time, including (sorted from smaller to wider area): the *Chandra* Deep Field South (CDFS,[4] Giacconi et al. 2002; Luo et al. 2008; Xue et al. 2011; Luo et al. 2017), the Extended *Chandra* Deep Field South (Lehmer et al. 2005), the *Chandra*

[4] First CDFS press release: http://chandra.harvard.edu/press/01_releases/press_031301.html.

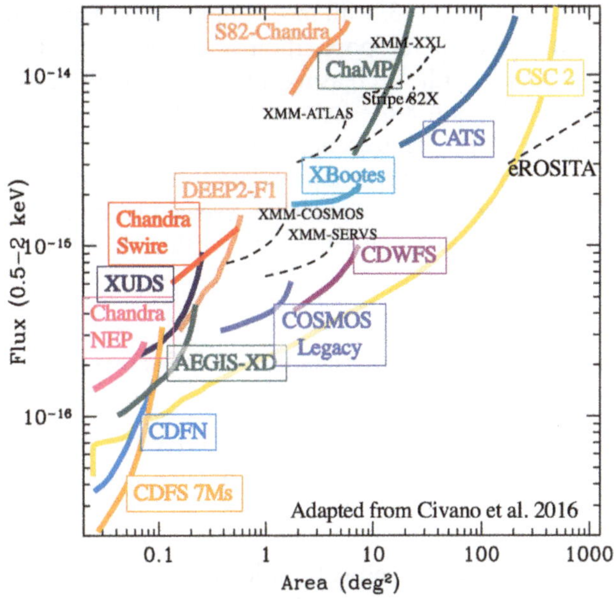


**Figure 8.19.** The sensitivity plot: area–flux curves for the *Chandra* (color solid) and *XMM-Newton* (black dashed) surveys. Each survey is represented using their sensitivity curve starting from the flux corresponding to 80% of the maximum area for that survey to the flux corresponding to the 20% of the total area. Reproduced from Civano et al. (2016) © 2016. The American Astronomical Society. All rights reserved. In the interactive figure, the labels link to the webpage for each survey, including information, images, and references to relevant publications. Additional materials available online at http://iopscience.iop.org/book/978-0-7503-2163-1.

Deep Field North[5] (Alexander et al. 2003), the AEGIS-X Deep survey (Nandra et al. 2015), the survey of the DEEP2 Galaxy Redshift Survey Fields (Goulding et al. 2012), SEXSI (Harrison et al. 2003), the *Chandra* survey of the UKIDSS field (Kocevski et al. 2018), the SWIRE/*Chandra* survey (Wilkes et al. 2009), the *Chandra* COSMOS Legacy Survey[6] (Elvis et al. 2009; Civano et al. 2016), the NDWFS XBootes (Murray et al. 2005), the CDWFS (A. Masini et al., 2019, in preparation), the ChaMP survey (Kim et al. 2007a, 2007b), the Cluster AGN Topography Survey (R. Canning et al., 2019, in preparation), the Stripe 82-X (LaMassa et al. 2013). Finally, the *Chandra* Source Catalog[7] (CSC 2.0), which now includes 15 years of data, is the *Chandra* largest area survey, covering both galactic and extragalactic fields (Evans et al. 2010; I. N. Evans et al., 2019, in preparation; see Chapter 2, Section 2.4 on CSC 2.0).

These *Chandra* surveys produced catalogs of X-ray-emitting sources, with a total of ~25,000 sources (excluding the CSC 2.0), mostly AGN. The extensive multi-wavelength spectroscopic and photometric data available in the same fields allow for studies of AGN demography. A number of key discoveries were made: (1) AGN

---

[5] First CDFN press release: http://chandra.harvard.edu/photo/2003/goods/.
[6] Mosaic of the COSMOS field from: http://chandra.harvard.edu/press/18_releases/press_080918.html.
[7] http://cxc.cfa.harvard.edu/csc2/.

number densities peak at $z \sim 2 - 3$, with more luminous sources dominant earlier in cosmic time and with most of the population being obscured; (2) the fraction of obscured AGN increases with decreasing luminosity with most sources being optically obscured (see Section 8.6.2 for a definition) below $L_X \sim 10^{43}$ erg s$^{-1}$; (3) correlations exist between SMBH accretion and galaxy properties, particularly with stellar/dark matter halo mass and (perhaps indirectly) star formation, although due to AGN stochasticity, these trends can only be revealed by large statistical samples; and (4) the space density of X-ray-selected high-redshift AGN is much higher than that measured from optical surveys alone.

### 8.6.2 Populations Studies: The Discoveries in X-Ray Surveys

*Source Number Counts*

Based on X-ray catalogs alone, it is possible to derive the number of sources per square degree (named number counts) detected above a given flux, the log $N$–log $S$ relation, which provides a rough estimate of the source space density and information on the cosmic population required for population synthesis models. The number counts also serve as a self-calibration tool, which validates the detection methods applied by different surveys. In Figure 8.20, we report a compilation of number counts for several *Chandra* surveys. The excellent statistics available in the surveys result in small uncertainties on the number counts over a broad range of fluxes, down to the very faint flux limits of the CDFS ($1.9 \times 10^{-17}$ erg cm$^{-2}$ s$^{-1}$, $6.4 \times 10^{-18}$ erg cm$^{-2}$ s$^{-1}$, and $2.7 \times 10^{-17}$ erg cm$^{-2}$ s$^{-1}$ in the respective energy ranges of 0.5–7.0 keV, 0.5–2.0 keV, and 2–7 keV in the central 1 arcmin$^2$ region). Overall, the shape of the measured log $N$–log $S$ is best represented by a double power law with a break at $\sim 10^{-15}$ erg cm$^{-2}$ s$^{-1}$ (see Luo et al. 2017 for the most recent modeling

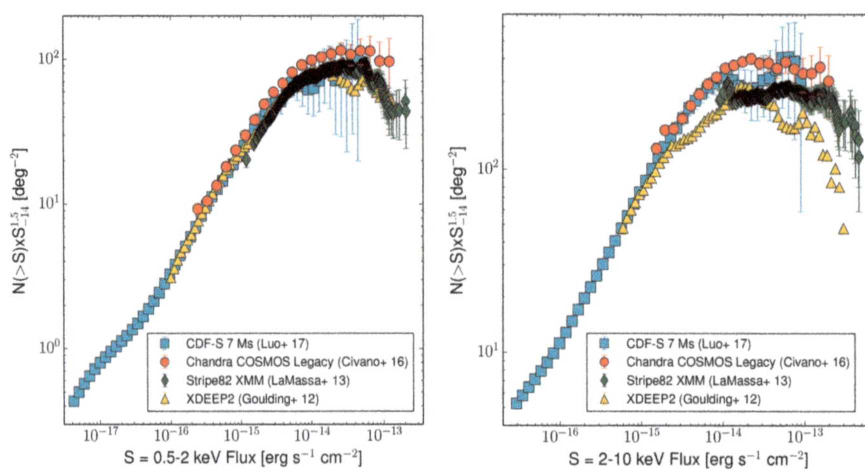


**Figure 8.20.** Euclidean-normalized log $N$–log $S$ curves in 0.5–2 keV (left) and 2–10 keV (right) bands for the surveys labeled in the figure, covering from the brightest to the faintest fluxes. The source number counts are multiplied by $(S/10^{14})^{1.5}$ to highlight the deviations from Euclidean behavior. Adapted from Civano et al. (2016) © 2016. The American Astronomical Society. All rights reserved.

of log $N$–log $S$). By including further information on source properties, it is possible to use log $N$–log $S$ to inform models of the population synthesis. For example, deriving the log $N$–log $S$ relation of the obscured and unobscured sources (by using hardness ratios or the intrinsic column density measured from X-ray spectroscopy) would inform on the relative fraction at each X-ray flux. In general, the population of unobscured AGN dominates at fluxes brighter than $\sim 10^{-16}$ erg cm$^{-2}$ s$^{-1}$ in the 0.5–2 keV band, while in the 2–10 keV band the contribution of obscured and unobscured AGN is roughly equal at all flux levels (see Civano et al. 2016). The optical classification of X-ray sources in surveys allows us to distinguish the relative contribution to the total number counts from AGN and sources for which the X-ray emission is not nuclear but produced by either hot gas or star formation (normal galaxies). As observed in the deep CDFS 7 Ms survey, the AGN contribution to the number counts is dominant at fluxes above $\sim 6.0 \times 10^{-18}$ erg cm$^{-2}$ s$^{-1}$, with the contribution of normal galaxies increasing at lower fluxes and overtaking the AGN number counts at this value (see Figure 31 in Luo et al. 2017).

*Stellar Content in Extragalactic Surveys*
A small percentage (3%–5%) of sources in the *Chandra* extragalactic surveys are foreground stars in the Milky Way. The detection of X-rays from stars in the Galactic halo allows the study of low-metallicity stars, probing the close binary populations. The number counts of the stellar component in these surveys have been presented in several works, including a detailed investigation of the stellar properties (Feigelson et al. 2004; Covey et al. 2008; Wright et al. 2010; Luo et al. 2017).

*Optical Classification of AGN in Surveys*
Historically, various classes of X-ray emitters have been characterized by different values of the X-ray to optical flux ratio, which provides an initial indication of the source classification (Maccacaro et al. 1988). This approach was used in the early years of the *Chandra* mission for the first X-ray surveys (e.g., Giacconi et al. 2002; Alexander et al. 2003; Civano et al. 2005; see Figure 8.21) and also allowed the classification of peculiar sources like X-ray-bright optically normal galaxies (XBONGs; see e.g., Civano et al. 2007). More recently, thanks to the advent of large aperture optical telescope with multiobject spectrographs and to the multiwavelength data available in most of the fields, source classification, and redshift measurements are done using both optical/NIR spectra and broadband spectral energy distributions (SEDs).

Traditionally, AGN classification was based on the optical spectroscopy allowing sources with broad and narrow emission lines, and emission lines ratios (e.g., BPT diagrams; Baldwin et al. 1981), to be distinguished. Optical spectroscopic follow-up observations of X-ray sources can determine redshift to 1% accuracy for the brightest ($i < 22$ mag) AGN, typically achieving a completeness of 50%–60% for medium-faint ($\sim 10^{-16}$ erg cm$^{-2}$ s$^{-1}$ in the 0.5–2 keV band) X-ray surveys. Moreover, infrared spectroscopic campaigns with both ground- (e.g., Subaru, Keck, and VLT) and space-based telescopes (e.g., 3D-HST survey; Momcheva et al. 2016) were designed to target the highest redshift sources. However, while extremely good for redshift measurements, the optical spectra of the AGN in surveys are often

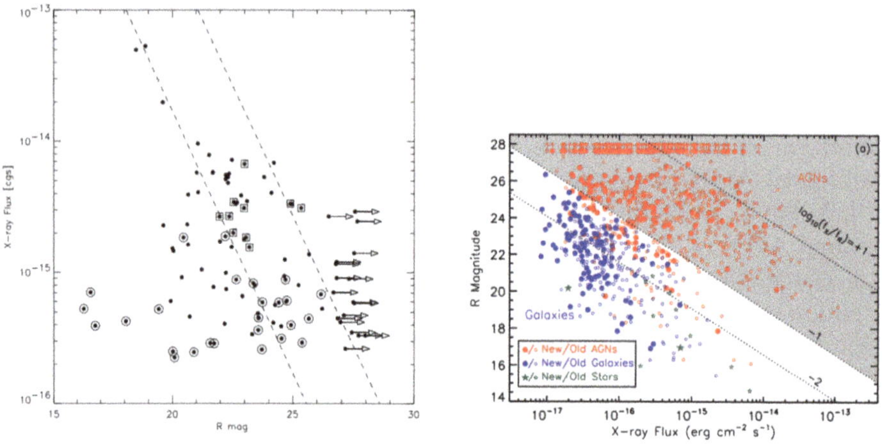


**Figure 8.21.** The X-ray to optical flux ratio plot for the CDFS survey for the first 120 ks data set as presented by Giacconi et al. (2001) © 2001. The American Astronomical Society. All rights reserved, and the same plot for the 7 Ms survey as in Luo et al. (2017) © 2017. The American Astronomical Society. All rights reserved. In both cases, the *R*-band magnitude is used.

characterized by low signal-to-noise ratios, and thus not accurate for using emission-line diagnostic diagrams.

With the rich multiwavelength data available, SEDs can be used for the source classification to derive the properties of the host galaxy (e.g., mass and star formation rate) as well as the relative contribution at all wavelengths of the nuclear and the stellar component (e.g., see Suh et al. 2017). Moreover, the SEDs can be used to measure the source redshift with an accuracy as high as 5% (see Salvato et al. 2018 for a review) down to very faint magnitudes (~24 mag in the *i*-band).

Overall, at the bright fluxes/luminosities ($>10^{-14}$ erg cm$^{-2}$ s$^{-1}$ in 0.5–2 keV or $L_X > 10^{44}$ erg s$^{-1}$ at $z \sim 1$), the source population is dominated by unobscured AGN (see, e.g., Stripe 82 and ChaMP; LaMassa et al. 2013; Trichas et al. 2012). The fraction of unobscured sources decreases in favor of sources with SED best fitted by obscured AGN templates (e.g., optical/UV power law with reddening applied) and/or normal galaxies templates, which are the most common in medium-deep and deep surveys (see, e.g., Figure 9 in Civano et al. 2012; also Figure 8.22). The X-ray luminosities of the galaxy-dominated sources are fairly high ($>10^{42}$ erg s$^{-1}$), consistent with being produced by nuclear emission and not just hot gas and/or star formation, suggesting their nuclear emission in the optical band is either diluted or mildly obscured. This difference between optical classification (obscured, not obscured, or galaxy dominated) and X-ray classification (obscured and not obscured) is, in general, true at all luminosities, and can be observed also when making use of the spectroscopic classification from the standard emission-line ratio diagnostics (see, e.g., Bongiorno et al. 2012 and discussion therein).

*Obscuration and Evolution of Obscured Fraction*

The main goals of directly measuring the intrinsic equivalent hydrogen column density, $N_H$, for AGN in X-ray surveys are to derive the obscuration distribution as

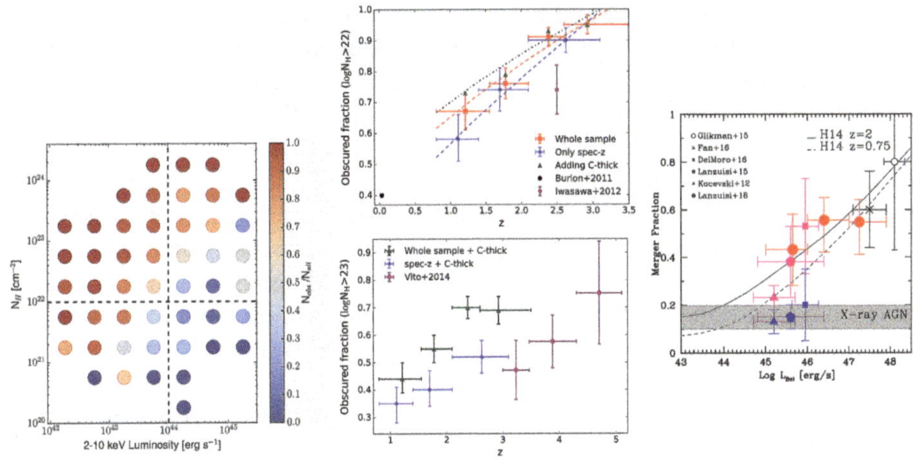


**Figure 8.22.** Left: measured rest-frame column density as a function of 2–10 keV absorption-corrected luminosity for the *Chandra* COSMOS Legacy sample from Marchesi et al. (2016a) © 2016. The American Astronomical Society. All rights reserved. The color map shows the ratio between the number of optical type 2/obscured AGN (N2) against all sources (Nall), in each bin of column density $N_H$, redshift, and luminosity (N2/Nall = 1 is plotted in red, N2/Nall = 0 in blue). The dashed lines mark log $N_H$ = 22 and X-ray luminosity $10^{44}$ erg cm$^{-2}$ s$^{-1}$ to divide obscured and luminous sources. Middle: fraction of AGN with log $N_H > 22$ (top) and log $N_H > 23$ (bottom) in redshift and luminosity bins ($0.8 < z < 3.5$, $43.5 < \log L < 44.2$) from CDFS spectral analysis from Liu et al. (2017) © 2017. The American Astronomical Society. All rights reserved. The best-fit lines are plotted in the top panel. Data from the entire complete sample (red) and the spectroscopic-*z* subsample (blue) and the Compton-thick AGN (green) are reported in the top panel. Comparison samples are reported as labeled in the figure from Burlon et al. (2011), Iwasawa et al. (2012), and Vito et al. (2014). Right: fraction of merger/disturbed morphology for several samples of CT (Compton-thick) and non-CT AGN at different bolometric luminosities. The red circles show the measurements for the *Chandra* COSMOS Legacy sample from Lanzuisi et al. (2018), by permission of Oxford University Press on behalf of the Royal Astronomical Society. The magenta points show fractions from different literature for CT AGN samples (Glikman et al. 2015; Fan et al. 2016; Lanzuisi et al. 2015; Kocevski et al. 2012; Del Moro et al. 2016) as labeled in the figure. Blue points are the parent sample of non-CT AGN in each of these studies. The green thick (dashed) line shows the prediction of the Hickox et al. (2014) model.

a function of redshift and luminosity, to determine the periods of obscured black hole growth and how this growth phase is related to the AGN trigerring and fueling mechanisms, and also to inform population synthesis models.

Obscuration can be estimated with different techniques, such as spectral analysis, hardness, or flux ratio analysis (i.e., the ratio between counts or fluxes in different bands). Unfortunately, the hardness ratio alone cannot be used as a reliable proxy of obscuration because it is degenerate with redshift, and it also depends on the *Chandra* effective area, which is changing over time.

The most reliable method is based on the X-ray spectral analysis; however, the majority of survey sources have typically fewer than 100 net counts[8] (see, e.g.,

[8] The narrow *Chandra* PSF and the low background allow the detection of sources with five total counts at the *Chandra* aimpoint.

Figure 9 in Civano et al. 2016 and Figure 24 in Luo et al. 2017), while a minimum of ~70–80 counts in the 0.5–7 keV energy band is required for any meaningful spectral constraint. This greatly limits the number of AGN for which spectral analysis can be performed (e.g., only ~270 sources were analyzed using the CDFS 7 Ms data in Liu et al. 2017 and 923 for the COSMOS field in Marchesi et al. 2016a). Based on spectral analysis of AGN in the COSMOS and CDFS surveys, the median power-law photon index is $\Gamma \sim 1.7 \pm 0.02$. This value varies slightly when selecting sources based on their unobscured and obscured optical classification with $\Gamma \sim 1.75$ ($\sigma = 0.31$) and $\Gamma = 1.61$ ($\sigma = 0.47$), respectively (Marchesi et al. 2016a). No significant correlation is found between the photon index and luminosity or redshift, while interesting correlations between obscuration and host galaxy properties were found in several samples (see Section 8.6.3).

According to the traditional unification-by-orientation schemes (see Section 8.4), obscured and unobscured AGN are postulated to be intrinsically the same objects seen from different angles with respect to a dusty "torus" (Antonucci & Miller 1985; Barthel 1989; Antonucci 1993; Urry & Padovani 1995; Netzer 2015). Under such unification schemes, there should be no difference between optical and X-ray classification, and also no dependence of the fraction of obscured AGN on intrinsic luminosity and/or redshift. On the contrary, AGN samples from the *Chandra* surveys are revealing a different picture.

First, while at high luminosities optical and X-ray classification are generally in agreement (see Figure 8.22, left), at lower luminosities a significant number of optically obscured AGN are classified as unobscured in the X-rays, with $N_H < 10^{22}$ cm$^{-2}$. This number strongly decreases with increasing (2–10 keV) luminosity and can be explained by the lack of broad emission lines in intrinsically unobscured, low-accretion AGN (Trump et al. 2011). Similarly, some optically classified unobscured sources have high X-ray obscuration ($N_H > 10^{22}$ cm$^{-2}$) and could be either broad absorption-line quasars or possibly sources with dust-free gas surrounding the inner part of the nuclei, therefore causing obscuration in the X-rays and not in the optical band (Risaliti et al. 2002; Maiolino et al. 2010; Fiore et al. 2012; Merloni et al. 2014).

Second, a significant decrease in the fraction of obscured AGN with increasing X-ray luminosity has been found in extragalactic surveys prior to *Chandra* (Lawrence & Elvis 1982; Steffen et al. 2003; Ueda et al. 2003) and then confirmed with both *Chandra* and *XMM-Newton* surveys (Hasinger 2008; Bongiorno et al. 2012; Merloni et al. 2014; Marchesi et al. 2016a; Liu et al. 2017), breaking down the simplest unification schemes.

The evolution of the obscured fraction with redshift was for a while a matter of debate, with claims favoring such evolution (La Franca et al. 2005; Treister & Urry 2006; Hasinger 2008), and as well as opposite conclusions of no significant evolution (Ueda et al. 2003; Tozzi et al. 2006; Gilli et al. 2007). When the selection biases and the k-correction are accounted for, the obscured fraction increases with redshift only for the most-luminous sources (Merloni et al. 2014; Buchner et al. 2015; Aird et al. 2015). However, combining the deepest CDFS 7 Ms data with the *Chandra* COSMOS Legacy survey, Liu et al. (2017) find a significant increase of the obscured fraction with redshift (up to $z = 3$) at any luminosity (see Figure 8.22, middle). No increase is instead measured at redshift $z > 3$ (see Figure 8.25, right; Vito et al. 2018).

*Compton-thick AGN*

As predicted by cosmic X-ray background synthesis models (see Section 8.6.7), a large fraction ( >60%) of AGN must be obscured. Constraining the fraction of the most obscured, Compton-thick (CT) AGN ($N_H > 10^{24}\ cm^{-2}$) beyond the local universe has been a challenge. While the shape of the X-ray CT AGN spectrum makes their detection favorable for high-redshift AGN, as the Compton reflection hump at 20–30 keV and the iron emission line at 6.4 keV become observable in the soft-energy band of *Chandra*; nonetheless, CT AGN tend to be very faint. Thus, modeling their spectrum is challenging because of the low number of counts available in surveys, and they can also be easily misclassified (see Castelló-Mor et al. 2013; Liu et al. 2017 and the discussion therein).

Thanks to a combination of new, dedicated models and advanced analysis techniques specifically developed for low-count statistics, recent studies have found a number of CT AGN at high redshift in deep X-ray surveys (Brightman & Ueda 2012; Georgantopoulos et al. 2013; Tozzi et al. 2006; Buchner et al. 2015; Brightman et al. 2014; Lanzuisi et al. 2013; Marchesi et al. 2016a; Liu et al. 2017; Lanzuisi et al. 2018).

The main conclusions made by combining several methods and AGN samples are as follows: the fraction of CT AGN increases with redshift, from approximately 20% to 50% at $z = 2$–$3$ (Figure 8.22 middle-bottom; Liu et al. 2017; Lanzuisi et al. 2018), in agreement with the CXRB synthesis models (see e.g., Ananna et al. 2019). These results are consistent with the fraction of CT AGN of 20%–25% at $z = 1$–$2$ estimated from *Chandra* observations of the low-frequency radio-selected sample of 3CRR sources (Wilkes et al. 2013).

CT AGNs seem to be preferentially hosted by galaxies with signs of recent merging (see Figure 8.22, right), in that there is an increase in merger fraction in CT AGN with respect to the Compton-thin AGN at the same luminosity (Kocevski et al. 2012; Lanzuisi et al. 2015; Del Moro et al. 2016; Lanzuisi et al. 2018). The survey results also support AGN–galaxy coevolutionary models according to which galaxy mergers play a prominent role in triggering the most-luminous and obscured AGN at $z \sim 2$, and the high obscuration is associated with the merger itself (Blecha et al. 2018).

### 8.6.3 SMBH and Host Galaxy Coevolution

One of the outstanding issues for understanding the formation and evolution of galaxies is how the presence of an SMBH affects its host galaxy. While correlations are found between host galaxy properties and SMBH mass in the local universe, it is not clear whether these correlations have been in place since the early universe, and to date, the evolution of scaling relations remains a matter of debate.

According to the generally accepted scenario of galaxy evolution, AGN activity can suppress star formation either by heating or removing the cold gas in the host galaxies (a.k.a. AGN feedback; e.g., Silk & Rees 1998; King 2003; Di Matteo et al. 2005; Hopkins et al. 2008). The samples of X-ray AGN in surveys, covering a wide range of luminosities, have provided a great test data set to understand how the AGN fueling mechanism is affecting star-forming activities. The low ratio between the nuclear emission and the stellar emission in the optical and UV band allows clean measurements of the host galaxy properties for most X-ray AGN. Contradictory results have been reported in the literature, finding either equivalent or enhanced star formation compared to normal star-forming galaxies (Silverman et al. 2009; Mullaney et al. 2012; Rosario et al. 2012; Santini et al. 2012; Juneau et al. 2013) or suppressed star formation (e.g., Barger et al. 2015; Mullaney et al. 2015; Riguccini et al. 2015; Shimizu et al. 2015). Similarly, contradictory results have been reported for the correlation between AGN accretion and SFR, which can give a crucial hint as to whether AGN activity can significantly enhance or quench star formation in galaxies. When measuring average properties of the population, it is important to account for varying time-scales for different processes, AGN intrinsic variability and diversity of accretion modes (Hickox et al. 2014; Volonteri et al. 2015), which can result in false correlations.

When accounting for selection biases, the star formation rates of X-ray-selected AGN are those expected for typical star-forming galaxies (Silverman et al. 2011; Suh et al. 2017), implying that the nuclear activity and star formation seem to coexist, fueling black hole accretion and star formation simultaneously.

Interestingly enough, no correlation between X-ray obscuration and star formation rate is found (Lanzuisi et al. 2017), while a correlation between X-ray obscuration and host galaxy stellar mass is found both for detections (Rodighiero et al. 2015; Marchesi et al. 2016a; Lanzuisi et al. 2017) and stacking analysis (Mezcua et al. 2016; Paggi et al. 2016; Fornasini et al. 2018). These findings are in agreement with the AGN obscuration model proposed by Buchner & Bauer (2017), in which galactic-scale obscuration depends on stellar mass and a additional nuclear obscuration is related to accretion rates.

Using the deepest *Hubble* data available on most of *Chandra* fields, it was found that most AGN host galaxies are likely to be structurally disk-dominated galaxies (Kocevski et al. 2012; Cisternas et al. 2011). The majority of AGN host galaxies do not show evidence of merger (aside from the most-luminous and obscured ones; see Section 8.6.2.5), suggesting that most AGN activity is not triggered by major mergers.

The above findings suggest that moderate-luminosity AGN are the products of modest accretion, rather than major mergers, in which case the gas accretion via secular processes trigger the AGN activity (see Figure 2 of Alexander & Hickox 2012).

### 8.6.4 X-Ray Luminosity Function

The observed luminosity function and its evolution with cosmic time is important for understanding the origin of SMBHs, their growth, and the processes that drive their accretion. The luminosity function represents the number of AGN per given luminosity range as a function of redshift, and its shape reflects the distribution of SMBH masses and accretion rates of the AGN population. As stated in Section 8.6.2.4, most AGN are obscured; therefore, understanding the distribution of obscuration is essential in deriving the true, intrinsic luminosity function. Additional factors that become important when modeling the luminosity function include classification and redshift completeness, and the Eddington bias.

The first X-ray luminosity function (XLF) was derived in the early 1990s using samples detected in the soft X-ray band dominated by unobscured (Type 1) AGN. With the advent of 2–10 keV surveys, *Chandra* and *XMM-Newton* data extended the XLF study to higher energies, allowing the population of obscured (Type 2) AGN detected in the 2–10 keV band, the dominant AGN population at low luminosities, to be accounted for.

The most recent XLF (Ueda et al. 2014; Miyaji et al. 2015; Ananna et al. 2019) compiles several thousand AGN, the majority from the *Chandra* surveys covering the luminosities of $10^{42}$–$10^{43}$ erg $s^{-1}$ up to $z = 2$, and $10^{45}$–$10^{46}$ erg $s^{-1}$ up to $z = 5$ (see Figure 8.23 from Miyaji et al. 2015). The *Chandra* surveys are critical in detecting the lowest luminosity AGN at each redshift. Based on the XLF, the AGN population is found to be strongly evolving with redshift, with the AGN space density peak located at $z = 2 - 3$ for the most-luminous sources ($L_X > 10^{45}$ erg $s^{-1}$), similarly to the optically selected quasars. The XLF peak for the low-luminosity AGN is found at lower redshift, $z = 1 - 2$, with a visible milder decline in a number of sources toward the present epoch. This anti-hierarchical evolution, termed "downsizing," also observed in galaxies (Cowie et al. 1996), is not expected by models predicting SMBH activity triggerd by hierarchical merging, but it is more in line with a differential AGN triggering mechanism for high- and low-luminosity AGN. According to this differential mechanism hypothesis, AGN may be triggered by secular processes (e.g., minor mergers, disk instabilities, bars, disks) or galaxy major mergers. Several cosmological simulations have been able to reproduce this AGN downsizing effect (e.g., Marulli et al. 2008; Fanidakis et al. 2012) by using the XLF as an important constraint.

To represent the evolving shape of the XLF, the function is modeled with a smoothed double power law that is steeper at high luminosities and flatter at low luminosities with a break point (knee) that changes as a function of redshift. This is also known as the luminosity-dependent density evolution (LDDE) model (e.g., Ueda et al. 2003, 2014; Hasinger et al. 2005; Miyaji et al. 2015; Buchner et al. 2015). Using a different approach, Aird et al. (2010, 2015) proposed a model where the shape of the luminosity function is not redshift dependent (i.e., does not change shape), but evolves in both luminosity and density. Recently, it was found that the XLF for obscured AGN has a lower break luminosity, higher normalization, and a steeper faint-end slope than the XLF of unabsorbed AGN up to $z \sim 2.5$ (Miyaji et al. 2015; Aird et al. 2015).

Integrating the XLF and converting into the total bolometric luminosity (using an appropriate correction; e.g., Hopkins et al. 2007; Lusso et al. 2012), it is possible to derive the overall AGN luminosity density. Furthermore, using the Soltan argument (Soltan 1982), it is possible to estimate the total SMBH mass density built up from

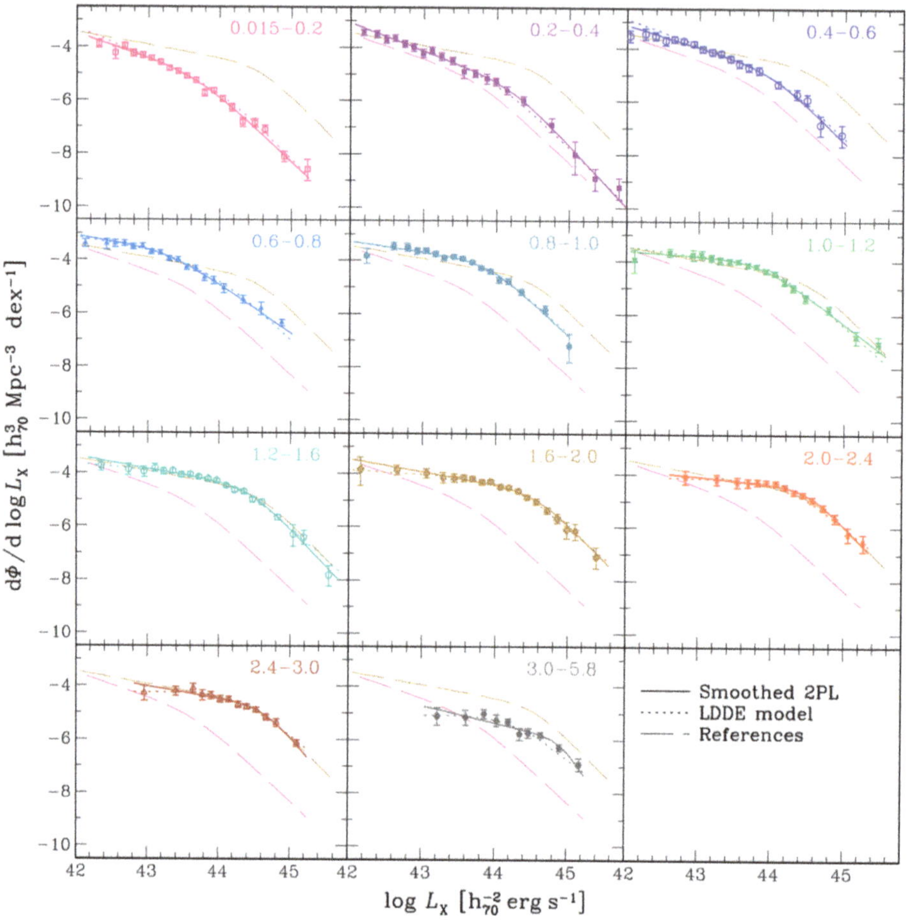


**Figure 8.23.** X-ray luminosity function in the 2–10 keV band as a function of redshift (marked in the upper corner of each panel) from Miyaji et al. (2015) © 2015. The American Astronomical Society. All rights reserved, computed including a sample of 3200 AGNs, covering six orders of magnitude in flux and drawn from mainly deep and wide *Chandra* and *XMM-Newton* surveys, but also *ASCA* and *Swift*. The solid line is the fit with a smoothed double power law, the dotted line is the luminosity-dependent density evolution model, and the dashed line is the same reference curve in each panel.

accretion (e.g., Figure 8.24, from Aird et al. 2015; see also Comastri et al. 2015 on the issue of the large local mass density that can be explained with the presence of a large population of heavily obscured AGN). Comparing with the star formation (SFR) density reveals that both SFR and SMBH accretion peak at $z \sim 2$, but have different slopes at low and high redshift. SMBH accretion might be more efficient than star formation at low-$z$. Conversely, at high redshift. SMBH accretion efficiency drops faster than SFR density (even though the slope and normalization of the SMBH accretion density is still a matter of debate; see, e.g., Madau & Haardt 2015; Ricci et al. 2015), signaling that the Galaxy build up might lead the formation

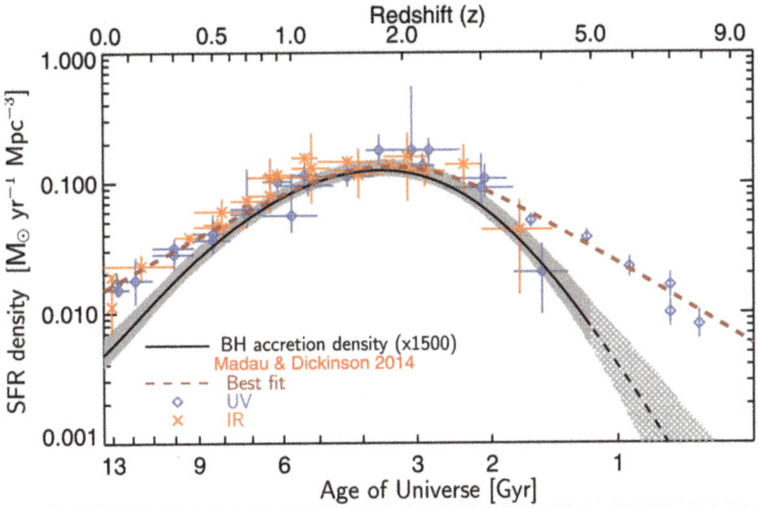


**Figure 8.24.** Total SFR density compared to the estimate of the total SMBH accretion density by Aird et al. (2015). The best overall fit (dashed dark-red line) for the evolution of the SFR density from the recent review by Madau & Dickinson (2014) along with their compilation of measurements based on rest-frame ultraviolet (blue diamonds) and infrared (orange crosses) observations. The estimate of the SMBH accretion density (solid black line) is scaled up by an arbitrary factor of 1500. Both galaxy and SMBH growth peak at $z \sim 2$. The SMBH accretion density appears to evolve more rapidly, with a stronger decline to higher redshifts. Figure is reproduced from Aird et al. (2015), by permission of Oxford University Press on behalf of the Royal Astronomical Society.

process in the early universe and precede the SMBH formation (see discussion in Volonteri 2012).

### 8.6.5 Probing Lower-luminosity Populations with Stacking Analysis

X-ray stacking techniques (for both imaging and spectroscopy) have been used with the goal of probing the average properties of the lowest luminosity AGNs, below the sensitivity limit of current surveys. Both X-ray-undetected sources (for imaging) and/or very faint sources at the detection limit (for spectroscopy) are grouped according to different multiwavelength properties, and their X-ray data are combined (and redshifted) to properly compute their average X-ray luminosities. This technique has been extremely successful, allowing studies to reach effective exposures of hundreds of millions of seconds and fluxes that are 2–3 orders of magnitude fainter than the nominal flux limits of the surveys, depending on the sample size of the stacked sources, as well as the depth of the survey and area covered.

Such studies have measured the unresolved X-ray background (see Section 8.6.7; e.g., Worsley et al. 2005, 2006; Hickox & Markevitch 2006, 2007; Cappelluti et al. 2017). Stacking of spectra has been also used to measure average spectral properties (e.g., Rosati et al. 2002; Civano et al. 2005) and the average profile of the 6.4 keV iron emission line in X-ray spectra (Brusa et al. 2005), which is found to be narrow and unchanging with redshift or luminosity.

X-ray image stacking of infrared-selected sources in *Chandra* surveys has revealed, in infrared- and near-infrared-selected galaxies, the ubiquitous presence of extremely obscured and even Compton-thick AGN not individually detected (Daddi et al. 2007; Fiore et al. 2009; Donley et al. 2012). Moreover, making use of the deepest data and largest samples available, multiple studies have put constraints on the AGN emission in normal elliptical, spiral, and dwarf galaxies beyond the local universe to $z \sim 3$ (see Chapter 7 and references therein). X-ray stacking has also been used to measure the evolution of X-ray binary populations out to $z \approx 2$ (Lehmer et al. 2016) and to study hot gas in the galaxies beyond the local universe (Paggi et al. 2016). Low-accretion and obscured AGN were discovered through X-ray stacking of galaxies spanning a large range of stellar masses and star formation rates, probing above and below the main sequence of star-forming galaxies up to $z = 4–5$ (see, e.g., Yang et al. 2018; Fornasini et al. 2018). In star-forming galaxies, the AGN X-ray luminosity, which is proportional to the SMBH accretion rate, is found to be more correlated with stellar mass than star formation rate. X-ray stacking studies have also revealed the presence of obscured AGN in early-type galaxies (Paggi et al. 2016), and detected X-ray emission likely arising from intermediate-mass black holes residing in dwarf galaxies (Mezcua et al. 2016).[9]

### 8.6.6 The High-rdshift Universe as Seen in Surveys

To develop a complete understanding of the coevolution of SMBHs and galaxies, it is important to explore the epoch of their formation and early growth. The shape of the rest-frame 2–10 keV energy comoving space density at $z > 3$ is linked to the timescale of SMBH accretion, and therefore, it provides a way to investigate the SMBH formation and growth scenario, and eventually distinguishing between major-merger driven accretion and secular accretion models. Optical surveys are limited to the highest luminosity population of quasars at high-$z$ ($M_{1450} < -24.5$ corresponding to $L_x > 10^{45}$ erg s$^{-1}$; e.g., Richards et al. 2006; Masters et al. 2012; Ross et al. 2013), while X-ray surveys allow us to probe more common, lower-luminosity AGN.

Two pioneering studies of the high-$z$ ($z = 3–6$) AGN population were performed in the COSMOS field, using *XMM-Newton* sources first (Brusa et al. 2009, $N_{\rm AGN} = 40$) and then *Chandra* (Civano et al. 2011, $N_{\rm AGN} = 81$; Marchesi et al. 2016b, $N_{\rm AGN} = 174$), reaching X-ray luminosities as low as $3 \times 10^{43}$ erg s$^{-1}$. Later, several samples were combined to decrease the statistical uncertainty of the space density measurements and extend the luminosity range covered: Kalfountzou et al. (2014) combined medium and wide *Chandra* surveys (C-COSMOS and ChaMP; $N_{\rm AGN} = 211$ at $z > 3$ and $N_{\rm AGN} = 27$ at $z > 4$), and Georgakakis et al. (2015) combined deep and medium *Chandra* surveys (CDFS, CDFN, AEGIS, ECDF-S, and C-COSMOS, and the shallow, wide-area XMM-XXL survey) to obtain a

[9] http://chandra.harvard.edu/press/18_releases/press_021518.html.

sample of 340 AGN at $z > 3$, spanning three orders of magnitude in luminosity. Only the CDFS 7 Ms data can reach the lower luminosities of $3 \times 10^{42}$ erg s$^{-1}$ at high redshift (Vito et al. 2018). However, this deep sample suffers from low statistics, consisting of only ~102 sources at $z > 3$ (combining both CDFS and CDFN, and integrating the contribution of all the sources' redshift probability distribution functions).

These high-redshift studies show that the X-ray AGN space density declines at $z > 3$ with the same slope as the optical AGN density, except for low-luminosity X-ray AGN, which exhibit a marginally faster decline. The X-ray space density normalization is higher than that for the optical AGN as the X-ray selection includes the majority of the obscured sources missed by optical surveys (see Figure 8.25, left). By comparing the space density evolution of AGN and galaxies, Vito et al. (2018) suggested that the decline of high-luminosity AGN is consistent with that of the underlying galaxy population, while for the low-luminosity AGN, there are hints of an intrinsic redshift evolution of the factors driving nuclear activity.

Marchesi et al. (2016b) found a good agreement between the observed space density of high-luminosity AGN and models of quasar activation via mergers (e.g., Shen 2009; Figure 13 of Marchesi et al. 2016b), but these models do not adequately represent the low-luminosity AGN data. Indeed, at these low luminosities, mergers may not be the unique driver for the evolution of AGN, and disk instabilities and/or clumpy accretion may trigger BH accretion by channeling gas down toward the host galaxy center (e.g., Bower et al. 2006; Bournaud et al. 2011; Di Matteo et al. 2012).

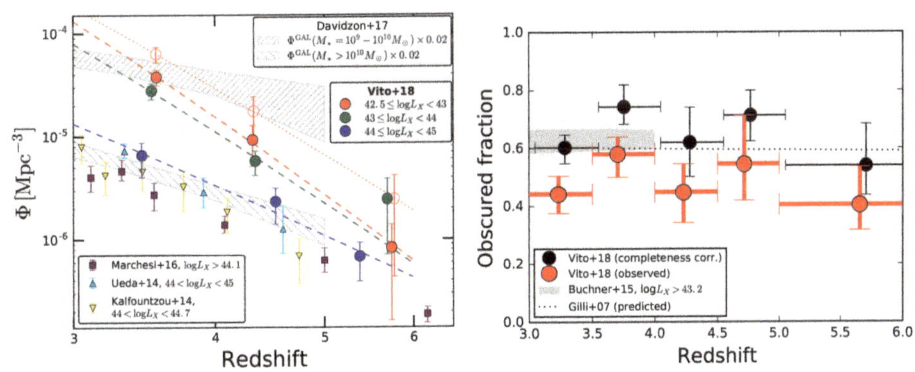


**Figure 8.25.** Left: comoving space density of three luminosity classes of AGN, divided into three redshift bins ($z$ = 3–4, 4–5, 5–6) from the *Chandra* deep field data (red, green, and blue circles from Vito et al. 2018) and wider area surveys results from COSMOS (Marchesi et al. 2016b), and Champ and COSMOS (Kalfountzou et al. 2014; Ueda et al. 2014). The dashed curves are the best-fitting models described in Vito et al. (2018, Section 6). The dotted curve is the best-fitting model for the low-luminosity bin, after the correction for incompleteness. The space density of galaxies in two different mass regimes from Davidzon et al. (2017), rescaled by an arbitrary factor of 0.02, is shown as a gray stripe. Right: obscured AGN fraction as a function of redshift from Vito et al. (2018). Gray region is from Buchner et al. (2015) for luminous AGN only. The dotted line shows the predictions from the X-ray background synthesis model of Gilli et al. (2007). Figure is adapted from Vito et al. (2018), by permission of Oxford University Press on behalf of the Royal Astronomical Society.

These results are in agreement with the findings from AGN clustering analysis (e.g., Allevato et al. 2016).

*Constraints on the Cosmic Dawn*

The main issue of the high-redshift X-ray *Chandra* studies is in the extreme paucity of $z > 4$ samples, with only three detections reported by Vito et al. (2018) and 27 in the COSMOS field (Marchesi et al. 2016b), but no spectroscopically confirmed (or with a strong photometric redshift) AGN at $z > 6$. On the contrary, in the optical bands, several hundreds of luminous and active SMBHs have been detected in the range $z = 5$–6.6, but only a handful above that redshift, with the record set at $z = 7.54$ (Bañados et al. 2018; see Section 8.2.4). These detections are just the tip of the iceberg as they are missing the obscured AGN, but they are already posing puzzling questions on SMBH formation models (e.g., Volonteri 2012) and the AGN contribution to cosmological reionization (see, e.g., Madau & Haardt 2015; Ricci et al. 2017; Kakiichi et al. 2018).

The discovery of SMBHs at $z > 6$ with masses $>10^9\ M_\odot$ (see Section 8.2.4). Section 8.2.4 implies that SMBHs were already in place and extremely massive when galaxies were only a few hundred megayears old. To reach the observed masses, constant Eddington-limit accretion from $z \sim 20$ down to $z = 7$ or sporadic episodes of super-Eddington accretion are required. To sustain such a high accretion rate requires a large amount of infalling material, which, at the same time, may screen and obscure the emitted radiation. As a consequence, X-ray observations should be the best available method to probe the existence of these seed black holes: the redshift effect combined with the shape of an extremely obscured spectrum favor their detection in the 2–10 keV observed energy band (corresponding to rest-frame energies of $>15$–70 keV for $z > 6$ sources). Observational searches are scarce: two $z \sim 6$ direct collapse BH candidates were proposed in the CANDELS/CDFS field (Pacucci et al. 2016); however, their X-ray spectra are consistent with the lower photometric redshift solution of Hsu et al. (2014). Looking for the obscured and low-accretion rate black holes in the highest redshift galaxies, Vito et al. (2016) performed X-ray stacking analysis of several thousands of galaxies in the CDFS field at $3.5 < z < 6.5$, accumalating many tens of megaseconds, finding that their X-ray emission is consistent with solely star formation processes.

While optical searches of high-redshift AGN will improve with the launch of the *James Webb Space Telescope* in 2021, a major leap forward in the characterization of the X-ray emission from high-redshift AGN will be possible with future X-ray missions—the upcoming *Athena* observatory (Nandra et al. 2013; see also Section 11.1.4), which will detect hundreds of AGN at $z > 6$ to luminosities of $10^{43}$ erg s$^{-1}$, and the mission concept *Lynx* (Gaskin et al. 2018; see also Chapter 11.2.1) that will probe the smallest black holes ($\sim 10^5\ M_\odot$) up to $z = 10$ (see also Haiman et al. 2019; Civano et al. 2019 for a discussion on future X-ray detections in the high-redshift universe).

### 8.6.7 Resolving the Cosmic X-Ray Background

The main goal of focusing X-ray telescopes has been to resolve and understand the origin of the cosmic X-ray background (CXRB; Giacconi et al. 1962), a diffuse X-ray radiation over the entire sky (Figure 8.26). Answering the question whether all of the CXRB radiation is due to emission of many faint individual sources or it is truly diffuse requires accurate detection of very faint sources. The density of X-ray sources increases with decreasing flux, and there are many more faint sources than bright ones. The calculations of the source contribution to the X-ray background requires detection of sources at the faintest end ($<10^{-15}$ erg cm$^{-2}$ s$^{-1}$) of the source flux distribution. The first *Chandra* deep fields observations detected many thousands of faint sources identified with distant galaxies and quasars (Mushotzky et al. 2000; Giacconi et al. 2001). The integrated contributions from these sources accounted for more than 90% of the CXRB radiation observed before *Chandra*.

It is now well established that the main contribution to the CXRB radiation is given by the AGN population. However, there is a large uncertainty on the normalization of the CXRB spectrum, which varies strongly between different measurements (Cappelluti et al. 2017). This uncertainty impacts the population synthesis models reproducing the CXRB global shape and intensity (Gilli et al. 2001, 2007; Treister & Urry 2006; Treister et al. 2009; Ananna et al. 2019). Most of the sources contributing to the CXRB are obscured AGN with X-ray spectral peak at ~30 keV. The least obscured sources dominate the emission at lower energies, while a large fraction, ~30% as measured in the local universe, needs to be extremely obscured and Compton thick. A comparably large fraction of obscured sources is

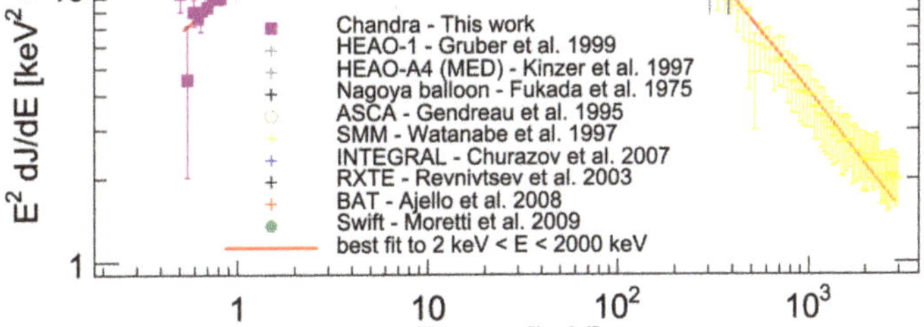


**Figure 8.26.** Compilation of the measurements of the cosmic X-ray background spectrum in the 0.5–1000 keV energy range from Cappelluti et al. (2017) © 2017. The American Astronomical Society. All rights reserved. Data points with different colors come from different combinations of missions and instruments as labeled and referenced. The *Chandra* data from Cappelluti et al. (2017) are in magenta (with the local soft component subtracted). The best fit is from Ajello et al. (2008).

inferred to exist at higher redshift to reproduce the shape of the cosmic X-ray background hump at 20–30 keV. However, constraints on the fraction of Compton-thick AGN as a function of redshift are still uncertain (see discussion in Section 8.6.2) because of the spectral degeneracies involved in modeling the X-ray spectra (photon index, reflection fraction, column density, and high-energy cutoff).

By adding together the contribution of the sources in the *Chandra* extragalactic surveys, it is possible to measure the fraction of the resolved X-ray background, which is now around 80–85% in the 0.5–2 keV energy range (Hickox & Markevitch 2006; Cappelluti et al. 2017). The spectrum of the extragalactic X-ray background observed by *Chandra* within the energy range 0.3–7 keV is a power law with a photon index of $\Gamma = 1.45 \pm 0.02$ (see Figure 8.26). Removing all of the X-ray detected sources in the *Chandra* COSMOS field and faint galaxies leaves diffuse emission with a hard X-ray background spectrum with a photon index $\Gamma \sim 1.2$ (Cappelluti et al. 2017). Understanding this spectrum requires studies of broadband properties of the sources detected in surveys, in particular the hard ($>$10 keV) X-rays.

## 8.7 Final Remarks

Looking back at the 20 years of *Chandra* studies, one can reflect on all the astonishing results. It is hard to decide which *Chandra* observation had the most impact on SMBH science, as it is a combination of many individual investigations that lead to progress. The recent EHT image of the black hole "shadow" in the millimeter waveband confirmed our theoretical understanding of the SMBH impact on the surrounding matter, an accretion flow and the radiation at the immediate vicinity of the event horizon. The X-ray radiation, unresolved on these scales, signals the presence of high-energy processes that are inherent to the existence and evolution of SMBHs. Our understanding of the BH phenomena, including the BH's significance in the formation and evolution of structures in the universe, requires multiwavelength and multimessenger experiments, which need to include high-quality X-ray observations. *Chandra* will provide crucial data for future decades of research, but we also look into the future of new technologies that will allow us to continue making progress in this field beyond *Chandra*.

## References

Abramowicz, M. A. 2005, Growing Black Holes: Accretion in a Cosmological Context, ed. A. Merloni, S. Nayakshin, & R. A. Sunyaev, 257–73

Abramowicz, M. A., Czerny, B., Lasota, J. P., & Szuszkiewicz, E. 1988, ApJ, 332, 646

Abramowicz, M. A., & Zurek, W. H. 1981, ApJ, 246, 314

Aird, J., Coil, A. L., Georgakakis, A., et al. 2015, MNRAS, 451, 1892

Aird, J., Nandra, K., Laird, E. S., et al. 2010, MNRAS, 401, 2531

Ajello, M., Greiner, J., Sato, G., et al. 2008, ApJ, 689, 666

Alexander, D. M., & Hickox, R. C. 2012, NewAR, 56, 93

Alexander, D. M., Bauer, F. E., Brandt, W. N., et al. 2003, AJ, 126, 539

Allevato, V., Civano, F., Finoguenov, A., et al. 2016, ApJ, 832, 70

Alonso-Herrero, A., Pereira-Santaella, M., García-Burillo, S., et al. 2018, ApJ, 859, 144
Ananna, T. T., Treister, E., Urry, C. M., et al. 2019, ApJ, 871, 240
Antonucci, R. 1993, ARA&A, 31, 473
Antonucci, R. R. J., & Miller, J. S. 1985, ApJ, 297, 621
Bañados, E., Connor, T., Stern, D., et al. 2018, ApJ, 856, L25
Baganoff, F. K., Bautz, M. W., Brandt, W. N., et al. 2001, Natur, 413, 45
Baganoff, F. K., Maeda, Y., Morris, M., et al. 2003, ApJ, 591, 891
Baldwin, J. A., Phillips, M. M., & Terlevich, R. 1981, PASP, 93, 5
Barger, A. J., Cowie, L. L., Owen, F. N., et al. 2015, ApJ, 801, 87
Barthel, P. D. 1989, ApJ, 336, 606
Bassett, L. C., Brandt, W. N., Schneider, D. P., et al. 2004, AJ, 128, 523
Bechtold, J., Elvis, M., Fiore, F., et al. 1994, AJ, 108, 759
Bechtold, J., Siemiginowska, A., Shields, J., et al. 2003, ApJ, 588, 119
Begelman, M. C. 1985, Astrophysics of Active Galaxies and Quasi-Stellar Objects, ed. J. S. Miller, 411–52
Begelman, M. C., & Cioffi, D. F. 1989, ApJ, 345, L21
Begelman, M. C., Volonteri, M., & Rees, M. J. 2006, MNRAS, 370, 289
Blandford, R. D., & Rees, M. J. 1974, MNRAS, 169, 395
Blandford, R. D., & Znajek, R. L. 1977, MNRAS, 179, 433
Blecha, L., Brisken, W., Burke-Spolaor, S., et al. 2019, arXiv: 1903.09301
Blecha, L., Civano, F., Elvis, M., & Loeb, A. 2013, MNRAS, 428, 1341
Blecha, L., Snyder, G. F., Satyapal, S., & Ellison, S. L. 2018, MNRAS, 478, 3056
Bondi, H. 1952, MNRAS, 112, 195
Bongiorno, A., Merloni, A., Brusa, M., et al. 2012, MNRAS, 427, 3103
Bournaud, F., Dekel, A., Teyssier, R., et al. 2011, ApJ, 741, L33
Bower, R. G., Benson, A. J., Malbon, R., et al. 2006, MNRAS, 370, 645
Brandt, W. N., & Alexander, D. M. 2015, A&ARv, 23, 1
Brandt, W. N., & Hasinger, G. 2005, ARA&A, 43, 827
Breiding, P., Meyer, E. T., Georganopoulos, M., et al. 2017, ApJ, 849, 95
Brightman, M., Nandra, K., Salvato, M., et al. 2014, MNRAS, 443, 1999
Brightman, M., & Ueda, Y. 2012, MNRAS, 423, 702
Brusa, M., Gilli, R., & Comastri, A. 2005, ApJ, 621, L5
Brusa, M., Comastri, A., Gilli, R., et al. 2009, ApJ, 693, 8
Buchner, J., & Bauer, F. E. 2017, MNRAS, 465, 4348
Buchner, J., Georgakakis, A., Nandra, K., et al. 2015, ApJ, 802, 89
Burlon, D., Ajello, M., Greiner, J., et al. 2011, ApJ, 728, 58
Cappelluti, N., Li, Y., Ricarte, A., et al. 2017, ApJ, 837, 19
Carilli, C. L., & Barthel, P. D. 1996, A&ARv, 7, 1
Carilli, C. L., Perley, R. A., & Harris, D. E. 1994, MNRAS, 270, 173
Castelló-Mor, N., Carrera, F. J., Alonso-Herrero, A., et al. 2013, A&A, 556, A114
Celotti, A., Ghisellini, G., & Chiaberge, M. 2001, MNRAS, 321, L1
Chartas, G., Gupta, V., Garmire, G., et al. 2002, ApJ, 565, 96
Chartas, G., Kochanek, C. S., Dai, X., et al. 2012, ApJ, 757, 137
Chartas, G., Kochanek, C. S., Dai, X., Poindexter, S., & Garmire, G. 2009, ApJ, 693, 174
Chartas, G., Krawczynski, H., Pooley, D., Mushotzky, R. F., & Ptak, A. J. 2019, arXiv: 1904.02018

Chartas, G., Krawczynski, H., Zalesky, L., et al. 2017, ApJ, 837, 26
Cheung, C. C. 2004, ApJ, 600, L23
Cheung, C. C., Stawarz, Ł., Siemiginowska, A., et al. 2012, ApJ, 756, L20
Cisternas, M., Jahnke, K., Inskip, K. J., et al. 2011, ApJ, 726, 57
Civano, F., Comastri, A., & Brusa, M. 2005, MNRAS, 358, 693
Civano, F., Mignoli, M., Comastri, A., et al. 2007, A&A, 476, 1223
Civano, F., Elvis, M., Lanzuisi, G., et al. 2010, ApJ, 717, 209
Civano, F., Brusa, M., Comastri, A., et al. 2011, ApJ, 741, 91
Civano, F., Elvis, M., Lanzuisi, G., et al. 2012, ApJ, 752, 49
Civano, F., Marchesi, S., Comastri, A., et al. 2016, ApJ, 819, 62
Civano, F., Cappelluti, N., Hickox, R., et al. 2019, BAAS, 51, 429
Clarke, D. A., Harris, D. E., & Carilli, C. L. 1997, MNRAS, 284, 981
Comastri, A., Gilli, R., Marconi, A., Risaliti, G., & Salvati, M. 2015, A&A, 574, L10
Comerford, J. M., Pooley, D., Barrows, R. S., et al. 2015, ApJ, 806, 219
Covey, K. R., Agüeros, M. A., Green, P. J., et al. 2008, ApJS, 178, 339
Cowie, L. L., Songaila, A., Hu, E. M., & Cohen, J. G. 1996, AJ, 112, 839
Croston, J. H., Kraft, R. P., & Hardcastle, M. J. 2007, ApJ, 660, 191
Croston, J. H., Kraft, R. P., Hardcastle, M. J., et al. 2009, MNRAS, 395, 1999
Czerny, B., & Elvis, M. 1987, ApJ, 321, 305
Czerny, B., & Naddaf, M. -H. 2018, arXiv: 1811.04326
Czerny, B., Nikołajuk, A., et al. 2003, A&A, 412, 317
Czerny, B., Siemiginowska, A., Janiuk, A., Nikiel-Wroczyński, B., & Stawarz, L. 2009, ApJ, 698, 840
Daddi, E., Alexander, D. M., Dickinson, M., et al. 2007, ApJ, 670, 173
Dai, X., Kochanek, C. S., Chartas, G., et al. 2010, ApJ, 709, 278
Danehkar, A., Nowak, M. A., Lee, J. C., et al. 2018, ApJ, 853, 165
Davidzon, I., Ilbert, O., Laigle, C., et al. 2017, A&A, 605, A70
Del Moro, A., Alexander, D. M., Bauer, F. E., et al. 2016, MNRAS, 456, 2105
Dexter, J., & Begelman, M. C. 2019, MNRAS, 483, L17
Di Matteo, T., Allen, S. W., Fabian, A. C., Wilson, A. S., & Young, A. J. 2003, ApJ, 582, 133
Di Matteo, T., Khandai, N., DeGraf, C., et al. 2012, ApJ, 745, L29
Di Matteo, T., Springel, V., & Hernquist, L. 2005, Natur, 433, 604
Donley, J. L., Koekemoer, A. M., Brusa, M., et al. 2012, ApJ, 748, 142
Duffy, R. T., Worrall, D. M., Birkinshaw, M., et al. 2018, MNRAS, 476, 4848
Edelson, R., Gelbord, J. M., Horne, K., et al. 2015, ApJ, 806, 129
Edelson, R., Gelbord, J., Cackett, E., et al. 2017, ApJ, 840, 41
Edelson, R., Gelbord, J., Cackett, E., et al. 2019, ApJ, 870, 123
Elvis, M. 2000, ApJ, 545, 63
Elvis, M. 2017, ApJ, 847, 56
Elvis, M., Wilkes, B. J., McDowell, J. C., et al. 1994, ApJS, 95, 1
Elvis, M., Civano, F., Vignali, C., et al. 2009, ApJS, 184, 158
Evans, D. A., Lee, J. C., Kamenetska, M., et al. 2006, ApJ, 653, 1121
Evans, I. N., Primini, F. A., Glotfelty, K. J., et al. 2010, ApJS, 189, 37
Event Horizon Telescope Collaboration, Akiyama, K., Alberdi, A., et al. 2019a, ApJ, 875, L1
Event Horizon Telescope Collaboration, Akiyama, K., Alberdi, A., et al. 2019b, ApJ, 875, L4
Event Horizon Telescope Collaboration, Akiyama, K., Alberdi, A., et al. 2019c, ApJ, 875, L5

Fabbiano, G., Paggi, A., Siemiginowska, A., & Elvis, M. 2018, ApJ, 869, L36
Fabbiano, G., Siemiginowska, A., Paggi, A., et al. 2019, ApJ, 870, 69
Fabian, A. C., Sanders, J. S., Allen, S. W., et al. 2003, MNRAS, 344, L43
Fabian, A. C., Sanders, J. S., Ettori, S., et al. 2000, MNRAS, 318, L65
Fan, L., Han, Y., Fang, G., et al. 2016, ApJ, 822, L32
Fanidakis, N., Baugh, C. M., Benson, A. J., et al. 2012, MNRAS, 419, 2797
Feigelson, E. D. 1980, PhD thesis, Harvard Univ., Cambridge, MA
Feigelson, E. D., Hornschemeier, A. E., Micela, G., et al. 2004, ApJ, 611, 1107
Ferrarese, L., & Ford, H. 2005, SSRv, 116, 523
Feruglio, C., Fabbiano, G., Bischetti, M., et al. 2019, arXiv: 1904.01483
Feruglio, C., Fiore, F., Carniani, S., et al. 2015, A&A, 583, A99
Fiore, F., Puccetti, S., Brusa, M., et al. 2009, ApJ, 693, 447
Fiore, F., Puccetti, S., Grazian, A., et al. 2012, A&A, 537, A16
Forman, W., Nulsen, P., Heinz, S., et al. 2005, ApJ, 635, 894
Forman, W., Jones, C., Churazov, E., et al. 2007, ApJ, 665, 1057
Fornasini, F. M., Civano, F., Fabbiano, G., et al. 2018, ApJ, 865, 43
Gaskin, J. A., Dominguez, A., Gelmis, K., et al. 2018, Proc. SPIE, 10699, 106990N
Gebhardt, K., Adams, J., Richstone, D., et al. 2011, ApJ, 729, 119
Georgakakis, A., Aird, J., Buchner, J., et al. 2015, MNRAS, 453, 1946
Georgantopoulos, I., Comastri, A., Vignali, C., et al. 2013, A&A, 555, A43
Ghez, A. M., Salim, S., Weinberg, N. N., et al. 2008, ApJ, 689, 1044
Giacconi, R., Gursky, H., Paolini, F. R., & Rossi, B. B. 1962, PhRvL, 9, 439
Giacconi, R., Rosati, P., Tozzi, P., et al. 2001, ApJ, 551, 624
Giacconi, R., Zirm, A., Wang, J., et al. 2002, ApJS, 139, 369
Gillessen, S., Eisenhauer, F., Trippe, S., et al. 2009, ApJ, 692, 1075
Gilli, R., Comastri, A., & Hasinger, G. 2007, A&A, 463, 79
Gilli, R., Salvati, M., & Hasinger, G. 2001, A&A, 366, 407
Glikman, E., Simmons, B., Mailly, M., et al. 2015, ApJ, 806, 218
Goodger, J. L., Hardcastle, M. J., Croston, J. H., et al. 2010, ApJ, 708, 675
Goulding, A. D., Forman, W. R., Hickox, R. C., et al. 2012, ApJS, 202, 6
Haardt, F., & Maraschi, L. 1991, ApJ, 380, L51
Haardt, F., & Maraschi, L. 1993, ApJ, 413, 507
Haiman, Z., Brandt, W. N., Vikhlinin, A., et al. 2019, BAAS, 51, 557
Halpern, J. P. 1984, ApJ, 281, 90
Hardcastle, M. 2015, in Astrophysics and Space Science Library, The Formation and Disruption of Black Hole Jets, Vol. 414, ed. I. Contopoulos, D. Gabuzda, & N. Kylafis, 83
Hardcastle, M. J., Massaro, F., & Harris, D. E. 2010, MNRAS, 401, 2697
Hardcastle, M. J., Worrall, D. M., Kraft, R. P., et al. 2003, ApJ, 593, 169
Hardcastle, M. J., Kraft, R. P., Sivakoff, G. R., et al. 2007, ApJ, 670, L81
Hardcastle, M. J., Massaro, F., Harris, D. E., et al. 2012, MNRAS, 424, 1774
Hardcastle, M. J., Lenc, E., Birkinshaw, M., et al. 2016, MNRAS, 455, 3526
Harris, D. E., Carilli, C. L., & Perley, R. A. 1994, Natur, 367, 713
Harris, D. E., Cheung, C. C., Stawarz, Ł., Biretta, J. A., & Perlman, E. S. 2009, ApJ, 699, 305
Harris, D. E., & Krawczynski, H. 2002, ApJ, 565, 244
Harris, D. E., & Krawczynski, H. 2006, ARA&A, 44, 463
Harrison, F. A., Eckart, M. E., Mao, P. H., Helfand, D. J., & Stern, D. 2003, ApJ, 596, 944

Harwood, J. J., & Hardcastle, M. J. 2012, MNRAS, 423, 1368
Hasinger, G. 2008, A&A, 490, 905
Hasinger, G., Miyaji, T., & Schmidt, M. 2005, A&A, 441, 417
Heinz, S., Reynolds, C. S., & Begelman, M. C. 1998, ApJ, 501, 126
Hickox, R. C., & Alexander, D. M. 2018, ARA&A, 56, 625
Hickox, R. C., & Markevitch, M. 2006, ApJ, 645, 95
Hickox, R. C., & Markevitch, M. 2007, ApJ, 661, L117
Hickox, R. C., Mullaney, J. R., Alexander, D. M., et al. 2014, ApJ, 782, 9
Hlavacek-Larrondo, J., Gandhi, P., Hogan, M. T., et al. 2017, MNRAS, 464, 2223
Hogan, B. S., Lister, M. L., Kharb, P., Marshall, H. L., & Cooper, N. J. 2011, ApJ, 730, 92
Hopkins, P. F., & Elvis, M. 2010, MNRAS, 401, 7
Hopkins, P. F., Hernquist, L., Cox, T. J., & Kereš, D. 2008, ApJS, 175, 356
Hopkins, P. F., Richards, G. T., & Hernquist, L. 2007, ApJ, 654, 731
Hsu, L. -T., Salvato, M., Nandra, K., et al. 2014, ApJ, 796, 60
Iwasawa, K., Gilli, R., Vignali, C., et al. 2012, A&A, 546, A84
Jamrozy, M., Konar, C., Saikia, D. J., et al. 2007, MNRAS, 378, 581
Janiuk, A., Czerny, B., & Siemiginowska, A. 2002, ApJ, 576, 908
Juneau, S., Dickinson, M., Bournaud, F., et al. 2013, ApJ, 764, 176
Kaastra, J. S., Mewe, R., Liedahl, D. A., Komossa, S., & Brinkman, A. C. 2000, A&A, 354, L83
Kaastra, J. S., Steenbrugge, K. C., Raassen, A. J. J., et al. 2002, A&A, 386, 427
Kakiichi, K., Ellis, R. S., Laporte, N., et al. 2018, MNRAS, 479, 43
Kalfountzou, E., Civano, F., Elvis, M., Trichas, M., & Green, P. 2014, MNRAS, 445, 1430
Kara, E., Alston, W. N., Fabian, A. C., et al. 2016, MNRAS, 462, 511
Kaspi, S., Brandt, W. N., Netzer, H., et al. 2000, ApJ, 535, L17
Kelly, B. C., Bechtold, J., Trump, J. R., Vestergaard, M., & Siemiginowska, A. 2008, ApJS, 176, 355
Kim, D. C., Yoon, I., Privon, G. C., et al. 2017, ApJ, 840, 71
Kim, M., Wilkes, B. J., Kim, D. -W., et al. 2007a, ApJ, 659, 29
Kim, M., Kim, D.-W., Wilkes, B. J., et al. 2007b, ApJS, 169, 401
King, A. 2003, ApJ, 596, L27
King, A., & Pounds, K. 2015, ARA&A, 53, 115
King, A. L., Miller, J. M., & Raymond, J. 2012, ApJ, 746, 2
Kocevski, D. D., Faber, S. M., Mozena, M., et al. 2012, ApJ, 744, 148
Kocevski, D. D., Hasinger, G., Brightman, M., et al. 2018, ApJS, 236, 48
Kraft, R. P., Forman, W., Jones, C., et al. 2000, ApJ, 531, L9
Krongold, Y., Nicastro, F., Brickhouse, N. S., et al. 2003, ApJ, 597, 832
Krongold, Y., Nicastro, F., Brickhouse, N. S., Elvis, M., & Mathur, S. 2005, ApJ, 622, 842
Kubota, A., & Done, C. 2018, MNRAS, 480, 1247
Kunert-Bajraszewska, M., Siemiginowska, A., & Labiano, A. 2013, ApJ, 772, L7
La Franca, F., Fiore, F., Comastri, A., et al. 2005, ApJ, 635, 864
LaMassa, S. M., Urry, C. M., Glikman, E., et al. 2013, MNRAS, 432, 1351
Lanzuisi, G., Civano, F., Elvis, M., et al. 2013, MNRAS, 431, 978
Lanzuisi, G., Ranalli, P., Georgantopoulos, I., et al. 2015, A&A, 573, A137
Lanzuisi, G., Delvecchio, I., Berta, S., et al. 2017, A&A, 602, A123
Lanzuisi, G., Civano, F., Marchesi, S., et al. 2018, MNRAS, 480, 2578
Lawrence, A., & Elvis, M. 1982, ApJ, 256, 410

Lee, J. C., Ogle, P. M., Canizares, C. R., et al. 2001, ApJ, 554, L13
Lee, J. C., Kriss, G. A., Chakravorty, S., et al. 2013, MNRAS, 430, 2650
Lehmer, B. D., Brandt, W. N., Alexander, D. M., et al. 2005, ApJS, 161, 21
Lehmer, B. D., Basu-Zych, A. R., Mineo, S., et al. 2016, ApJ, 825, 7
Lightman, A. P., & Eardley, D. M. 1974, ApJ, 187, L1
Liu, T., Tozzi, P., Wang, J. -X., et al. 2017, ApJS, 232, 8
Liu, X., Civano, F., Shen, Y., et al. 2013, ApJ, 762, 110
Luo, B., Bauer, F. E., Brandt, W. N., et al. 2008, ApJS, 179, 19
Luo, B., Brandt, W. N., Xue, Y. Q., et al. 2017, ApJS, 228, 2
Lusso, E., Comastri, A., Simmons, B. D., et al. 2012, MNRAS, 425, 623
Maccacaro, T., Gioia, I. M., Wolter, A., Zamorani, G., & Stocke, J. T. 1988, ApJ, 326, 680
Madau, P., & Dickinson, M. 2014, ARA&A, 52, 415
Madau, P., & Haardt, F. 2015, ApJ, 813, L8
Maiolino, R., Risaliti, G., Salvati, M., et al. 2010, A&A, 517, A47
Marchesi, S., Lanzuisi, G., Civano, F., et al. 2016a, ApJ, 830, 100
Marchesi, S., Civano, F., Salvato, M., et al. 2016b, ApJ, 827, 150
Marinucci, A., Risaliti, G., Wang, J., et al. 2013, MNRAS, 429, 2581
Marshall, H. L., Miller, B. P., Davis, D. S., et al. 2002, ApJ, 564, 683
Marshall, H. L., Harris, D. E., Grimes, J. P., et al. 2001, ApJ, 549, L167
Marshall, H. L., Hardcastle, M. J., Birkinshaw, M., et al. 2010, ApJ, 714, L213
Marshall, H. L., Gelbord, J. M., Schwartz, D. A., et al. 2011, ApJS, 193, 15
Marshall, H. L., Gelbord, J. M., Worrall, D. M., et al. 2018, ApJ, 856, 66
Marulli, F., Bonoli, S., Branchini, E., Moscardini, L., & Springel, V. 2008, MNRAS, 385, 1846
Maselli, A., Kraft, R. P., Massaro, F., & Hardcastle, M. J. 2018, A&A, 619, A75
Massaro, F., Harris, D. E., & Cheung, C. C. 2011, ApJS, 197, 24
Massaro, F., Chiaberge, M., Grandi, P., et al. 2009, ApJ, 692, L123
Masters, D., Capak, P., Salvato, M., et al. 2012, ApJ, 755, 169
McKeough, K., Siemiginowska, A., Cheung, C. C., et al. 2016, ApJ, 833, 123
McKernan, B., Yaqoob, T., & Reynolds, C. S. 2007, MNRAS, 379, 1359
McNamara, B. R., & Nulsen, P. E. J. 2007, ARA&A, 45, 117
Melia, F., & Falcke, H. 2001, ARA&A, 39, 309
Merloni, A., Bongiorno, A., Brusa, M., et al. 2014, MNRAS, 437, 3550
Meyer, E. T., Breiding, P., Georganopoulos, M., et al. 2017, ApJ, 835, L35
Meyer, E. T., & Georganopoulos, M. 2014, ApJ, 780, L27
Meyer, E. T., Georganopoulos, M., Sparks, W. B., et al. 2015, ApJ, 805, 154
Mezcua, M., Civano, F., Fabbiano, G., Miyaji, T., & Marchesi, S. 2016, ApJ, 817, 20
Miller, B. P., Brandt, W. N., Schneider, D. P., et al. 2011, ApJ, 726, 20
Miyaji, T., Hasinger, G., Salvato, M., et al. 2015, ApJ, 804, 104
Momcheva, I. G., Brammer, G. B., van Dokkum, P. G., et al. 2016, ApJS, 225, 27
Morgan, C. W., Kochanek, C. S., Dai, X., Morgan, N. D., & Falco, E. E. 2008, ApJ, 689, 755
Mościbrodzka, M., Falcke, H., & Shiokawa, H. 2016, A&A, 586, A38
Mościbrodzka, M., Falcke, H., Shiokawa, H., & Gammie, C. F. 2014, A&A, 570, A7
Mullaney, J. R., Daddi, E., Béthermin, M., et al. 2012, ApJL, 753, L30
Mullaney, J. R., Alexander, D. M., Aird, J., et al. 2015, MNRAS, 453, L83
Muno, M. P., Baganoff, F. K., Bautz, M. W., et al. 2003, ApJ, 589, 225
Muno, M. P., Baganoff, F. K., Bautz, M. W., et al. 2004, ApJ, 613, 326

Muno, M. P., Bauer, F. E., Baganoff, F. K., et al. 2009, ApJS, 181, 110
Murray, S. S., Kenter, A., Forman, W. R., et al. 2005, ApJS, 161, 1
Mushotzky, R. F., Cowie, L. L., Barger, A. J., & Arnaud, K. A. 2000, Natur, 404, 459
Nandra, K., Barret, D., Barcons, X., et al. 2013, arXiv: 1306.2307
Nandra, K., Laird, E. S., Aird, J. A., et al. 2015, ApJS, 220, 10
Nanni, R., Vignali, C., Gilli, R., Moretti, A., & Brandt, W. N. 2017, A&A, 603, A128
Neilsen, J., Nowak, M. A., Gammie, C., et al. 2013, ApJ, 774, 42
Neilsen, J., Markoff, S., Nowak, M. A., et al. 2015, ApJ, 799, 199
Netzer, H. 2015, ARA&A, 53, 365
Nowak, M. A., Neilsen, J., Markoff, S. B., et al. 2012, ApJ, 759, 95
Nulsen, P., Johnson, A., Snios, B., et al. 2019, AAS/HEAD 17, 106.21
O'Dea, C. P. 1998, PASP, 110, 493
Owen, F. N., Ledlow, M. J., Morrison, G. E., & Hill, J. M. 1997, ApJ, 488, L15
Pacucci, F., Ferrara, A., Grazian, A., et al. 2016, MNRAS, 459, 1432
Paczyński, B., & Wiita, P. J. 1980, A&A, 88, 23
Padovani, P., Alexander, D. M., Assef, R. J., et al. 2017, A&ARv, 25, 2
Paggi, A., Fabbiano, G., Civano, F., et al. 2016, ApJ, 823, 112
Pellegrini, S., Siemiginowska, A., Fabbiano, G., et al. 2007, ApJ, 667, 749
Perlman, E. S., Biretta, J. A., Sparks, W. B., Macchetto, F. D., & Leahy, J. P. 2001, ApJ, 551, 206
Perlman, E. S., Meyer, E., Eilek, J., et al. 2019, arXiv: 1903.03657
Pezzulli, E., Volonteri, M., Schneider, R., & Valiante, R. 2017, MNRAS, 471, 589
Pooley, D., Blackburne, J. A., Rappaport, S., & Schechter, P. L. 2007, ApJ, 661, 19
Pounds, K., Lobban, A., Reeves, J., & Vaughan, S. 2016, MNRAS, 457, 2951
Pounds, K. A., Reeves, J. N., King, A. R., et al. 2003, MNRAS, 345, 705
Pringle, J. E., Rees, M. J., & Pacholczyk, A. G. 1973, A&A, 29, 179
Proga, D. 2003, ApJ, 585, 406
Pudritz, R. E., Hardcastle, M. J., & Gabuzda, D. C. 2012, SSRv, 169, 27
Pyrzas, S., Steenbrugge, K. C., & Blundell, K. M. 2015, A&A, 574, A30
Reeves, J. N., Gofford, J., Braito, V., & Sambruna, R. 2010, ApJ, 725, 803
Regan, J. A., Downes, T. P., Volonteri, M., et al. 2019, MNRAS, 486, 3892
Reid, M. J., Readhead, A. C. S., Vermeulen, R. C., & Treuhaft, R. N. 1999, ApJ, 524, 816
Reis, R. C., & Miller, J. M. 2013, ApJ, 769, L7
Reynolds, C. S., Heinz, S., & Begelman, M. C. 2001, ApJ, 549, L179
Ricci, C., Ueda, Y., Koss, M. J., et al. 2015, ApJ, 815, L13
Ricci, F., Marchesi, S., Shankar, F., La Franca, F., & Civano, F. 2017, MNRAS, 465, 1915
Richards, G. T., Strauss, M. A., Fan, X., et al. 2006, AJ, 131, 2766
Riguccini, L., Le Floc'h, E., Mullaney, J. R., et al. 2015, MNRAS, 452, 470
Risaliti, G., Elvis, M., Bianchi, S., & Matt, G. 2010, MNRAS, 406, L20
Risaliti, G., Elvis, M., Fabbiano, G., Baldi, A., & Zezas, A. 2005, ApJ, 623, L93
Risaliti, G., Elvis, M., & Nicastro, F. 2002, ApJ, 571, 234
Risaliti, G., & Lusso, E. 2015, ApJ, 815, 33
Rodighiero, G., Brusa, M., Daddi, E., et al. 2015, ApJ, 800, L10
Rosario, D. J., Togi, A., Burtscher, L., et al. 2019, ApJ, 875, L8
Rosario, D. J., Santini, P., Lutz, D., et al. 2012, A&A, 545, A45
Rosati, P., Tozzi, P., Giacconi, R., et al. 2002, ApJ, 566, 667
Ross, N. P., McGreer, I. D., White, M., et al. 2013, ApJ, 773, 14

Różańska, A., & Czerny, B. 2000, A&A, 360, 1170
Russell, H. R., Fabian, A. C., McNamara, B. R., et al. 2018, MNRAS, 477, 3583
Russell, H. R., Fabian, A. C., Sanders, J. S., et al. 2010, MNRAS, 402, 1561
Russell, H. R., Fabian, A. C., Taylor, G. B., et al. 2012, MNRAS, 422, 590
Salvato, M., Buchner, J., Budavári, T., et al. 2018, MNRAS, 473, 4937
Sambruna, R. M., Gambill, J. K., Maraschi, L., et al. 2004, ApJ, 608, 698
Sanders, J. S., Fabian, A. C., Taylor, G. B., et al. 2016, MNRAS, 457, 82
Santini, P., Rosario, D. J., Shao, L., et al. 2012, A&A, 540, A109
Scheuer, P. A. G. 1974, MNRAS, 166, 513
Scheuer, P. A. G. 1982, in IAU Symp. 97, Extragalactic Radio Sources, ed. D. C. Heeschen, & C. M. Wade, 163–5
Schreier, E. J., Feigelson, E., Delvaille, J., et al. 1979, ApJ, 234, L39
Schreier, E. J., Gorenstein, P., & Feigelson, E. D. 1982, ApJ, 261, 42
Schwartz, D., & Harris, D. 2006, ChNew, 13, 1
Schwartz, D. A., Marshall, H. L., Lovell, J. E. J., et al. 2000, ApJ, 540, 69
Shakura, N. I., & Sunyaev, R. A. 1973, A&A, 24, 337
Shemmer, O., Brandt, W. N., Schneider, D. P., et al. 2006, ApJ, 644, 86
Shen, Y. 2009, ApJ, 704, 89
Shimizu, T. T., Mushotzky, R. F., Meléndez, M., Koss, M., & Rosario, D. J. 2015, MNRAS, 452, 1841
Siemiginowska, A. 2009, AN, 330, 264
Siemiginowska, A. 2014, ChNew, 21, 3
Siemiginowska, A., Bechtold, J., Aldcroft, T. L., et al. 2002, ApJ, 570, 543
Siemiginowska, A., Burke, D. J., Aldcroft, T. L., et al. 2010, ApJ, 722, 102
Siemiginowska, A., Cheung, C. C., LaMassa, S., et al. 2005, ApJ, 632, 110
Siemiginowska, A., Smith, R. K., Aldcroft, T. L., et al. 2003a, ApJ, 598, L15
Siemiginowska, A., Sobolewska, M., Migliori, G., et al. 2016, ApJ, 823, 57
Siemiginowska, A., Stanghellini, C., Brunetti, G., et al. 2003b, ApJ, 595, 643
Siemiginowska, A., Stawarz, Ł., Cheung, C. C., et al. 2007, ApJ, 657, 145
Siemiginowska, A., Stawarz, Ł., Cheung, C. C., et al. 2012, ApJ, 750, 124
Silk, J., & Rees, M. J. 1998, A&A, 331, L1
Silverman, J. D., Lamareille, F., Maier, C., et al. 2009, ApJ, 696, 396
Silverman, J. D., Kampczyk, P., Jahnke, K., et al. 2011, ApJ, 743, 2
Simionescu, A., Stawarz, Ł., Ichinohe, Y., et al. 2016, ApJ, 816, L15
Smith, D. A., Wilson, A. S., Arnaud, K. A., Terashima, Y., & Young, A. J. 2002, ApJ, 565, 195
Snios, B., Nulsen, P. E. J., Kraft, R. P., et al. 2019a, arXiv: 1905.04330
Snios, B., Nulsen, P. E. J., Kraft, R. P., et al. 2019, arXiv: 1905.04330
Snios, B., Nulsen, P. E. J., Wise, M. W., et al. 2018, ApJ, 855, 71
Snios, B., Wykes, S., Nulsen, P. E. J., et al. 2019c, ApJ, 871, 248
Sobolewska, M., Siemiginowska, A., Guainazzi, M., et al. 2019, ApJ, 871, 71
Sobolewska, M. A., Siemiginowska, A., Migliori, G., et al. 2012, ApJ, 758, 90
Sobolewska, M. A., Siemiginowska, A., & Życki, P. T. 2004a, ApJ, 608, 80
Sobolewska, M. A., Siemiginowska, A., & Życki, P. T. 2004b, ApJ, 617, 102
Soltan, A. 1982, MNRAS, 200, 115
Stawarz, Ł., Aharonian, F., Kataoka, J., et al. 2006, MNRAS, 370, 981
Stawarz, Ł., Cheung, C. C., Harris, D. E., & Ostrowski, M. 2007, ApJ, 662, 213

Stawarz, Ł., Siemiginowska, A., Ostrowski, M., & Sikora, M. 2005, ApJ, 626, 120
Steenbrugge, K. C., Blundell, K. M., & Duffy, P. 2008, MNRAS, 388, 1465
Steenbrugge, K. C., Kaastra, J. S., Crenshaw, D. M., et al. 2005, A&A, 434, 569
Steffen, A. T., Barger, A. J., Cowie, L. L., Mushotzky, R. F., & Yang, Y. 2003, ApJ, 596, L23
Stein, N. M., van Dyk, D. A., Kashyap, V. L., & Siemiginowska, A. 2015, ApJ, 813, 66
Suh, H., Civano, F., Hasinger, G., et al. 2017, ApJ, 841, 102
Tananbaum, H., Avni, Y., Branduardi, G., et al. 1979, ApJ, 234, L9
Tchekhovskoy, A. 2015, in Astrophysics and Space Science Library, The Formation and Disruption of Black Hole Jets, Vol. 414, ed. I. Contopoulos, D. Gabuzda, & N. Kylafis, 45
Tchekhovskoy, A., Narayan, R., & McKinney, J. C. 2011, MNRAS, 418, L79
Tombesi, F., Cappi, M., Reeves, J. N., et al. 2013, MNRAS, 430, 1102
Tombesi, F., Cappi, M., Reeves, J. N., et al. 2010, A&A, 521, A57
Tombesi, F., Reeves, J. N., Kallman, T., et al. 2016, ApJ, 830, 98
Tonry, J. L., Dressler, A., Blakeslee, J. P., et al. 2001, ApJ, 546, 681
Tozzi, P., Gilli, R., Mainieri, V., et al. 2006, A&A, 451, 457
Trebitsch, M., Volonteri, M., & Dubois, Y. 2019, MNRAS, 487, 819
Treister, E., & Urry, C. M. 2006, ApJ, 652, L79
Treister, E., Urry, C. M., & Virani, S. 2009, ApJ, 696, 110
Tremblay, S. E., Taylor, G. B., Ortiz, A. A., et al. 2016, MNRAS, 459, 820
Trichas, M., Green, P. J., Silverman, J. D., et al. 2012, ApJS, 200, 17
Trump, J. R., Impey, C. D., Kelly, B. o. C., et al. 2011, ApJ, 733, 60
Turner, T. J., & Miller, L. 2009, A&ARv, 17, 47
Ueda, Y., Akiyama, M., Hasinger, G., Miyaji, T., & Watson, M. G. 2014, ApJ, 786, 104
Ueda, Y., Akiyama, M., Ohta, K., & Miyaji, T. 2003, ApJ, 598, 886
Urry, C. M., & Padovani, P. 1995, PASP, 107, 803
Uttley, P., Cackett, E. M., Fabian, A. C., Kara, E., & Wilkins, D. R. 2014, A&ARv, 22, 72
Vignali, C., Brandt, W. N., Schneider, D. P., & Kaspi, S. 2005, AJ, 129, 2519
Vito, F., Gilli, R., Vignali, C., et al. 2014, MNRAS, 445, 3557
Vito, F., Gilli, R., Vignali, C., et al. 2016, MNRAS, 463, 348
Vito, F., Brandt, W. N., Yang, G., et al. 2018, MNRAS, 473, 2378
Vito, F., Brandt, W. N., Bauer, F. E., et al. 2019, A&A, 628, L6
Volonteri, M., Capelo, P. R., Netzer, H., et al. 2015, MNRAS, 452, L6
Volonteri, M. 2010, A&ARv, 18, 279
Volonteri, M. 2012, Sci., 337, 544
Volonteri, M., & Rees, M. J. 2005, ApJ, 633, 624
Wagner, A. Y., Bicknell, G. V., & Umemura, M. 2012, ApJ, 757, 136
Wagner, A. Y., Bicknell, G. V., Umemura, M., Sutherland, R. S., & Silk, J. 2016, AN, 337, 167
Walsh, J. L., Barth, A. J., Ho, L. C., & Sarzi, M. 2013, ApJ, 770, 86
Wang, Q. D., Nowak, M. A., Markoff, S. B., et al. 2013, Sci, 341, 981
Wilkes, B. J., & Elvis, M. 1987, ApJ, 323, 243
Wilkes, B. J., Kilgard, R., Kim, D. -W., et al. 2009, ApJS, 185, 433
Wilkes, B. J., Kuraszkiewicz, J., Haas, M., et al. 2013, ApJ, 773, 15
Wilkins, D. R., & Fabian, A. C. 2012, MNRAS, 424, 1284
Wilkins, D. R., & Fabian, A. C. 2013, MNRAS, 430, 247
Williams, A. G., & Gull, S. F. 1985, Natur, 313, 34
Wilson, A. S., Smith, D. A., & Young, A. J. 2006, ApJ, 644, L9

Wilson, A. S., Young, A. J., & Shopbell, P. L. 2000, ApJ, 544, L27
Worrall, D. M. 2009, A&ARv, 17, 1
Worsley, M. A., Fabian, A. C., Bauer, F. E., et al. 2006, MNRAS, 368, 1735
Worsley, M. A., Fabian, A. C., Bauer, F. E., et al. 2005, MNRAS, 357, 1281
Wright, N. J., Drake, J. J., & Civano, F. 2010, ApJ, 725, 480
Wu, J., Brandt, W. N., Miller, B. P., et al. 2013, ApJ, 763, 109
Xue, Y. Q., Luo, B., Brandt, W. N., et al. 2011, ApJS, 195, 10
Yang, G., Brandt, W. N., Vito, F., et al. 2018, MNRAS, 475, 1887
Zhu, S. F., Brandt, W. N., Wu, J., Garmire, G. P., & Miller, B. P. 2019, MNRAS, 482, 2016

# Chapter 9

# Groups and Clusters of Galaxies

**Paul Nulsen and Brian McNamara**

## 9.1 Introduction

Prior to the launch of *Chandra*, the X-ray-emitting hot intergalactic medium that fills groups and clusters of galaxies appeared to be largely smooth and structureless. The hierarchical collapse that forms groups and clusters was expected to drive turbulence, while more violent mergers could cause shocks (e.g., Sarazin 1988), but little detailed consideration had been given to other impacts of structure formation on the intracluster medium (ICM[1]). The improved combination of effective area and spatial and spectral resolution provided by *Chandra* has radically revised our view of the ICM, revealing that complex substructure is the norm in both galaxy groups and clusters. In turn, this has led to an appreciation that the substructure can be used as a powerful probe of the physical processes taking place in these systems.

The material here focuses on advances that have relied on *Chandra* data, with limited discussion of the related theory. Space does not permit an exhaustive review of the past 20 years, so that some topics worthy of review are not covered and those that have been are presented through illustrative examples.

## 9.2 Basic Properties of Clusters and the ICM

Galaxy clusters are the most massive, gravitationally collapsed structures in the universe, and they are still in the process of forming (Sarazin 1988; Kravtsov & Borgani 2012). They have total masses of up to $\sim 2 \times 10^{15}\ M_{\odot}$, mainly in the form of dark matter, and radii up to ~2 Mpc. The second most massive constituent of clusters is the hot gas in the ICM, which contains up to ~12% of the mass in the most massive clusters (Mantz et al. 2014) and less in smaller systems. The ICM is a highly ionized, weakly magnetized plasma (Govoni & Feretti 2004; Bonafede et al. 2010), composed mainly of gas, together with a modest population of highly energetic particles (e.g., Churazov et al. 2010; Brunetti & Jones 2014). With gas temperatures

[1] In this chapter, ICM will refer to the medium filling both groups and clusters, unless noted otherwise.

doi:10.1088/2514-3433/ab43dcch9  

typically in the range $kT \sim$ 1–10 keV, the ICM produces thermal bremsstrahlung and line radiation mainly in the X-ray band accessible to *Chandra* (Sarazin 1988, Chapter 3). The composition of the gas is typical of cosmic plasma, predominantly hydrogen and helium, with heavier element abundances in the range of ~0.3–1 times the solar values (e.g., Mernier et al. 2017). For gas temperatures, $kT$, in the range 1–10 keV, the sound speed in the gas, $\sqrt{\gamma kT/(\mu m_{\rm H})}$, where the ratio of specific heats is $\gamma = 5/3$ and the mean mass per particle is $\mu m_{\rm H} \simeq 1.0 \times 10^{-24}$ g, lies in the range 520–1640 km s$^{-1}$, comparable to the thermal speed of a typical proton in the gas.

Apart from a few low-energy lines in the dense cores of some clusters (Churazov et al. 2004), the gas is optically thin, so that X-ray spectra provide readily used diagnostics for the physical state of the gas. Because collisions between gas particles are the main cause of its radiation, the power emitted per unit volume is proportional to the square of the gas density. It is traditionally expressed as $n_e n_{\rm H} \Lambda(T, Z)$, where $n_e$ is the electron number density, the hydrogen (i.e., proton) number density is $n_{\rm H} \simeq 0.85 n_e$, and the numerical factor depends weakly on the gas temperature and elemental abundances. The cooling function, $\Lambda$, depends on the gas temperature and its composition, signified here by $Z$.

Galaxy groups share many of the properties of galaxy clusters, but they are less massive. The dividing line between groups and clusters is not well defined, but systems with gas temperatures $kT \lesssim 1$ keV are generally regarded as groups. Another common distinction for more massive clusters is between "cool-core" and "noncool-core" systems. The temperature generally rises inward from the virial radius of a cluster, but in some clusters it declines again in the central region, giving rise to the term "cool core." More significantly, coupled to the temperature decline, the gas density rises steeply, making cool cores very bright in X-rays. A number of different criteria have been used to distinguish cool cores (e.g., see Hudson et al. 2010). The physical significance of cool cores is that the heat lost from the gas there over the life of a cluster can exceed the thermal energy of the gas, in which case we should expect it to cool significantly. Roughly one-third of galaxy clusters have a cool core (e.g., Andrade-Santos et al. 2017).

## 9.3 AGN Feedback in Groups and Clusters

### 9.3.1 The Need for Feedback

The cooling time of the X-ray-emitting gas is the time required for it to radiate its thermal energy,

$$t_{\rm cool} = \frac{3p}{2 n_e n_{\rm H} \Lambda}, \tag{9.1}$$

where $p$ is the pressure. Prior to the launch of *Chandra*, the cooling time of the gas was known to be significantly shorter than the age in cluster cool cores. Large quantities of the intracluster gas were thought to be cooling to low temperatures in these systems, causing hundreds of solar masses, or more, per year to flow toward the cluster center in what is known as a cooling flow (Fabian 1994). In the Perseus cluster, for example, ~300 $M_\odot$ yr$^{-1}$ were thought to be cooling from the within the central ~100 kpc. However, there was little evidence for the copious quantities of cooled gas and star formation expected, a conundrum known as the cooling flow problem (e.g., McNamara & Nulsen 2007). Many heating mechanisms had been proposed to prevent the gas from cooling, including thermal conduction from larger radii (e.g., Zakamska & Narayan 2003; Voigt & Fabian 2004), dynamical friction (Miller 1986), shocks and turbulence generated by ongoing infall and mergers (Gómez et al. 2002; O'Hara et al. 2006), heating by central active galactic nuclei (AGN; Pedlar et al. 1990; Binney & Tabor 1995; Tucker & David 1997), and many others.

On all scales and in almost all known cases, the time required for sound to propagate to a cluster center is shorter than the cooling time of the gas, so that the hot gas should remain close to local pressure equilibrium as it radiates. Thus, higher gas densities accompany lower temperatures and vice versa. Higher density and lower temperature both tend to reduce the cooling time of the gas, whereas lower density and higher temperature generally prolong cooling. By contrast, heating processes tend to be more effective when the gas density is lower. As a result, when heating and cooling processes compete, one tends to end up dominating. This leads to local thermal instability. If cooling dominates, large quantities of gas are expected to cool, leaving the cooling flow problem unresolved. If heating dominates, it will tend to drive the cooling time upwards toward values comparable to a cluster's age, inconsistent with the high proportion of short central cooling times found in clusters. The prevalence of short central cooling times (Hudson et al. 2010; Andrade-Santos et al. 2017) combined with the shortage of cooled and cooling gas (e.g., McDonald et al. 2018) cannot be explained by thermally unstable heating mechanisms.

Although some locally stable heating mechanisms have been proposed (e.g., Dennis & Chandran 2005), they are not supported by observations. The likely explanation for the persistence of short cooling times is that heating and cooling rates are linked in a negative feedback loop.[2] This can occur naturally if cooled or cooling gas is the fuel for a central AGN that provides the power to inhibit further cooling. Some review articles on AGN feedback in clusters and less massive hot atmospheres include McNamara & Nulsen (2007, 2012), Fabian (2012), Gitti et al. (2012), and Werner et al. (2019).

### 9.3.2 The Case for Feedback

An extragalactic radio source is powered by a pair of opposed jets that convey power from the vicinity of the event horizon of a supermassive black hole into the surrounding medium, on galactic scales. The power deposited via the jets inflates radio lobes with relativistic particles and magnetic field. The expanding radio lobe drives cocoon shocks into the surrounding medium (Scheuer 1974; Begelman et al. 1984). In a prescient paper, Gull & Northover (1973) argued that radio lobes will displace the hot ICM to create X-ray cavities, or "bubbles," in galaxy clusters. This

[2] Astronomers often use "feedback" to mean the impact of AGN on their environments; here, it has the engineering sense of a closed cycle in which the output is coupled to the input.

was confirmed 20 years later by the discovery, with the *ROSAT* High Resolution Imager, of X-ray cavities in three systems, the Perseus cluster (Boehringer et al. 1993), Cygnus A (Carilli et al. 1994), and Abell 4059 (Huang & Sarazin 1998).

Commencing early in the mission, *Chandra* observations started revealing systems of X-ray cavities in cool-core clusters, e.g., Hydra A (McNamara et al. 2000), Perseus (Figure 9.1; Fabian et al. 2000), Abell 2052 (Figure 9.2; Blanton et al. 2001), Abell 2597 (McNamara et al. 2001), RBS 797 (Schindler et al. 2001), and Abell 4059 (Heinz et al. 2002). In many cases, the cavities coincide with the lobes of a radio source hosted by the central galaxy. Cavities have also been found associated with radio lobes in many lower mass systems, such as M84 (Finoguenov & Jones 2001), NGC 4636 (Jones et al. 2002), NGC 4552 (Machacek et al. 2006), Centaurus A (Figure 9.6; Croston et al. 2009), and NGC 5813 (Figure 9.3; Randall et al. 2011). Although fewer radio lobe cavities are known in lower mass systems, this is chiefly because cavities embedded in the dense, X-ray-bright gas of a cool-core cluster are more readily detected. *Chandra* observations have established that cavities are ubiquitous around the lobes of radio sources in X-ray-emitting hot atmospheres. The total number of hot atmospheres shown by *Chandra* to have radio lobe cavities is now of order 100 (e.g., Shin et al. 2016). The great majority of the associated radio sources are Fanaroff–Riley class I (FRI), although a small number of cavities are associated with Fanaroff–Riley class II (FRII; Fanaroff & Riley 1974). Some "ghost" cavities, lacking detected radio emission, have been found, although deep low-frequency radio observations have tended to reveal the radio emission expected for an aging radio lobe (e.g., Fabian et al. 2002; Clarke et al. 2005). The X-ray cavities are often referred to as bubbles, reflecting the low density of the radio plasma that fills them compared to the surrounding ICM.

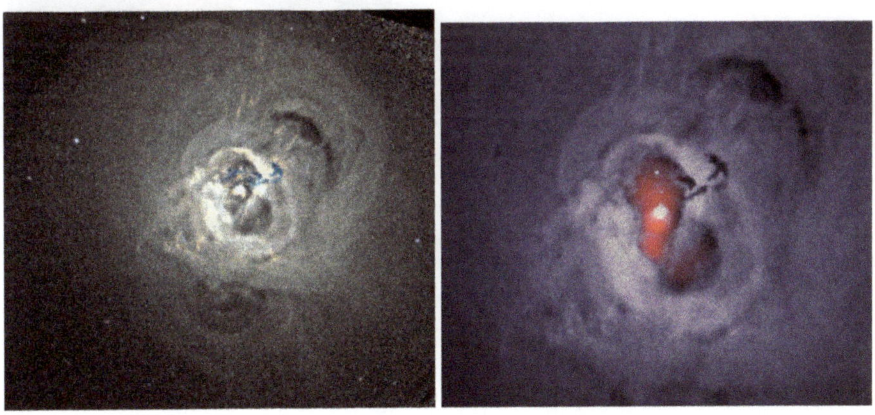

**Figure 9.1.** Left: unsharp mask color *Chandra* image of the central region of the Perseus cluster (Fabian et al. 2006, by permission of Oxford University Press on behalf of the Royal Astronomical Society). A color image combining 0.3–1.2 keV (red), 1.2–2 keV (green), and 2–7 keV (blue) images was smoothed, multiplied by 0.8, and subtracted from the unsmoothed image to obtain this image highlighting small-scale structure. The field of view is $\simeq 6.3' \times 5.7'$. The AGN (bright spot near the center) lies between cavities $\sim 0.5'$ (11 kpc) to the north and south, with a crescent-shaped cavity $\sim 1.5'$ to the northwest. Right: radio image overlaid in red, filling the inner cavities. The interactive figure shows separately the X-ray emission (blue), radio emission (pink), and the composite image, as shown in the right hand panel. Additional materials available online at http://iopscience.iop.org/book/978-0-7503-2163-1.

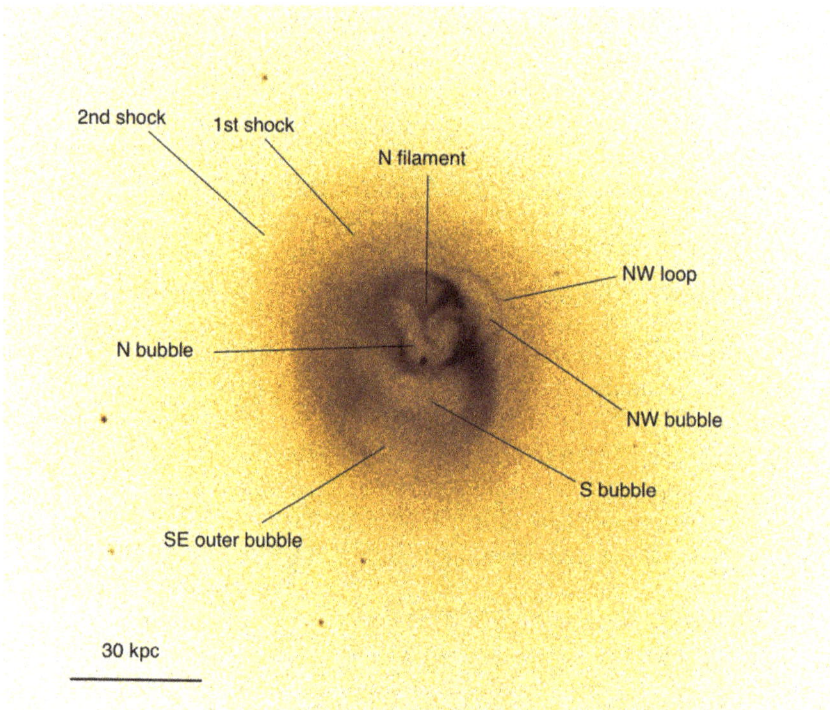


**Figure 9.2.** 0.5–3 keV *Chandra* image of Abell 2052, with cavities (bubbles) and shocks marked (Blanton et al. 2011) 

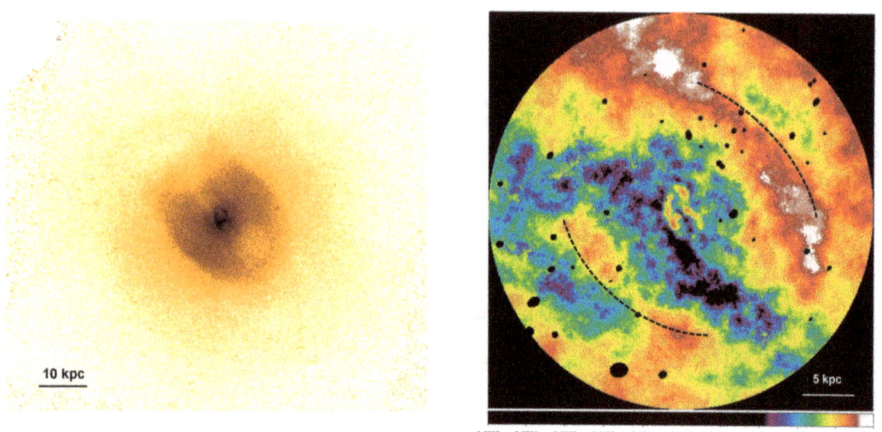


**Figure 9.3.** NGC 5813 (Randall et al. 2015)  Left: 0.3–3 keV *Chandra* image, showing three pairs of cavities, at 1–2 kpc from the center, 5–10 kpc out, and, faintly visible, at 20–30 kpc. Right: temperature map (note the different spatial scale). Black dashed lines mark the positions of the ~10 kpc shocks. Temperature rises can be seen behind those and the ~1 kpc shocks.

AGN feedback is thought to play a broad role in galaxy formation. Evidence is accumulating that, at early times, when the supermassive black holes in galactic nuclei were growing rapidly, the radiation emerging from matter accreting at close to the Eddington limit can drive away surrounding gas, throttling back star formation and further accretion (reviewed by Fabian 2012; King & Pounds 2015). At later times, when accretion rates are lower, a substantial fraction of the accretion power emerges in jets (Churazov et al. 2002). The former process is called "quasar mode" feedback, while the latter is known as "radio mode" feedback (Croton et al. 2006; Sijacki & Springel 2006). *Chandra* findings on radio mode feedback are discussed here.

### 9.3.3 Cavity Calorimetry

Observations of X-ray cavities have provided a valuable tool for the study of AGN feedback in clusters. Churazov et al. (2000) were the first to recognize that the cavities in the Perseus cluster could be used to estimate the power of the AGN jets. The total energy required to inflate a radio lobe is the sum of the work done on the surrounding gas and the internal energy stored within it. If a cavity is inflated against a constant external pressure, $p$, the work required is just $pV$, where $V$ is the volume of the cavity. The internal energy of a cavity at pressure $p$ can be expressed as $E = pV/(\gamma - 1)$, where $\gamma$ is the ratio of specific heats of the fluid within the cavity. Thus, Churazov et al. (2000) estimated the energy required to inflate a cavity as the sum of these, which is the enthalpy,

$$H = E + pV = \frac{\gamma}{\gamma - 1} pV. \tag{9.2}$$

Because radio synchrotron emission from a lobe requires the presence of relativistic electrons and a magnetic field, $\gamma$ is generally assumed to be close to 4/3, giving an enthalpy of $H = 4pV$. If the cavity is filled predominantly with nonrelativistic, hot cosmic gas, as may be the case in older radio lobes (e.g., Bîrzan et al. 2008), we would have instead $\gamma = 5/3$ and $H = 2.5pV$.

Combining this with the age of a cavity, we can estimate the mean jet power of the radio source that inflated the lobe. To estimate the age of a cavity, Churazov et al. (2000) assumed that it has risen to its current position at its terminal speed due to buoyancy, i.e.,

$$v_{\rm b} \simeq K_{\rm b} v_{\rm K} \sqrt{r/R}, \tag{9.3}$$

where the constant $K_{\rm b}$ is of order unity, $r$ is the radius of the cavity, $R$ is its distance from the cluster center, the *Kepler* speed is $v_{\rm K} = \sqrt{GM(R)/R}$, and $M(R)$ is the total gravitating mass enclosed within the radius $R$. The buoyant rise time would then be $t_{\rm b} = R/v_{\rm b}$, and the mean jet power over the lifetime of the AGN outburst can be estimated as

$$P_{\rm cav} = \frac{4pV}{t_{\rm b}}. \tag{9.4}$$

Subsequently, Churazov et al. (2002) introduced the term "cavity calorimetry" for this approach to estimating jet power. This estimate of the jet power is often referred to as the "cavity power."

*Calorimetry and Feedback*

In the first systematic study, Bîrzan et al. (2004) used cavity calorimetry to estimate jet powers for 16 galaxy clusters observed with *Chandra*. To address uncertainties in the ages of the radio lobes, they used three different estimates for cavity ages—the sonic rise time, the buoyant rise time, and the refill time—but found that all gave similar results. This is because the sound speed is generally similar to the *Kepler* speed, and the size of a cavity, $r$, is comparable to its distance from the AGN, $R$, making the buoyant speed (Equation (9.3)) close to the sound speed. The relatively narrow distribution of $r/R$ (e.g., Shin et al. 2016) is a selection effect, at least in part, as larger values make the X-ray decrement over a cavity larger, favoring its detection.

For feedback from a radio AGN to prevent the gas from cooling and forming stars at a cluster center, the power it injects into the ICM must, at least, be sufficient to replace the power lost by radiation from the region that would otherwise cool. Within that region, the cooling time is shorter than the time since the last major merger. For their low-redshift sample, Bîrzan et al. (2004) assumed that the last major merger was at redshift unity, approximately 7.7 Gyr ago, so they defined the cooling radius to be where the cooling time is 7.7 Gyr. The cooling power, $L_{\rm cool}$, is then the power radiated from within the cooling radius. Bîrzan et al. (2004) found that the jet power, estimated as the cavity power, is comparable to the cooling power for the systems in their sample, although with significant scatter. A more recent compilation of results is shown in Figure 9.4. Observed values of $P_{\rm cav}$ range from more than an order of magnitude greater than $L_{\rm cool}$ in some systems down to zero in effect for systems where no cavities are detected. However, there is a better match in the mean relationship between $P_{\rm cav}$ and $L_{\rm cool}$. If, as seems likely, the jet power is insufficient to prevent cooling in some systems, the jet powers of the central AGN must be variable, while maintaining an average heating rate sufficient to balance cooling in each core.

Subsequent studies have confirmed and strengthened these results. Rafferty et al. (2006) extended the sample of Bîrzan et al. (2004) to 33 systems and confirmed their conclusions. For the 55 brightest galaxy clusters, Dunn & Fabian (2006) found that 20 would be prone to cool at the center. Of those, at least 14 have radio lobe cavities, implying a duty cycle of 70% or more for the presence of X-ray cavities. Similarly, for lower mass systems, AGN jets have been found to inject sufficient power to prevent the gas from cooling (Nulsen et al. 2009). The mean value of $P_{\rm cav}/L_{\rm cool}$ appears to be larger for the low-mass systems, and a smaller fraction of them have cavities, suggesting that the duty cycle of their AGN outbursts is lower. However, Dong et al. (2010) argued that this may still be a selection effect, as cavities are harder to detect in less massive halos.

Cavity calorimetry will be one of the significant legacies of *Chandra*. The sensitivity and high spatial resolution have been critical for the detection of cavities

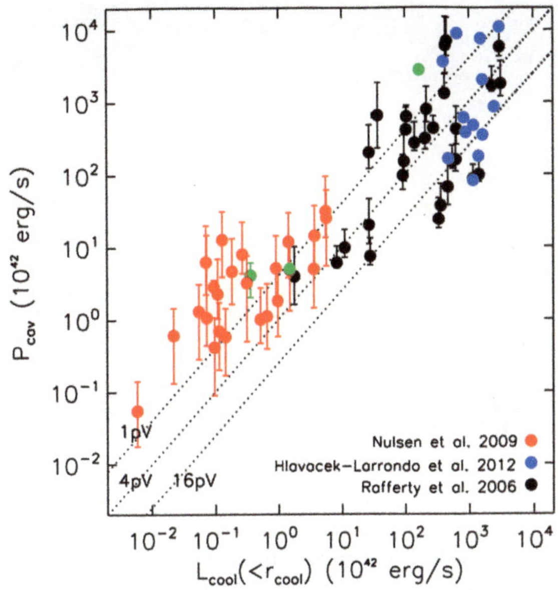


**Figure 9.4.** Cavity power versus cooling power. This compilation includes results from Rafferty et al. (2006) with those for a lower mass sample (Nulsen et al. 2009) and a sample of very massive clusters in the redshift range $0.3 < z < 0.7$ (Hlavacek-Larrondo et al. 2012).

and measurement of their properties, particularly the pressure and volume. Cavity calorimetry has made a strong empirical case for radio mode feedback.

*Limitations of Cavity Calorimetry*

Cavities can be difficult to identify (e.g., Bîrzan et al. 2012), and many cavities are surrounded by X-ray-bright, lower entropy gas (e.g., Nulsen et al. 2002; Rafferty et al. 2006), making it hard to determine their properties accurately. Many assumptions underlie cavity calorimetry. For example, if the X-ray-emitting gas is not displaced completely by a radio lobe, the volume in the expression for the enthalpy (Equation (9.2)) should be reduced by the filling factor of the radio plasma. X-ray emission from the remaining gas would then partly fill in the cavity, reducing its contrast with the surrounding ICM. The difference between the X-ray emission over an empty cavity and an adjacent region can be used to constrain the filling factor, but it depends on the dimensions and location of the cavity in a cluster, and the placement along our line of sight is not known. Existing estimates of cavity-filling factors are consistent with unity (e.g., McNamara et al. 2000; Wise et al. 2007), but this issue has not be studied extensively. Working in the opposite sense, inverse-Compton X-rays from the radio lobes of Cygnus A (Figure 9.5), and in other powerful radio galaxies, more than make up for the thermal emission of the gas displaced by the lobes, causing the lobe volumes to be significantly underestimated (de Vries et al. 2018). Uncertainties in cavity volumes have a significant impact on the results of cavity calorimetry (McNamara & Nulsen 2012).

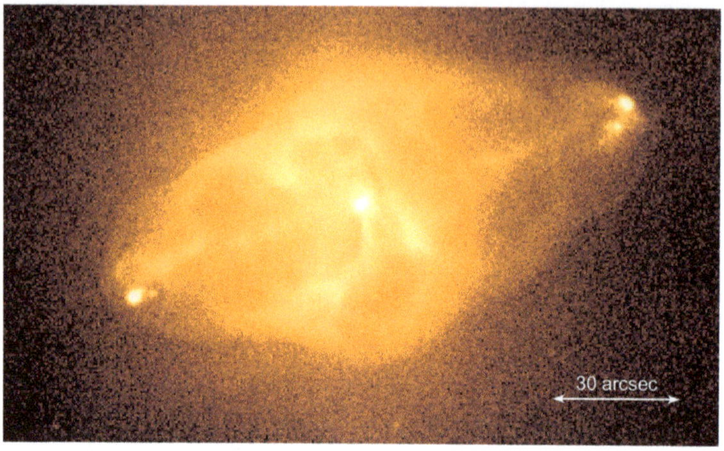


**Figure 9.5.** 0.5–5 keV *Chandra* image of the radio cocoon of Cygnus A. The AGN is the bright spot near the center, and the radio hotspots can be seen near the eastern and western ends of the cocoon. The cocoon shock appears as the sharp edge enveloping the whole of the bright cocoon. The image scale is 1.09 kpc $\text{arcsec}^{-1}$. The interactive figure shows separately the X-ray emission (blue, shown in the figure), optical star field (yellow), radio emission (red), and the composite image of Cygnus A. Additional materials available online at http://iopscience.iop.org/book/978-0-7503-2163-1

**Figure 9.6.** *Chandra* false-color image of Centaurus A. Courtesy of NASA/CXC/CfA/R.Kraft et al. The color scheme makes thermal emission from hot gas red, while non-thermal emission appears yellow/green to white. The absorbing dust lane is dark green/gray. The southwest shock front is $\simeq 5.7'$ or $\simeq 6.0$ kpc from the AGN, near the center. The interactive figure shows separately the X-ray high energy, non-thermal emission (blue), medium energy emission (green), low energy thermal emission (red), and the composite image of Cen A, similar to that shown in the figure. Additional materials available online at http://iopscience.iop.org/book/978-0-7503-2163-1

The work done on its surroundings by an expanding radio lobe is $W = \int p\, dV$, so the rate of work scales as $pR^2\dot{R}$, where $\dot{R}$ is the speed at which the lobe expands into a cluster atmosphere. In cluster cores, $pR^2$ generally decreases inward, so that, under constant jet power, a lobe will expand faster, at higher Mach number, early in its life. A higher Mach number increases the ratio of the pressure inside the cocoon shock to that in the ambient ICM. Thus, earlier in the life of an expanding radio lobe, the pressure in the lobe will have been substantially greater than it is today, causing the total work done by the expanding lobe to exceed the $pV$ evaluated today by a substantial margin. This means that the enthalpy represents the minimum energy required to create a cavity. In reality, the total energy deposited by a radio outburst will usually exceed the lobe enthalpy by a significant margin. The division of the energy deposited via the jets between the enthalpy of the radio lobe and the excess work, $\int p\, dV - pV$, depends on the history of the AGN outburst. For example, Forman et al. (2017) found that roughly half of the total energy deposited by the jets of M87 remains as the enthalpy of its lobes today. While it is generally assumed that the energy input represented by a cavity is $4pV$, a more realistic value could easily be twice this.

The pressure within a cavity is usually assumed to match that in the surrounding hot atmosphere at the radius of the cavity center, being a relatively accessible quantity. While this approximation is good for a slowly expanding lobe, it will significantly underestimate the pressure in a young, rapidly expanding radio lobe with a moderately strong cocoon shock. The Rankine–Hugoniot jump condition for the pressure increase behind a shock with Mach number $\mathcal{M}$ can be expressed as

$$\frac{\delta p}{p} = \frac{2\gamma}{\gamma + 1}(\mathcal{M}^2 - 1), \tag{9.5}$$

where $p$ is the preshock pressure, $p + \delta p$ is the postshock pressure, and $\gamma = 5/3$ for the ICM. Thus, even for a shock Mach number of $\simeq 1.34$, the pressure behind the shock is doubled. Even for the weakest known cocoon shocks in a cluster, with $\mathcal{M} \simeq 1.2$, the cocoon pressure would be 55% higher than in the adjacent ICM. Although the majority of the observed cocoon shocks for radio sources hosted by cluster central galaxies are weak (Section 9.3.4), this correction has an appreciable effect. These weak shocks are difficult to detect, and many more are likely to be revealed in future data.

In summary, the conventional assumption of $4pV$ of energy injected per cavity is almost certainly an underestimate. More realistic values are probably several times greater, but detailed radio source models and deeper data will be required to resolve this.

### 9.3.4 AGN Shocks

Expanding radio lobes have long been expected to drive "cocoon" shocks into the surrounding gas (Scheuer 1974). Prior to the launch of *Chandra*, the shocks were expected to be strong enough to be readily detected in the X-ray (e.g., Heinz et al. 1998), but very few of the early observations of radio lobe cavities revealed signs of shocks. Perhaps the clearest evidence of shocks in early *Chandra* observations was found in the powerful radio galaxy, Cygnus A, which is hosted by the central galaxy of a cool-core cluster. Smith et al. (2002) detected a band of enhanced emission around the radio cavities of Cygnus A from gas that is hotter than the adjacent ICM. A shock boosts both the density and temperature of the gas, so this feature is consistent with expectations for the shocked gas being pushed aside by a rapidly expanding radio lobe. More recent, very deep observations of Cygnus A clearly confirm the nature of this feature, but even for Cygnus A, the measured shock strength only ranges between Mach 1.2 and 1.7 (Snios et al. 2018). Figure 9.5 shows a *Chandra* image of the radio cocoon of Cygnus A, with a sharp shock front enveloping the X-ray-bright radio cocoon. The cocoon is approximately 140 kpc in length. The shock strength varies over the cocoon, implying Mach numbers approaching 3 at the eastern and western extremities. although such high Mach numbers are yet to be measured from X-ray data. Detectability is sensitive to the preshock density and the radius of curvature of the front, both of which are minimized where the shock is fastest. About 30″ east of the AGN, narrow bright layers of shock-compressed gas can be seen along the southern and northern boundaries of the cocoon, with a fainter region in between, where the radio lobe has expelled the ICM to create an X-ray cavity.

Another likely system of cocoon shocks, detected in early *Chandra* observations, is in NGC 4636, the dominant galaxy of a small subgroup on the outskirts of the Virgo cluster (Jones et al. 2002; Baldi et al. 2009). It is noteworthy that the Mach number of the shocks in NGC 4636 is $\simeq$1.7, a far less powerful radio galaxy than Cygnus A ($P_{1.4} \simeq 10^{22}$W Hz$^{-1}$ versus $6 \times 10^{27}$ W Hz$^{-1}$ for Cygnus A). Principally, this reflects the difference in central pressure between a massive cool-core cluster and a small group (see Equation (9.5)).

While cocoon shocks are not detected around most radio lobe cavities, they have been detected in a number of cases, including Centaurus A (Figure 9.6; Kraft et al. 2003), Perseus A (Figure 9.1; Fabian et al. 2003a), Hercules A (Nulsen et al. 2005a), M87 (Figure 9.7; Forman et al. 2005, 2007), MS0735.6+7421 (McNamara et al. 2005), Hydra A (Nulsen et al. 2005b), and NGC 5813 (Figure 9.3; Randall et al.

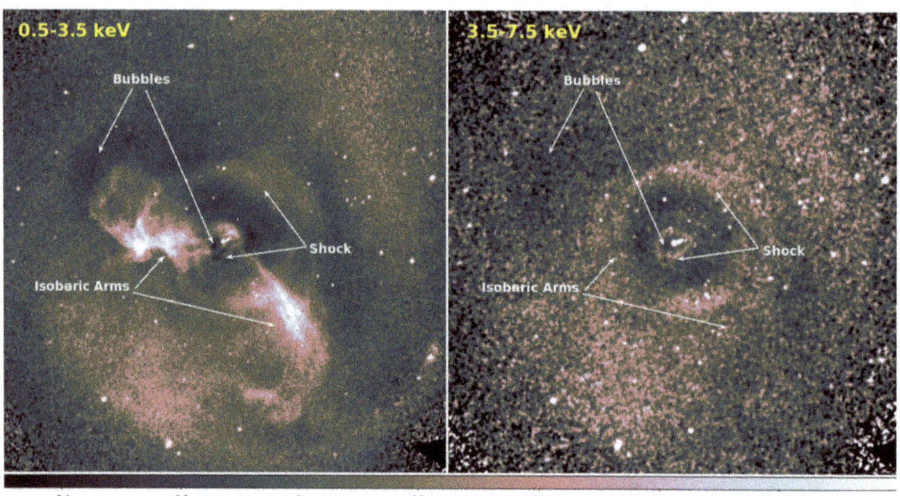


**Figure 9.7.** *Chandra* images of M87 divided by the best-fitting beta model (Arévalo et al. 2016). © 2016. The American Astronomical Society. All rights reserved. Left: the 0.5–3.5 keV image is most sensitive to density fluctuations, highlighting the X-ray cavities (bubbles) and the "arms" of low-entropy gas lifted by the radio lobes. Right: the 3.5–7.5 keV image is sensitive to the gas pressure and highlights the shocks around the inner radio lobe and at 13 kpc.

2011). By far the strongest known cocoon shock delimits the southwest lobe of Centaurus A (Croston et al. 2009), which can be observed in far greater detail than any other radio galaxy, due to its proximity. Centaurus A resides in what is probably the poorest X-ray-detected group environment (Figure 9.6). Along the northern edge of the southwest radio lobe, where the shock Mach number is estimated to be $\simeq 2.8$, emission from the shocked gas is predominantly thermal. At the southwestern edge of the lobe, farthest from the AGN, the Mach number is estimated to be closer to 8, resulting in predominantly non-thermal postshock X-ray emission.

The next strongest X-ray-detected cocoon shock surrounds NGC 5813 (Figure 9.3) at the center of a modest group (Randall et al. 2015), with a Mach number of $\simeq 1.8$. The remaining cocoon shocks detected thus far are weaker. The scarcity of stronger shocks is likely to be a selection effect. As discussed in Section 9.3.3, lobes expand faster early in their lives, so that they spend substantially longer in their later stages, when the rate of expansion is slowest and any shocks will be weak. Furthermore, the largest intact lobes will generally be the most easily detected, increasing the likelihood of finding them in their later stages. The ultimate fate is not well determined. Simulated lobes are prone to instabilities that cause them to disintegrate quickly, so that viscosity (Reynolds et al. 2005), magnetic field (Kaiser et al. 2005), or some other physical processes (e.g., Pizzolato & Soker 2006) are required to slow their disruption. If, as argued by Pfrommer (2013), cosmic rays stream out of lobes to heat that ICM, the lobes would tend to deflate. However, at least some cavities appear to have detached from the jet and now survive as free-floating bubbles. Also, as most of the known shocks are weak, a substantial fraction of them must survive at least until their pressure approaches that of the surrounding ICM. In short, it remains unclear whether the absence of cocoon shocks in the majority of cases is primarily due to inadequate observations, or because the cavities survive until after the cocoon shocks detach.

*Shock Calorimetry*

Cocoon shocks provide an alternative means of estimating jet powers that relies on distinct data, thus providing a check on the assumptions of cavity calorimetry. The physics of hydrodynamic shocks is relatively simple, and spherical models are adequate for the spherical parts of a shock front, which is usually the closest to the AGN. This is also the region where the X-ray data provide the best information, making shocks more likely to be detected. Fully hydrodynamic, spherical, numerical models of an explosive outburst have been used to estimate the total energies and ages for several shocks (e.g., Nulsen et al. 2005b; McNamara et al. 2005; Forman et al. 2005; Randall et al. 2011). Allowing for the shortcomings of these models, they have provided results consistent with cavity-based estimates of the ages and powers of the AGN outbursts.

Improved modeling will provide more accurate tests of radio source and feedback models in the future, and some steps have already been taken in that direction. For a fixed total energy, explosive injection maximizes the shock speed, therefore minimizing the total energy required to obtain an observed Mach number. The age of the outburst is also minimized, along with the remaining cavity enthalpy. More realistic models, with continuous energy injection, leave a greater proportion to the total energy as cavity enthalpy. To match the Mach number and radius of the shock in M87, as well as the size of the cavities, Forman et al. (2017) concluded that the shock at 13 kpc from the AGN was driven by an outburst that started $\simeq$12 Myr ago and had a duration of $\simeq$2 Myr (Figure 9.7). There is considerable potential for further improvements in shock models, using existing data to better constrain the properties of radio outbursts.

### 9.3.5 Feeding the AGN

Initially, Bondi accretion (see Section 8.2.2) from the hot ICM was considered the natural candidate to fuel AGN feedback. The Bondi accretion rate is sensitive to the entropy of the hot gas, which is modified directly by heating and cooling, providing a simple feedback mechanism. However, although Bondi accretion would be sufficient to power many of the observed AGN outbursts, it falls short for the most powerful ones (Rafferty et al. 2006). Recently, Russell et al. (2018) used carefully planned *Chandra* observations to show that the flow speed of gas at the Bondi radius in M87 falls well short of that expected for Bondi flow, largely ruling out Bondi accretion in M87.

The alternative fuel source is cold gas (Pizzolato & Soker 2005; Gaspari et al. 2013; Voit et al. 2015). Cool-core clusters can contain substantial masses of cold molecular gas (Edge 2001; Salomé & Combes 2003), some exhibit significant star formation (e.g., Johnstone et al. 1987; McNamara & O'Connell 1989), and many have extended emission-line nebulae in their cores (e.g., Heckman 1981; Johnstone et al. 1987; Crawford et al. 1999). The presence of these components is related to the thermal state of the hot ICM. Cavagnolo et al. (2008) showed that emission-line nebulae are found almost exclusively in cluster cores with a central entropy index of $kT/n_e^{2/3} \lesssim 30$ keV cm$^2$. Rafferty et al. (2008) only found star formation in systems with central cooling times $\lesssim 10^9$ yr, which amount to much the same criterion as the entropy threshold for emission-line nebulae. There is also a strong physical association between the various cooler gas phases. For example, Fabian et al. (2003b) found that soft X-ray filaments, revealing the presence of a cooler phase in the hot ICM, accompany H$\alpha$-emitting gas in the core of the Perseus cluster. The H$\alpha$-emitting gas in Perseus is also closely associated with warm molecular gas at $\sim$1000 K (Lim et al. 2012) and cold molecular gas. Of the cooler phases, the great bulk of the mass resides in the cold molecular gas (Salomé et al. 2011). ALMA observations are confirming the association between cold molecular gas and emission-line nebulae in other cool-core clusters (e.g., Russell et al. 2017a, 2017b).

The strong physical association between the cold molecular gas and intermediate phases up to the temperature of the hot ICM is a clear signature of interaction between the cold gas and the hot ICM. The segregation of cold gas to systems with short cooling times, i.e., those that are most prone to cooling, suggests strongly that the cold gas has cooled out of the hot ICM. Although individual cold clouds have too much angular momentum to reach the AGN, models for the accretion of cold clouds rely on collisions between the clouds on a hierarchy of scales to feed the gas inward and, eventually, via an accretion disk to the AGN (Gaspari et al. 2012, 2013).

*Thermal Instability*

Feedback models generally assume that a heating process balances the heat lost by radiation in some average sense. In local pressure equilibrium, gas that is cooler than average is also denser and so radiates faster. Because the average heating rate cannot keep up, this leads to net heat loss from the cooler gas. Similarly, gas that is hotter than average will be less dense, leading it to gain heat. In a stably stratified atmosphere, a gas "blob" that is denser than average also experiences a net downward force that will set it moving toward the place where its entropy matches the local average. At that location, the average heating rate would also match its radiative heat loss, because gas with the same entropy and pressure will have the same density. The fate of the blob thus depends on whether it can reach that location before cooling to a low temperature.

Disregarding heating and cooling, in the linear approximation, an adiabatic blob will oscillate about its equilibrium position at the Brunt–Väisälä frequency, with angular frequency, $\omega_{\rm BV}$, given by

$$\omega_{\rm BV}^2 = \frac{1}{\gamma}\frac{d\ln K}{d\ln R}\frac{g}{R}, \tag{9.6}$$

where $g$ is the acceleration due to gravity, $K = kT/n_{\rm e}^{2/3}$ is the entropy index, and the remaining symbols are defined in Section 9.3.3. The first two factors on the right are both of order unity. Starting from rest, a blob would take a quarter of a period to reach its equilibrium position, and this is close to the free-fall time, defined to be

$$t_{\rm ff} = \sqrt{2R/g}\,. \tag{9.7}$$

The timescale for the growth of thermal instability is generally the cooling time, so, in the linear approximation, the fate of an overdense gas blob is largely determined by the ratio of the cooling time to the free-fall time, $t_{\rm cool}/t_{\rm ff}$. For $t_{\rm cool}/t_{\rm ff} \lesssim 1$, thermally unstable cooling is expected to be rampant (McCourt et al. 2012). The central role of $t_{\rm cool}/t_{\rm ff}$ in the linear instability has driven the discussion of the more general case in terms of this ratio. Based on numerical simulations of AGN feedback, several authors have argued for a threshold on $t_{\rm cool}/t_{\rm ff}$ of 10 or somewhat greater for thermally unstable cooling (McCourt et al. 2012; Sharma et al. 2012; Gaspari et al. 2013; Voit et al. 2015). More recently, Choudhury & Sharma (2016) used numerical simulations to determine a threshold of $t_{\rm cool}/t_{\rm ff} < 10$. However, such a threshold is not supported by observations. There is no evidence for such low values in clusters, with observed minimum values in the range 10–30 (Figure 9.8, Hogan et al. 2017b).

In practice, nonlinear effects almost certainly come into play for thermally unstable cooling. Allowing for buoyancy, the net weight of a gas blob of volume $V$ is $gV\delta\rho$, where $\delta\rho = \rho - \rho_{\rm e}$ is the difference between the densities of the blob and the ambient gas. If the blob moves through the ambient gas at speed $v$, presenting a cross-sectional area $A$ to it, the drag on the blob is approximately $\rho_{\rm e} v^2 A$. Equating the net weight to the drag gives the terminal speed of a blob,

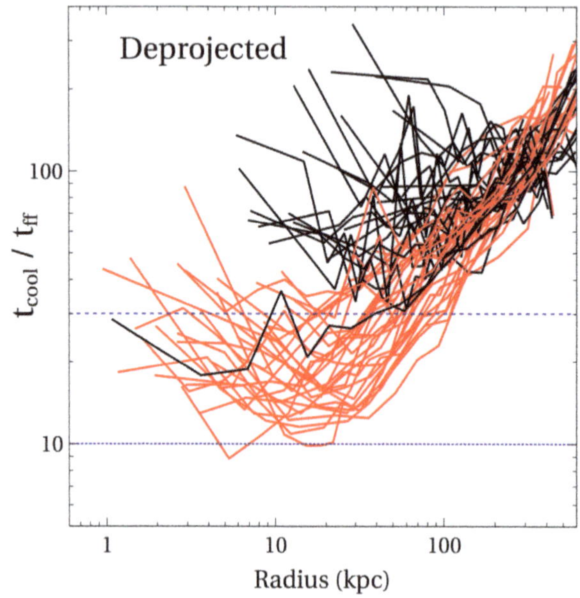


**Figure 9.8.** Ratio of cooling time to free-fall time for the sample of Hogan et al. (2017b).  Systems with H$\alpha$ detections, indicating the presence of cold gas, are plotted in red. Error bars are omitted, but there is little evidence for values of $t_{\rm cool}/t_{\rm ff} < 10$.

$$v_{\rm t} \simeq \sqrt{gr\delta\rho/\rho_{\rm e}} = v_{\rm K}\sqrt{\frac{r}{R}\frac{\delta\rho}{\rho_{\rm e}}}, \tag{9.8}$$

where the blob is assumed to be a sphere of radius $r$ and some factors of order unity are omitted. Because the drag is fundamentally nonlinear, if a blob undergoing Brunt–Väisälä oscillations reaches terminal speed, its motion cannot be treated in the linear approximation. For small oscillations, the amplitude, $A$, is related to the density perturbation by $A \simeq R\delta\rho/\rho_{\rm e}$, so the condition for nonlinearity, $\omega_{\rm BV}A \gtrsim v_{\rm t}$, gives $\delta\rho/\rho_{\rm e} \gtrsim r/R$. In other words, the motion of a small blob becomes nonlinear if it is displaced from its equilibrium position by a distance exceeding its size. This makes it very easy to generate nonlinear perturbations, limiting the value of the linear analysis.

For a small blob moving at its terminal speed, the time required for it to return to its equilibrium position is $\simeq t_{\rm ff}\sqrt{(\delta\rho/\rho_{\rm e})/(r/R)}$, where the factor under the square root exceeds unity. The condition for thermal instability would then be $t_{\rm cool} \lesssim t_{\rm ff}\sqrt{(\delta\rho/\rho_{\rm e})/(r/R)}$, so the blob may be able to cool, even if $t_{\rm cool} > t_{\rm ff}$. In reality, nonlinear perturbations generally cannot be treated as isolated blobs and even an isolated blob is likely to be pulled apart before it cools. Numerical simulations are required to deal with such complications (e.g., Choudhury & Sharma 2016).

*Entropy Profiles and Thermal Instability*

A large body of data on the thermal properties of the ICM has been assembled with *Chandra*. During its first decade, the large-scale entropy profile of the ICM was found to be a power law, $K \sim r^{\eta}$, with $\eta \simeq 1.1$, consistent with expectations for hierarchical gravitational collapse (e.g., Voit et al. 2005; Cavagnolo et al. 2009). Very high-quality X-ray data are required to resolve gas temperatures on small scales, so that central entropy profiles of cool cores were not well determined before the work of Panagoulia et al. (2014), who found that the entropy continues to decline to the smallest scales resolved. The core profiles follow a power law with $\eta \simeq 2/3$.

Subsequently, Hogan et al. (2017b) measured entropy profiles for 57 clusters, finding that the systems with H$\alpha$ line emission, signaling the presence of cold gas, have lower central entropies than those without. In order to obtain realistic estimates of the free-fall time, they included an isothermal component in the gravitational potential to represent the mass of the central galaxy, which generally dominates within the central ~10 kpc (Hogan et al. 2017a). Their deprojected profiles of $t_{\rm cool}/t_{\rm ff}$ mostly show a minimum at 10–20 kpc. However, cooling times for the innermost regions are biased upward, due to the challenges of making measurements on such small scales. If the entropy index, $K \propto R^{2/3}$, as found by Panagoulia et al. (2014), and the gas is isothermal, the cooling time will vary as $t_{\rm cool} \propto R$. Where the isothermal component dominates the potential, the free-fall time scales as $t_{\rm ff} \propto R$, so that $t_{\rm cool}/t_{\rm ff}$ would approach a constant near the cluster center. This suggests that $t_{\rm cool}/t_{\rm ff}$ actually approaches a lower limit in the range 10–30 in cores of cluster with cold gas (Figure 9.8). Voit et al. (2015) have suggested that the apparent minimum at 10 seen in this figure is a limit set by AGN feedback, where atmospheres stabilize and are thus most likely to reside. However, $t_{\rm cool}/t_{\rm ff}$ is expected to drop below 10 during cooling episodes, but this is rarely seen, even in clusters where cooling is occurring. The floor may also be an observational selection effect (Hogan et al. 2017b). Nevertheless, the effect is potentially important and should be explored further in the future. Hogan et al. (2017b) also underlined that the range of free-fall times at any fixed radius in this region is too small in their sample to show that the free-fall time has any impact on the ratio. Babyk et al. (2018) measured entropy profiles of 40 early-type galaxies with X-ray-emitting hot atmospheres, including six lenticular and spiral galaxies. Their entropy profiles at small radii are consistent with those of the cool-core clusters. In particular, the dependence of the median entropy index on the radius is $K \propto R^{2/3}$.

For a sample of 55 clusters with molecular gas masses in the range $10^8$–$10^{11}\ M_{\odot}$, Pulido et al. (2018) found that the molecular gas mass is correlated with star formation, H$\alpha$ line luminosity, and central gas density. In these systems, they found that the minimum ratio of $t_{\rm cool}/t_{\rm ff}$ lies in the range $10 \lesssim (t_{\rm cool}/t_{\rm ff})_{\rm min} \lesssim 25$. They also argued that the limited range of $(t_{\rm cool}/t_{\rm ff})_{\rm min}$ could be a selection effect.

*Driving Unstable Cooling*

Although not conclusive, the observational results are a strong indication that entropy profiles are governed by thermally unstable cooling. Thermal instability is almost inevitable when $t_{\rm cool}/t_{\rm ff}$ is smaller than about unity. Numerical simulations find a margin of stability for linear perturbations closer to $t_{\rm cool}/t_{\rm ff} \lesssim 10$ (Choudhury & Sharma 2016) as outlined above (Section 9.3.5). For values in the observed range of $10 \lesssim t_{\rm cool}/t_{\rm ff} \lesssim 30$, larger disturbances are required to trigger cooling (Pizzolato & Soker 2005). Such disturbances might include a radio outburst that lifts low-entropy gas well out beyond its thermally stable location (McNamara et al. 2016), or perhaps, a minor merger (Russell et al. 2017b). Voit (2018) argued that moderately strong turbulence can be sufficient.

The process of thermally unstable cooling is complex, and the details of how the hot ICM gives rise to cold gas have yet to be settled. Numerical models of the cycle of heating and cooling are getting increasingly sophisticated, and their predictions more realistic. For example, the models of Li & Bryan (2014) produce cycles in which AGN outbursts lead to a period of thermally unstable cooling, hence further outbursts. By contrast, in the model of Gaspari et al. (2017), cooling of the hot gas is sustained by interaction with cold clouds. Future observations will be able to distinguish such models. For example, when the hot ICM cools by radiation, there is a well-defined relationship between the rate of cooling and the strengths of some soft X-ray lines (e.g., McNamara & Nulsen 2007). Even when cooling is intermittent, the mean cooling rate for a sample of systems should be sufficient to account for the total amount of cold gas present. Thus, observations of the soft X-ray lines for sufficiently large samples should settle the question of whether the ICM is cooling by radiation.

### 9.3.6 AGN Heating

*Heating is Gentle*

Among the surprising results from *Chandra* is that radio-mechanical feedback heats gently. Early hydrodynamic models for radio mode feedback tended to overheat the central parts of hot atmospheres (e.g., Sijacki & Springel 2006). These models were essentially bombs triggered by radio jets that drive the density down and the temperature up in the inner 10 or so kiloparsecs. *Chandra* has shown that despite ongoing and powerful AGN feedback, atmospheres have remarkably similar density and temperature profiles (McNamara et al. 2016). Detailed studies of clusters such as Virgo (Forman et al. 2017) and Perseus (Fabian et al. 2006; Sanders & Fabian 2007) have shown that X-ray bubbles are inflated slowly, driving weak shocks and sound waves, not strong shocks. For example, the AGN outburst that produced the 13 kpc shock in M87 deposited an energy of about $5 \times 10^{57}$ erg over about 2 Myr. Forman et al. (2017) found that only 22% of the energy is carried away in the Mach 1.2 shocks, while more than half of the energy is contained in bubble enthalpy. This enthalpy is available to heat the atmosphere as the bubble rises.

*Shock and Sound Heating*

The Mach 8 shock in Centaurus A is an isolated example of a strong shock driven by a radio outburst in an X-ray-emitting atmosphere (Figure 9.6; Croston et al. 2009). Centaurus A resides in a poor group and demonstrates that AGN shocks can play a significant role in the energetics of a poor group atmosphere. The next strongest shocks, in NGC 5813, are significantly weaker, with a maximum Mach number of $\simeq$1.8. Three generations of shocks are observed in NGC 5813, and, if they continue to recur, Randall et al. (2015) have shown that their net heating rate is sufficient to prevent the group atmosphere from cooling (Figure 9.3). Million et al. (2010) also argued that repeated weak shocks can prevent cooling in M87. Although sometimes effective, weak shocks are an inefficient heating mechanism. The transient impact of a shock may be substantial, but after it moves onward, the residual heating is determined by the entropy increase. The heat input per unit mass is $\Delta Q = T\Delta S$, and the entropy increase per unit mass, $\Delta S$, is cubic in the shock strength for weak shocks. Repeated weak shocks provide gentle heating, but the evidence does not support them being an effective source of heating in massive clusters.

The difference between sound waves and shocks is mainly that the pressure disturbance, $\delta p/p$, in a sound wave must be small. An expanding radio lobe will drive sound waves, particularly if the power of the AGN jet is variable. Sound waves generated by the radio lobes propagate into the ICM, where they are dissipated by viscosity and thermal conduction, distributing heat throughout a cluster core (Fabian et al. 2005). Furthermore, ripples in deep X-ray images of the Perseus cluster are suggestive of sound waves with sufficient intensity to provide the necessary level of heating (Figure 9.1; Fabian et al. 2006; Sanders & Fabian 2007). Unfortunately, the transport properties of the ICM in Perseus are poorly determined, leaving the significance of sound heating uncertain (e.g., Zweibel et al. 2018). From their analysis of fluctuations in the Perseus ICM, Zhuravleva et al. (2016) also concluded that the ripples in Perseus are mainly isobaric, not sound waves. They argue that these features are the result of turbulence in a stably stratified atmosphere.

*Turbulent Heating*

If a radio lobe cavity remains intact or, more precisely, adiabatic, as it rises through the ICM, the ultimate fate of its enthalpy will mostly be as turbulence in the ICM (e.g., McNamara & Nulsen 2007). For hydrodynamic turbulence, the velocity distribution follows the Kolmogorov spectrum, which only depends on the dissipation rate, so that a measurement of the velocity distribution gives the dissipation rate. By analyzing the distribution of emissivity fluctuations in the central region of the Perseus cluster, Zhuravleva et al. (2016) inferred the turbulent spectrum and, hence, determined that the turbulent dissipation rate is close to the heating rate required to prevent the gas from cooling. Their results are in reasonable agreement with the line-of-sight velocity dispersion of 164 km $s^{-1}$, determined from the *Hitomi* soft X-ray spectrometer (SXS) spectrum (Hitomi Collaboration et al. 2016), for a region lying in the range 30–60 kpc from the center of Perseus.

Turbulent heating is gentle, and the results of Zhuravleva et al. (2016) lend it strong support. However, it is not clear that radio outbursts can inject sufficient power in the form of turbulence for them to be the primary heating channel. Although cavity enthalpy is efficiently converted to turbulence, for an adiabatic "bubble," the total enthalpy scales with the confining pressure as $p^{(\gamma-1)/\gamma}$, where the exponent would be 1/4 for relativistic bubble plasma. As a result, for half of the cavity enthalpy to be converted to turbulent energy, the external pressure must drop by a factor of 16. In a typical cluster core, the bubble would need to rise a long way for that to happen, causing the turbulent heat input to be diluted. Separately, using numerical hydrodynamic simulations, Reynolds et al. (2015) found that less than 1% of the energy injected by AGN activity ends up in turbulence. While that may be an underestimate in the long run, it does nothing to explain how jet power could be efficiently converted to ICM turbulence.

*Cosmic-Ray Heating*

On the basis of radio, X-ray, and gamma-ray observations of M87, Pfrommer (2013) has argued that streaming cosmic rays can balance radiative cooling in its core. Cosmic rays escaping from the radio lobes can stream out of the core along magnetic field lines. If they stream faster than the Alfvén speed, they will generate Alfvén waves that then dissipate and heat the gas. If sufficient cosmic rays escape from the lobes, this mechanism has the potential to dissipate the energy stored in a lobe as heat in a more centrally concentrated manner than the adiabatic loss discussed above.

The key to this mechanism is the gradient of cosmic-ray pressure. Jacob & Pfrommer (2017) have constructed steady-state models for cosmic-ray heating to match the properties of 39 galaxy clusters from the ACCEPT database (Cavagnolo et al. 2009). Their models needed to include thermal conduction in order to balance cooling at larger radii. In a second paper, they found that, for the systems with radio mini halos (Section 9.4.2), the predicted non-thermal emission due to cosmic rays exceeds the observed limits in the radio and gamma-ray ranges. However, their models for the systems lacking mini halos are consistent with observational limits.

*Heating by Mixing*

Early hydrodynamic models of radio mode feedback simulated jets, hence radio lobes, as very hot gas (e.g., Brüggen & Kaiser 2002). The resulting lobes were prone to fluid instabilities, helping them to disrupt and mix quickly with the ICM and causing the energy of the radio lobe fluid to instantly "thermalize" in the ICM under the model assumptions. This made the simulated AGN outbursts very effective at heating the ICM. Hillel & Soker (2017) have again proposed that mixing of radio lobe plasma into the surrounding ICM is the main route by which the lobe energy is transformed into heat. Real radio plasma is probably dominated by relativistic particles and magnetic field, at least when it is young. Old radio lobes do tend to be dominated by nonradiating matter (e.g., Bîrzan et al. 2008), but the nature and state of this matter are unclear. If it is hot gas, that should be revealed in the forseeable future by mapping the Sunyaev–Zel'dovich effect over the lobes (e.g.Abdulla et al. 2019). If the lobe plasma is dominated by relativistic particles, cosmic-ray streaming, as discussed above, is likely the most effective way of transforming the particle energy into heat in the ICM.

### 9.3.7 Growth of Cool Cores

The primary challenges to studying cluster evolution are identifying a representative and unbiased sample in redshift space and obtaining enough photons to make a measurement. X-ray surveys, such as the *ROSAT* All Sky Survey, successfully avoided many of the biases inherent in optical surveys, such as the tendency to identify projected mass concentrations with rich clusters. However, the cosmological dimming of X-ray atmospheres at higher redshifts increasingly favors the detection of systems with bright, cool cores. Clusters identified using the Sunyaev–Zel'dovich effect avoid these biases, as sensitivity is independent of distance. Although this technique tends to find hotter clusters, selection is well understood (McDonald et al. 2013).

The most comprehensive studies involve clusters identified with the South Pole Telescope (SPT). Using samples designed for cosmological studies, McDonald and collaborators have pioneered studies of evolution of the gas density and temperature in cool cores (McDonald et al. 2017, 2019). The clusters evolve in a way that maintains the gas entropy profiles constant over the redshift range 1–0. Distant cool cores tend to be cooler by ~30%. The sample spans a timescale of 5 Gyr, giving the cores adequate time to cool and the atmospheres to adjust to lower entropy,but they do not. This is a strong indication they are maintained by AGN feedback (Hlavacek-Larrondo et al. 2015). The central pressure in clusters has increased over the past 5 Gyr due to the increase in average atmospheric gas density. The outer parts of clusters beyond $0.3R_{500}$ apparently evolve self-similarly. This indicates that cooling cores are growing with time and are stabilized by heating by the AGN in the central galaxy.

### 9.3.8 Synopsis

Much of what we know about radio mode AGN feedback in galaxy clusters has been assembled from *Chandra* data. Starting from the few isolated examples known before the launch of *Chandra*, enormous progress has been made. The relationship between cooling power and cavity power makes a strong empirical case for a feedback cycle (Section 9.3.3). The large masses of cold gas and star formation only present in clusters with short central cooling times (Section 9.3.5) and entropy profiles that make $t_{\rm cool}/t_{\rm ff}$ roughly constant indicate that the central density of the ICM is limited by thermally unstable cooling and feedback. The constancy of the core profiles indicates that this has held for a substantial fraction of the age of the universe. Nevertheless, a lot of the details of the feedback process remain to be settled. The precise manner in which the hot ICM cools to fuel the AGN and the primary channel by which radio outbursts heat the ICM have yet to be determined. Although shock heating is ruled out in more massive clusters, the shock in Centaurus A clearly has a major impact on its group medium. Together with the repeated weak shocks in NGC 5813 (Section 9.3.6), it indicates that radio outbursts can heat clusters via mechanisms that vary with cluster mass. We should be wary of trying to interpret all of the data in terms of a single heating model. *Chandra* data will continue to play a key role in answering the outstanding questions.

## 9.4 Atmospheric Dynamics

This section covers almost everything about *Chandra* observations of the hot ICM, apart from AGN feedback. Acknowledging the marvelous spatial resolution of

*Chandra*, we concentrates on imaging results. One way or another, the main subject is the growth of structure by hierarchical gravitational collapse. The review of Markevitch & Vikhlinin (2007) remains an excellent source for deeper discussion of a number of topics covered here.

Early in the mission, *Chandra* observations of some clusters revealed an abrupt surface brightness feature along a well-defined line, or “edge” (Markevitch et al. 2000; Vikhlinin et al. 2001b). *Chandra* was widely anticipated to detect merger shocks, but contrary to expectation for shock fronts, the gas density was higher on the opposite side of the edge to the temperature. It was quickly realized that these features are the projected images of contact discontinuities in the ICM. The edge occurs along a line where our lines of sight are tangent to the surface of the discontinuity. By analogy with the terrestrial weather phenomenon, they are called “cold fronts.” They are formed by at least two quite distinct physical mechanisms, as outlined below.

The surface defined by a contact discontinuity or shock front is always curved, so that on one side of the edge our sight lines do not intersect the front, but pass through a single medium and the X-ray surface brightness varies smoothly. On the opposite side of the edge, our sight lines intersect the front twice and their length within the second medium is given approximately by $2\sqrt{2rx}$, where $x \ll r$ is the distance from the front, and $r$ is its radius of curvature (relative to our line of sight). If the emission per unit volume on the first side of the front is $A_1$ and that on the side where our sight lines intersect both media is $A_2$, then, on that side of the edge, the surface brightness will vary as $(A_2 - A_1)2\sqrt{2rx}$, in addition to the underlying smooth variation. Thus, the surface brightness is not smooth at the edge, but it is continuous. This is commonly misstated, in part because the surface brightness can appear to be discontinuous if it is sampled too coarsely. Also, if the radius of curvature of the front is known, the change in emission per unit volume, $A_2 - A_1$, can be determined from the surface brightness profile of the front.

### 9.4.1 Merger Cold Fronts

Galaxy clusters are at the pinnacle of the hierarchy of gravitational collapse, still assembling at the present (e.g., Kravtsov & Borgani 2012), so that many are undergoing mergers as we observe them. As a previously known merger, Abell 3667 was targeted early in the *Chandra* mission, but a bright feature that was thought to be a shock turned out to be a cold front. This was the first instance found of a “merger” cold front (Vikhlinin et al. 2001b), which is formed by the contact discontinuity between the remnant of the dense gas core of the smaller subcluster undergoing a merger and the ICM of the more massive subcluster. The speed of the low-entropy gas in the remnant core through the ICM is comparable to the sound speed, so that ram pressure shapes the the leading edge of the remnant, sweeping back the gas that is farther from its center and giving it a roughly conical or rounded shape. From the pressure change across the front, Vikhlinin et al. (2001b) estimated that the remnant core in Abell 3667 is moving at close to the sound speed. Using a deeper *Chandra* observation to better resolve the front, Owers et al. (2009)

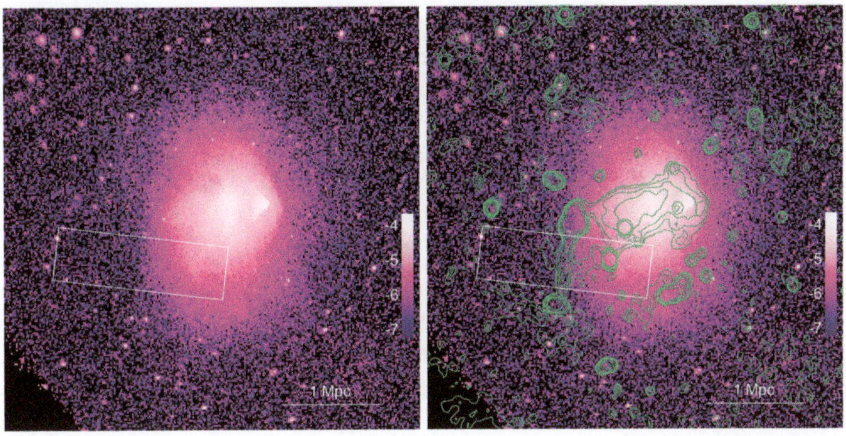


**Figure 9.9.** Left: Bullet cluster 0.8–4 keV X-ray image from Shimwell et al. (2015), by permission of Oxford University Press on behalf of the Royal Astronomical Society. The remnant core of the infalling cluster is the brightest, roughly conical region to the right of the center. The Mach 3 shock front is the edge partly enfolding that on the right. Right: 1.3–3.1 GHz radio contours overlaid on the X-ray image (for details, see Shimwell et al. 2015). The eastward traveling shock front coincides with the radio relic approximately at the middle of the rectangular box.

determined that the pressure is continuous across this front, as expected for a contact discontinuity.

In order to explain the absence of Kelvin–Helmholtz instability (KHI) on the merger cold front in Abell 3667, Vikhlinin et al. (2001c) estimated that the magnetic field strength there needs to be $\sim 10\ \mu$G. The ICM is weakly magnetized, with typical fields of order $3\ \mu$G (Govoni & Feretti 2004; Bonafede et al. 2010), somewhat weaker than this. A body moving through the ICM will tend to sweep up the field, causing field lines to be stretched as the bulk of the ICM flows on around the body and away in its wake. The stretching will align field lines tangent to the surface of the body and increase the field strength. Although field lines cannot normally be broken, as the field strengthens and the magnetic tension increases, they can slip past the body, limiting the growth of the field (Dursi & Pfrommer 2008). This "magnetic draping" could play a significant role in the interaction of moving bodies with the ICM. However, recent analysis of the deep *Chandra* image of Abell 3667 has revealed evidence of KHI, suggesting that the ICM is nearly inviscid, and removing the need for amplified magnetic field to suppress the instability (Ichinohe et al. 2017).

Perhaps the best-known merger cold front is in 1E 0657−56, better known as the Bullet cluster (Markevitch et al. 2002 Figure 9.9). In this merger, the conically shaped remnant core can be seen close behind a Mach $\sim$3 shock. The merger direction is almost perpendicular to our line of sight, presenting us with an optimal view of the merger. The remnant core has passed through the center of the larger cluster and is approaching complete disruption. Among the significant features of the bullet cluster, Clowe et al. (2006) demonstrated significant offsets between the peaks in the mass distribution determined from weak lensing and the peaks of the X-ray emission. For example, the lensing mass peak associated with the infalling subcluster lies close to the BCG from the subcluster, $\simeq 0.5'$ ahead of the X-ray peak of the remnant gas core. Because the great bulk of the baryons are in the X-ray-emitting gas, this shows that the gravitating mass must be dominated by dark matter. Using these data and *N*-body simulations, Randall et al. (2008b) were able to place a lower limit on the self-interaction cross section per unit mass of the dark matter of 1.25 $cm^2$ $g^{-1}$.

*Slingshot Tails*
Because the gas is subject to the drag of the ICM, while the galaxies and dark matter are not, it has generally been assumed that the galaxies and dark matter lead the gas of a subcluster on its orbit during a merger. However, as discussed in the review of Markevitch & Vikhlinin (2007), simulations have shown that the gas may appear to lead at times, a situation they refer to as a "ram pressure slingshot." More generally, as a subcluster nears its apocenter after at least one pericenter passage, the directions of motion of the gas and galaxies will often change with respect to their relative positions, so that the relative positions do not reveal their direction of motion. Sheardown et al. (2019) used the term "slingshot tail" to describe this situation. In a series of simulations, they have investigated observable properties that may be used to distinguish slingshot tails from cases where the galaxies do lead the gas. They identify three systems as slingshot tails, LEDA 87445 in Hydra A, the merging groups NGC 7618 and UGC 12491, and the NGC 4839 group in Coma. Lyskova et al. (2019) independently arrived at the same conclusion for NGC 4389 in Coma. Another likely candidate for a slingshot tail is the NGC 4472 group in Virgo (Su et al. 2019). Slingshot tails may prove to be fairly common, due to the relatively long time a body spends near the apocenter of its orbit. It is important to recognize their nature, as otherwise attempts to determine properties such as the speed of the cold front through the ICM will lead to spurious results.

### 9.4.2 Sloshing Cold Fronts

The first cluster observed by *Chandra* to have cold fronts was Abell 2142 (Figure 9.10, Markevitch et al. 2000), with fronts at $\simeq 40''$ southeast and $160''$ northwest of the center, but the cause of the fronts was not immediately understood. Modeling the Perseus cluster, Churazov et al. (2003) showed that shocks generated by an infalling subcluster can induce oscillatory motion in the gas, causing sharp edges to develop on opposite sides of the central galaxy. A realistic model for Abell 2142 was proposed by Tittley & Henriksen (2005), who found similar structures in hydrodynamic simulations of off-center cluster mergers. These mergers can produce sets of fronts at increasing radii, with successive fronts positioned on opposite sides of the cluster center. Tittley & Henriksen (2005) applied their model to several clusters in addition to Abell 2142. Subsequently, Ascasibar & Markevitch (2006) introduced the term "sloshing cold front" for this type of front, because the primary cause is the sloshing motion of the ICM in the cluster potential initiated by a merger. Ascasibar & Markevitch (2006) used a series of hydrodynamic simulations to examine the phenomenon in more detail, finding that the sloshing motion develops

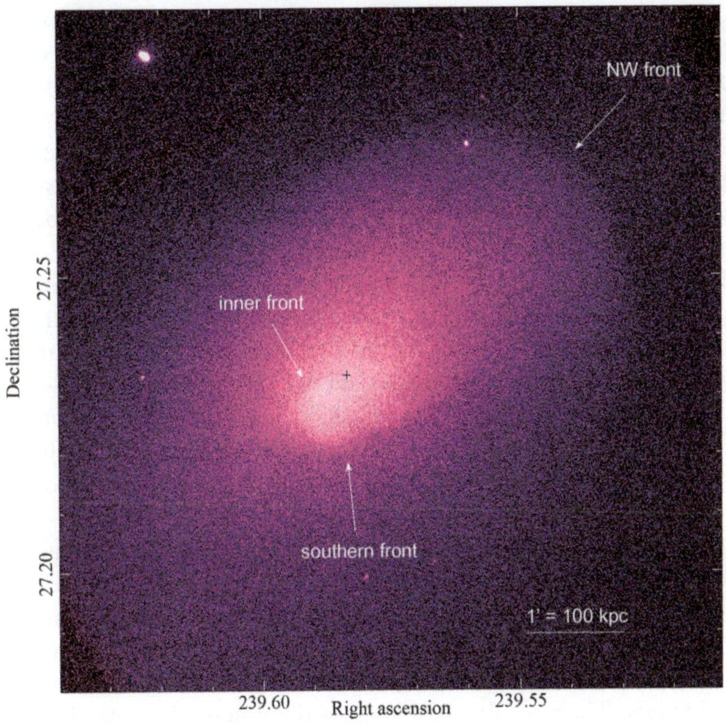


**Figure 9.10.** 0.8–4 keV *Chandra* image of cold fronts in Abell 2142 (from Wang & Markevitch 2018 © 2018. The American Astronomical Society. All rights reserved).

into a spiral-shaped density enhancement on which the fronts appear. The appearance of the fronts is affected by the direction from which the spiral is viewed, but their interleaved arrangement is visible from a wide range of viewing directions.

The main properties of sloshing cold fronts can be understood by approximating the gas motion as a superposition of dipolar g-modes, or internal gravity waves (P. E. J. Nulsen & E. Roediger, 2020, in preparation). In a spherical cluster, these modes come in pairs that rotate rigidly in the opposite sense about the $z$-axis. If the orbit of an infalling subcluster is in the $x$–$y$ plane, the modes that rotate in the same sense as the orbit are excited more than those that rotate in the opposite sense. The angular frequency of the modes is related to the Brunt–Väisälä frequency (Equation (9.6)), which decreases with distance from the cluster center, roughly as $\omega_{\rm BV} \sim 1/R$. A merger excites a broad spectrum of modes initially, with the ICM moving together in much the same direction. As time passes, the modes at smaller radii rotate increasingly farther ahead of those at larger radii, creating the characteristic spiral pattern. Interference between the counter-rotating pairs causes the disturbance to be greatest along one axis, which is where the fronts appear. Fronts form at places where fluid elements initially from different radii would need to pass through one another in the linear approximation—which would then fail. Being in essence a wave, the contact discontinuity moves through the ICM, so that the fluid on either side of the front is continually changing (Roediger et al. 2012). As the spiral tightens, the separation between regions of fluid moving in opposite directions decreases, so that remarkably small initial perturbations can lead to the formation of such fronts.

At a sloshing cold front, samples of ICM from different radii in the unperturbed cluster atmosphere are brought into contact by the sloshing motion. At the same pressure, low-entropy gas from deeper within a cluster has a higher density and lower temperature than higher-entropy gas from farther out in the cluster, producing the front seen in X-rays. The gas from deeper in a cluster may also be more metal-rich, creating a step in metallicity at the cold front (Simionescu et al. 2010; Blanton et al. 2011; Sanders et al. 2016). Sloshing fronts can persist for several gigayears (Ascasibar & Markevitch 2006) and, coupled with the high rate of minor mergers in clusters, they are expected to be a common feature in the ICM (Markevitch et al. 2003). Signatures of sloshing are detected in many nearby cool-core clusters, including the best observed ones, such as Virgo (Simionescu et al. 2010) and Perseus (Churazov et al. 2003). The sloshing spiral in Abell 2029 extends from the cluster center out to ~400 kpc (Paterno-Mahler et al. 2013). The one in Perseus extends to ~700 kpc (Simionescu et al. 2012).

From hydrodynamic simulations of sloshing fronts in the Virgo cluster, Roediger et al. (2011) found that the individual fronts move outward at nearly constant speeds. By matching the locations and density jumps of the fronts in Virgo and also in Abell 496 (Roediger et al. 2012), they were able to place fairly narrow limits on the mass, orbit, and time of core passage of the subcluster that initiated the sloshing to produce the observed fronts. Their results demonstrate the potential to use sloshing fronts to determine the recent merger history of a cluster. Roediger et al. (2011) found that the perturber would probably have been stripped of its hot atmosphere, but they showed that the shock front of a fast-moving galaxy could also have initiated the sloshing in Virgo, as initially proposed for Perseus by Churazov et al. (2003).

*Sloshing Fronts and ICM Physics*

Significant interest in sloshing cold fronts is motivated by their potential to be used as probes of ICM physics (see the review of Zuhone & Roediger 2016). In particular, the structure of a sloshing front is affected by magnetic field in the ICM. Strong shear across a front tends to align the field parallel to the front and to strengthen it, in effect producing magnetic draping by a different mechanism from that operating in merger cold fronts (Section 9.4.1). High-resolution simulations show that the magnetic pressure at the front can become comparable to the ICM pressure and suppress the growth of KHI (ZuHone et al. 2011). The magnetic field gives rise to fronts that are more abrupt across the edge and smoother along the edge, better matching observed fronts.

In deep *Chandra* images of the sloshing cold fronts in each of the nearby galaxy clusters Perseus, Centaurus, and Abell 1795, Walker et al. (2017) found a similar, large "bay," where the front curves inwards toward the cluster center, rather than following the outward curvature of the sloshing spiral. They argued that the bays are formed by KHI on the front. In particular, they found a good match for the ~50 kpc bay about 100 kpc south of the center of the Perseus cluster in magnetohydrodynamic (MHD) simulations when the initial value of $\beta$, the ratio of thermal to magnetic pressure, is 200. Increasing the initial magnetic field to make $\beta = 100$ prevents a bay from developing, while decreasing it to make $\beta = 500$ lets KHI develop too far.

Farther out in the Perseus cluster, 730 kpc from the cluster center, Walker et al. (2018) found an unusual "hook" feature, where the front splits into two separate parts, with a band of hotter gas intruding between them. They found that similar features only occur in MHD simulations if the strength of the magnetic field lies in a relatively narrow range. The hook results from the partial development of Rayleigh–Taylor instability in the presence of the magnetic draping layer. Walker et al. (2018) found this feature develops for an initial value of $\beta = 200$, but not for $\beta = 100$ or 500. They estimated the age of the feature to be ~5 Gyr, demonstrating the remarkable longevity of sloshing cold fronts.

The mean free path for Coulomb collisions between protons is

$$\lambda_{\rm pp} \simeq 3.2\left(\frac{kT}{10\ {\rm keV}}\right)^2\left(\frac{n_{\rm e}}{0.01\ {\rm cm}^{-3}}\right)^{-1} {\rm kpc}, \tag{9.9}$$

and the electron–electron mean free path is a bit smaller. From the outset, it was recognized that the widths of some cold fronts are comparable to or smaller than the particle mean free paths (e.g.Vikhlinin et al. 2001b). The Larmor radius of a thermal proton is only

$$r_{\rm L} \simeq 1.4 \times 10^4\left(\frac{kT}{10\ {\rm keV}}\right)^{1/2}\left(\frac{B}{10\ \mu{\rm G}}\right)^{-1} {\rm km} \tag{9.10}$$

and smaller for electrons, many orders of magnitude less than the mean free paths. In effect, the particles can only diffuse along the magnetic field, so that a draped magnetic field seems a plausible cause for the narrow fronts.

Anisotropic diffusion has a significant impact on transport processes. For thermal conduction in a magnetized plasma, it is often assumed that heat only flows parallel to the magnetic field, driven by the component of the temperature gradient in that direction, and with a thermal conductivity equal to the field-free value. ZuHone et al. (2013b) incorporated such a model into a numerical MHD simulation to examine the effect of conduction on the structure of sloshing fronts. Uninhibited, isotropic thermal conduction would quickly erase the cold fronts in many clusters. Although the shear at a sloshing front combs the field parallel to the front, sloshing is a dynamic process, and ZuHone et al. (2013b) found that sufficient field still penetrates the front to allow thermal conduction to smooth the temperature jump. They found that, unless the conductivity along the field is smaller than the field-free value, the temperature jump is smoothed too much to be consistent with *Chandra* observations. Plasma processes may have a significant impact on heat flow. Fluid motion tends to make the particle velocity distributions anisotropic, which can drive plasma instabilities that may dramatically reduce the effective particle mean free paths (Schekochihin et al. 2010). Komarov et al. (2016) found that when a plasma is subject to mirror instability (i.e., the difference between the particle pressures perpendicular and parallel to the field exceed the magnetic pressure), the conductivity is reduced by a factor of up to 5.

Roberg-Clark et al. (2018) found that plasma processes can place an upper limit on the heat flux along the field that scales as $\beta^{-1}$.

The anisotropic form of the viscous stress tensor in the presence of a magnetic field is called Braginskii viscosity. ZuHone et al. (2015) have examined the impact of viscosity on the structure of sloshing cold fronts. They found that the Braginskii viscosity suppresses the growth of KHI, in agreement with observations, although they found that they could not readily determine whether viscosity or magnetic fields are the dominant suppression mechanism. They also found that including Braginskii viscosity did not alter the need to suppress thermal conduction along the field. In deep *Chandra* observations of a cold front in the Virgo cluster, ~90 kpc northwest of M87, the closest cold front known, Werner et al. (2016) found evidence of KHI, which places similar constraints on the transport properties of the ICM. There are also linear features that are ~10% brighter than adjacent regions. Based on their tailored simulations, Werner et al. (2016) proposed that these are due to the enhanced magnetic field in adjacent regions, where the magnetic pressure reaches 5%–10% of the thermal pressure, further evidence that magnetic field has a marked effect on the structure of sloshing cold fronts.

*Sloshing and Radio Mini Halos*

Radio mini halos are diffuse synchrotron sources with steep radio spectra found in the central regions of some cool-core galaxy clusters (reviewed in Feretti et al. 2012). Mazzotta & Giacintucci (2008) found two cool-core clusters that show a striking correlation between the X-ray structure and the diffuse radio emission. In brief, both mini halos seem to be bounded on the outer edge by sloshing cold fronts. Subsequent observations have confirmed the relationship between the X-ray and radio emission in several more cases (e.g., Giacintucci et al. 2014b, 2014a). In a promising model for mini halos, relativistic electrons with long synchrotron cooling times remain in the core region from earlier radio outbursts of a central AGN. Turbulence generated by the sloshing can boost the energies of these electrons enough for them to produce observable radio synchrotron emission again. Numerical simulations of this process confirm the expectation of a tight correlation between the radio and X-ray properties, with results that are insensitive to the initial spatial distribution of the relativistic electrons (ZuHone et al. 2013a). Although the model is successful, the substantial uncertainties in the details of the particle acceleration process mean that other mechanisms cannot be ruled out (Brunetti & Jones 2014).

### 9.4.3 Merger Shocks

The best known and probably the clearest example of a merger shock is in the Bullet cluster (Section 9.4.1; Markevitch et al. 2002). The remnant gas core of the infalling subcluster is driving a relatively strong shock, at Mach 3.0, through the ICM of the cluster (Markevitch 2006). Initially, the speed of the subcluster relative to the cluster was assumed equal to the speed of the shock, at $\simeq$4700 km $s^{-1}$, but such a high speed would create tension with the expectations of structure formation in the standard ΛCDM cosmological model. Springel & Farrar (2007) showed that that the shock speed can be substantially greater than the speed of the merging subcluster. Two effects are primarily responsible for the difference. First, the gravity of the subcluster accelerates the ICM toward itself so that, on the outgoing part of its orbit, the speed of the subcluster is significantly greater relative to the gas than to the cluster. Second, the shock expands away from the slowing subcluster. In the simulation of Springel & Farrar (2007), the speed of the subcluster is only $\sim$2700 km s$^{-1}$.

Another noteworthy feature of the Bullet cluster merger is the presence of a second shock front on the opposite side of the cluster center (Figure 9.9). Shimwell et al. (2015) showed that relic radio emission about 1 Mpc east of the cluster center coincides with an edge in the X-ray surface brightness characteristic of a shock with a Mach number of about 2.5. Such a feature is seen in a number of cluster simulations (e.g., Roettiger et al. 1999). In a symmetric merger, this is largely inevitable. In an asymmetric merger, the infalling subcluster drives a shock ahead of itself. If it is not too small, the subcluster also draws a wake composed of the gas stripped from it and ICM that falls into the wake after being displaced by the passage of the subcluster. When it is stopped by the dense central atmosphere of the cluster, a reverse shock propagates back through the wake to the cluster outskirts (e.g., Poole et al. 2006). Opposed pairs of shocks formed by mergers give rise to the characteristic pairs of radio relics (see the subsection on particle acceleration below) found on opposite sides of some clusters in the later stages of a merger (e.g., van Weeren et al. 2012; Storm et al. 2018).

*Shock Structure*

By analogy with shocks in supernova remnants (Chapter 5), moderately strong shocks in galaxy clusters are likely to be "collisionless." In that case, strong plasma waves scatter the ions as they enter the shock, effectively thermalizing their kinetic energy and producing a shock front that is much thinner than the Coulomb mean free path. Being much less massive than the ions, the electrons are hardly affected by the plasma waves, but electrostatic forces maintain near-perfect charge neutrality, so that the electrons are compressed by the same amount as the ions in the shock. The rise in electron temperature at the shock is mainly due to adiabatic compression. The velocity, density, and pressure jumps match the usual Rankine–Hugoniot jump conditions, but immediately behind the shock, the ion temperature can be higher than the electron temperature, particularly in a strong shock. The time required for the electrons and ions to return to thermal equilibrium due to Coulomb collisions,

$$t_{\rm ei} \simeq 28\left(\frac{kT}{10\ \rm keV}\right)^{3/2}\left(\frac{n_{\rm e}}{0.01\ \rm cm^{-3}}\right)^{-1}\ \rm Myr, \qquad (9.11)$$

is long enough for the gas to travel a resolvable distance behind the shock in some cases. The X-ray continuum spectrum radiated by the gas is largely determined by the electron temperature (nonequilibrium ionization may affect line emission, but the continuum dominates the spectrum in most clusters). If our understanding of collisionless shocks is correct, it should, therefore, be possible to resolve the gradual rise in the electron temperature behind some shocks.

For the Mach 3 shock in the Bullet cluster, Markevitch (2006) found that the postshock temperature is consistent with a constant value, in which case the temperature rise appears to be faster than expected if the rate of equilibration was determined by Coulomb collisions. For the Mach 2.3 shock in Abell 2146, Russell et al. (2012) found that the postshock temperature profile is consistent with expectations for equilibration via Coulomb collisions. In fact, neither measurement is sufficiently accurate to leave the result ambiguous. However, these observations do demonstrate that it should be possible to resolve the issue with deeper data. *Chandra* observations of Abell 2146 currently underway should settle the matter for that case.

*Particle Acceleration*

Large-scale, diffuse radio synchrotron emission has long been associated with mergers in galaxy clusters (see the reviews of Feretti et al. 2012; van Weeren et al. 2019). Shocks accelerate particles to relativistic energies in supernova remnants, and merger shocks are expected to do the same in galaxy clusters, although the merger shocks are not as strong. Mergers also stir the ICM, producing turbulence that can provide energy to maintain the population of synchrotron-emitting electrons too. Recent improvements in sensitivity and spatial resolution, especially at low radio frequencies, have yielded images and spectra that, in combination with *Chandra* data, make a clear case for particle acceleration being powered by mergers.

A spectacular example of a radio “relic” resulting from a merger shock is seen in the “Toothbrush,” in the merging cluster RX J0603.3+4214. Figure 9.11 shows a low-frequency radio image of the cluster from LOFAR. The Toothbrush relic is the radio-bright region ~2 Mpc long, stretching along the northern edge of the cluster. In the *Chandra* image (upper-right panel), there are two brighter regions within the cluster, capped at the southern end by a merger cold front, and also shock fronts farther to the north and south, both roughly coincident with the outermost radio contours. This X-ray structure matches the later stages of a merger outlined above. The shock to the north coincides with the northern limit of the Toothbrush’s head, where it is brightest. The lower-right panel of the figure shows a map of the radio spectral index, $\alpha$, defined so that the dependence of the radio flux on the frequency, $\nu$, is $F_\nu \propto \nu^\alpha$. Relativistic electrons are accelerated by the shock, causing synchrotron emission with a spectral index of $-0.8$ immediately behind the shock. The timescale for energy loss is shortest for the most-energetic electrons, so that the spectrum steepens and synchrotron emission fades with increasing distance behind the shock. At least qualitatively, this is a textbook illustration of expectations for particle acceleration by a shock.

The mechanism usually assumed to be responsible for particle acceleration at shocks is diffusive shock acceleration (DSA). A relativistic particle trapped near a nonrelativistic shock can scatter back and forth across the shock many times. In the upstream region, the fast particles generate plasma waves, while in the downstream region waves can be generated by plasma turbulence. The waves scatter the particles strongly, slowing their diffusion and so keeping them trapped near the shock for a relatively long time. From a frame at rest in the downstream flow, the gas upstream of the shock is approaching, so that a particle scattered back and forth across the

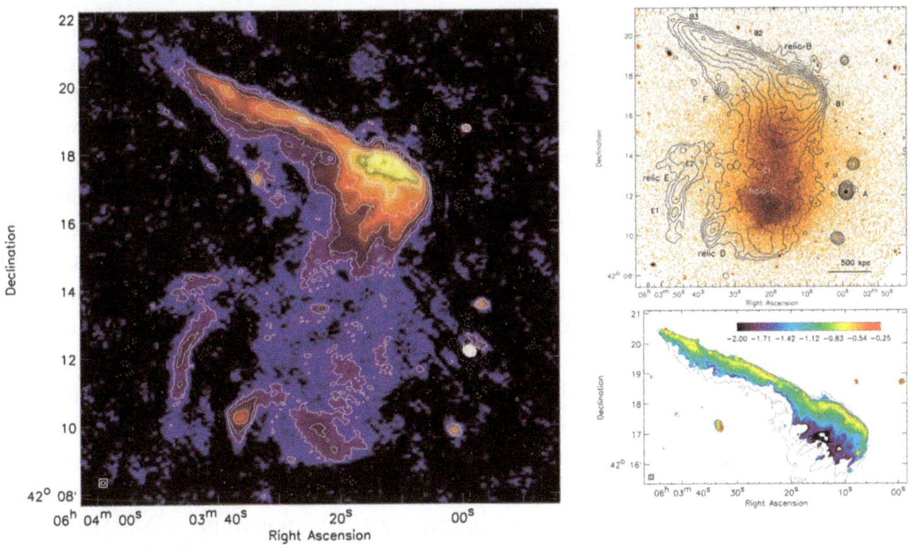


**Figure 9.11.** Left: LOFAR 120–181 MHz image of the Toothbrush and its host cluster. The 8″ × 6.5″ beam is shown in the lower-left corner. Contours are plotted for [1, 2, 4, 8, ...] × $4\sigma_{\rm rms}$ and, dotted, for $-3\sigma_{\rm rms}$, where $\sigma_{\rm rms} = 93\ \mu$ Jy beam$^{-1}$. Upper right: radio contours plotted on a 0.5–2 keV *Chandra* image. Lower right: 151–610 MHz radio spectral index for the Toothbrush, showing the spectra steepen from the northern edge to the south. Images from van Weeren et al. (2016) © 2016. The American Astronomical Society. All rights reserved.

shock experiences first-order Fermi acceleration. Under the usual assumptions of DSA (e.g., Brunetti & Jones 2014), the energy distribution of relativistic electrons emerging from the shock will be a power law, $dN/dE \propto E^{-s}$, where $s = (r + 2)/(r - 1)$ and $r$ is the factor by which the ICM is compressed in the shock. The corresponding spectrum of synchrotron radiation is a power law, with spectral index $\alpha = -(s - 1)/2 = -3/[2(r - 1)]$. From the Rankine–Hugoniot shock jump conditions, the Mach number is related to the shock compression by

$$M^2 = \frac{2r}{(\gamma + 1) - (\gamma - 1)r} = \frac{-4\alpha + 6}{-4\alpha - 3(\gamma - 1)}, \tag{9.12}$$

where $r$ has been expressed in terms of the spectral index. For the Toothbrush, with $\alpha \simeq -0.8$ and $\gamma = 5/3$, this would require a shock Mach number of $M \simeq 2.8$. However, from the X-ray surface brightness profile of the shock, van Weeren et al. (2016) determined that the Mach number cannot exceed 1.5. On the face of it, DSA cannot account for particle acceleration in the Toothbrush and some other relics. There are observational issues that might account for this discrepancy, so the issue has yet to be settled (discussed in van Weeren et al. 2019).

DSA is only effective at accelerating particles with highly suprathermal speeds from the outset. It also depends on the orientation of the magnetic field swept into the shock front. Because particles must diffuse back and forth across the front many times to gain energy, DSA is most effective when the magnetic field is nearly perpendicular to the shock front. As the angle between the field and the front decreases, the efficiency of particle acceleration decreases (demonstrated for ions by Caprioli & Spitkovsky 2014). Magnetic flux conservation means that compression in the shock will cause randomly oriented field before a shock to be preferentially parallel to the front after the shock. This is why the preferred synchrotron polarization of relics is perpendicular to the shock front (e.g., van Weeren et al. 2010). DSA is not the only possible shock acceleration mechanism. Using plasma PIC simulations, Guo et al. (2014) have demonstrated moderately efficient shock drift acceleration (SDA) of electrons in cases where the magnetic field is almost perpendicular to the shock front. In this process, the abrupt change in magnetic field at the shock creates an electric field parallel to the front, while the steep field gradient there causes a drift current parallel to the field. Electrons contributing to the drift current gain energy from the electric field. It has yet to be determined how significant SDA is for particle acceleration in relics.

Reacceleration of old populations of energetic particles can significantly augment the effectiveness of DSA (Kang & Ryu 2011; Kang et al. 2012). Near the outskirts of a cluster, the energy density of the cosmic microwave background (CMB) dominates the magnetic energy density, and the cooling time due to inverse-Compton scattering of the CMB exceeds a Hubble time for electron energies $E/(mc^2) \lesssim 200$. At these energies, the Coulomb loss timescale is also well in excess of a Hubble time, due to the low density of the ICM. Thus, moderately relativistic electrons can accumulate over time from earlier radio outbursts, merger shocks, or other events that accelerate particles. More energetic electrons can remain from more recent events, too. In the Toothbrush cluster, for example, it is noteworthy that the northern relic is bright, while there is only a small relic at the eastern end of the southern shock, even though it is stronger (van Weeren et al. 2016). A likely reason for the discrepancy is that the northern shock has overrun a region with a substantial remnant population of energetic electrons, probably from an old extragalactic radio source.

In observations of the Abell 3411–Abell 3412 merger, van Weeren et al. (2017) found a compelling case for reacceleration in a merger shock. Figure 9.12 shows the region around a tailed radio galaxy in this system. The radio tail extends ~90 kpc (~30″) southeast from its AGN host to where it meets a relic radio source at the merger shock, which was identified in a *Chandra* image and is marked in the right panel of the figure. The map of the radio spectral index in the right panel shows the radio synchrotron spectrum steepening along the tail, in the direction away from the AGN, as expected due to aging of the radio plasma. However, where the tail meets the relic, the spectrum flattens again. Furthermore, as in the Toothbrush, we see that the spectrum steepens across the relic away from the shock, due to aging. The striking feature is that the relic is so much brighter where the merger shock has encountered the aging radio tail, almost certainly because of the pre-existing population of energetic particles in the tail available to be reaccelerated in the shock.

### 9.4.4 Turbulence in the ICM

Turbulence in the ICM is driven primarily by the continuing growth of clusters through mergers and infall. A powerful central radio AGN can have a significant

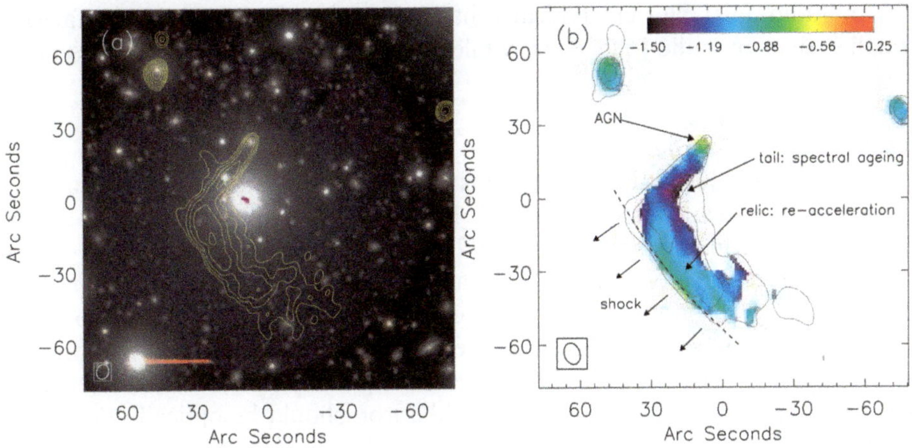


**Figure 9.12.** Left: Subaru *gri* image of the region around a tailed radio galaxy in the Abell 3411–Abell 3412 merger. Contours from a 610 MHz GMRT image are shown in yellow. The galaxy hosting the radio tail is at the northern end of the contours, near the center of the image. The radio tail extends from it ~90 kpc to the southeast, where it meets a radio relic that extends to the southwest. Right: map of the radio spectral index (color) from fits to 0.325, 0.61, 1.5, and 3.0 GHz radio data, with contours from a 325 MHz GMRT image. The dashed line marks the position of the merger shock determined from *Chandra* data. For more details, see van Weeren et al. (2017) © Springer Nature 2017.

impact on the ICM in that region, but the energy released by continuing collapse dominates on large scales in clusters. Merger shocks, merger cold fronts, sloshing, and AGN outbursts are all processes that drive turbulence. Even in a cluster that has not undergone a recent merger, the gravity of the surviving subhalos continues to stir the gas as they orbit within the cluster.

*Radio Halos*

Radio halos are diffuse, unpolarized, synchrotron radio sources that occupy a substantial fraction of the volume of a cluster (reviewed by Feretti et al. 2012). They are very likely powered by turbulence, probably through turbulent reacceleration of energetic electrons (Brunetti & Jones 2014). In Figure 9.11, the faint radio emission that can be seen bridging the entire region from the northern relic to the merger shock in the south is a radio halo. Compared to the mini halos mentioned above (Section 9.4.2), radio halos are more extended, with lower radio surface brightness. For the halo in the Toothbrush cluster, van Weeren et al. (2016) found that the spectral index is fairly uniform over the region, varying in the range $-1.2 \lesssim \alpha \lesssim -1.1$. They proposed that the turbulence is homogeneous in this region, where the two dark matter cores have crossed recently. Other systems show greater variations in the spectral index (e.g., Abell 2744, Orrú et al. 2007), presumably due to differences in the dynamical state of the ICM. Strengthening the case that radio halos are powered by ICM turbulence, Eckert et al. (2017) have found a correlation between the turbulent velocity dispersion and halo radio power. Using X-ray surface brightness fluctuations to estimate turbulent speeds (see below), they found that the

ICM in radio-quiet clusters typically has a velocity dispersion, $\sigma_V$, a factor of 2 smaller than in radio-loud ones, while the radio power of the halos scales as $\sigma_V^{3.3\pm0.7}$.

*Measurements of Turbulent Velocities*

The limited energy resolution of X-ray instruments has generally frustrated attempts to measure turbulent velocities in the ICM. For example, in one of the best measurements available, Sanders et al. (2010) placed an upper limit of 274 km $s^{-1}$ on the turbulent velocity dispersion in the central 30 kpc of Abell 1835 using the *XMM-Newton* Reflection Grating Spectrometer. Indirect methods can also be used to constrain the velocity dispersion of the gas. Churazov et al. (2004) showed that some resonant absorption lines should become optically thick at cluster centers. In particular, the 6.7 keV $K_\alpha$ line of helium-like iron should be optically thick at the center of the Perseus cluster, but not the 7.9 keV $K_\beta$ line. However, in a spectrum of the region 0.5′–2.0′ from the cluster center, they found the line ratios are consistent with both lines being optically thin. This can be explained by line broadening due to gas motions at speeds of about half of the sound speed.

In an important development, using the SXS on the *Hitomi* mission (Mitsuda et al. 2014), Hitomi Collaboration et al. (2016) made a direct measurement of the turbulent line broadening for the central Perseus cluster. They found a turbulent velocity width of 164 km $s^{-1}$ in a ~60 kpc region near the cluster center, and also measured a velocity gradient of ≃150 km $s^{-1}$ over ~60 kpc. These results show the great promise of calorimeter spectrometers on future X-ray missions, but the Perseus measurement is an isolated case at present. Although turbulent speeds in Perseus were lower than anticipated by many, ZuHone et al. (2018) used mock observations to show that they are consistent with expectations from hydrodynamic simulations of sloshing. Their result is also insensitive to the viscosity of the ICM, because the driving scale for the turbulence is significantly larger than the dissipation scale.

Because the ICM is optically thin and the broadband response of X-ray detectors is usually insensitive to the gas temperature, density in the ICM can be determined from surface brightness fluctuations (e.g., Churazov et al. 2012). In the absence of direct measurements of turbulent velocities, Zhuravleva et al. (2014a) have argued that the relationship between velocity perturbations and density fluctuations in the ICM of relaxed clusters is linear, with Fourier components of the density fluctuations and velocity power spectrum related by $(\delta\rho_k/\rho)^2 = \eta_1^2 v_k^2/(3c_s^2)$, where $c_s$ is the sound speed and $k$ is the wave vector. From numerical simulations, they found $\eta_1 \simeq 1$. This has enabled measurements of the turbulent power spectrum for the ICM for several nearby galaxy clusters. Zhuravleva et al. (2014b) measured turbulent velocity spectra for central regions in the Perseus and Virgo clusters, finding that they are consistent with the Kolmogorov spectrum, $|v_k| \propto k^{-1/3}$. In that case, the normalization of the turbulent spectrum is directly related to the turbulent dissipation rate, allowing Zhuravleva et al. (2014b) to determine that turbulent dissipation is sufficient to make up for the energy lost by radiation in these two clusters, although their method does leave substantial uncertainty in the actual dissipation rate.

Applying their method of measuring the turbulent velocity spectrum to a region a few hundred kiloparsecs from the center of the Coma cluster, for a $\simeq$1 Ms *Chandra* exposure, Zhuravleva et al. (2019) were able to resolve scales down to $\simeq$35 kpc, comparable to the ~30 kpc Coulomb mean free path of the protons and electrons. They found that the turbulent velocity spectrum has the Kolmogorov form all the way to their resolution limit. Even if the thermal conductivity is suppressed significantly, they find that for a viscosity determined by Coulomb collisions, i.e., the "Spitzer" viscosity (Spitzer 1956), the velocity spectrum should turn over at scales about an order of magnitude greater than their smallest resolved scale. Although there is other evidence to show that conduction and viscosity are suppressed well below the levels expected if transport in the ICM was determined solely by Coulomb collisions, this measurement applies to the bulk ICM, rather than a particular flow, such as a sloshing cold front. It shows that viscous stresses in the bulk of the ICM are suppressed relative to the values expected in an unmagnetized plasma. However, these results do not rule out the anisotropic Braginskii viscosity for a magnetized plasma.

### 9.4.5 Large-scale Abundance Distribution

The spatial distribution of the elements in the ICM is determined by the composition of the gas that falls into clusters as they build up over time and by the composition of the gas shed into the ICM by the stars and galaxies within the cluster. A great deal of work has been done on measuring the distribution of the elements in the ICM, which we do not attempt to do justice here. For a much more comprehensive discussion, see the review of Böhringer & Werner (2010). One important recent contribution to this field from *Chandra* data is the work of Mantz et al. (2017). For a sample of 245 of the most massive clusters with redshifts out to 1.2, they measured the metallicity in the three ranges of radii, $<0.1 r_{500}$, $(0.1–0.5)\ r_{500}$, and $(0.5–1.0)\ r_{500}$, where $r_{500}$ is the radius within which the mean density is equal to 500 times the critical density. For clusters, the metallicity is determined mainly by the iron abundance. Perhaps the most significant result of these measurements is that the metallicity in the outermost range of radius does not evolve with redshift, implying that the gas in the ICM was enriched to about 0.3 times the solar abundance before it fell into clusters, in agreement with findings from *XMM-Newton* (Ettori et al. 2015). Mantz et al. (2017) also found evidence of an increase in metallicity at intermediate radii, $(0.1–0.5)\ r_{500}$, at late times, i.e., at redshifts $\lesssim 0.4$, which might be due to mass shed by evolving stars, or mixing with more enriched gas from cluster cores. For small radii, they find a large scatter in metallicity and no evidence of evolution with redshift. The large scatter is mainly due to cool-core clusters, distinguished by their "peakiness" in this study.

### 9.4.6 Galaxy Stripping

*Elliptical Galaxies*

One of the early surprises from *Chandra* was the discovery of small "coronae" of hot gas bound to each of the two giant elliptical galaxies, NGC 4874 and NGC 4889, that orbit one another at the center of the Coma cluster (Vikhlinin et al. 2001a). Each corona has a radius of ≃3kpc and a temperature of ≃1–2 keV, much cooler than the Coma ICM, and they have masses of $\simeq 1.1 \times 10^8\ M_\odot$ and $1.6 \times 10^8\ M_\odot$, respectively. The mass of each can be built up in $\sim 10^9$ yr by stellar mass loss from the host. Cooling times in the coronae are in the range $(1–3) \times 10^8$ yr and, if thermal conduction was uninhibited, they would be evaporated within a few million years. Heat conduction into the coronae needs to be suppressed by factors of ~30–100, or more, to explain their survival. It is also surprising that they are not being stripped, but that may be explained if each host carries with it a large mass of hotter gas at close to the temperature of the ICM. A fine balance is required between heating and cooling rates to maintain the gas as it is, without either cooling or being evaporated. Sun et al. (2007b) found that similar coronae occur in at least 60% of bright ($L_{K_s} > 2L_*$) early-type galaxies in clusters. They argued that the coronae are probably maintained by supernova feedback. However, their principal argument is that AGN feedback is likely to destroy the small coronae. It is noteworthy that, apart from being embedded in a hot ICM, the coronae match the properties of the cores in the sample of Babyk et al. (2018), suggesting that they, too, might be maintained by AGN feedback (Section 9.3.5).

M86 is close to pericenter on its orbit through the Virgo cluster, approaching Earth at ≃1550 km $s^{-1}$. It is the dominant galaxy of a former group that is in the process of merging with the Virgo cluster. The ram-pressure-stripped tail of M86 was discovered in observations of the Virgo cluster with the *Einstein* X-ray observatory (Forman et al. 1979). Observations with *Chandra* show that the tail extends a projected distance of ≃150 kpc behind M86, or closer to 400 kpc deprojected. The structure of the stripped gas is complex. In addition to the gas still bound to M86, there is a "plume" projected ≃20 kpc northwest of M86 and a split tail behind that, each feature containing more gas than is left in M86. The split in the tail may be due to the aspherical gravitational potential of M86 (Randall et al. 2008a).

One of the best studied cases of an elliptical galaxy in the process of being stripped is NGC 1404 in the Fornax cluster (Figure 9.13, Su et al. 2017a). The hot corona of NGC 1404 was seen in *ROSAT* PSPC images of the Fornax cluster (Jones et al. 1997) and targeted for deeper observations early in the *Chandra* mission. Machacek et al. (2005) showed that the hot interstellar medium (ISM) of NGC 1404 is truncated at a cold front ≃8 kpc northwest of the galaxy center, facing approximately toward NGC 1399, the central galaxy of the Fornax cluster. Based on the ratio of pressures in the free stream ahead of NGC 1404 and in its ISM just inside the cold front, they estimated that it is moving through the ≃1.5 keV Fornax ICM with a Mach number near unity. Using deeper *Chandra* observations, Su et al. (2017b) measured the distribution of pressure along the cold front. Fitting that to a hydrodynamic model with an assumed form for the velocity variation, they found that NGC 1404 is close to the plane of the sky through the center of the Fornax cluster and moving at Mach ≃1.3 through the ICM, on a velocity vector inclined ≃33° to our line of sight.

Su et al. (2017a) studied the effect of transport processes on the flow around NGC 1404. The cold front is not resolved by *Chandra*, placing an upper limit of ~45 pc on

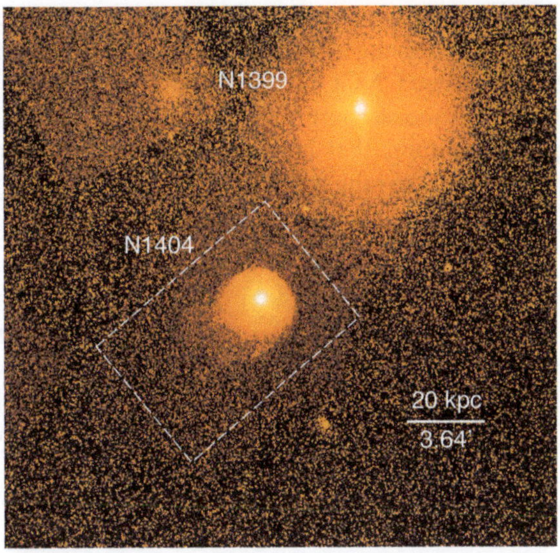


**Figure 9.13.** 0.5–2 keV *Chandra* image of the region around NGC 1404 (Su et al. 2017a © 2017. The American Astronomical Society. All rights reserved). NGC 1399 is the central galaxy of the Fornax cluster. NGC 1404 has a cold front facing northwest and, opposite that, a short wake of stripped gas trailing to the southeast. The movie shows density and temperature slices from a simulation of the elliptical galaxy NGC 1404 merging with the Fornax cluster. The galaxy loses its gas and causes sloshing in the cluster. The images in the movie are from Sheardown et al. (2018) and Su et al. (2017a, 2017c). Additional materials available online at http://iopscience.iop.org/book/978-0-7503-2163-1.

its width, which requires thermal conduction to be heavily suppressed. The cold front is surrounded by a band of diffuse emission, due to KHI eddies. Formation of the eddies requires the effective viscosity of the gas to be smaller than 5% of the Spitzer viscosity for an unmagnetized plasma (Spitzer 1956). Su et al. (2017a) also estimated that the magnetic field in the gas is $\lesssim 5\ \mu$G to allow the KHI to grow. NGC 1404 has a cometary stripped tail on the opposite side to the cold front, where Su et al. (2017a) found gas temperatures that are intermediate between the 0.6 keV ISM and the 1.5 keV ICM. This is evidence of rapid mixing between the ISM and ICM and another signature of a low viscosity. Lastly, the short, $\sim$8 kpc, length of the stripped tail also requires a low effective viscosity (Roediger et al. 2015).

Su et al. (2017c) identified a set of sloshing cold fronts in the Fornax cluster, 3 kpc west, 10 kpc northeast, 30 kpc southwest, and 200 kpc east of the central galaxy, NGC 1399. They are spread over a single spiral pattern that is likely to have been initiated by the infall of NGC 1404. Based on all the observational data, Sheardown et al. (2018) used numerical hydrodynamic simulations to model the infall of NGC 1404 into the Fornax cluster. In accordance with the findings above, the simulation included no viscosity or thermal conduction. They concluded that, if NGC 1404 was on its first core passage, it would have a much larger gas tail and be led by a prominent bow shock, neither of which is present. On second core passage, they found that NGC 1404 would still be somewhat too fast, with a more prominent stripped tail, and also that the locations and strengths of the cold fronts would not match as well as on the third core passage. They most favored a model in which NGC 1404 is on its third core passage. Whether it is on its second or third core passage, NGC 1404 would have passed NGC 1399 traveling from northeast to southwest 1.1–1.3 Gyr ago, triggering the sloshing that has produced the observed cold fronts. The models also predict features that have yet to be observed. In particular, there should be a detached shock in the range 450–750 kpc south of NGC 1404, with a Mach number in the range 1.3–1.5 and a turbulent wake remaining to the south.

*Spiral Galaxies*

Ram pressure stripping in clusters (Gunn et al. 1972) can convert a late-type, star-forming, spiral galaxy into a quiescent S0 galaxy (e.g., reviewed by Boselli & Gavazzi 2014). The first tail of X-ray-emitting gas associated with a spiral galaxy was found using *Chandra* in Abell 2125, behind the galaxy C153 (Wang et al. 2004). This galaxy also has tails of radio and [O II] emission (Owen et al. 2006). In retrospect, C153 was the first jellyfish galaxy observed with *Chandra.*

The galaxy UGC 6697, which is falling nearly edge on into Abell 1367, was already a well-known case of stripping before it was observed with *Chandra* by Sun & Vikhlinin (2005). They found lopsided thermal X-ray emission with a temperature of ~0.7 keV around the disk of the galaxy. On the leading, southeastern edge, nearer to the cluster center, the X-ray-emitting gas is truncated abruptly at ~11 kpc from the galactic center, in the same place as the H$\alpha$ emission, well inside the edge of the stellar disk. On the opposite, trailing, side of the galaxy, the hot gas extends to ~60 kpc, also similar to the extent of the H$\alpha$ emission. Because the X-ray emission is confined to a relatively narrow layer above and below the disk, Sun & Vikhlinin (2005) suggested that the X-ray-emitting gas has been heated by supernovae accompanying the star formation in this galaxy. However, now, with extensive data on jellyfish galaxies, such as ESO 137−001 and D-100 discussed below, it appears more likely that the hot gas results from mixing or conduction between the 5.7 keV ICM and cold gas swept from the galaxy.

In extreme cases of ram pressure stripping, the stripped gas can form tentacle-like trails of filaments and star-forming knots displaced to one side of the disk (e.g., Kenney & Koopmann 1999, Owers et al. 2012). Such systems have been dubbed jellyfish galaxies. One of the surprises from *Chandra* was the discovery by Sun et al. (2006) of a relatively narrow, ~80 kpc long tail of enhanced X-ray emission behind the galaxy ESO 137−001 in Abell 3627 (Figure 9.14). Despite being embedded in a 6.3 keV ICM, the temperature of the tail was found to be 0.7 keV along its whole length. The most plausible explanation is that there is a stream of cold gas swept from the galaxy that extends over the length of the tail. As the cold gas is mixed into or evaporated by the hot ICM, the interface between the cold stream and hot ICM develops an approximately steady state. Subsequent observations have found H$\alpha$ emission along the tail, up to 40 kpc from the galaxy, and H II regions that are sites of extragalactic star formation (Sun et al. 2007a). There is also a tail of $4 \times 10^7 M_\odot$ of warm molecular gas extending 20 kpc behind the galaxy (Sivanandam et al. 2010). Deeper *Chandra* observations show that the tail is bifurcated, with a fainter

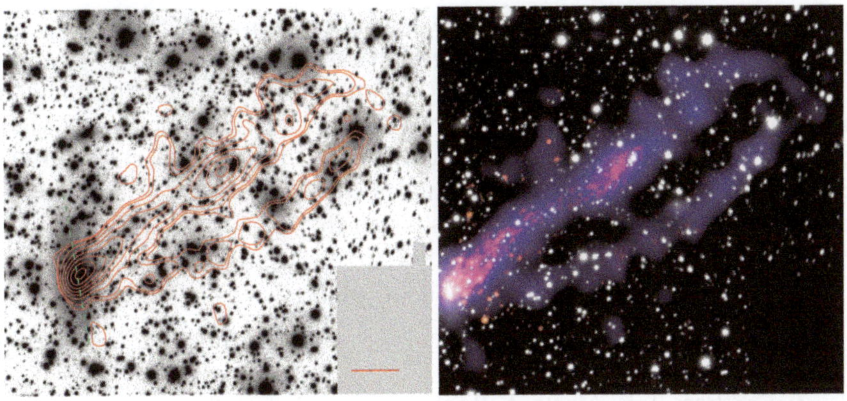

**Figure 9.14.** Left: optical image of ESO 137–001 from the Southern Observatory for Astrophysical Research (SOAR) in H$\alpha$ and continuum, with contours from an adaptively smoothed 0.6–2 keV *Chandra* image in red. The location of the galaxy is marked by a green dashed line at the lower left. The red scale bar is 10 kpc long. Right: SOAR image with the adaptively smoothed *Chandra* image in blue and H$\alpha$ in red (for details, see Sun et al. 2010 © 2010. The American Astronomical Society. All rights reserved).

secondary tail roughly parallel to the main tail (Figure 9.14; Sun et al. 2010). Optical spectroscopy of H II regions in the wake shows that they preserve the rotation of the disk to ~20 kpc downstream, making it even clearer that they formed from material originating in the disk (Sun et al. 2010; Fumagalli et al. 2014).

Some other examples of jellyfish galaxies with X-ray tails include ESO 137–002, which is moving almost edge on through Abell 3627 with a 40 kpc X-ray tail and a bifurcated H$\alpha$ tail (Zhang et al. 2013), CGCG254-021 in Zwicky 8338, with a ~76 kpc long tail (Schellenberger & Reiprich 2015), and D-100 in the Coma cluster (Jáchym et al. 2017). The tail of D-100 is ~60 kpc long. Assuming the standard conversion from CO to molecular gas mass, this tail contains $\sim 10^9\, M_\odot$ of cold molecular gas. This massive stream of cold gas supports the case that the X-ray tail is due to cold gas mixing with or evaporating into the hot ICM.

## 9.5 The Future

The gain in observing capability provided by *Chandra* has had a dramatic impact on the understanding of physical processes in the ICM. A great deal has been learned about AGN feedback in hot atmospheres. Merger shocks and cold fronts provide diagnostics for the collapse history of clusters. Observations of galaxy stripping allow us to see the transformation of galaxy types in progress. Closer examination of these processes can also teach us about the physical properties of the ICM. Together with these observations, advances in our ability to simulate gas flows and steps toward understanding the basic physics of transport in the intracluster plasma have led to enormous progress in the past 20 years.

Even so, large gaps remain in the basic understanding of the ICM. The focus on viscosity and conduction here reflects the critical part they play in governing flows in the ICM. There is a long way to go in both theory and observation before transport processes stop being an issue. For stripping of spiral galaxies and the AGN feedback cycle, we need to understand in detail how the kiloelectronvolt ICM plasma interacts with cold molecular gas clouds and all gas phases in between.

A great deal remains to be done with *Chandra*. Larger, better defined, and better observed samples are required to make progress in understanding how clusters and the galaxies within them grow, evolve, and interact. In this respect, the eROSITA survey promises to provide a greatly expanded cluster catalog to be explored. Other planned missions promise improvements in sensitivity and spectral resolution that will enable direct measurement of flow speeds in the gas. Such a capability will take us into a new era of observations of galaxy clusters.

## References

Abdulla, Z., Carlstrom, J. E., Mantz, A. B., et al. 2019, ApJ, 871, 195
Andrade-Santos, F., Jones, C., Forman, W. R., et al. 2017, ApJ, 843, 76
Arévalo, P., Churazov, E., Zhuravleva, I., Forman, W. R., & Jones, C. 2016, ApJ, 818, 14
Ascasibar, Y., & Markevitch, M. 2006, ApJ, 650, 102
Babyk, I. V., McNamara, B. R., Nulsen, P. E. J., et al. 2018, ApJ, 862, 39
Baldi, A., Forman, W., Jones, C., et al. 2009, ApJ, 707, 1034
Begelman, M. C., Blandford, R. D., & Rees, M. J. 1984, RvMP, 56, 255
Binney, J., & Tabor, G. 1995, MNRAS, 276, 663
Bîrzan, L., McNamara, B. R., Nulsen, P. E. J., Carilli, C. L., & Wise, M. W. 2008, ApJ, 686, 859
Bîrzan, L., Rafferty, D. A., McNamara, B. R., Wise, M. W., & Nulsen, P. E. J. 2004, ApJ, 607, 800
Bîrzan, L., Rafferty, D. A., Nulsen, P. E. J., et al. 2012, MNRAS, 427, 3468
Blanton, E. L., Randall, S. W., Clarke, T. E., et al. 2011, ApJ, 737, 99
Blanton, E. L., Sarazin, C. L., McNamara, B. R., & Wise, M. W. 2001, ApJ, 558, L15
Boehringer, H., Voges, W., Fabian, A. C., Edge, A. C., & Neumann, D. M. 1993, MNRAS, 264, L25
Böhringer, H., & Werner, N. 2010, A&ARv, 18, 127
Bonafede, A., Feretti, L., Murgia, M., et al. 2010, A&A, 513, A30
Boselli, A., & Gavazzi, G. 2014, A&ARv, 22, 74
Brüggen, M., & Kaiser, C. R. 2002, Natur, 418, 301
Brunetti, G., & Jones, T. W. 2014, IJMPD, 23, 1430007
Caprioli, D., & Spitkovsky, A. 2014, ApJ, 783, 91
Carilli, C. L., Perley, R. A., & Harris, D. E. 1994, MNRAS, 270, 173
Cavagnolo, K. W., Donahue, M., Voit, G. M., & Sun, M. 2008, ApJ, 683, L107
Cavagnolo, K. W., Donahue, M., Voit, G. M., & Sun, M. 2009, ApJS, 182, 12
Choudhury, P. P., & Sharma, P. 2016, MNRAS, 457, 2554
Churazov, E., Forman, W., Jones, C., & Böhringer, H. 2000, A&A, 356, 788
Churazov, E., Forman, W., Jones, C., & Böhringer, H. 2003, ApJ, 590, 225
Churazov, E., Forman, W., Jones, C., Sunyaev, R., & Böhringer, H. 2004, MNRAS, 347, 29
Churazov, E., Sunyaev, R., Forman, W., & Böhringer, H. 2002, MNRAS, 332, 729
Churazov, E., Tremaine, S., Forman, W., et al. 2010, MNRAS, 404, 1165
Churazov, E., Vikhlinin, A., Zhuravleva, I., et al. 2012, MNRAS, 421, 1123
Clarke, T. E., Sarazin, C. L., Blanton, E. L., Neumann, D. M., & Kassim, N. E. 2005, ApJ, 625, 748
Clowe, D., Bradač, M., Gonzalez, A. H., et al. 2006, ApJ, 648, L109
Crawford, C. S., Allen, S. W., Ebeling, H., Edge, A. C., & Fabian, A. C. 1999, MNRAS, 306, 857

Croston, J. H., Kraft, R. P., Hardcastle, M. J., et al. 2009, MNRAS, 395, 1999
Croton, D. J., Springel, V., & White, S. D. M. 2006, MNRAS, 365, 11
de Vries, M. N., Wise, M. W., Huppenkothen, D., et al. 2018, MNRAS, 478, 4010
Dennis, T. J., & Chandran, B. D. G. 2005, ApJ, 622, 205
Dong, R., Rasmussen, J., & Mulchaey, J. S. 2010, ApJ, 712, 883
Dunn, R. J. H., & Fabian, A. C. 2006, MNRAS, 373, 959
Dursi, L. J., & Pfrommer, C. 2008, ApJ, 677, 993
Eckert, D., Gaspari, M., Vazza, F., et al. 2017, ApJ, 843, L29
Edge, A. C. 2001, MNRAS, 328, 762
Ettori, S., Baldi, A., Balestra, I., et al. 2015, A&A, 578, A46
Fabian, A. C. 1994, ARAA, 32, 277
Fabian, A. C. 2012, ARAA, 50, 455
Fabian, A. C., Celotti, A., Blundell, K. M., Kassim, N. E., & Perley, R. A. 2002, MNRAS, 331, 369
Fabian, A. C., Reynolds, C. S., Taylor, G. B., & Dunn, R. J. H. 2005, MNRAS, 363, 891
Fabian, A. C., Sanders, J. S., Allen, S. W., et al. 2003a, MNRAS, 344, L43
Fabian, A. C., Sanders, J. S., Crawford, C. S., et al. 2003b, MNRAS, 344, L48
Fabian, A. C., Sanders, J. S., Ettori, S., et al. 2000, MNRAS, 318, L65
Fabian, A. C., Sanders, J. S., Taylor, G. B., et al. 2006, MNRAS, 366, 417
Fanaroff, B. L., & Riley, J. M. 1974, MNRAS, 167, 31P
Feretti, L., Giovannini, G., Govoni, F., & Murgia, M. 2012, A&ARv, 20, 54
Finoguenov, A., & Jones, C. 2001, ApJ, 547, L107
Forman, W., Churazov, E., Jones, C., et al. 2017, ApJ, 844, 122
Forman, W., Jones, C., Churazov, E., et al. 2007, ApJ, 665, 1057
Forman, W., Nulsen, P., Heinz, S., et al. 2005, ApJ, 635, 894
Forman, W., Schwarz, J., Jones, C., Liller, W., & Fabian, A. C. 1979, ApJ, 234, L27
Fumagalli, M., Fossati, M., & Hau, G. K. T. 2014, MNRAS, 445, 4335
Gaspari, M., Ruszkowski, M., & Oh, S. P. 2013, MNRAS, 432, 3401
Gaspari, M., Ruszkowski, M., & Sharma, P. 2012, ApJ, 746, 94
Gaspari, M., Temi, P., & Brighenti, F. 2017, MNRAS, 466, 677
Giacintucci, S., Markevitch, M., Brunetti, G., et al. 2014a, ApJ, 795, 73
Giacintucci, S., Markevitch, M., Venturi, T., et al. 2014b, ApJ, 781, 9
Gitti, M., Brighenti, F., & McNamara, B. R. 2012, AdAst, 2012, 950641
Gómez, P. L., Loken, C., Roettiger, K., & Burns, J. O. 2002, ApJ, 569, 122
Govoni, F., & Feretti, L. 2004, IJMPD, 13, 1549
Gull, S. F., & Northover, K. J. E. 1973, Natur, 244, 80
Gunn, J. E., Gott, J., & Richard, I. 1972, ApJ, 176, 1
Guo, X., Sironi, L., & Narayan, R. 2014, ApJ, 794, 153
Heckman, T. M. 1981, ApJ, 250, L59
Heinz, S., Choi, Y.-Y., Reynolds, C. S., & Begelman, M. C. 2002, ApJ, 569, L79
Heinz, S., Reynolds, C. S., & Begelman, M. C. 1998, ApJ, 501, 126
Hillel, S., & Soker, N. 2017, ApJ, 845, 91
Hitomi CollaborationAharonian, F., Akamatsu, H., et al. 2016, Natur, 535, 117
Hlavacek-Larrondo, J., Fabian, A. C., Edge, A. C., et al. 2012, MNRAS, 421, 1360
Hlavacek-Larrondo, J., McDonald, M., Benson, B. A., et al. 2015, ApJ, 805, 35
Hogan, M. T., McNamara, B. R., Pulido, F., et al. 2017a, ApJ, 837, 51

Hogan, M. T., McNamara, B. R., Pulido, F. A., et al. 2017b, ApJ, 851, 66
Huang, Z., & Sarazin, C. L. 1998, ApJ, 496, 728
Hudson, D. S., Mittal, R., Reiprich, T. H., et al. 2010, A&A, 513, A37
Ichinohe, Y., Simionescu, A., Werner, N., & Takahashi, T. 2017, MNRAS, 467, 3662
Jáchym, P., Sun, M., Kenney, J. D. P., et al. 2017, ApJ, 839, 114
Jacob, S., & Pfrommer, C. 2017, MNRAS, 467, 1449
Johnstone, R. M., Fabian, A. C., & Nulsen, P. E. J. 1987, MNRAS, 224, 75
Jones, C., Forman, W., Vikhlinin, A., et al. 2002, ApJ, 567, L115
Jones, C., Stern, C., Forman, W., et al. 1997, ApJ, 482, 143
Kaiser, C. R., Pavlovski, G., Pope, E. C. D., & Fangohr, H. 2005, MNRAS, 359, 493
Kang, H., & Ryu, D. 2011, ApJ, 734, 18
Kang, H., Ryu, D., & Jones, T. W. 2012, ApJ, 756, 97
Kenney, J. D. P., & Koopmann, R. A. 1999, AJ, 117, 181
King, A., & Pounds, K. 2015, ARAA, 53, 115
Komarov, S. V., Churazov, E. M., Kunz, M. W., & Schekochihin, A. A. 2016, MNRAS, 460, 467
Kraft, R. P., Vázquez, S. E., Forman, W. R., et al. 2003, ApJ, 592, 129
Kravtsov, A. V., & Borgani, S. 2012, ARAA, 50, 353
Li, Y., & Bryan, G. L. 2014, ApJ, 789, 54
Lim, J., Ohyama, Y., Chi-Hung, Y., Dinh-V-Trung,, & Shiang-Yu, W. 2012, ApJ, 744, 112
Lyskova, N., Churazov, E., Zhang, C., et al. 2019, MNRAS, 485, 2922
Machacek, M., Dosaj, A., Forman, W., et al. 2005, ApJ, 621, 663
Machacek, M., Nulsen, P. E. J., Jones, C., & Forman, W. R. 2006, ApJ, 648, 947
Mantz, A. B., Allen, S. W., Morris, R. G., et al. 2014, MNRAS, 440, 2077
Mantz, A. B., Allen, S. W., Morris, R. G., et al. 2017, MNRAS, 472, 2877
Markevitch, M. 2006, in ESA Special Publication, Vol. 604, The X-ray Universe 2005, ed. A. Wilson, 723
Markevitch, M., Gonzalez, A. H., David, L., et al. 2002, ApJ, 567, L27
Markevitch, M., Ponman, T. J., Nulsen, P. E. J., et al. 2000, ApJ, 541, 542
Markevitch, M., & Vikhlinin, A. 2007, PhR, 443, 1
Markevitch, M., Vikhlinin, A., & Forman, W. R. 2003, in ASP Conf. Ser. 30137, Matter and Energy in Clusters of Galaxies, ed. S. Bowyer, & C.-Y. Hwang (San Francisco, CA: ASP), 37
Mazzotta, P., & Giacintucci, S. 2008, ApJL, 675, L9
McCourt, M., Sharma, P., Quataert, E., & Parrish, I. J. 2012, MNRAS, 419, 3319
McDonald, M., Allen, S. W., Bayliss, M., et al. 2017, ApJ, 843, 28
McDonald, M., Allen, S. W., Hlavacek-Larrondo, J., et al. 2019, ApJ, 870, 85
McDonald, M., Benson, B. A., Vikhlinin, A., et al. 2013, ApJ, 774, 23
McDonald, M., Gaspari, M., McNamara, B. R., & Tremblay, G. R. 2018, ApJ, 858, 45
McNamara, B. R., & Nulsen, P. E. J. 2007, ARAA, 45, 117
McNamara, B. R., & Nulsen, P. E. J. 2012, NJPh, 14, 055023
McNamara, B. R., Nulsen, P. E. J., Wise, M. W., et al. 2005, Natur, 433, 45
McNamara, B. R., & O'Connell, R. W. 1989, AJ, 98, 2018
McNamara, B. R., Russell, H. R., Nulsen, P. E. J., et al. 2016, ApJ, 830, 79
McNamara, B. R., Wise, M., Nulsen, P. E. J., et al. 2000, ApJ, 534, L135
McNamara, B. R., Wise, M. W., Nulsen, P. E. J., et al. 2001, ApJ, 562, L149
Mernier, F., de Plaa, J., Kaastra, J. S., et al. 2017, A&A, 603, A80
Miller, L. 1986, MNRAS, 220, 713

Million, E. T., Werner, N., Simionescu, A., et al. 2010, MNRAS, 407, 2046
Mitsuda, K., Kelley, R. L., Akamatsu, H., et al. 2014, Proc. SPIE, 9144, 91442A
Nulsen, P., Jones, C., Forman, W., et al. 2009, in AIP Conf. Ser. 1201, ed. S. Heinz, & E. Wilcots (Melville, NY: AIP), 198–201
Nulsen, P. E. J., David, L. P., McNamara, B. R., et al. 2002, ApJ, 568, 163
Nulsen, P. E. J., Hambrick, D. C., McNamara, B. R., et al. 2005a, ApJ, 625, L9
Nulsen, P. E. J., McNamara, B. R., Wise, M. W., & David, L. P. 2005b, ApJ, 628, 629
O'Hara, T. B., Mohr, J. J., Bialek, J. J., & Evrard, A. E. 2006, ApJ, 639, 64
Orrú, E., Murgia, M., Feretti, L., et al. 2007, A&A, 467, 943
Owen, F. N., Keel, W. C., Wang, Q. D., Ledlow, M. J., & Morrison, G. E. 2006, AJ, 131, 1974
Owers, M. S., Couch, W. J., Nulsen, P. E. J., & Rand all, S. W. 2012, ApJL, 750, L23
Owers, M. S., Nulsen, P. E. J., Couch, W. J., & Markevitch, M. 2009, ApJ, 704, 1349
Panagoulia, E. K., Fabian, A. C., & Sanders, J. S. 2014, MNRAS, 438, 2341
Paterno-Mahler, R., Blanton, E. L., Randall, S. W., & Clarke, T. E. 2013, ApJ, 773, 114
Pedlar, A., Ghataure, H. S., Davies, R. D., et al. 1990, MNRAS, 246, 477
Pfrommer, C. 2013, ApJ, 779, 10
Pizzolato, F., & Soker, N. 2005, ApJ, 632, 821
Pizzolato, F., & Soker, N. 2006, MNRAS, 371, 1835
Poole, G. B., Fardal, M. A., Babul, A., et al. 2006, MNRAS, 373, 881
Pulido, F. A., McNamara, B. R., Edge, A. C., et al. 2018, ApJ, 853, 177
Rafferty, D. A., McNamara, B. R., & Nulsen, P. E. J. 2008, ApJ, 687, 899
Rafferty, D. A., McNamara, B. R., Nulsen, P. E. J., & Wise, M. W. 2006, ApJ, 652, 216
Randall, S. W., Forman, W. R., Giacintucci, S., et al. 2011, ApJ, 726, 86
Randall, S. W., Markevitch, M., Clowe, D., Gonzalez, A. H., & Bradač, M. 2008b, ApJ, 679, 1173
Randall, S., Nulsen, P., Forman, W. R., et al. 2008a, ApJ, 688, 208
Randall, S. W., Nulsen, P. E. J., Jones, C., et al. 2015, ApJ, 805, 112
Reynolds, C. S., Balbus, S. A., & Schekochihin, A. A. 2015, ApJ, 815, 41
Reynolds, C. S., McKernan, B., Fabian, A. C., Stone, J. M., & Vernaleo, J. C. 2005, MNRAS, 357, 242
Roberg-Clark, G. T., Drake, J. F., Swisdak, M., & Reynolds, C. S. 2018, ApJ, 867, 154
Roediger, E., Brüggen, M., Simionescu, A., et al. 2011, MNRAS, 413, 2057
Roettiger, K., Burns, J. O., & Stone, J. M. 1999, ApJ, 518, 603
Roediger, E., Kraft, R. P., Nulsen, P. E. J., et al. 2015, ApJ, 806, 103
Roediger, E., Lovisari, L., Dupke, R., et al. 2012, MNRAS, 420, 3632
Russell, H. R., Fabian, A. C., McNamara, B. R., et al. 2018, MNRAS, 477, 3583
Russell, H. R., McDonald, M., McNamara, B. R., et al. 2017a, ApJ, 836, 130
Russell, H. R., McNamara, B. R., Fabian, A. C., et al. 2017b, MNRAS, 472, 4024
Russell, H. R., McNamara, B. R., Sanders, J. S., et al. 2012, MNRAS, 423, 236
Salomé, P., & Combes, F. 2003, A&A, 412, 657
Salomé, P., Combes, F., Revaz, Y., et al. 2011, A&A, 531, A85
Sanders, J. S., & Fabian, A. C. 2007, MNRAS, 381, 1381
Sanders, J. S., Fabian, A. C., Smith, R. K., & Peterson, J. R. 2010, MNRAS, 402, L11
Sanders, J. S., Fabian, A. C., Taylor, G. B., et al. 2016, MNRAS, 457, 82
Sarazin, C. L. 1988, X-ray Emission from Clusters of Galaxies (Cambridge: Cambridge Univ. Press)

Schekochihin, A. A., Cowley, S. C., Rincon, F., & Rosin, M. S. 2010, MNRAS, 405, 291
Schellenberger, G., & Reiprich, T. H. 2015, A&A, 583, L2
Scheuer, P. A. G. 1974, MNRAS, 166, 513
Schindler, S., Castillo-Morales, A., De Filippis, E., Schwope, A., & Wambsganss, J. 2001, A&A, 376, L27
Sharma, P., McCourt, M., Quataert, E., & Parrish, I. J. 2012, MNRAS, 420, 3174
Sheardown, A., Fish, T. M., Roediger, E., et al. 2019, ApJ, 874, 112
Sheardown, A., Roediger, E., Su, Y., et al. 2018, ApJ, 865, 118
Shimwell, T. W., Markevitch, M., Brown, S., et al. 2015, MNRAS, 449, 1486
Shin, J., Woo, J.-H., & Mulchaey, J. S. 2016, ApJS, 227, 31
Sijacki, D., & Springel, V. 2006, MNRAS, 366, 397
Simionescu, A., Werner, N., Forman, W. R., et al. 2010, MNRAS, 405, 91
Simionescu, A., Werner, N., Urban, O., et al. 2012, ApJ, 757, 182
Sivanandam, S., Rieke, M. J., & Rieke, G. H. 2010, ApJ, 717, 147
Smith, D. A., Wilson, A. S., Arnaud, K. A., Terashima, Y., & Young, A. J. 2002, ApJ, 565, 195
Snios, B., Nulsen, P. E. J., Wise, M. W., et al. 2018, ApJ, 855, 71
Spitzer, L. 1956, Physics of Fully Ionized Gases (New York: Interscience)
Springel, V., & Farrar, G. R. 2007, MNRAS, 380, 911
Storm, E., Vink, J., Zandanel, F., & Akamatsu, H. 2018, MNRAS, 479, 553
Su, Y., Kraft, R. P., Roediger, E., et al. 2017a, ApJ, 834, 74
Su, Y., Kraft, R. P., Nulsen, P. E. J., et al. 2017b, ApJ, 835, 19
Su, Y., Nulsen, P. E. J., Kraft, R. P., et al. 2017c, ApJ, 851, 69
Su, Y., Kraft, R. P., Nulsen, P. E. J., et al. 2019, AJ, 158, 6
Sun, M., Donahue, M., Roediger, E., et al. 2010, ApJ, 708, 946
Sun, M., Donahue, M., & Voit, G. M. 2007a, ApJ, 671, 190
Sun, M., Jones, C., Forman, W., et al. 2006, ApJ, 637, L81
Sun, M., Jones, C., Forman, W., et al. 2007b, ApJ, 657, 197
Sun, M., & Vikhlinin, A. 2005, ApJ, 621, 718
Tittley, E. R., & Henriksen, M. 2005, ApJ, 618, 227
Tucker, W., & David, L. P. 1997, ApJ, 484, 602
van Weeren, R. J., Andrade-Santos, F., Dawson, W. A., et al. 2017, NatAs, 1, 0005
van Weeren, R. J., Bonafede, A., Ebeling, H., et al. 2012, MNRAS, 425, L36
van Weeren, R. J., Brunetti, G., Brüggen, M., et al. 2016, ApJ, 818, 204
van Weeren, R. J., de Gasperin, F., Akamatsu, H., et al. 2019, SSRv, 215, 16
van Weeren, R. J., Röttgering, H. J. A., Brüggen, M., & Hoeft, M. 2010, Sci., 330, 347
Vikhlinin, A., Markevitch, M., Forman, W., & Jones, C. 2001a, ApJ, 555, L87
Vikhlinin, A., Markevitch, M., & Murray, S. S. 2001b, ApJ, 551, 160
Vikhlinin, A., Markevitch, M., & Murray, S. S. 2001c, ApJ, 549, L47
Voigt, L. M., & Fabian, A. C. 2004, MNRAS, 347, 1130
Voit, G. M. 2018, ApJ, 868, 102
Voit, G. M., Donahue, M., Bryan, G. L., & McDonald, M. 2015, Natur, 519, 203
Voit, G. M., Kay, S. T., & Bryan, G. L. 2005, MNRAS, 364, 909
Walker, S. A., Hlavacek-Larrondo, J., Gendron-Marsolais, M., et al. 2017, MNRAS, 468, 2506
Walker, S. A., ZuHone, J., Fabian, A., & Sand ers, J. 2018, NatAs, 2, 292
Wang, Q. D., Owen, F., & Ledlow, M. 2004, ApJ, 611, 821
Wang, Q. H. S., & Markevitch, M. 2018, ApJ, 868, 45

Werner, N., McNamara, B. R., Churazov, E., & Scannapieco, E. 2019, SSRv, 215, 5
Werner, N., ZuHone, J. A., Zhuravleva, I., et al. 2016, MNRAS, 455, 846
Wise, M. W., McNamara, B. R., Nulsen, P. E. J., Houck, J. C., & David, L. P. 2007, ApJ, 659, 1153
Zakamska, N. L., & Narayan, R. 2003, ApJ, 582, 162
Zhang, B., Sun, M., Ji, L., et al. 2013, ApJ, 777, 122
Zhuravleva, I., Churazov, E., Arévalo, P., et al. 2016, MNRAS, 458, 2902
Zhuravleva, I., Churazov, E., Schekochihin, A. A., et al. 2019, NatAs, 3, 832
Zhuravleva, I., Churazov, E. M., Schekochihin, A. A., et al. 2014a, ApJ, 788, L13
Zhuravleva, I., Churazov, E., Schekochihin, A. A., et al. 2014b, Natur, 515, 85
ZuHone, J. A., Kunz, M. W., Markevitch, M., Stone, J. M., & Biffi, V. 2015, ApJ, 798, 90
ZuHone, J. A., Markevitch, M., Brunetti, G., & Giacintucci, S. 2013a, ApJ, 762, 78
ZuHone, J. A., Markevitch, M., & Lee, D. 2011, ApJ, 743, 16
ZuHone, J. A., Markevitch, M., Ruszkowski, M., & Lee, D. 2013b, ApJ, 762, 69
ZuHone, J. A., Miller, E. D., Bulbul, E., & Zhuravleva, I. 2018, ApJ, 853, 180
Zuhone, J. A., & Roediger, E. 2016, JPlPh, 82, 535820301
Zweibel, E. G., Mirnov, V. V., Ruszkowski, M., et al. 2018, ApJ, 858, 5

# Chapter 10

# Galaxy Cluster Cosmology

**Steven W Allen and Adam B Mantz**

## 10.1 Introduction

Observations of galaxy clusters have played a key role in helping to establish the standard ΛCDM model of cosmology, with a universe dominated by dark matter and dark energy (see Allen et al. 2011 and Weinberg et al. 2013 for reviews). The birth of cluster cosmology dates back almost a century to Zwicky's discovery of dark matter in the Coma Cluster (Figure 10.1(a)) using pioneering measurements of the luminosities and velocities of cluster galaxies (Zwicky 1933; note that these measurements were made just a few years after Hubble's discovery of the expansion of the universe; Hubble 1929). At the time, optical wavelengths were the only part of the electromagnetic spectrum available to astronomers. However, as the century progressed, additional wavebands would became available and, in the latter part of the 20th century, X-ray observations would move to the fore.

By the start of the 1990s, observations with the *Einstein Observatory* and *EXOSAT* had established that the baryonic matter content of galaxy clusters is dominated by the hot intracluster medium (ICM; see e.g., Hughes 1989; David et al. 1990; Edge & Stewart 1991; Fabian 1991; and references therein). While (very) hard to observe in other wavebands, the ICM emits brightly in X-rays through bremsstrahlung and line radiation (Chapter 3). Figure 10.2 illustrates our modern understanding of the typical matter content of galaxy clusters: for the most massive systems like the Coma Cluster, approximately 5/6 of the matter content is dark matter and close to 1/6 is in the ICM, with only about 1% of the mass in the form of stars.[1]

The first major result of cluster cosmology in the modern era was presented by White et al. (1993a) using measurements of the baryonic mass fraction in the Coma Cluster (Figure 10.1(b)) and recently improved constraints on big bang

[1] Planets and other "cold" baryons make negligible contributions to the overall matter content of galaxy clusters.

doi:10.1088/2514-3433/ab43dcch10  

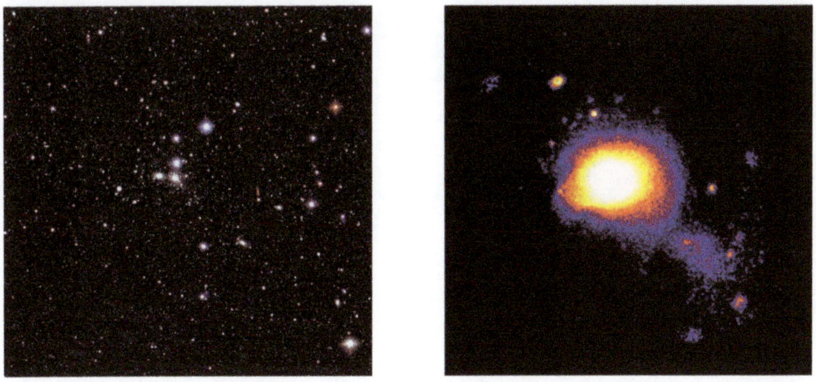

**Figure 10.1.** Left: optical image of the Coma Cluster from the Sloan Digital Sky Survey. It was in this cluster that Zwicky discovered the existence of dark matter (Zwicky 1933). Right: *ROSAT* X-ray image of the Coma Cluster covering the same spatial region. The hot ICM emits brightly in X-rays and permeates the entire cluster volume. The interactive figure fades between the optical (SDSS) and X-ray (*ROSAT*) images of the Coma Cluster shown in the left and right panels of the figure. Additional materials available online at http://iopscience.iop.org/book/978-0-7503-2163-1

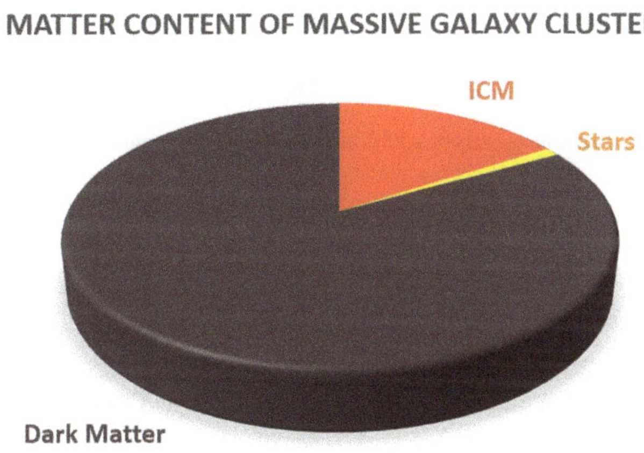


**Figure 10.2.** For the most massive galaxy clusters, approximately 5/6 of the matter content is dark matter and roughly 1/6 is in the ICM. Only ~1% of the mass is in the form of stars.

nucleosynthesis models from quasar absorption-line spectroscopy (Kolb & Turner 1990). Recognizing that the largest galaxy clusters should provide approximately fair samples of the matter content of the universe, White et al. (1993b) used these measurements to infer that the mean matter density of the universe, $\Omega_m$, must have a value significantly less than one. While considered at the time a "challenge to cosmological orthodoxy," this result provided arguably the first compelling evidence that we live in a low matter density universe. Later in the decade, early measurements of the evolution of cluster number counts, i.e., the number density of clusters as a function of mass and redshift, would provide further evidence for a low value of $\Omega_m$ (e.g., Bahcall et al. 1997). In particular, the discovery of the first massive clusters

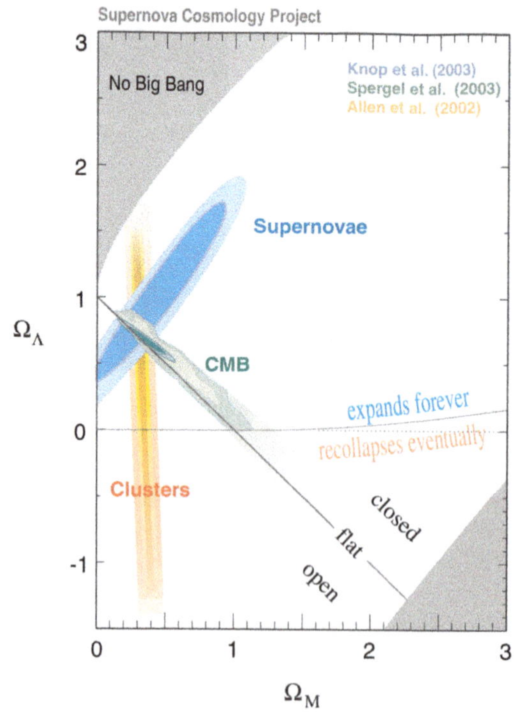


**Figure 10.3.** Cosmological measurements circa 2003, showing the complementary nature of constraints on the mean matter density, $\Omega_m$, and dark energy density, $\Omega_\Lambda$, from Type Ia supernovae, cosmic microwave background (CMB) data, and early *Chandra* cluster baryon fraction constraints. Figure credit http://supernova.lbl.gov/. Perlmutter et al., The Supernova Cosmology Project

at intermediate to high redshift ($0.5 < z < 0.9$) with the *Einstein Observatory* Extended Medium Sensitivity Survey (EMSS; Henry et al. 1992) provided powerful confirmatory evidence for $\Omega_m < 1$ (Bahcall & Fan 1998; Donahue et al. 1998).

The launch of the *Chandra* X-ray Observatory in 1999 would lead to rapid advances in the precision and accuracy of cluster baryon fraction measurements, and *Chandra* measurements of low-to-intermediate redshift clusters would quickly tighten the constraints on the mean matter density to $\Omega_m = 0.30 \pm 0.04$ (Allen et al. 2002). The late 1990s had also seen the discovery of cosmic acceleration using measurements of Type 1a supernovae (Riess et al. 1998; Perlmutter et al. 1999). The complementary nature of the supernova and cluster baryon fraction constraints, and of then rapidly improving cosmic microwave background measurements (Spergel et al. 2003) was recognized, and contour plots like the one shown in Figure 10.3 from the Supernova Cosmology Team (Knop et al. 2003) would soon become a familiar tool for representing our understanding of the mass–energy content of the universe.

By 2004, *Chandra* had extended measurements of the cluster baryon fraction out to $z \sim 1$ (Allen et al. 2004). Critically, these measurements would provide the first opportunity to study dark energy at X-ray wavelengths: *Chandra* measurements of the apparent evolution of the cluster baryon fraction (which is expected to be an approximately standard quantity) would provide the first direct confirmation of the finding from Type Ia supernova studies that the expansion of the universe is accelerating (at comparable statistical significance to the supernovae results; Allen et al. 2004). This work, and later studies using this technique, is discussed in Section 10.2.

During the early 2000s, work within the cosmology community would also return to measurements of cluster counts. By 2004, studies utilizing improved measurements of mass–observable scaling relations had made good progress, forging a path toward our modern understanding of the amplitude of structure formation, encapsulated by the parameter $\sigma_8$ (e.g., Seljak 2002; Pierpaoli et al. 2003; Schuecker et al. 2003). While controversial at the time, being in tension with early weak-lensing work (Van Waerbeke et al. 2002; Refregier et al. 2002; Bacon et al. 2003) and also, to some degree, the first results from the *WMAP* satellite (Spergel et al. 2003), these measurements would stand the test of time. Cluster count measurements would really come into their own toward the end of the decade, however, when two independent teams (Vikhlinin et al. 2009; Mantz et al. 2010c) would develop and apply new methodology that would, for the first time, unlock the full power of this technique. These two studies, utilizing many megaseconds of *Chandra* observations, would confirm and refine the value of $\sigma_8$ and, more significantly, provide the first robust detection of the effects of dark energy in slowing the growth of cosmic structure. These studies, and more recent work using the cluster count technique, are discussed in Section 10.3.

Throughout the 2000s, *Chandra* would also facilitate a series of stringent tests of the standard cold dark matter (CDM) paradigm, the highlight being the reported "direct empirical proof of the existence of dark matter" obtained using *Chandra* X-ray and ground- and space-based gravitational lensing measurements of the Bullet Cluster (Clowe et al. 2006; Bradač et al. 2006). This work, and the findings from other *Chandra* studies of dark matter, is discussed in Section 10.4.

One of the science goals that originally motivated the *Chandra* mission (Weisskopf 2010) was the recognition that, for dynamically relaxed clusters, measurements of their X-ray properties and Sunyaev–Zel'dovich effects (Sunyaev & Zeldovich 1972) could be combined to determine the distances to the systems, independent of redshift, and constrain the Hubble constant. Such measurements are discussed in Section 10.5.

*Chandra* observations of galaxy clusters also enabled other important measurements of fundamental physics, including tests of gravity on cosmological scales, measurements of neutrino properties, and constraints on the physics of cosmic inflation. These results are discussed in Section 10.6.

Finally, in Section 10.7, we discuss how, though now 20 years old, *Chandra* looks likely to remain vital to cosmological studies through its third decade, providing measurements that continue to boost the constraining power of cluster surveys and enhance the robustness of cosmological measurements.

## 10.2 Cosmology with the $f_{gas}$ Test

Because the most massive clusters are expected to provide approximately fair samples of the matter content of the universe (Section 10.1), and their baryonic matter content is dominated by the ICM (Figure 10.2), the mass fraction of hot gas, $f_{gas} = M_{gas}/M_{tot}$, measured within a characteristic radius, $r_\Delta$,[2] of such a cluster at redshift $z$, can be written as $f_{gas}(r_\Delta, z) = \Upsilon(r_\Delta, z)(\Omega_b/\Omega_m)$. Here, $\Upsilon(r_\Delta, z)$ is the gas depletion parameter, which accounts for the combined effects of a host of baryonic physics, including shocks, cooling, active galactic nucleus (AGN) feedback and star formation, the net effect of which is to inflate the ICM slightly with respect to the dark matter distribution. For the most massive halos and typical measurement radii, however, the effect is modest, and $\Upsilon \sim 0.8$–$0.9$, independent of redshift, is expected (e.g., Battaglia et al. 2013; Planelles et al. 2013).

Cluster $f_{gas}$ values can be measured robustly with *Chandra* X-ray data for dynamically relaxed clusters under the assumption of hydrostatic equilibrium (Chapter 9). The apparent $f_{gas}(z)$ value measured within a given aperture will depend on the assumed reference cosmology as $f_{gas}(z)^{ref} \propto [d_L(z)d_A^{0.5}(z)]^{ref}$, where $d_L^{ref}$ and $d_A^{ref}$ are the luminosity and angular diameter distances in that reference cosmology, respectively. However, the true, intrinsic $f_{gas}$ value should be an approximately constant, standard quantity, dependent only on the detailed form of $\Upsilon(r_\Delta, z)$. A complete expression for the expected dependence of the measured $f_{gas}$ values for a given reference cosmology is given by Allen et al. (2008; see also Mantz et al. 2014),

$$f_{gas}^{ref}(r_\Delta, z) = AK(r_\Delta, z)\Upsilon(r_\Delta, z)\left(\frac{\Omega_b}{\Omega_m}\right)\left[\frac{d^{ref}(z)}{d(z)}\right]^{3/2}, \quad (10.1)$$

where

$$A = \left(\frac{\theta_{2500}^{ref}}{\theta_{2500}}\right)^{\eta} \approx \left(\frac{H(z)\,d(z)}{[H(z)\,d(z)]^{ref}}\right)^{\eta}. \quad (10.2)$$

Here, the factor $A$ accounts for the change in angle subtended by the measurement aperture as the test cosmology is varied, while $\eta$ is the slope of the $f_{gas}(r_\Delta, z)$ data in the region of the measurement aperture for the reference cosmology. $\Upsilon(r_\Delta, z)$, the gas depletion parameter, can be predicted by hydrodynamical simulation: generally, the form of this prior is insensitive to cosmology, though dependent to some degree on the baryonic physics prescription in the simulations. $K(r_\Delta, z)$ describes the expected deviation from hydrostatic equilibrium as a function of radius and redshift, which can be predicted by simulations or calibrated directly with weak lensing measurements (e.g., Applegate et al. 2016). Constraints on $\Omega_b$ are incorporated

[2] As the virial radii of clusters are not readily observable, characteristic radii and masses corresponding to smaller scales are jointly defined by convention in terms of an "overdensity," $\Delta$, as $M_\Delta = 4/3\pi\Delta\rho_{cr}(z)r_\Delta^3$, where $\rho_{cr}(z)$ is the critical density at a cluster's redshift.

independently from CMB data or using external priors on $\Omega_b h^2$ and $h$ from other external measurements.

Examining the form of Equation (10.1), we see how the normalization of the $f_{gas}(z)$ curve provides a constraint on $\Omega_m$, while the shape of $f_{gas}(z)$ constrains both $\Omega_m$ and the effects of dark energy through the distance dependence of the $f_{gas}$ measurements. Benefits of this cosmological test include the ability to operate with a relatively small number of clusters and the relative insensitivity of the technique to the details of the surveys from which the clusters are found.[3]

In principle, $f_{gas}$ measurements can be made at radii corresponding to any overdensity. In practice, though, this overdensity should be low enough (i.e., the radius should be large enough) that the uncertainties in the depletion parameter do not dominate the overall error budget, but not so low (i.e., the radius not so large) that the measurements become compromised by systematic limitations at faint surface brightness levels. Hydrodynamical simulations indicate that radii $\gtrsim r_{2500}$ are sufficiently large to benefit from robust predictions for $\Upsilon(z)$ (Battaglia et al. 2012; Planelles et al. 2013). These radii are also small enough that astrophysical uncertainties associated with the effects of gas clumping, and systematic uncertainties in the background modeling, should generally not be significant.

Using this technique, Allen et al. (2004) presented *Chandra* $f_{gas}$ measurements for 26 hot, dynamically relaxed clusters spanning the redshift range $0.07 < z < 0.9$. This analysis provided the first direct confirmation of the results from Type Ia supernova studies that the expansion of the universe is accelerating, presenting a clear detection of the effects of dark energy at >99.9% confidence. The marginalized constraints on the mean matter and dark energy densities from this work for nonflat ΛCDM models (i.e., ΛCDM models with curvature included as a free parameter) were $\Omega_m = 0.25 \pm 0.04$ and $\Omega_\Lambda = 0.96 \pm 0.21$, with the best-fit model providing statistically acceptable description of the data.

This work was expanded by Allen et al. (2008) using *Chandra* observations for 42 hot ($kT_X > 5$ keV), dynamically relaxed clusters extending out to $z < 1.1$. (This study utilized approximately twice the total *Chandra* exposure time of the Allen et al. 2004 work.) The constraints from these improved measurements, marginalizing conservatively over all residual systematic uncertainties, are shown in Figure 10.4. For nonflat ΛCDM models, the results give constraints of $\Omega_m = 0.27 \pm 0.06$ and $\Omega_\Lambda = 0.86 \pm 0.19$. For flat, constant-$w$ models, the measurements imply a dark energy equation of state, $w = -1.14^{+0.27}_{-0.35}$. This analysis also confirmed, quantitatively, that the observed systematic, cluster-to-cluster scatter in $f_{gas}$ for such massive, relaxed clusters is small, <7% (corresponding to only 5% in distance), in agreement with the predictions from numerical simulations.

Ettori et al. (2009) presented $f_{gas}$ measurements for 52 clusters spanning the redshift range $0.3 < z < 1.3$. However, their study was not restricted to dynamically

[3] It is the small intrinsic scatter in $f_{gas}$ measurements for the most massive, dynamically relaxed clusters that leads to the insensitivity of the technique to details of survey selection functions.

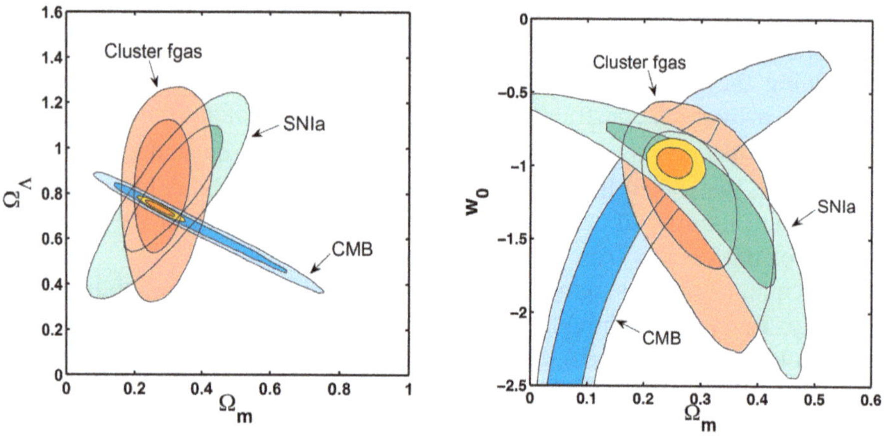


**Figure 10.4.** Cosmological constraints from the $f_{gas}$ test circa 2008. Left: joint 68.3% and 95.4% confidence constraints on the mean matter density, $\Omega_m$, and the dark energy density, $\Omega_\Lambda$, for ΛCDM models with curvature included as a free parameter. Right: constraints on $\Omega_m$ and the dark energy equation of state, $w$, for flat, constant-$w$ models. Contemporary constraints from CMB data (Spergel et al. 2007) and Type Ia supernova (SN Ia; Davis et al. 2007) measurements are also shown. Reproduced from Allen et al. (2008), by permission of Oxford University Press on behalf of the Royal Astronomical Society.

relaxed systems, which resulted in significantly larger system-to-system scatter and weaker cosmological constraints.

The most recent application of the $f_{gas}$ technique was presented by Mantz et al. (2014). This study employed approximately twice the total *Chandra* exposure of the Allen et al. (2008) work (and approximately four times that of Allen et al. 2004) and incorporated several modeling advances. Most notable among these were the application of an automated, algorithmic selection of the most dynamically relaxed systems from *Chandra* images (Mantz et al. 2015a); the use of $f_{gas}$ measurements made within a spherical shell, spanning radii $0.8 < r/r_{2500} < 1.2$, rather than a complete spherical volume (which results in slightly larger $f_{gas}$ measurement errors but minimizes the uncertainties in the gas depletion parameter); and, for the first time, the incorporation of a direct, robust constraint on $K(r_\Delta, z)$ from high-quality weak-lensing measurements (Applegate et al. 2016). A comparison of cosmological constraints from the Mantz et al. (2014) study with those from the earlier work of Allen et al. (2008) is shown in Figure 10.5.

Improvements in the constraints on $\Omega_m$ from the $f_{gas}$ experiment, at approximately the factor two level, should be possible over the next couple of years simply by expanding the number of clusters in the $f_{gas}$ sample with high-quality weak-lensing measurements (Mantz et al. 2014 included weak-lensing measurements for only 12/40 targets; work to deliver such gains is underway). Improvements in the dark energy constraints, however, will require an expansion of the cluster sample. Mantz et al. (2014) described how, with an investment of ~10 Ms of *Chandra* time over the coming decade, targeting the hottest, most-relaxed clusters at intermediate to high redshifts discovered with next-generation X-ray and Sunyaev–Zel'dovich

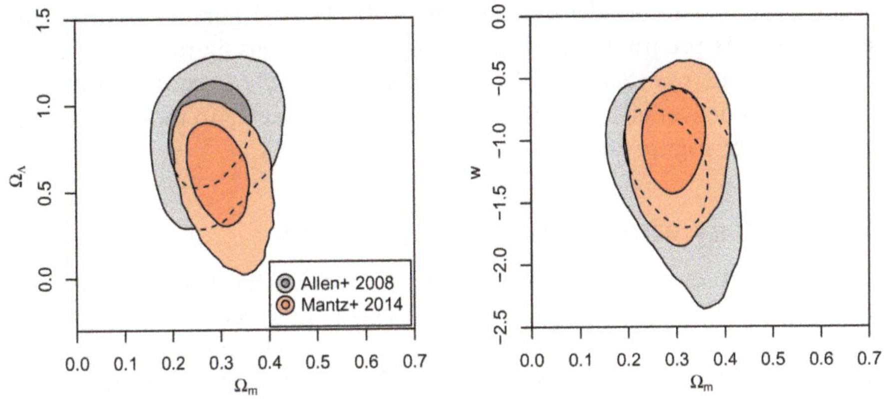


**Figure 10.5.** Left: a comparison of the joint 68.3% and 95.4% confidence constraints on $\Omega_m$ and and $\Omega_\Lambda$ from the most recent $f_{gas}$ study of Mantz et al. (2014; red curves) with those from the earlier study of Allen et al. (2008). Right: a similar comparison for flat, constant-$w$ models. *Chandra* measurements with the $f_{gas}$ technique have shown excellent stability over time. Figures adapted from Mantz et al. (2014), by permission of Oxford University Press on behalf of the Royal Astronomical Society.

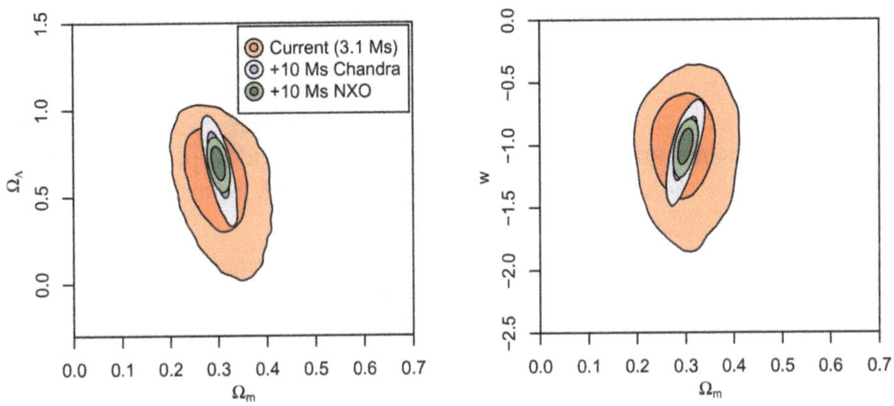


**Figure 10.6.** Projected joint 68.3% and 95.4% confidence constraints from the $f_{gas}$ test with current (red contours) and future data for (left) nonflat ΛCDM and (right) flat, constant-$w$ cosmologies. Blue contours show the predicted constraints that could be obtained with the addition of 10 Ms of *Chandra* data, providing $f_{gas}$ measurements for an additional ~50 clusters at intermediate to high redshifts (to a precision of ~15%). Green contours show the predicted constraints from a future data set of 400 clusters with $f_{gas}$ measured with a new, next-generation X-ray observatory (NXO) with capabilities similar to the *Lynx* concept (Gaskin et al. 2018) being considered by the 2020 Decadal Survey. Figures adapted from Mantz et al. (2014), by permission of Oxford University Press on behalf of the Royal Astronomical Society.

(hereafter SZ) surveys (Section 10.7), substantial improvements in the dark energy constraints from the $f_{gas}$ experiment could be obtained. The projected constraints from adding 50 new clusters at $0.3 < z < 1.75$, with accompanying weak-lensing measurements, are presented in Figure 10.6 (blue contours). Mantz et al. (2014) also explored the improvements that could be obtained with a next-generation X-ray observatory (NXO) with capabilities similar to the *Lynx* concept (Gaskin et al.

2018) currently under consideration by the 2020 Decadal Survey[4] (Astro2020). In this case, a 10 Ms program, with accompanying weak-lensing measurements from optical/near-IR surveys (Section 10.7), could expand the $f_{\rm gas}$ sample to more than 400 clusters and deliver cosmological constraints comparable to those plotted in the green contours of Figure 10.6.

## 10.3 Cosmology with Cluster Number Counts

The large-scale structure of our universe was generated by the growth, under gravity, of initially small density perturbations, with clusters of galaxies being the largest virialized structures to have formed (Peebles 1980; Dodelson 2003). The gravitational collapse that results in clusters, unlike the larger-scale ($\gtrsim$100 Mpc) distribution of matter, has reached the nonlinear regime. As a result, the number of massive clusters formed as a function of mass and time is a sensitive probe of both cosmological initial conditions (the spectrum of initial density perturbations) and the growth process itself (gravity, as influenced by the presence and nature of dark matter and dark energy).

From a theoretical or simulation perspective, the analog of the (observationally defined) galaxy cluster is the halo—a massive, gravitationally bound structure. The key theoretical prediction needed to constrain cosmological models from cluster counts is the density of halos as a function of mass and redshift, known as the mass function. While the mass function is dependent on the cosmological model, theoretical arguments and simulations show that, to good approximation, it can be expressed in a cosmology-independent form (e.g., Press & Schechter 1974; Sheth & Tormen 1999; Jenkins et al. 2001),

$$\frac{dn}{d\ln M} = \frac{\bar{\rho}_{\rm m}}{M}\left|\frac{d\ln\sigma}{d\ln M}\right| f(\sigma), \tag{10.3}$$

where $\bar{\rho}_{\rm m} = \Omega_{\rm m}\rho_{\rm cr}$ is the comoving mean matter density. In this formulation, all of the cosmology dependence of the mass function is captured by the function $f$ and the parameter $\sigma$, the latter being itself a function of mass, defined as the standard deviation of the linearly evolved cosmic matter density after smoothing by a mass-dependent spherical top-hat filter:

$$\sigma^2(M, z) = \int \frac{d^3k}{(2\pi)^3}\, W^2(kR_M)\, P_{\rm m}(k, z), \tag{10.4}$$

where $P_{\rm m}(k, z)$ is the matter power spectrum and $M = (4\pi/3)\rho_{\rm cr}(z)R_M^3$ defines the smoothing scale, $R_M$, in terms of the halo mass. The cosmological parameter $\sigma_8$, used to normalize the power spectrum, corresponds to evaluating Equation (10.4) with a smoothing scale of $R_M = 8h^{-1}$ comoving Mpc at $z = 0$. The form of $f(\sigma)$ can be calibrated from simulations, which have shown it to be approximately invariant for a range of cosmologies near the concordance model (Jenkins et al. 2001; Tinker

[4] https://sites.nationalacademies.org/DEPS/Astro2020/index.htm

et al. 2008; Bocquet et al. 2016). Similarly, the impact of nonzero neutrino mass on the mass function can be approximately accounted for by ignoring neutrino contributions in Equations (10.3)–(10.4), i.e., considering only baryons and dark matter (Costanzi et al. 2013; LoVerde 2014). While this approach has proven sufficient for the limited precision of current data, the field is currently moving toward the use of "emulators" that accurately interpolate the mass function between points in cosmological parameter space where detailed simulations have been performed (e.g., Heitmann et al. 2016; McClintock et al. 2019).

With this theoretical input, we can straightforwardly compute the expected number of clusters detected in a mass-limited survey according to a given cosmological model. Broadly speaking, a constraint on the total mass contained in clusters at low redshifts translates to a constraint on a degenerate combination of the cosmic matter density, $\Omega_{\mathrm{m}}$, and the power spectrum amplitude, $\sigma_8$. The degeneracy between these parameters can be broken either by measuring the shape of the mass function or its evolution with redshift. The evolution furthermore constrains additional cosmological parameters that dictate the rate of growth of structure, namely $\Omega_{\mathrm{m}}$, $\Omega_\Lambda$, and $w$ in the most common parameterizations (see Section 10.6.1 for a discussion of models in which modifications to gravity are also considered).

In practice, cluster surveys necessarily rely on observable signals for source detection, so real cluster catalogs are selected based on astrophysical properties rather than mass (which is not directly observable). Cosmologically relevant catalogs have been produced by surveys in three primary wavelength regimes: X-rays, detecting clusters through their ICM emission; optical/IR wavelengths, by identifying local overdensities of red galaxies (richness); and millimeter wavelengths, through the SZ effect imprinted on the CMB by inverse-Compton scattering with the ICM. While a survey operating in any single waveband will have certain limitations, the complementary nature of X-ray, optical, and millimeter-wavelength measurements provides direct, observational solutions to most issues. X-ray surveys, for example, can provide clean, complete catalogs of clusters; their primary disadvantages are the need to perform them from space (which brings associated cost and risk), the impact of surface brightness dimming, and the need for adequate spatial resolution to distinguish X-ray emission from the ICM and contaminating point sources. SZ surveys provide a more uniform selection in redshift, with only their sensitivity determining the mass down to which clusters can in principle be detected. This technique provides our best route for finding clusters at high redshifts. Challenges again include the need for spatial resolution and multifrequency coverage to discriminate the emission from contaminating sources. Like X-ray surveys, optical and near-IR surveys are most effective at low-to-intermediate redshifts, but have the benefit of finding larger numbers of clusters down to lower masses. The primary challenge for optical cluster selection is projection effects, which lead to overestimated richnesses for some clusters. Nonetheless, optical surveys provide an essential complement to X-ray and SZ data in cluster identification, in absolute mass calibration (using galaxy weak-lensing measurements) and, uniquely, in providing essential redshift information (from precise multiband photometry or spectroscopy).

Because the selection function of a given survey can only be cleanly defined (i.e., without invoking cosmological or astrophysical model assumptions) in terms of observable signals, the task of constraining cosmology from cluster counts necessarily requires us to also model and constrain the scaling relations relating observable signals to cluster mass and redshift. In principle, these relations could be "self-calibrated" from the survey data alone using information present in the shape and evolution of the mass function. However, pure self-calibration requires strong assumptions about the form of the scaling relations and their intrinsic scatter that cannot be directly verified; in addition, information from the shape and evolution of the mass function would no longer be available to break other cosmological parameter degeneracies. Hence, the availability of additional follow-up information for individual clusters provides an invaluable complement to the survey data for cluster cosmology. It is in providing such information that *Chandra* has excelled.

This follow-up information comes in two forms; the first provides estimates of cluster masses based on the hydrostatic assumption, which can be used to provide an absolute mass calibration for the experiment (i.e., the normalization of the observable–mass scaling relations). This approach is necessarily restricted to dynamically relaxed clusters and has been superseded in recent years by the use of weak gravitational lensing measurements, which are approximately unbiased, independent of dynamical state. Nonetheless, the use of hydrostatic mass measurements enabled the first generation of cosmology constraints, including the first competitive constraints on dark energy, from cluster counts (see below). Second, and more critical to modern cluster count measurements, X-ray data can also provide precise mass proxies–observables, other than those used to detect clusters in the survey, that scale tightly (i.e., with $\lesssim$15% intrinsic scatter) with the total cluster mass; these include the gas mass (Section 10.2) and the product of gas mass and temperature ($Y_X$). Given a separate absolute mass calibration, these additional proxies act as precise relative mass estimates, providing the means to constrain the scaling relations and intrinsic scatters of the more complex observables used to detect clusters in surveys.

Through the early 2000s, measurements of cluster counts, utilizing observationally derived mass–observable scaling relations calibrated with hydrostatic mass measurements, were used to place interesting constraints on a degenerate combination of $\Omega_m$ and $\sigma_8$ (e.g., Borgani et al. 2001; Seljak 2002; Allen et al. 2003; Pierpaoli et al. 2003; Schuecker et al. 2003), as well as the first, relatively weak constraints on dark energy parameters from cluster counts (Mantz et al. 2008).

This landscape was transformed later in the decade, however, when two teams independently brought substantial amounts of *Chandra* mass proxy information to bear on measurements of cluster counts (Vikhlinin et al. 2009; Mantz et al. 2010c). This enabled far stronger constraints on dark energy to be placed, constraints comparable to (or even exceeding) those from the best other cosmological probes, providing the first clear detection of the effects of dark energy in slowing the growth of cosmic structure. These works exemplified two different approaches to the modeling of follow-up data: the first restricted its analysis to a relatively small catalog where complete follow-up was available; the second introduced a method to statistically leverage incomplete follow-up for a subset of clusters to improve constraints from a larger parent catalog, simultaneously solving for both cosmological and scaling relation parameters.

The cluster catalog used by Vikhlinin et al. (2009) is X-ray flux selected and consists of disjoint low-redshift and high-redshift samples, with 49 clusters at redshifts $0.025 < z < 0.22$ (nearly all $<0.1$) and an additional 36 clusters serendipitously detected in the *ROSAT* 400 Square Degree Survey (Burenin et al. 2007) at $0.35 < z < 0.9$. This survey strategy, covering only a modest sky area in the higher redshift catalog but to significant depths, finds very few of the most massive clusters in the universe but extends down to relatively low masses ($M_{500} > 1.3 \times 10^{14}\ M_{\odot}$). All 85 clusters in the full data set were followed up with *Chandra*, and their masses were estimated using either $M_{gas}$ or $Y_X$ as a proxy; the proxy–mass relations were calibrated using hydrostatic mass estimates for a sample of well-observed, low-redshift clusters. Figure 10.7 illustrates schematically how these data provide cosmological constraints by comparing low- and high-redshift empirical and predicted mass functions for two different cosmological models.

The work of Mantz et al. (2010c, 2015b) employed the BCS (Ebeling et al. 1998), REFLEX (Böhringer et al. 2004), and Bright MACS (Ebeling et al. 2010) cluster samples compiled from the *ROSAT* All-Sky Survey (RASS). This survey strategy, covering a large fraction of the sky to relatively shallow depth in X-rays, was optimized to the task of finding the largest clusters at the expense of depth in redshift. This data set consists of >200 clusters with masses $M_{500} > 2.7 \times 10^{14}\ M_{\odot}$, distributed over the redshift range $0 < z < 0.5$. Pointed *Chandra* or *ROSAT* follow-up observations were used to measure $M_{gas}$ for a subset of clusters in this catalog, with the remaining clusters having measurements of only the cluster redshift and survey flux. The $M_{gas} - M_{tot}$ relation was marginalized over using either hydrostatic mass estimates for relaxed clusters Mantz et al. (2010c) or weak-lensing mass measurements (Mantz et al. 2015b).

The same RASS-based cluster catalog was also used in the earlier work of Mantz et al. (2008). In this previous generation of analysis, no mass proxy information for

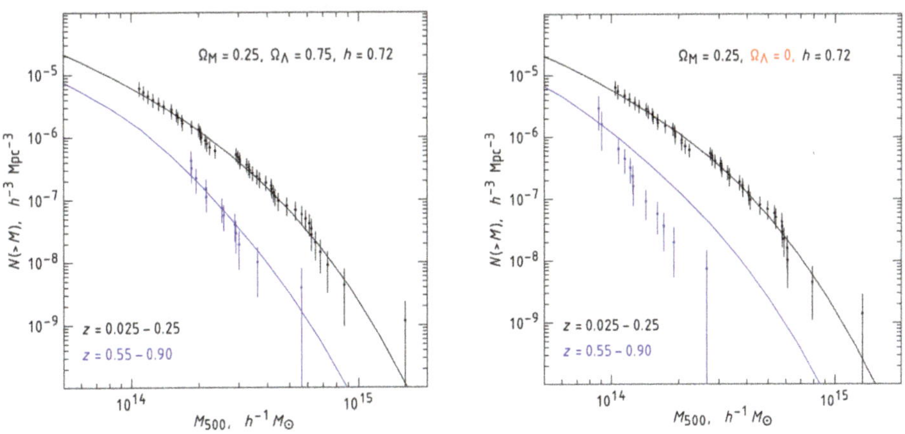


**Figure 10.7.** Measured mass functions of clusters at low and high redshifts are compared with predictions of a flat, ΛCDM model (left) and an open model without dark energy (right). For the purpose of illustration, cosmology-dependent masses are shown; in practice, model predictions are compared with cosmology-independent measurements. Reproduced from Vikhlinin et al. (2009). 

the cosmological sample was used, but instead weak constraints on the X-ray luminosity–mass scaling relation and its scatter were derived from an external data set. This is responsible for the elongated "banana"-shaped constraints on $\Omega_m$ and $\sigma_8$ seen in the left panel of Figure 10.8. The 2010 Mantz et al. (2010c) study improved on this by employing a hydrostatic absolute mass calibration for relaxed clusters from Allen et al. (2008), via *Chandra* $M_{gas}$ mass proxy measurements for 94 clusters in the survey data set (the relaxed clusters also directly constrain $\Omega_m$, as in Section 10.2). Combined with improvements in the statistical analysis, this development significantly reduced both the statistical and systematic uncertainties, placing the constraints on a firmer footing. From 2010 to 2015, the principal changes were an expansion in the amount of *Chandra* mass proxy data (both deeper data in some cases and an expansion to 139 followed-up clusters) and the use of an absolute mass calibration from weak-lensing measurements (von der Linden et al. 2014a; Applegate et al. 2014), both of which contributed to tighter constraints on the parameters shown. The importance of each type of follow-up information is illustrated in the right panel of Figure 10.8, which shows how the Mantz et al. (2015b) constraints vary when only subsets of the data are employed. We see that the weak-lensing data, while critical for establishing an unbiased absolute mass calibration, are not precise enough to significantly reduce the degeneracy between $\Omega_m$ and $\sigma_8$. The addition of *Chandra* $f_{gas}$ (Section 10.2) measurements provides a precise external constraint on $\Omega_m$. The constraints are tightened still further by incorporating precise *Chandra* X-ray mass proxy information for approximately 60% of the clusters.

The past decade has also seen the emergence of cosmological constraints from SZ-selected cluster counts (Benson et al. 2013; Hasselfield et al. 2013; Planck Collaboration 2014, 2016a; deHaan et al. 2016; Bocquet et al. 2019) with three projects simultaneously leading the way: the South Pole Telescope (Carlstrom et al.

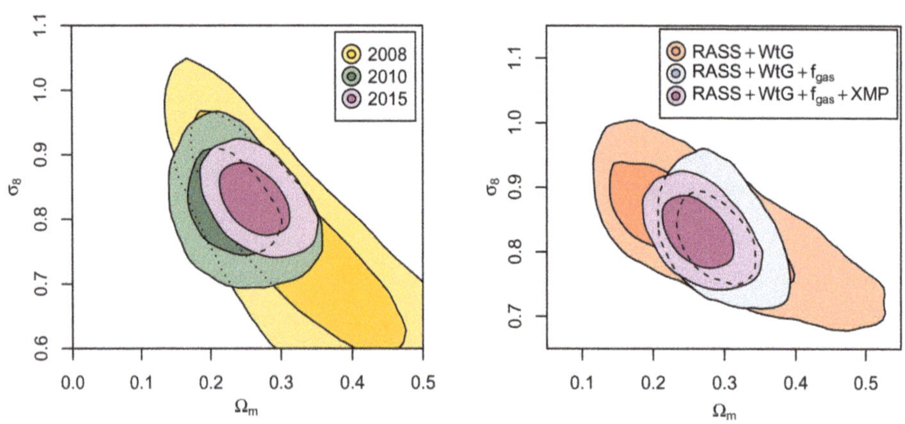


**Figure 10.8.** Left: joint 68.3% and 95.4% confidence constraints on $\Omega_m$ and $\sigma_8$ from cluster number counts (Mantz et al. 2008, 2010c, 2015b), showing the improvement as additional follow-up data, are incorporated (see text for details). Right: similarly, constraints from Mantz et al. (2015b) when only subsets of the available follow-up data are used to supplement the cluster survey (RASS): weak-lensing (WtG), gas-mass fractions of relaxed clusters ($f_{gas}$), and X-ray mass proxies (XMP). Figures adapted from Mantz et al. (2015b), by permission of Oxford University Press on behalf of the Royal Astronomical Society.

2011), the Atacama Cosmology Telescope (Swetz et al. 2011), and the *Planck* satellite (Planck Collaboration 2011a). As noted previously, the primary strength of SZ surveys is in finding massive clusters out to high redshifts. However, X-ray measurements have also played an important role in the full utilization of these surveys for cluster cosmology. The cosmological analyses of the 2500 deg$^2$ SPT-SZ survey by deHaan et al. (2016) and Bocquet et al. (2019) incorporated *Chandra* X-ray measurements for more than 80 of the >350 systems found, primarily to provide precise $Y_X$ measurements. The *Planck* cluster cosmology analysis (Planck Collaboration 2016a) is also notable for its extensive use of *XMM-Newton* observations, both to aid with cluster identification (Planck Collaboration 2011b) and in constraining the key mass–observable scaling relations (Planck Collaboration 2011c, 2011d).

Figure 10.9 compares the cosmological constraints reported by some of the works discussed above. The left panel shows constraints on $\Omega_m$ and $\sigma_8$, where we see the impact of different assumptions or measurements of the cluster-mass scale. For the Vikhlinin et al. (2009) work, the brown-shaded regions show 68.3% and 95.4% confidence regions reflecting statistical uncertainties only and notably excluding any systematic uncertainty in the absolute mass calibration. (Note also that this analysis was limited to $0.15 \leqslant \Omega_m \leqslant 0.4$, within which the 95.4% contours do not quite close.) The Vikhlinin et al. (2009) contours illustrate the main parameter degeneracy that exists natively in cluster counts data if the mass calibration is assumed to be known precisely. The other works shown marginalize over various degrees of uncertainty in the mass calibration and thus are consistent with a wider range of $\sigma_8$ at fixed $\Omega_m$. From the background forward, the results plotted are from SPT-SZ (deHaan et al. 2016, orange shading), with a mass calibration motivated by weak-lensing measurements from von der Linden et al. (2014a) and Hoekstra et al. (2015);

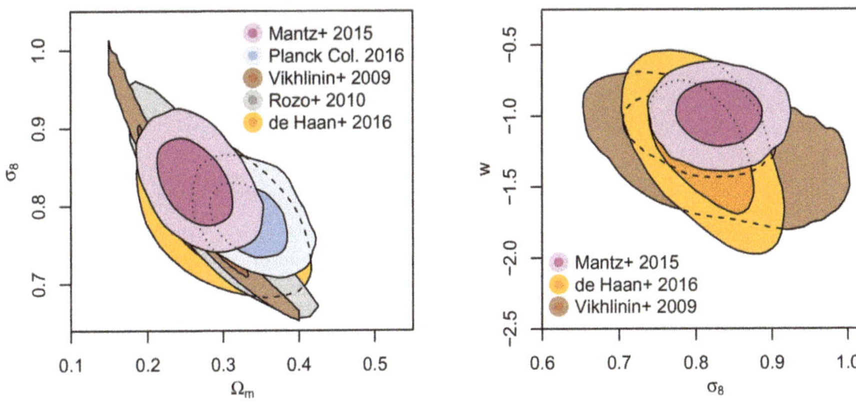


**Figure 10.9.** Joint 68.3% and 95.4% confidence constraints on (left) $\Omega_m$ and $\sigma_8$ from cluster number counts for flat ΛCDM models and (right) $\sigma_8$ and $w$ for flat, constant-$w$ models (Vikhlinin et al. 2009; Rozo et al. 2010; Mantz et al. 2015b; deHaan et al. 2016; Planck Collaboration 2016a). (See the text for a detailed description.) With the exception of Vikhlinin et al. (2009), the results shown are marginalized over systematic allowances, in particular on the absolute mass calibration, which strongly impacts the constraints on $\sigma_8$ at fixed $\Omega_m$ (see text for details). The Planck Collaboration (2016a) results incorporate an external prior on $\Omega_m$ from BAO data.

Rozo et al. (2010, gray shading), from optically selected clusters, with masses calibrated using stacked weak lensing and priors from *Chandra* X-ray measurements (Vikhlinin et al. 2009); Planck Collaboration (2016a, green shading), from *Planck* SZ-selected clusters plus a constraint on $\Omega_m$ from baryon acoustic oscillations (BAO), with a cluster-mass calibration motivated by von der Linden et al. (2014a, 2014b); and Mantz et al. (2015b, purple shading), the results from RASS cluster counts combined with *Chandra* X-ray mass proxies, weak-lensing (von der Linden et al. 2014a), and $f_{gas}$ data. While these results are not each independent in terms of the absolute mass calibration employed, they nevertheless show the concordance achieved by different cluster surveys and analysis teams in the past decade.

The right panel of Figure 10.9 similarly compares joint constraints on $\sigma_8$ and $w$ (for constant-$w$ models) from cluster counts. Note that constraints on $w$ are essentially independent of the absolute mass calibration, deriving primarily from the evolution of the mass function rather than its normalization. The first tight constraints on $w$ from cluster counts, at the ±0.2 level, using extensive *Chandra* follow-up measurements (Vikhlinin et al. 2009; Mantz et al. 2010c) remain competitive with those from all other leading individual cosmological probes even 10 years later (see below). The consistent constraints on dark energy from these two different X-ray-selected samples, and with later work based on SZ-selected catalogs from SPT and *Planck* (deHaan et al. 2016; Planck Collaboration 2016a), speak to the robustness of the technique.

The tightest cosmological constraints from clusters at the time of this writing, obtained by combining counts and $f_{gas}$ data, are those of Mantz et al. (2015b). The left panel of Figure 10.10 shows how these compare to contemporaneous constraints from primary CMB anisotropies, Type Ia supernovae, and BAO for constant-$w$

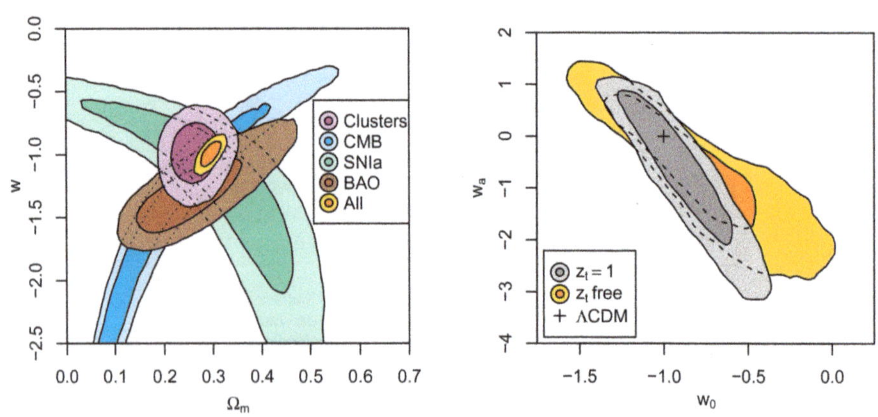


**Figure 10.10.** Cosmological measurements circa 2015 showing joint 68.3% and 95.4% confidence constraints on (left) $\Omega_m$ and $w$ from cluster counts and $f_{gas}$ (Mantz et al. 2015b), alongside contemporaneous constraints from CMB (Hinshaw et al. 2013), SNe Ia (Suzuki et al. 2012), and BAO (Beutler et al. 2011; Padmanabhan et al. 2012; Anderson et al. 2014); and (right) evolving-$w$ models (see text for details) from the combination of the above data sets. The results are consistent with a cosmological constant (ΛCDM; $w_0 = -1$ and $w_a = 0$). Figures adapted from Mantz et al. (2015b), by permission of Oxford University Press on behalf of the Royal Astronomical Society.

models. The counts + $f_{gas}$ data yield $\Omega_m = 0.26 \pm 0.03$ and $\sigma_8 = 0.83 \pm 0.04$ essentially independent of the dark energy model assumed. For constant-$w$ models, they constrain $w$ to be $-0.98 \pm 0.15$, while for nonflat ΛCDM models they yield $\Omega_\Lambda = 0.728 \pm 0.115$. Beyond the agreement of the cluster constraints with other probes and with the flat ΛCDM model, it is worth noting that this combination of counts and $f_{gas}$ data does not suffer from strong parameter degeneracies in these models; this is the result of internally having multiple constraints on the cosmic expansion history, as well as an independent constraint on $\Omega_m$ from the normalization of $f_{gas}(z)$. The right panel of Figure 10.10 shows constraints on evolving-$w$ models, with $w(z) = w_0 + w_a\, z/(z + z_t)$, which are consistent with the simple cosmological constant model, from the combination of all the probes mentioned.

As is discussed in Section 10.7, the coming decade will see a vast expansion in the size of available cluster catalogs through new surveys at X-ray, optical, and millimeter wavelengths. The role of *Chandra* in exploiting these catalogs will remain significant. In particular, *Chandra* will be needed to provide the robust, low-scatter mass proxies needed to pin down the mass–observable scaling relations and their scatter in the greatly expanded intermediate- to high-redshift regime.

Mantz et al. (2018) presented a quantitative exploration of the potential impact of *Chandra* and *XMM-Newton* follow-up measurements on future studies with cluster counts. By way of example, they consider a fiducial survey with a mass limit of $2 \times 10^{14}\, M_\odot$ spanning 2000 deg$^2$; such a survey is broadly comparable to that expected to be produced by the union of the current RASS and ongoing SPT-3G surveys (Benson et al. 2014), delivering a catalog of ~5000 clusters extending out to redshifts $z < 2$ (Section 10.7; we note, however, that the forecasts are not especially sensitive to the details of the cluster catalog, and the findings discussed below should be broadly applicable in a qualitative sense). Figure 10.11 shows the predicted improvements in the standard dark energy figure of merit (FoM; Albrecht et al. 2006) from this fiducial future survey as a function of the number of follow-up mass proxy measurements, assuming an intrinsic scatter in these measurements of 15%. The upper edge of the shaded region corresponds to the case where the follow-up targets are chosen to optimize the dark energy FoM (assuming a power-law form for the scaling relation and its evolution; see Wu et al. 2010), while the lower edge corresponds to a representative follow-up strategy. In the optimized case, roughly half of the mass proxy measurements gathered would be for relatively low-redshift, high-mass clusters; as such observations already populate the *Chandra* and *XMM-Newton* archives, it is unlikely that many additional observations of this type would be required. The new follow-up observations should focus on relatively high-redshift, low-mass clusters. In this regime, *Chandra* measurements will be especially important to cleanly discriminate the X-ray emission from point-like AGNs in the clusters, which are expected to be enhanced in high-$z$, low-mass systems (e.g., Ehlert et al. 2015; Koulouridis et al. 2018). Depending on whether center-excised X-ray luminosity (discussed by Maughan 2007; Mantz et al. 2018), $M_{gas}$, or $Y_X$ is to be measured (the efficacy of each mass proxy will need to be verified in this high-$z$ regime), follow-up observations might require ~100–1000 source counts.

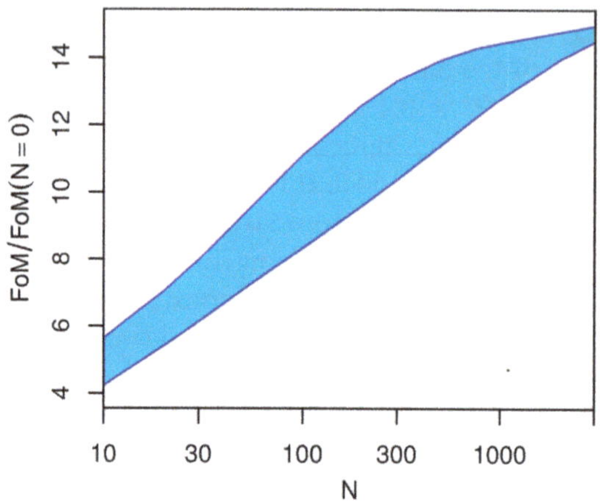


**Figure 10.11.** Predicted improvements in the standard dark energy FoM (Albrecht et al. 2006) for a fiducial survey of massive clusters extending to high redshifts, as a function of the number of clusters followed-up to obtain low-scatter ($\leqslant$15%) mass proxies. The upper edge of the shaded region corresponds to follow-up targets chosen to optimize the FoM, while the lower edge corresponds to the representative follow-up of detected clusters. Both cases assume power-law scaling relations with constant scatter. In practice, the best strategy would not rely on such strong assumptions about the form of the scaling relations, but would spread follow-up observations more evenly in mass and redshift. Figure from Mantz et al. (2018), by permission of Oxford University Press on behalf of the Royal Astronomical Society.

Accounting for the expected mass and redshift distribution of targets, this corresponds to *Chandra* exposure times of approximately 1–10 Ms per 50 new cluster observations.

Arguably the most beneficial overall follow-up strategy in terms of the potential for discovery is not one that is optimized assuming a particular dark energy model, as above, but rather one that spreads the follow-up observations throughout the interesting redshift and mass ranges. The FoM improvement per target for such an "evenly sampled" program would lie between the extremes represented by the optimal and representative follow-up cases, i.e., order-of-magnitude improvements in the FoM should be obtained for ~100–300 mass proxy measurements in total (with about half of the total being drawn from existing archival data). In this context, we note that the exposure time required per 50 clusters uniformly distributed in redshift and $\log(M)$, for $0.3 < z < 1.5$ and $10^{14} < M/M_{\odot} < 6 \times 10^{14}$ (i.e., the regime not well represented in archival data), remains similar to the optimized case above. The rough scale of a follow-up program is thus not too sensitive to the choice of targets, with *Chandra*-equivalent investments of ~10 Ms over the next decade potentially producing order-of-magnitude improvements in the FoM with respect to no follow-up for the fiducial case. Finally, we note that there is a great deal of science beyond cosmology that such observations would uniquely enable, such as studies of the evolution of cool cores (McDonald et al. 2019), cluster morphologies (Nurgaliev et al. 2017; Mantz et al. 2015a), high-$z$ metal enrichment (Ettori et al. 2015; McDonald et al. 2016; Mantz et al. 2017), and AGNs in cluster environments (Ehlert et al. 2015; Koulouridis et al. 2018; and references therein).

The most interesting clusters identified from these initial exposures will also be exceptional candidates for deeper follow-up *Chandra* observations, including the most dynamically relaxed systems, which would be the primary targets for the $f_{\rm gas}$ test (Section 10.2).

## 10.4 Dark Matter

While providing an excellent model for the large-scale structure of the universe, the standard ΛCDM (cosmological constant plus cold dark matter) paradigm says little about the physical nature of dark matter. It assumes only that dark matter is nonbaryonic, that it emits and absorbs no detectable electromagnetic radiation, that it interacts with itself and baryonic matter effectively only via gravity, and that the dark matter particles move at subrelativistic speeds. *Chandra* observations of galaxy clusters have provided some of the most stringent tests to date of these key assumptions.

### 10.4.1 Constraints on Dark Matter from Merging Clusters

Within the CDM paradigm, as galaxy clusters merge under the pull of gravity, their (collisionless) dark matter halos and (collisional) ICM should become separated temporarily, as the X-ray-emitting gas experiences ram pressure and is slowed. This effect was first observed in the Bullet Cluster (1E 0657−558; Bradač et al. 2006; Clowe et al. 2006; left panel of Figure 10.12), for which gravitational lensing measurements with the *Hubble Space Telescope* and ground-based observatories were used to independently trace both the stellar and total (dark plus baryonic) matter distributions in the cluster, while *Chandra* X-ray imaging measured the ICM (the dominant baryonic mass component; Figure 10.2). The results showed a clear separation between the dark and baryonic matter peaks, confirming that the dark matter particles have a much smaller interaction cross section than the ICM. Confirmation soon followed with a combined *Hubble Space Telescope* and *Chandra* study of MACS J0025.4−1222, a second massive merging cluster with a simple plane-of-the-sky merger geometry (Bradač et al. 2008; right panel of Figure 10.12).

Markevitch et al. (2004) and Bradač et al. (2008) used the observed separations between the lensing and X-ray peaks in the Bullet Cluster and MACS J0025.4−1222 to derive limits on the velocity-independent dark matter self-interaction cross section per unit mass of $\sigma/m < 5\ \mathrm{cm^2\ g^{-1}}$ and $\sigma/m < 4\ \mathrm{cm^2\ g^{-1}}$, respectively. Randall et al. (2008) used the nondetection of an offset between the lensing peaks and the galaxy centroids for the Bullet Cluster to refine this constraint to $\sigma/m < 1.25\ \mathrm{cm^2\ g^{-1}}$.

This work was followed by studies for other massive, merging systems, including Abell 2744 (Merten et al. 2011) and Abell 520 (Clowe et al. 2012; Jee et al. 2014). More recently, Harvey et al. (2015) and Wittman et al. (2018) presented results for dozens of massive clusters, seeking to measure similar effects. However, limitations in the data quality and the more complex merger geometries of most systems have to date prevented significantly tighter constraints on the dark matter self-interaction cross section from being obtained (Wittman et al. 2018).

Upcoming surveys (Section 10.7) will provide high-quality lensing data for hundreds of merging clusters. In combination with new *Chandra* observations and

**Figure 10.12.** *Hubble Space Telescope* optical images of the massive, merging clusters 1E0657−558 (a.k.a. the Bullet Cluster; $z = 0.30$) and MACS J0025.4−1222 ($z = 0.54$), with the X-ray emission measured with *Chandra* overlaid in pink and total mass reconstructions from gravitational lensing data in blue. The separations of the X-ray and lensing peaks, and the coincidence of the lensing and optical centroids, imply that the dark matter has a small self-interaction cross section. Figure credits. (Left) X-ray: NASA/CXC/CfA/M. Markevitch et al.; optical: NASA/STScI, Magellan/U. Arizona/D. Clowe et al.; lensing map: NASA/STScI, ESO WFI, *Magellan*/U. Arizona/D. Clowe et al. (Right) X-ray: NASA/CXC/Stanford/S. Allen; optical/lensing: NASA/STScI/UC Santa Barbara/M. Bradač. The interactive figure shows separately the X-ray/optical emission (red/yellow), lensing map/optical emission (blue/yellow), and the composite image of the bullet cluster, shown in the left hand panel. The movie shows a simulation of the collision taking place between the hot gas (red) and the dark matter (blue) in the bullet cluster. The hot gas (red) is slowed by the impact, forming the bullet-shaped structure, but the dark matter (blue) is not, demonstrating its lack of interaction with itself or the hot gas. The movie fades into a still of the composite image, made up of X-ray, optical, and lensing map data, as shown in the left hand panel. The MACS J0025.4-1222 interactive figure shows separately the X-ray emission (red), optical emission (yellow), lensing map (blue), and the composite image of MACS J0025.4-1222, as shown in the right hand panel. Additional materials available online at http://iopscience.iop.org/book/978-0-7503-2163-1

improved numerical simulations (e.g., Forero-Romero et al. 2010), this should allow the properties of dark matter to be constrained with improved precision.

### 10.4.2 Constraints on Dark Matter from Dynamically Relaxed Clusters

One of the most remarkable predictions of the CDM model is that the density profiles of relaxed dark matter halos, on all resolvable mass scales, can be approximated by a simple, universal profile,

$$\rho(r) = \frac{\rho_{\rm cr}(z)\, A_c}{(r/r_{\rm s})^{\beta}\, (1 + r/r_{\rm s})^{(3-\beta)}}, \tag{10.5}$$

where $\rho$ is the total matter density, $\beta$ is the asymptotic inner density slope, $r_{\rm s}$ is the scale radius, $c$ the concentration parameter (with $c = r_{200}/r_{\rm s}$), and $A_c = 200c^3/3[\ln(1 + c) - c/(1 + c)]$. CDM simulations predict $\beta \sim 1$. The special case of $\beta = 1$ is known as the Navarro–Frenk–White (NFW) profile (Navarro et al. 1995), while the more general form of Equation (10.5) is known as the generalized NFW or GNFW model.

CDM simulations additionally predict that concentration and mass should be weakly correlated. Gao et al. (2008) found $c \propto M^{-\zeta}$, with $\zeta \sim 0.14$ at low redshift, trending to $\zeta \rightarrow 0$ at $z \gtrsim 3$. For the most massive, most relaxed clusters, a typical value of $c \sim 4$–5 is predicted, independent of redshift (Gao et al. 2008; Ludlow et al. 2012).

In contrast, for dark matter models with significant self-interaction cross sections, halos are expected to exhibit flattened, quasi-isothermal cores (Spergel & Steinhardt 2000; Yoshida et al. 2000) and enhanced sphericity with respect to CDM models.

The most recent, comprehensive test of these predictions, using *Chandra* observations for 40 of the most massive, most dynamically relaxed galaxy clusters (the same systems used for the $f_{\rm gas}$ test; Section 10.2), is presented by Mantz et al. (2016a). (For examples of earlier work, see, e.g., Vikhlinin et al. 2006; Voigt & Fabian 2006; Zhang et al. 2006; Schmidt & Allen 2007; Host & Hansen 2011; Amodeo et al. 2016). For their sample, Mantz et al. (2016b) measured a mean central density slope, $\beta = 1.02 \pm 0.08$, and a power-law mass dependence of the concentration–mass relation of $\zeta = 0.16 \pm 0.07$.[5] An analysis of weak-lensing data for a subset of these clusters provides an ensemble average concentration of $c = 3.0^{+4.4}_{-1.8}$ (Applegate et al. 2016). These results, and those of previous X-ray studies of relaxed galaxy clusters, are generally in excellent agreement with the standard CDM model. Efforts to use such results to obtain complementary constraints on the dark matter self-interaction cross section are underway.

### 10.4.3 Constraints on Dark Matter from X-Ray Spectral-line Searches

Certain dark matter candidates, including sterile neutrinos, possess a two-body radiative decay channel that produces a photon with energy $E_\gamma = M_{\rm DM}/2$, where $M_{\rm DM}$ is the dark matter particle mass (e.g., Feng 2010). Galaxy clusters have been the targets of searches for emission lines associated with such decays. The soft X-ray (keV) regime is particularly interesting, marking the lower limit of masses consistent with constraints from large-scale structure formation. To date, all searches for monochromatic X-ray emission lines associated with nonbaryonic matter in clusters (as well as other dark-matter-rich objects) have failed to provide convincing evidence for such signatures. While considerable excitement was generated by reports of a detection of an unidentified emission line at an energy $E \sim 3.5$ keV in stacked X-ray spectra for 73 galaxy clusters observed with the *XMM-Newton* satellite, as well as *XMM-Newton* and *Chandra* data for the Perseus Cluster individually (Bulbul et al. 2014), later work, while confirming the presence of such a feature in CCD-quality data under certain modeling assumptions, has cast doubt on a dark matter origin (Urban et al. 2015). Moreover, high-spectral-resolution X-ray measurements of the central regions of the Perseus Cluster gathered with the *Hitomi* satellite failed to detect an emission-line signature at the energies and line strength predicted from the *XMM-Newton* data for this system, under the dark matter hypothesis (Aharonian et al. 2017).

[5] Note that the Mantz et al. (2016a) study excludes the central ~50 kpc regions of clusters, where the complicating effects of AGN feedback processes and residual dynamical activity commonly become important.

## 10.5 Measurements of the Hubble Constant

Silk & White (1978) showed that X-ray measurements can be combined with measurements of the SZ effect for clusters to determine the distances to these systems. The amplitude of the SZ effect is governed by the Compton $y$-parameter, which is a measure of the electron pressure along the line of sight, $y \propto \int n_e T dl$. Given an observed SZ signal, $y_{obs}$, and a predicted signal based on X-ray measurements of the ICM density and temperature of a cluster, $y_{pred}$, the angular diameter distance to the system scales as

$$d_A \propto \left( \frac{y_{obs}}{y_{pred}} \right)^2 . \quad (10.6)$$

The cosmological constraint originates from the distance dependence of the X-ray measurements, $y_{pred}(z) \propto d_A(z)^{1/2}$, and the requirement that $y_{pred} = y_{obs}$. The reader will note that this cosmological test (sometimes called the XSZ test) is intrinsically less sensitive to distance than $f_{gas}$ measurements, with the signal being proportional to $d_A(z)^{1/2}$ rather than $d_A(z)^{3/2}$. For this reason, to date, such measurements have only been able to constrain one free parameter, the Hubble constant, $H_0$ (or equivalently, the Hubble parameter, $h = H_0/100\ \mathrm{km\ s^{-1}\ Mpc^{-1}}$) to an interesting level. Examination of Equation (10.6) also shows the sensitivity of the measurements to the absolute calibration of the SZ and X-ray flux and X-ray temperature measurements, which provides additional challenges.

Bonamente et al. (2006) presented a realization of the XSZ experiment using *Chandra* and SZ measurements for 38 clusters at redshifts $0.14 < z < 0.89$. Assuming spatial flatness and fixing $\Omega_m = 0.3$, they measured a Hubble parameter, $h = 0.77^{+0.11}_{-0.09}$. Using a smaller sample of *Chandra* and SZ measurements restricted to dynamically relaxed clusters, for which systematic uncertainties associated with geometrical complexity are minimized, Schmidt et al. (2004) measured $h = 0.69 \pm 0.08$, in good agreement with results from the Hubble Key Project ($0.72 \pm 0.08$; Freedman et al. 2001) and more recent CMB measurements with *WMAP* ($h = 0.697 \pm 0.024$; Hinshaw et al. 2013) and *Planck* ($h = 0.674 \pm 0.005$; Planck Collaboration 2018).

## 10.6 Other Fundamental Physics

### 10.6.1 Gravity

Dark energy, though a central axiom of the ΛCDM model, is not the only possible explanation for cosmic acceleration. Various nonstandard gravity models can also produce acceleration on cosmological scales (see, e.g., Copeland et al. 2006; Frieman et al. 2008; Jain & Khoury 2010).[6] These include frameworks that

[6] A critical requirement for any modified gravity model is that it should mimic GR in the relatively small-scale, high-density regime where GR has been tested precisely.

consistently parameterize departures from general relativity (GR; Zhang et al. 2007; Hu & Sawicki 2007; Amin et al. 2008; Daniel et al. 2010)—full, alternative theories such as the Dvali–Gabadadze–Porrati (DGP) braneworld gravity (Dvali et al. 2000), $f(R)$ modifications of the Einstein–Hilbert action (Carroll et al. 2004), and modifications of gravity based on the mechanism of ghost condensation (Hamed et al. 2004). Thus, in addition to investigating whether dark energy is well described by a cosmological constant, we are also interested in asking whether GR provides the correct description of gravity and, within that broader context, whether dark energy is needed at all.

In discriminating among these possibilities, the combination of measurements of both the expansion history and (scale-dependent) growth of cosmic structure provides a particularly powerful approach. In essence, under the assumption that GR is correct, measurements of the expansion history (such as those from Type Ia supernova, BAO, and cluster $f_{\rm gas}$ measurements) will automatically specify the expected growth of cosmic structure. Thus, direct measurements of structure growth enable a powerful consistency test of the $\Lambda$CDM + GR paradigm.

As discussed in Section 10.3, galaxy clusters provide some of our strongest constraints on cosmological structure growth. To utilize these constraints robustly, however, accurate predictions for the halo mass function are required. Recently, mass functions for a few specific modified gravity models have been constructed and calibrated using $N$-body simulations. These include the self-accelerated branch (Chan & Scoccimarro 2009; Schmidt 2009a) and normal branch (Schmidt 2009b) of DGP gravity, and $f(R)$ models (Schmidt et al. 2009; Cataneo et al. 2016; Hagstotz et al. 2019). Constraints on the latter model using the observed local cluster abundance are presented by Cataneo et al. (2015).

An alternative to evaluating specific gravity theories is to adopt a convenient, parameterized description for the growth of structure. This can be used to constrain departures from the predictions of $\Lambda$CDM + GR (Nesseris & Perivolaropoulos 2008). At late times, the linear growth rate can be simply parameterized as

$$\frac{d\ln\delta}{d\ln a} = \Omega_{\rm m}(a)^{\gamma}, \tag{10.7}$$

where $\delta$ is the density contrast and $\gamma$ the growth index (Linder 2005). Conveniently, GR predicts a nearly constant and scale-independent value of $\gamma \approx 0.55$ for models consistent with current expansion data. Similarly to the case of $w$ for dark energy models, constraining $\gamma$ constitutes a phenomenological approach to studying gravity. Rapetti et al. (2013; see also Rapetti et al. 2009, 2010; Mantz et al. 2015b) reported constraints on departures from GR on cosmic scales for this parameterization using cluster data, CMB measurements, and galaxy redshift space distortions (see left panel of Figure 10.13). Combining the independent measurements increases the power of the constraints. In all cases, the results are simultaneously consistent with GR ($\gamma \sim 0.55$) and $\Lambda$CDM ($w = -1$), with $w = -0.987^{+0.054}_{-0.053}$ and $\gamma = 0.604 \pm 0.078$ for the full data combination shown in the right panel of Figure 10.13.

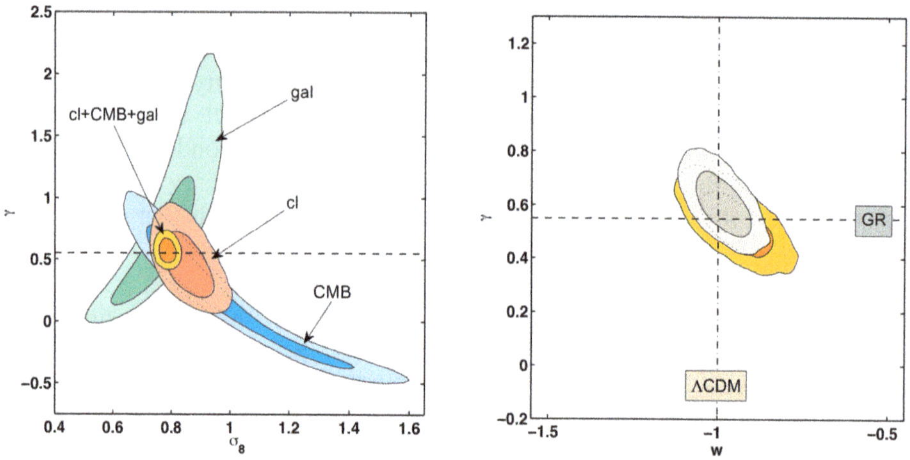


**Figure 10.13.** Constraints on modified gravity. Left: joint 68.3% and 95.4% confidence contours on $\sigma_8$ and the growth index, $\gamma$. The horizontal dashed line indicates a constant value of $\gamma = 0.55$, appropriate for a broad range of GR cosmologies. Results are shown separately for galaxy redshift space distortion data (gal; green contours; Blake et al. 2011; Beutler et al. 2011; Reid et al. 2012), CMB measurements (blue contours; Dunkley et al. 2009), and cluster counts (cl; red contours; Mantz et al. 2010a, 2010c), as well as the combination of these data (gold contours). Right: joint 68.3% and 95.4% confidence regions in the $\gamma$, $w$ plane, simultaneously constraining departures from GR and a cosmological constant model ($w = -1$). Gold contours show the results for cl + CMB + gal, while the platinum contours show the impact of also including contemporary Type Ia supernova data (Suzuki et al. 2012), BAO measurements (Percival et al. 2010; Reid et al. 2012), and external constraints on the Hubble Constant (Riess et al. 2011). Reproduced from Rapetti et al. (2013), by permission of Oxford University Press on behalf of the Royal Astronomical Society.

### 10.6.2 Neutrinos

The mass of neutrinos directly influences the growth of cosmic structure because any particle with nonzero mass at some point cools from a relativistic state, in which it effectively suppresses structure formation, to a nonrelativistic state, in which it actively participates in structure growth (for a review, see Lesgourgues & Pastor 2006). In the standard scenario, where the neutrino species have approximately degenerate mass, the species-summed mass, $\sum m_\nu$, is sufficient to describe their cosmological effects.

Although current data lack the precision to directly detect the effects of neutrino mass on the time-dependent growth of clusters, cluster data have played an important role over the past decade in constraining the value of $\sum m_\nu$. Until recently (Planck Collaboration 2016b, 2018), CMB data alone had only been able to place relatively weak upper bounds on the mass. For example, the constraints from five years of *WMAP* data for a flat, ΛCDM model were $\sum m_\nu < 1.3$ eV at 95% confidence (Dunkley et al. 2009). Incorporating cosmic distance measurements from cluster $f_{\rm gas}$, Type Ia supernovae, and BAO (Allen et al. 2008; Kowalski et al. 2008; Percival et al. 2007) improved this constraint to $\sum m_\nu < 0.61$ eV, although the results still displayed a strong degeneracy between $\sum m_\nu$ and $\sigma_8$ (see left panel of Figure 10.14). The introduction of cluster count measurements, which provide a direct constraint on $\sigma_8$,

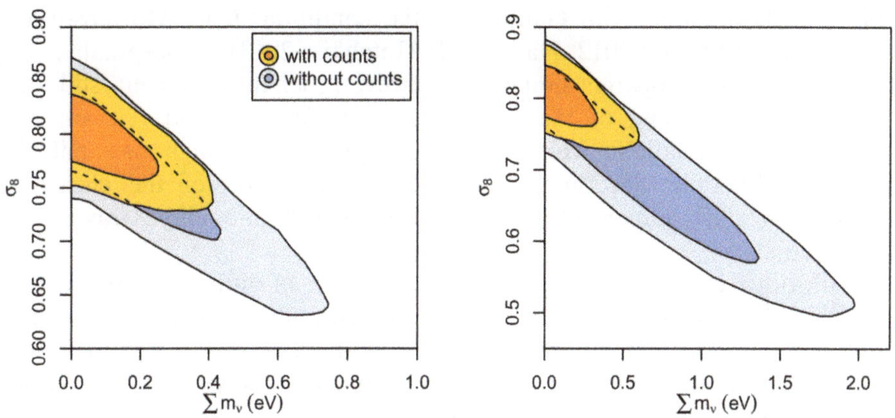


**Figure 10.14.** Constraints on the species-summed neutrino mass, $\sum m_\nu$, and the amplitude of density perturbations, $\sigma_8$, for (left) a basic flat ΛCDM model and (right) an extended model marginalized over global curvature, and the amplitude and spectral index of primordial tensor perturbations. Blue contours show the results from the combination of cluster $f_{\rm gas}$ (Allen et al. 2008), *WMAP* (Dunkley et al. 2009), SN Ia (Kowalski et al. 2008), and BAO (Percival et al. 2007) data. A clear degeneracy between the species-summed neutrino mass and $\sigma_8$ is observed. The gold contours show how this degeneracy is broken by the inclusion of cluster data from cluster counts (Mantz et al. 2010c). Figures adapted from Mantz et al. (2010b), by permission of Oxford University Press on behalf of the Royal Astronomical Society.

breaks this degeneracy (gold contours), improving the upper limit by a further factor of 2, $\sum m_\nu < 0.33$ eV (Mantz et al. 2010b; see also Vikhlinin et al. 2009; Reid et al. 2010). The degeneracy-breaking power of the cluster observations also significantly improves the robustness of neutrino mass limits to the assumed cosmological model (Mantz et al. 2010b; Reid et al. 2010). The right panel of Figure 10.14 shows the results for an extended cosmological model with free global curvature, also marginalizing over the amplitude and spectral index of primordial tensor perturbations. In this case, the addition of the cluster counts data improves the constraints from $\sum m_\nu < 1.6$ eV (CMB + BAO + SN Ia + $f_{\rm gas}$ only) to $\sum m_\nu < 0.49$ eV.

### 10.6.3 Inflation

Galaxy clusters trace the rare, high-mass tail of density perturbations in the universe, and thus provide an important probe of non-Gaussianity in the primordial density perturbation spectrum. In principle, a robust detection of such non-Gaussianity would open a new window into the physics of inflation (see, e.g., Chen 2010 & Komatsu 2010 for reviews). For example, any measurable non-Gaussianity of the "local" type (referring to particular configurations in Fourier space) would rule out not only slow roll but all classes of single field inflation models (Creminelli & Zaldarriaga 2004).

Galaxy clusters are so well suited to such studies that, in principle, the identification of even a single galaxy cluster of sufficiently high mass (for example, located within the gray region of Figure 10.16) would severely challenge the

Standard ΛCDM model with Gaussian initial conditions (e.g., Mortonson et al. 2011; Holz & Perlmutter 2012; Harrison & Hotchkiss 2013). Conceptually, such a test is attractive in its apparent simplicity. However, such measurements still require a detailed understanding of the survey selection function and relevant mass–observable scaling relations. Moreover, the steepness of the high-mass tail of the cluster-mass function (Section 10.3) makes the precision of the cluster-mass estimates particularly important, which in turn makes high-quality X-ray measurements essential.

As with other cosmological tests, the combination of different, complementary probes provides richer insights. In this case, CMB and/or galaxy autocorrelation data place tight constraints on the three-point (and eventually four-point) statistics of the perturbations, while the cluster-mass function is sensitive to moments extending to higher order (Barnaby & Shandera 2012). The combination of cluster data with the CMB and large-scale structure measurements can thus potentially distinguish between different mechanisms of generating non-Gaussianity during inflation.

The leading constraints on non-Gaussianity from cluster data to date have used X-ray-selected cluster catalogs at $z < 0.5$, coupled with extensive *Chandra* X-ray follow-up (Shandera et al. 2013; Mantz et al. 2015b). The results for two distinct classes of inflation models are shown in Figure 10.15. These measurements use an analytic prescription to include non-Gaussianity in the cluster-mass function, incorporating moments beyond the skewness. Figure 10.15 shows the current cluster

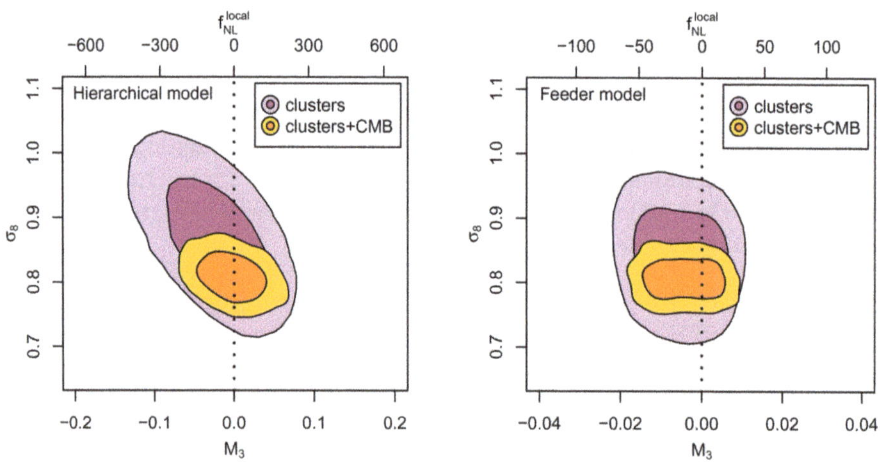


**Figure 10.15.** Joint constraints on $M_3$, the amplitude of non-Gaussian moments (and the corresponding $f_{\rm NL}^{\rm local}$, parameterizing the strength of "local" type non-Gaussianity), and $\sigma_8$ for single-parameter non-Gaussian models. Contours are drawn at the 68.3% and 95.4% confidence level, marginalizing over residual systematic uncertainties. For the clusters + CMB combinations, only the CMB power spectra (not bispectra or trispectra) are used in the combination; this tightens the constraints on the background cosmology but provides no additional, direct constraining power for non-Gaussianity. The results are consistent with Gaussianity in all cases, although their strength and character depend on whether the hierarchical or feeder scaling (left and right panels) are used (note that feeder models generate more non-Gaussianity for a given value of $M_3$). Reproduced from Shandera et al. (2013). 

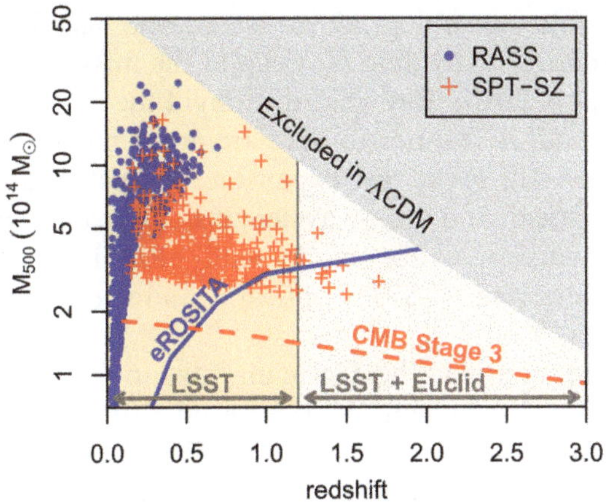


**Figure 10.16.** A mass–redshift plot showing some of the existing cluster catalogs that have been used extensively for cosmological studies: blue circles show clusters in the RASS X-ray catalogs of Ebeling et al. (1998, 2010) and Böhringer et al. (2004); red crosses mark clusters in the SPT-SZ millimeter-wavelength survey (Bleem et al. 2015). Also shown are the approximate expected reaches of the future cluster catalogs that will be produced by eROSITA (Merloni et al. 2012), LSST (LSST Dark Energy Science Collaboration 2012), and the various CMB Stage-3 projects (Benson et al. 2014; Henderson et al. 2016; Ade et al. 2019). In the standard cosmological model, clusters are not expected to exist in the gray "exclusion" region (see Section 10.6.3).

constraints on the amplitude of non-Gaussianity for two classes of inflation models, corresponding to single-field inflation (hierarchical) and inflation coupled to a spectator gauge field (feeder). In the former case, the leading non-Gaussian term (skewness) essentially characterizes the entire model and, as expected, the cluster constraints are not competitive with CMB data. However, the second scenario generates a significant contribution to the cluster signal at higher orders, where CMB data do not provide constraints. For this model, and particularly for equilateral and orthogonal configurations of the bispectrum, cluster data provide a powerful complement to CMB measurements of non-Gaussianity.

## 10.7 Conclusions and Future Prospects

We have summarized the ways in which, over the past 20 years, *Chandra* studies of galaxy clusters have enabled significant advances in cosmology and fundamental physics. *Chandra* measurements have been used to place competitive constraints on cosmological parameters, including the mean matter density, $\Omega_m$; the mean dark energy density, $\Omega_\Lambda$; the dark energy equation of state, $w$; the amplitude of matter fluctuations, $\sigma_8$; and the Hubble parameter, $h$. *Chandra* observations have additionally shed new light on the nature of dark matter and the physics of inflation, neutrinos, and gravity on cosmological scales. On virtually all fronts, the cluster constraints have proved highly complementary to those from other leading techniques; and, for those cases where cluster measurements initially led the way, for example with studies of dark matter and measurements of $\Omega_m$ and $\sigma_8$, the results have stood the test of time.

*Chandra*'s foray into cosmological measurements has been an unequivocal success, with achievements extending far beyond the mission's originally described science goals (Weisskopf 2010). But what role might the mission continue to play as it enters its third decade? A few possibilities are noted below.

The coming decade will see a vast expansion of cluster catalogs, generated by surveys at X-ray, optical and mm-wavelengths. Of particular note for cluster cosmology are the surveys to be made with eROSITA (X-rays), the Large Synoptic Survey Telescope (LSST; optical), and new millimeter-wavelength experiments.

At the time of this writing, the German–Russian *SRG* mission, bearing the eROSITA X-ray survey instrument (Merloni et al. 2012), has launched and is en route to L2, where it will perform its survey. The eROSITA all-sky survey will extend approximately 20 times fainter than the previous survey carried out by *ROSAT*, identifying vast numbers of galaxy groups (systems with $M_{500} < 10^{14}\ M_{\odot}$) out to $z \sim 0.3$, intermediate-mass clusters out to $\sim$0.6, and the most-massive clusters out to $z \lesssim 1.5$.

LSST will survey the southern sky in the optical *ugrizy* bands over a 10 year period. Beginning in 2022, it will identify clusters down to the group scale, constrain their redshifts photometrically, and provide precise, stacked weak-lensing mass measurements out to a redshift of $\sim$1.2 (LSST Dark Energy Science Collaboration 2012). Combining LSST data with near-IR data from *Euclid* (an ESA M-class mission scheduled for launch in 2021) will extend the redshift range for optical/near-IR cluster catalogs even further (Laureijs et al. 2011).

The most relevant, near-term millimeter-wavelength surveys for cluster science are those being carried out by SPT-3G, AdvACT, and the planned Simons Observatory, collectively referred to as the "Stage-3" CMB projects. Taking advantage of the SZ effect, these surveys will break new ground in providing the first large, robustly selected catalogs of clusters at $z > 1$, and the first informative absolute mass calibration from CMB-cluster lensing. Together, they should find >3000 clusters at $z > 1$ and $\sim$50 at $z > 2$ (Benson et al. 2014; Henderson et al. 2016; Ade et al. 2019). The expected reach of the eROSITA, LSST/Euclid, and the CMB Stage-3 surveys are illustrated in Figure 10.16.

*Chandra* will play multiple roles in enabling the exploitation of these surveys. For example, while the eROSITA survey will find tens of thousand of clusters down to (far) fainter fluxes than *ROSAT*, at intermediate to high redshifts, it will not have the spatial resolution to discriminate cleanly between the X-ray emission from the ICM and that from point-like AGN in the cluster fields. Short *Chandra* observations for representative samples of these clusters should allow this effect to be modeled. Cosmological studies based on the $f_{gas}$ test (Section 10.2) will continue to rely on *Chandra* measurements as their primary data source. For the new X-ray, optical, and millimeter-wavelength surveys, the availability of low-scatter X-ray mass proxies from targeted X-ray measurements will continue to enhance both the statistical power and robustness of the constraints from cluster counts; while the optical and millimeter-wavelength surveys will provide exquisite constraints on the mean masses of clusters using galaxy- and CMB-weak lensing methods,

X-ray follow-up observations will be required to provide the low-scatter mass proxies for individual systems necessary to pin down the mass–observable scaling relations and their scatter, as a function of mass and redshift. Among the most exciting targets from these new surveys will be systems at the highest redshifts, which Figure 10.16 shows is likely to be the primary discovery space. Observations of these new high-$z$ targets will hold the potential to deliver significant advances in our understanding of cosmology and fundamental physics, as well as a host of astrophysics. The faintness of these targets, however, will mean that substantial *Chandra* exposure times will be necessary. The success of this work will therefore continue to depend on the strong support of the broader X-ray astronomy community.

## References

Ade, P., Aguirre, J., Ahmed, Z., et al. 2019, JCAP, 2019, 056
Aharonian, F. A., Akamatsu, H., Akimoto, F., et al. 2017, ApJ, 837, L15
Albrecht, A., Bernstein, G., Cahn, R., et al. 2006, arXiv:astro-ph/0609591
Allen, S. W., Evrard, A. E., & Mantz, A. B. 2011, ARA&A, 49, 409
Allen, S. W., Rapetti, D. A., Schmidt, R. W., et al. 2008, MNRAS, 383, 879
Allen, S. W., Schmidt, R. W., & Bridle, S. L. 2003, MNRAS, 346, 593
Allen, S. W., Schmidt, R. W., Ebeling, H., Fabian, A. C., & van Speybroeck, L. 2004, MNRAS, 353, 457
Allen, S. W., Schmidt, R. W., & Fabian, A. C. 2002, MNRAS, 334, L11
Amin, M. A., Wagoner, R. V., & Blandford, R. D. 2008, MNRAS, 390, 131
Amodeo, S., Ettori, S., Capasso, R., & Sereno, M. 2016, A&A, 590, A126
Anderson, L., Aubourg, E., Bailey, S., et al. 2014, MNRAS, 439, 83
Applegate, D. E., Mantz, A., Allen, S. W., et al. 2016, MNRAS, 457, 1522
Applegate, D. E., von der Linden, A., Kelly, P. L., et al. 2014, MNRAS, 439, 48
Bacon, D. J., Massey, R. J., Refregier, A. R., & Ellis, R. S. 2003, MNRAS, 344, 673
Bahcall, N. A., & Fan, X. 1998, ApJ, 504, 1
Bahcall, N. A., Fan, X., & Cen, R. 1997, ApJ, 485, L53
Barnaby, N., & Shandera, S. 2012, JCAP, 1, 34
Battaglia, N., Bond, J. R., Pfrommer, C., & Sievers, J. L. 2012, ApJ, 758, 74
Battaglia, N., Bond, J. R., Pfrommer, C., & Sievers, J. L. 2013, ApJ, 777, 123
Benson, B. A., Ade, P. A. R., Ahmed, Z., et al. 2014, Proc. SPIE, 9153, 91531P
Benson, B. A., de Haan, T., Dudley, J. P., et al. 2013, ApJ, 763, 147
Beutler, F., Blake, C., Colless, M., et al. 2011, MNRAS, 416, 3017
Blake, C., Brough, S., Colless, M., et al. 2011, MNRAS, 415, 2876
Bleem, L. E., Stalder, B., de Haan, T., et al. 2015, ApJS, 216, 27
Bocquet, S., Dietrich, J. P., Schrabback, T., et al. 2019, ApJ, 878, 55
Bocquet, S., Saro, A., Dolag, K., & Mohr, J. J. 2016, MNRAS, 456, 2361
Böhringer, H., Schuecker, P., Guzzo, L., et al. 2004, A&A, 425, 367
Bonamente, M., Joy, M. K., LaRoque, S. J., et al. 2006, ApJ, 647, 25
Borgani, S., Rosati, P., Tozzi, P., et al. 2001, ApJ, 561, 13
Bradač, M., Allen, S. W., Treu, T., et al. 2008, ApJ, 687, 959
Bradač, M., Clowe, D., Gonzalez, A. H., et al. 2006, ApJ, 652, 937
Bulbul, E., Markevitch, M., Foster, A., et al. 2014, ApJ, 789, 13
Burenin, R. A., Vikhlinin, A., Hornstrup, A., et al. 2007, ApJS, 172, 561

Carlstrom, J. E., Ade, P. A. R., Aird, K. A., et al. 2011, PASP, 123, 568
Carroll, S. M., Duvvuri, V., Trodden, M., & Turner, M. S. 2004, PhRvD, 70, 043528
Cataneo, M., Rapetti, D., Lombriser, L., & Li, B. 2016, JCAP, 12, 024
Cataneo, M., Rapetti, D., Schmidt, F., et al. 2015, PhRvD, 92, 044009
Chan, K. C., & Scoccimarro, R. 2009, PhRvD, 80, 104005
Chen, X. 2010, AdAst, 2010, 638979
Clowe, D., Bradač, M., Gonzalez, A. H., et al. 2006, ApJ, 648, L109
Clowe, D., Markevitch, M., Bradač, M., et al. 2012, ApJ, 758, 128
Copeland, E. J., Sami, M., & Tsujikawa, S. 2006, IJMPD, 15, 1753
Costanzi, M., Villaescusa-Navarro, F., Viel, M., et al. 2013, JCAP, 12, 12
Creminelli, P., & Zaldarriaga, M. 2004, JCAP, 10, 6
Daniel, S. F., Linder, E. V., Smith, T. L., et al. 2010, PhRvD, 81, 123508
David, L. P., Arnaud, K. A., Forman, W., & Jones, C. 1990, ApJ, 356, 32
Davis, T. M., Mörtsell, E., Sollerman, J., et al. 2007, ApJ, 666, 716
deHaan, T., Benson, B. A., Bleem, L. E., et al. 2016, ApJ, 832, 95
Dodelson, S. 2003, Modern Cosmology (New York: Academic)
Donahue, M., Voit, G. M., Gioia, I., et al. 1998, ApJ, 502, 550
Dunkley, J., Komatsu, E., Nolta, M. R., et al. 2009, ApJS, 180, 306
Dvali, G., Gabadadze, G., & Porrati, M. 2000, PhLB, 485, 208
Ebeling, H., Edge, A. C., Bohringer, H., et al. 1998, MNRAS, 301, 881
Ebeling, H., Edge, A. C., Mantz, A., et al. 2010, MNRAS, 407, 83
Edge, A. C., & Stewart, G. C. 1991, MNRAS, 252, 428
Ehlert, S., Allen, S. W., Brandt, W. N., et al. 2015, MNRAS, 446, 2709
Ettori, S., Baldi, A., Balestra, I., et al. 2015, A&A, 578, A46
Ettori, S., Morandi, A., Tozzi, P., et al. 2009, A&A, 501, 61
Fabian, A. C. 1991, MNRAS, 253, 29P
Feng, J. L. 2010, ARA&A, 48, 495
Forero-Romero, J. E., Gottlöber, S., & Yepes, G. 2010, ApJ, 725, 598
Freedman, W. L., Madore, B. F., Gibson, B. K., et al. 2001, ApJ, 553, 47
Frieman, J. A., Turner, M. S., & Huterer, D. 2008, ARA&A, 46, 385
Gao, L., Navarro, J. F., Cole, S., et al. 2008, MNRAS, 387, 536
Gaskin, J. A., Dominguez, A., Gelmis, K., et al. 2018, Proc. SPIE, 10699, 106990N
Hagstotz, S., Costanzi, M., Baldi, M., & Weller, J. 2019, MNRAS, 486, 3927
Hamed, N. A., Cheng, H. S., Luty, M. A., & Mukohyama, S. 2004, JHEP, 5, 74
Harrison, I., & Hotchkiss, S. 2013, JCAP, 7, 022
Harvey, D., Massey, R., Kitching, T., Taylor, A., & Tittley, E. 2015, Sci., 347, 1462
Hasselfield, M., Hilton, M., Marriage, T. A., et al. 2013, JCAP, 7, 8
Heitmann, K., Bingham, D., Lawrence, E., et al. 2016, ApJ, 820, 108
Henderson, S. W., Allison, R., Austermann, J., et al. 2016, JLTP, 184, 772
Henry, J. P., Gioia, I. M., Maccacaro, T., et al. 1992, ApJ, 386, 408
Hinshaw, G., Larson, D., Komatsu, E., et al. 2013, ApJS, 208, 19
Hoekstra, H., Herbonnet, R., Muzzin, A., et al. 2015, MNRAS, 449, 685
Holz, D. E., & Perlmutter, S. 2012, ApJ, 755, L36
Host, O., & Hansen, S. H. 2011, ApJ, 736, 52
Hu, W., & Sawicki, I. 2007, PhRvD, 76, 104043
Hubble, E. 1929, PNAS, 15, 168

Hughes, J. P. 1989, ApJ, 337, 21
Jain, B., & Khoury, J. 2010, AnPhy, 325, 1479
Jee, M. J., Hoekstra, H., Mahdavi, A., & Babul, A. 2014, ApJ, 783, 78
Jenkins, A., Frenk, C. S., White, S. D. M., et al. 2001, MNRAS, 321, 372
Knop, R. A., Aldering, G., Amanullah, R., et al. 2003, ApJ, 598, 102
Kolb, E. W., & Turner, M. S. 1990, The Early Universe (Reading, MA: Addison-Wesley)
Komatsu, E. 2010, CQGra, 27, 124010
Koulouridis, E., Ricci, M., Giles, P., et al. 2018, A&A, 620, A20
Kowalski, M., Rubin, D., Aldering, G., et al. 2008, ApJ, 686, 749
Laureijs, R., Amiaux, J., Arduini, S., et al. 2011, arXiv:1110.3193
Lesgourgues, J., & Pastor, S. 2006, PhR, 429, 307
Linder, E. V. 2005, PhRvD, 72, 043529
LoVerde, M. 2014, PhRvD, 90, 083518
LSST Dark Energy Science Collaboration 2012, arXiv:1211.0310
Ludlow, A. D., Navarro, J. F., Li, M., et al. 2012, MNRAS, 427, 1322
Mantz, A., Allen, S. W., Ebeling, H., & Rapetti, D. 2008, MNRAS, 387, 1179
Mantz, A., Allen, S. W., Ebeling, H., Rapetti, D., & Drlica-Wagner, A. 2010a, MNRAS, 406, 1773
Mantz, A., Allen, S. W., & Rapetti, D. 2010b, MNRAS, 406, 1805
Mantz, A., Allen, S. W., Rapetti, D., & Ebeling, H. 2010c, MNRAS, 406, 1759
Mantz, A. B., Allen, S. W., & Morris, R. G. 2016a, MNRAS, 462, 681
Mantz, A. B., Allen, S. W., & Morris, R. G. 2016b, arXiv:1607.04686
Mantz, A. B., Allen, S. W., Morris, R. G., et al. 2014, MNRAS, 440, 2077
Mantz, A. B., Allen, S. W., Morris, R. G., et al. 2015a, MNRAS, 449, 199
Mantz, A. B., Allen, S. W., Morris, R. G., et al. 2017, MNRAS, 472, 2877
Mantz, A. B., Allen, S. W., Morris, R. G., & von der Linden, A. 2018, MNRAS, 473, 3072
Mantz, A. B., von der Linden, A., Allen, S. W., et al. 2015b, MNRAS, 446, 2205
Markevitch, M., Gonzalez, A. H., Clowe, D., et al. 2004, ApJ, 606, 819
Maughan, B. J. 2007, ApJ, 668, 772
McClintock, T., Rozo, E., Becker, M. R., et al. 2019, ApJ, 872, 53
McDonald, M., Allen, S. W., Hlavacek-Larrondo, J., et al. 2019, ApJ, 870, 85
McDonald, M., Bulbul, E., de Haan, T., et al. 2016, ApJ, 826, 124
Merloni, A., Predehl, P., Becker, W., et al. 2012, arXiv:1209.3114
Merten, J., Coe, D., Dupke, R., et al. 2011, MNRAS, 417, 333
Mortonson, M. J., Hu, W., & Huterer, D. 2011, PhRvD, 83, 023015
Navarro, J. F., Frenk, C. S., & White, S. D. M. 1995, MNRAS, 275, 720
Nesseris, S., & Perivolaropoulos, L. 2008, PhRvD, 77, 023504
Nurgaliev, D., McDonald, M., Benson, B. A., et al. 2017, ApJ, 841, 5
Padmanabhan, N., Xu, X., Eisenstein, D. J., et al. 2012, MNRAS, 427, 2132
Peebles, P. J. E. 1980, The Large-scale Structure of the Universe (Princeton, NJ: Princeton Univ. Press)
Percival, W. J., Nichol, R. C., Eisenstein, D. J., et al. 2007, ApJ, 657, 51
Percival, W. J., Reid, B. A., Eisenstein, D. J., et al. 2010, MNRAS, 401, 2148
Perlmutter, S., Aldering, G., Goldhaber, G., et al. 1999, ApJ, 517, 565
Pierpaoli, E., Borgani, S., Scott, D., & White, M. 2003, MNRAS, 342, 163
Planck Collaboration: Ade, P. A. R., Aghanim, N., et al. 2011a, A&A, 536, A1

Planck Collaboration: Ade, P. A. R., Aghanim, N., et al. 2011b, A&A, 536, A8
Planck Collaboration: Ade, P. A. R., Aghanim, N., et al. 2011c, A&A, 536, A11
Planck Collaboration: Aghanim, N., Arnaud, M., et al. 2011d, A&A, 536, A10
Planck Collaboration: Ade, P. A. R., Aghanim, N., et al. 2014, A&A, 571, A20
Planck Collaboration: Ade, P. A. R., Aghanim, N., et al. 2016a, A&A, 594, A24
Planck Collaboration: Ade, P. A. R., Aghanim, N., et al. 2016b, A&A, 594, A13
Planck Collaboration: Aghanim, N., Akrami, Y., et al. 2018, arXiv:1807.06209
Planelles, S., Borgani, S., Dolag, K., et al. 2013, MNRAS, 431, 1487
Press, W. H., & Schechter, P. 1974, ApJ, 187, 425
Randall, S. W., Markevitch, M., Clowe, D., Gonzalez, A. H., & Bradač, M. 2008, ApJ, 679, 1173
Rapetti, D., Allen, S. W., Mantz, A., & Ebeling, H. 2009, MNRAS, 400, 699
Rapetti, D., Allen, S. W., Mantz, A., & Ebeling, H. 2010, MNRAS, 406, 1796
Rapetti, D., Blake, C., Allen, S. W., et al. 2013, MNRAS, 432, 973
Refregier, A., Rhodes, J., & Groth, E. J. 2002, ApJ, 572, L131
Reid, B. A., Samushia, L., White, M., et al. 2012, MNRAS, 426, 2719
Reid, B. A., Verde, L., Jimenez, R., & Mena, O. 2010, JCAP, 1, 3
Riess, A. G., Filippenko, A. V., Challis, P., et al. 1998, AJ, 116, 1009
Riess, A. G., Macri, L., Casertano, S., et al. 2011, ApJ, 730, 119
Rozo, E., Wechsler, R. H., Rykoff, E. S., et al. 2010, ApJ, 708, 645
Schmidt, F. 2009a, PhRvD, 80, 043001
Schmidt, F. 2009b, PhRvD, 80, 123003
Schmidt, F., Lima, M., Oyaizu, H., & Hu, W. 2009, PhRvD, 79, 083518
Schmidt, R. W., & Allen, S. W. 2007, MNRAS, 379, 209
Schmidt, R. W., Allen, S. W., & Fabian, A. C. 2004, MNRAS, 352, 1413
Schuecker, P., Böhringer, H., Collins, C. A., & Guzzo, L. 2003, A&A, 398, 867
Seljak, U. 2002, MNRAS, 337, 769
Shandera, S., Mantz, A., Rapetti, D., & Allen, S. W. 2013, JCAP, 8, 4
Sheth, R. K., & Tormen, G. 1999, MNRAS, 308, 119
Silk, J., & White, S. D. M. 1978, ApJ, 226, L103
Spergel, D. N., & Steinhardt, P. J. 2000, PhRvL, 84, 3760
Spergel, D. N., Verde, L., Peiris, H. V., et al. 2003, ApJS, 148, 175
Spergel, D. N., Bean, R., Doré, O., et al. 2007, ApJS, 170, 377
Sunyaev, R. A., & Zeldovich, Y. B. 1972, CoASP, 4, 173
Suzuki, N., Rubin, D., Lidman, C., et al. 2012, ApJ, 746, 85
Swetz, D. S., Ade, P. A. R., Amiri, M., et al. 2011, ApJS, 194, 41
Tinker, J., Kravtsov, A. V., Klypin, A., et al. 2008, ApJ, 688, 709
Urban, O., Werner, N., Allen, S. W., et al. 2015, MNRAS, 451, 2447
Van Waerbeke, L., Mellier, Y., Pelló, R., et al. 2002, A&A, 393, 369
Vikhlinin, A., Kravtsov, A. V., Burenin, R. A., et al. 2009, ApJ, 692, 1060
Vikhlinin, A., Kravtsov, A., Forman, W., et al. 2006, ApJ, 640, 691
Voigt, L. M., & Fabian, A. C. 2006, MNRAS, 368, 518
von der Linden, A., Allen, M. T., Applegate, D. E., et al. 2014a, MNRAS, 439, 2
von der Linden, A., Mantz, A., Allen, S. W., et al. 2014b, MNRAS, 443, 1973
Weinberg, D. H., Mortonson, M. J., Eisenstein, D. J., et al. 2013, PhR, 530, 87
Weisskopf, M. C. 2010, PNAS, 107, 7135
White, S. D. M., Efstathiou, G., & Frenk, C. S. 1993a, MNRAS, 262, 1023

White, S. D. M., Navarro, J. F., Evrard, A. E., & Frenk, C. S. 1993b, Natur, 366, 429
Wittman, D., Golovich, N., & Dawson, W. A. 2018, ApJ, 869, 104
Wu, H., Rozo, E., & Wechsler, R. H. 2010, ApJ, 713, 1207
Yoshida, N., Springel, V., White, S. D. M., & Tormen, G. 2000, ApJ, 544, L87
Zhang, P., Liguori, M., Bean, R., & Dodelson, S. 2007, PhRvL, 99, 141302
Zhang, Y.-Y., Böhringer, H., Finoguenov, A., et al. 2006, A&A, 456, 55
Zwicky, F. 1933, AcHPh, 6, 110

# Chapter 11

## Future X-Ray Missions

**Belinda J Wilkes**

The past 20 years, since the launch of both NASA's *Chandra*[1] and ESA's *XMM-Newton*[2] in 1999, have seen a golden age of X-ray astronomy. These two observatories, complementary in many ways, have formed the backbone of the quest to understand the X-ray universe. *Chandra*'s high spatial resolution resolves sources in crowded fields, sharp structures, and looks deep into the universe, while *XMM-Newton*'s larger mirrors and field of view efficiently obtain high signal-to-noise spectra and light curves, and more efficient sky coverage at lower ($\sim 16''$, half-energy width) spatial resolution. Critically, working together, new phenomena have been independently confirmed and characterized by both observatories.

Working both independently and in consort with the large observatories, several smaller missions have significantly enhanced progress in understanding the X-ray universe. Those still in operation include: the *Neil Gehrels Swift*[3] *Observatory*, which constantly monitors the sky for gamma-ray bursts and other transients, often handing off monitoring to *Chandra* or *XMM-Newton* as the source fades below its flux limit; *NuStar*,[4] which images the sky at energies up to ~80 keV for the first time, and often works with *XMM-Newton* and/or *Chandra* to define spectra over a much broader band; and the youngest member of the team, NICER[5] (the Neutron star Interior Composition Explorer), mounted on the *International Space Station*, which studies neutron stars and other soft transients with ~10× the effective area of *Chandra* at low energies.

While the unexpected longevity of operations for many of these X-ray missions continues to forge new paths into X-ray studies of celestial sources, many are well beyond their original life span and are likely to cease operations over the next decade

[1] http://cxc.harvard.edu
[2] https://heasarc.gsfc.nasa.gov/docs/xmm/xmmgof.html
[3] https://swift.gsfc.nasa.gov/
[4] https://www.nasa.gov/mission_pages/nustar/main/index.html
[5] https://heasarc.gsfc.nasa.gov/docs/nicer/index.html

doi:10.1088/2514-3433/ab43dcch11

or so. With this reality, the X-ray community is developing new missions, including both expanded and new capabilities, that will facilitate more advanced X-ray exploration. This chapter summarizes several upcoming approved missions, along with an incomplete list of possible future US-based missions, including several NASA-funded mission concept studies, as the community prepares for the US 2020 Decadal Survey in Astronomy and Astrophysics (Astro2020).[6]

## 11.1 Approved Missions

### 11.1.1 *Spektr-RG/SRG*

*Spektr-RG/SRG*[7] (Figure 11.1) is a Russian mission with German participation. It was launched from the Baikonur Cosmodrome on 2019 July 12 and placed in an L2 orbit. *SRG* is a collaboration between the space agencies Roskosmos (Russia) and DLR (Germany). *SRG* contains two X-ray telescopes: eROSITA as the primary scientific payload, developed under the leadership of MPE (Max-Planck-Institute for Extraterrestrial Physics, Garching, Germany), and ART-XC, developed under the leadership of IKI RAS (Space Research Institute of the Russian Academy of Sciences, Moscow) with a contribution made by NASA/MSFC, Huntsville, AL, USA.

The overall objective of the mission is to conduct the first all-sky survey with an imaging telescope in the 0.5–11 keV band, and the first all-sky imaging X-ray time-variability survey.

The eRosita[8] instrument (MPE Garching) consists of seven identical Wolter I telescopes, small versions of the *XMM-Newton* mirrors but manufactured using the

**Figure 11.1.** The *SRG* spacecraft is based on the "Navigator" multiuse bus of the NPO Lavochkin Scientific Production Association, Khimki, Russia. The spacecraft is three-axis stabilized and has an estimated launch mass of ~2712.5 kg (total payload mass of 1165 kg). (Reproduced courtesy of Roscosmos TV.)

---

[6] https://sites.nationalacademies.org/DEPS/astro2020/index.htm
[7] http://srg.iki.rssi.ru/?lang=en
[8] https://www.mpe.mpg.de/455799/instrument

**Figure 11.2.** The front view of the eROSITA space telescope, consisting of seven identical mirror modules aligned in parallel. Each has a diameter of 36 centimeters and consists of 54 nested mirror shells, the surface of which is composed of a paraboloid and a hyperboloid (Wolter I optics). (Reproduced courtesy of MPE.)

same technology (Figure 11.2) . At each focus there are seven framestore pnCCDs, upgraded versions of the successful *XMM-Newton* cameras. These detectors permit accurate spectroscopy of X-rays as well as imaging with high (50 ms) timing resolution. All seven telescopes have a spatial resolution of 16″ (on-axis half-energy width) at 1 keV (~25″ in survey mode). The effective area is 2500 $cm^2$ at 1 keV over a 61′ field of view with 9″ pixels. The energy range is 0.3–10 keV with energy resolution of 138 eV at 6 keV. For the first four years of the mission, eROSITA will perform a deep survey of the entire X-ray sky. In the soft X-ray band (0.5–2 keV), this will be about 20 times more sensitive than the *ROSAT* all-sky survey (point-source flux limit $\sim 1.2 \times 10^{-14}$ erg $cm^{-2}$ $s^{-1}$ after four years), while in the hard band (2–10 keV) it will provide the first imaging survey of the sky (point-source flux limit $\sim 2.0 \times 10^{-13}$ erg $cm^{-2}$ $s^{-1}$ after four years). The German eROSITA survey data, of the western sky in Galactic coordinates, will be made public after a two-year proprietary period.

The ART-XC[9] (Astronomical Roentgen Telescope—X-ray Concentrator, Space Research Institute, Moscow) is an X-ray-energy survey instrument, providing higher energy coverage, 5–30 keV (with limited sensitivity above 12 keV). ART-XC also contains seven identical telescopes with mirrors fabricated at NASA/MSFC. The cadmium telluride (CdTe) double-sided strip focal plane detectors will operate over an energy range of ~5–30 keV, with an angular resolution of 30″ at 8 keV, a field of view of 36′ diameter, and an energy resolution of ~10% at 10 keV. It will carry out an all-sky X-ray survey in the 6–11 keV energy range with a sensitivity of $(2\text{–}20) \times 10^{-13}$ erg $cm^{-2}$ $s^{-1}$ $keV^{-1}$. The effective area is 410 $cm^{-2}$ at 8 keV.

The expected lifetime of SRG is 6.5 years. It is planned that the survey will be followed by pointed observations of selected fields, such as the Galactic Center and

[9] http://srg.iki.rssi.ru/?page_id=674&lang=en

the SMC, in the full, 0.3–30 keV energy range, and a search for transients. Beyond mapping the full high-energy sky, the mission will facilitate the detection of large samples of galaxy clusters and active galactic nuclei (AGNs), including obscured sources, as well as insights into many kinds of X-ray-emitting celestial sources such as X-ray binaries and supernova remnants.

### 11.1.2 The *X-Ray Imaging and Spectroscopy Mission*

The *X-Ray Imaging and Spectroscopy Mission*[10] (*XRISM*; Figure 11.3) is a JAXA/NASA collaborative mission, with ESA participation. It provides high-throughput imaging and high-resolution spectroscopy (Guainazzi & Tashiro 2018). *XRISM* is expected to launch in 2021 (TBR) on a JAXA H-2A rocket.

The *XRISM* payload consists of two instruments. A soft X-ray spectrometer, Resolve, which combines a lightweight X-Ray Mirror Assembly paired with an X-ray calorimeter spectrometer and provides nondispersive 5–7 eV energy resolution in the 0.3–12 keV bandpass with a field of view of about $3'$ . A soft X-ray imager, Xtend, with a CCD detector that extends the field of the observatory to $38'$ over the energy range 0.4–13 keV, using an identical lightweight X-Ray Mirror Assembly. The instruments' characteristics are similar to those of the Soft X-ray Spectrometer and Soft X-ray Imager, respectively, flown on *Hitomi*.[12]

*XRISM* is designed to cover the science capability lost with the *Hitomi* mishap, such as studying the structure of the universe, including the creation and evolution of galaxy clusters and supermassive black holes, and the physics in extreme conditions such as high density and strong magnetic fields.

**Figure 11.3.** The *XRISM* spacecraft. (Reproduced courtesy of JAXA.[11])

[10] https://heasarc.gsfc.nasa.gov/docs/xrism/

[11] https://global.jaxa.jp/projects/sas/xrism/

[12] https://heasarc.gsfc.nasa.gov/docs/hitomi/

### 11.1.3 The *Imaging X-Ray Polarimetry Explorer*

The NASA Explorer mission *Imaging X-Ray Polarimetry Explorer*[13] (*IXPE*; Figure 11.4) will be launched on or after 2021 April into a 540 km circular orbit at 0° inclination (Weisskopf et al. 2016). It will have three identical X-ray telescopes. The Gas Pixel Detectors (GPD) at each telescope focus are based on proportional counters. The mission will provide $\lesssim 30''$ angular resolution (half-power diameter) over a $9'$ field of view with an effective area of $>100$ cm$^2$ at 2.3 keV. The energy range will be 2–8 keV with energy resolution of $\sim$1.5 keV at 5.9 keV (FWHM $\propto 1/\sqrt{E}$) and $\leqslant$100 μs timing resolution. *IXPE* shall provide a minimum detectable polarization not to exceed 5.5% for a point source with an $E^{-2}$ photon spectrum, and reach a 2–8 keV flux of $10^{-11}$ ergs cm$^{-2}$ s$^{-1}$ in an integration time of 10 days.

*IXPE* shall obtain polarimetric images, including spatial, spectral, and timing data, of celestial sources such as AGN jets, pulsar wind nebulae, and shell-type supernova remnants. The data will make it possible to map the magnetic fields of the X-ray-emitting regions.

### 11.1.4 The Advanced Telescope for High ENergy Astrophysics (Athena)

The European Space Agency's (ESA) L-class mission *Advanced Telescope for High ENergy Astrophysics*[14] (*Athena*; Figure 11.5) was selected on 2014 June 27 as part of the Cosmic Vision 2015–2025 plan (Barcons et al. 2017). The mission is currently in the study phase: once the design and costing are complete, it will be proposed for "adoption" around 2021, starting the construction phase. It is currently slated for launch in 2031 into a halo orbit around the Earth–Sun L2 Lagrange point. The

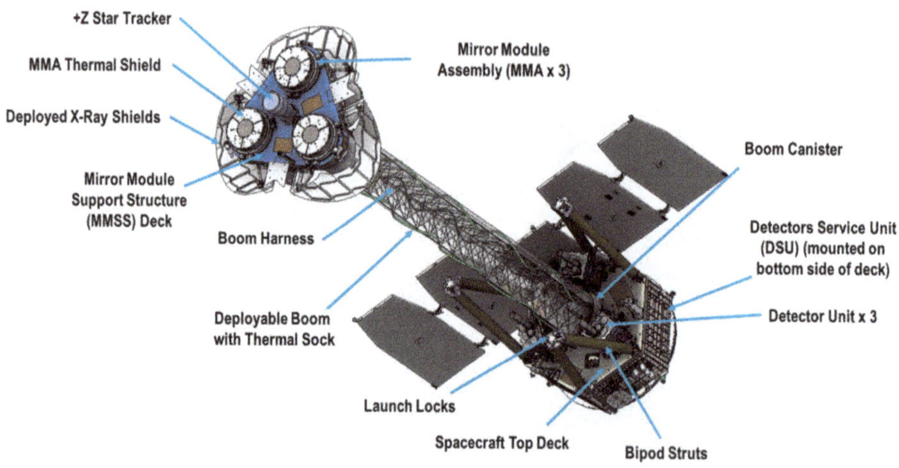


**Figure 11.4.** The *IXPE* spacecraft deployed. The total length is 5.2 m. (Courtesy of NASA.)

[13] https://ixpe.msfc.nasa.gov/

[14] http://sci.esa.int/athena/

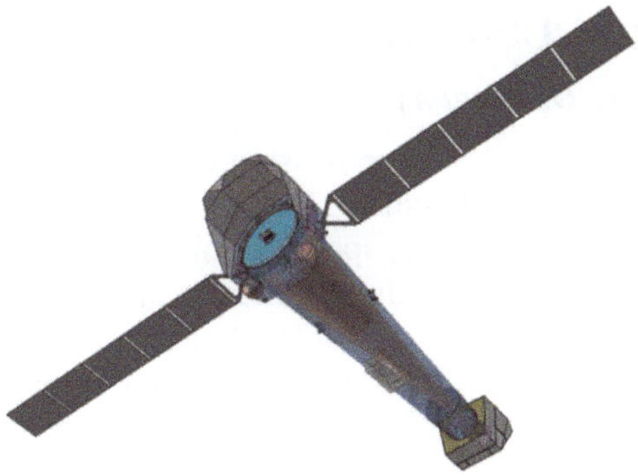

**Figure 11.5.** A conceptual design for the *Athena* spacecraft. (Reproduced from ESA CDF Study Report: CDF-150(A).[15])

baseline mission duration will be four years, with consumables sized for an extension to 10 years.

*Athena* will be a general-purpose observatory and is being designed to address the following key questions in astrophysics: how does ordinary matter assemble into the large-scale structures we see today? How do black holes grow and shape the universe?

The X-ray telescope design has a focal length of 12 m, with spatial resolution of $5''$ (half-energy width) on axis, and an effective area of $\sim$1.4 m$^2$ at 1 keV. Using silicon pore optics technology, around 1.5 million pores with cross-sections of a few millimeter squared will provide the required collecting area and angular resolution. Each pore acts as a very small section of a Wolter I telescope. The Mirror Assembly Module (MAM) will support the X-ray optics. The MAM is mounted on a hexapod system to move the focus between the two instruments: an X-ray Integral Field Unit (X-IFU) for high spectral-resolution imaging, and a Wide Field Imager (WFI) for imaging and moderate-resolution spectroscopy over a large field of view, and high count-rate capability.

The X-IFU[16] will be an advanced, actively shielded X-ray microcalorimeter spectrometer for high spectral-resolution imaging cooled to 100 mK. The instrument is based on a large array of Transition Edge Sensors (TES), with $5''$ pixels, over a field of view of $5'$ in equivalent diameter. It covers an energy range of 0.2–12 keV with 2.5 eV spectral resolution ($E < 7$ keV) and $E/\Delta E = 2800$ ($E > 7$ keV). The X-IFU also provides microsecond time resolution together with the capability to observe very bright X-ray sources with better than 10 eV spectral resolution between 5 and 8 keV using the defocusing capability of the *Athena* mirror.

The WFI[17] uses silicon-depleted p-channel field effect transistor (DEPFET) active pixel sensors with energy resolution of <170 eV (FWHM) at 7 keV. The Large Detector Array covers a $40' \times 40'$ field of view with $2.2'' \times 2.2''$ pixels to oversample the PSF. The separate Fast Detector is optimized for high count-rate applications with 80 μs time resolution.

[15] https://www.cosmos.esa.int/web/athena/study-documents

[16] http://x-ifu.irap.omp.eu/

[17] http://www.mpe.mpg.de/ATHENA-WFI/

## 11.2 Possible Future US-based X-Ray Missions

### 11.2.1 The *Lynx X-Ray Observatory*

The *Lynx X-ray Observatory* is a large mission concept study funded by NASA in advance of Astro2020. *Lynx*[18] (Figure 11.6) would be a true *Chandra* successor, aiming to achieve *Chandra*-like, ~0.5″ spatial resolution with a large, 22′ × 22′ field-of-view, (Gaskin et al. 2019). The mirror design is an assembly of densely packed, thin, grazing-incidence mirrors with an outer diameter of 3 m and a total effective area greater than 2 $m^2$ at 1 keV. The on-axis angular resolution will be 0.5″ (50% power diameter) with subarcsecond resolution out to 10′ off axis.

The instrument complement[19] will include a state-of-the-art microcalorimeter providing nondispersive imaging spectroscopy, a high-definition X-ray Imager (HDXI), and a X-ray Grating Spectrometer (HGS).

The microcalorimeter design calls for energy resolution ≲3 eV (0.2–7 keV) and the standard instrument setup will include 1″ pixels over a 5′ × 5′ field of view. Several subarrays will be available, optimized for (a) subarcsecond imaging, (b) 0.3 eV energy resolution, and (3) a large, 20′ × 20′, field of view.

The HDXI design comprises an array of silicon sensors with ~0.3″ pixels covering a field of view in excess of 20′ × 20′, providing moderate (~100 eV) spectral resolution over the full 0.1–10 keV energy band. High frame rates will minimize pileup and provide at least 100 μs time resolution. Technology prototypes for the

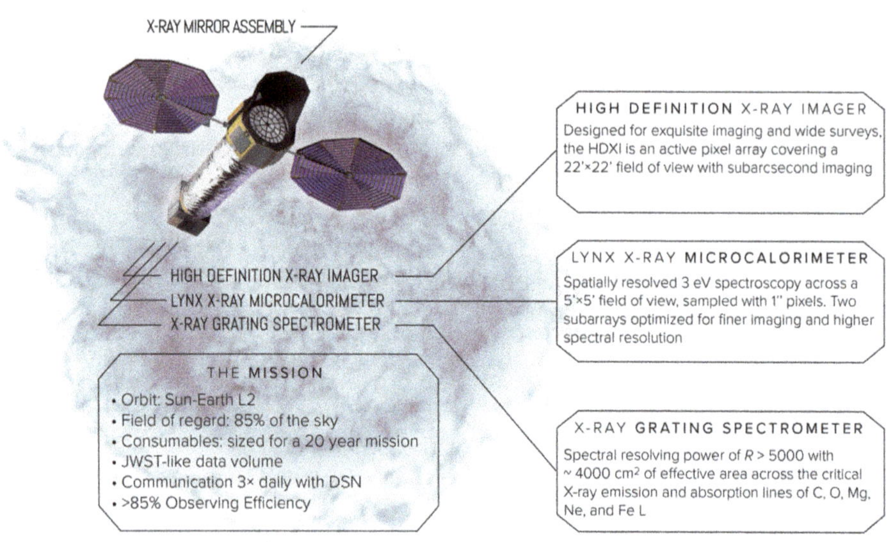


**Figure 11.6.** A conceptual design for the *Lynx* spacecraft. (Reproduced from the *Lynx* Mission Study Report 2019.[20])

[18] https://wwwastro.msfc.nasa.gov/lynx/docs/science/observatory.html
[19] https://www.lynxobservatory.com/mission
[20] https://wwwastro.msfc.nasa.gov/lynx/

detector include a hybrid complementary metal-oxide–semiconductor (CMOS), a digital CCD with CMOS readout, and a monolithic CMOS detector.

The XGS design provides extremely high resolving power, $R > 5000$, spectroscopy, combined with an effective area exceeding 4000 $cm^2$ over the astrophysically important X-ray emission and absorption lines of C, O, Mg, Ne, and Fe–L. Technology prototypes include critical angle transmission and off-plane gratings.

*Lynx* will be a general-purpose observatory, with the science driven by proposals from the astronomical community. Its sensitivity in one day of observing will be comparable with two weeks of *Chandra* observing time. *Lynx* is being designed to address the following key science[21] questions in current astrophysics. It will directly observe the birth and early growth of supermassive black holes, allowing us to understand how they formed and grew so quickly in the early universe. *Lynx* will reveal the drivers of galaxy formation by mapping hot gas around galaxies and in the cosmic web, and discovering how this baryonic component is heated and ionized to X-ray temperatures. *Lynx* will trace stellar activity including the effects on planet habitability and probe the entire mass scale of stars out to 5 kpc, transforming our knowledge of the endpoints of stellar evolution.

### 11.2.2 Arcus

*Arcus*[22] (Figure 11.7) is a high-resolution X-ray grating spectrometer mission that was proposed to NASA as a MIDEX in 2016 and selected for a Phase A concept study (Smith et al. 2017). It was not approved for Phase B, but will be reproposed for a future call.

In the 2016 proposal, *Arcus* had four identical, independent optical assemblies with a 12 m focal length feeding into two identical focal planes. Each assembly was composed of 34 coaligned, silicon pore optics mirror modules, each of which focused their collected X-rays through the Critical Angle Transmission grating windows that provided a minimum resolution ($\Delta\lambda/\lambda$) of 2500 over the 12–50 Å bandpass. The dispersed X-rays traveled through a boom onto a CCD-based detector array. The effective area was ~250 $cm^2$ over the key 16–29 Å bandpass, which, together with

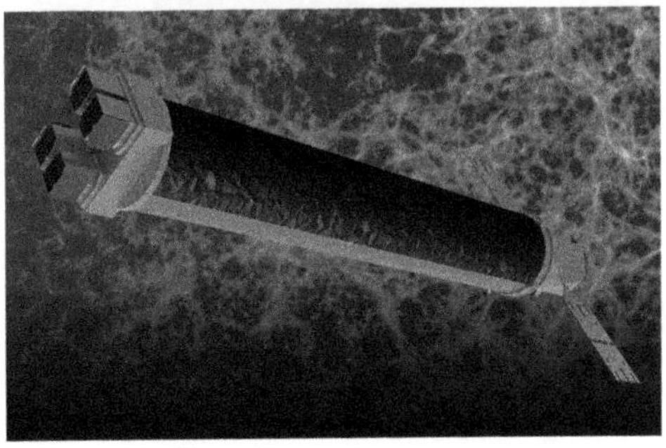

**Figure 11.7.** The *Arcus* spacecraft. (Credit: *Arcus* image courtesy of the *Arcus* team; background from the EAGLE simulation, a project of the VIRGO Consortium.)

[21] https://wwwastro.msfc.nasa.gov/lynx/docs/science/

[22] http://www.arcusxray.org/science.html

the high resolution, enabled sensitivity to absorption features as weak as 3 mÅ equivalent width.

The science[23] goals were aimed at 2010 decadal survey priorities, primarily community driven through a guest observer program. They included (1) determining how baryons cycle in and out of galaxies, by observing the distribution and temperature of hot gas around galaxies and clusters, and mapping the temperature and abundances of gas in and around the Galaxy; (2) understanding how feedback from a black hole system influences its surroundings by determining the mass, energy, and momentum in the accretion-driven, outflowing winds from both supermassive and stellar mass black holes; and (3) examining how stellar systems form and evolve by measuring the balance between accretion and outflow in young accreting stars, evolved coronal stars, and exoplanet atmospheres.

### 11.2.3 *Advanced X-Ray Imaging Satellite*

The NASA-funded probe concept mission study, in advance of Astro2020, *Advanced X-Ray Imaging Satellite*[24] (*AXIS*; Figure 11.8) employs lightweight, thin-shell mirror technology with expected ~0.3″ resolution being developed at GSFC, as also proposed for the *Lynx* mission (Section 11.2.1). The nested array of mirrors will be ~7000 $cm^2$, 10× the effective area of *Chandra* at 1 keV (15× at 10 keV). The energy range is ~0.3–12 keV with resolution 150 eV at 6 keV and timing resolution <50 ms.

The focal plane detector design includes an array of either CCDs or active pixel sensor (APS) devices covering a field of view of 15′, or larger. The spatial resolution will remain <1″ over the full 15′ field. Small pixels and fast readout allow for sampling the point-spread function and significantly reduce pileup from bright point sources.

*AXIS* will be a general-purpose X-ray CCD imaging, low-resolution spectroscopy, and timing mission. Science goals include galaxy growth and feedback, supermassive black holes, the high-redshift universe, and more.

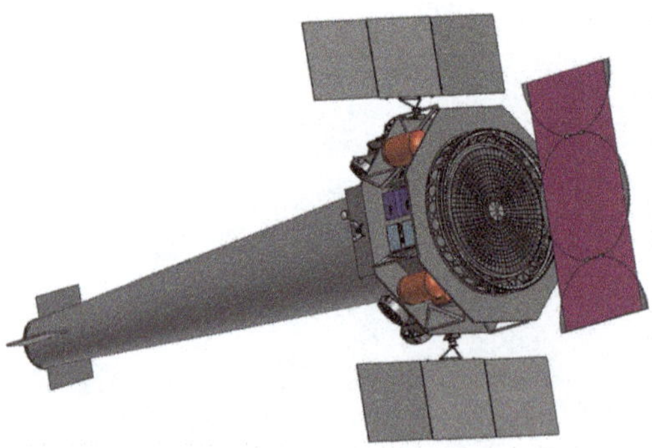

**Figure 11.8.** The *AXIS* spacecraft. (Courtesy of NASA/GSFC, Mission Design Lab; Mushotzky et al. 2019.)

[23] http://www.arcusxray.org/science.html
[24] http://axis.astro.umd.edu/

### 11.2.4 The *High-Energy X-Ray Probe*

The *High-Energy X-ray Probe* (*HEX-P*) is a next-generation, high-energy X-ray observatory concept that was submitted as a White Paper to Astro2020. The *HEX-P* design has broadband (2–200 keV) response with 40 times the sensitivity of any previous mission in the 10–80 keV band (e.g., *NuStar*; Harrison et al. 2013), and >100 times that in the 80–200 keV band. Relying on *NuStar*'s heritage and employing recent technological advances, *HEX-P* will achieve an angular resolution of 5″ (half-power diameter) over the full energy range, a 13′ × 13′ field of view, spectral resolution of 120 eV at 6 keV (0.8 keV at 60 keV), and timing resolution of 1 μs. The 5″ resolution is achieved using monosilicate mirror technology (Madsen et al. 2018), and the energy range is extended up to 200 keV by using a 20 m long focal length together with Ni/C multilayer coatings. In its baseline configuration, *HEX-P* will have three coaligned grazing-incidence optics modules, as being developed for missions such as *Lynx*, by the Next Generation X-ray Optics team at NASA's Goddard Space Flight Center (Zhang et al. 2019). Each module will include 389 multilayer-coated shells. *HEX-P* would be launched into a circular, low-Earth orbit with near-equatorial inclination. It would operate primarily as a pointed observatory, with a funded, competitive Guest Observer program.

*HEX-P* will probe the extreme environments around black holes and neutron stars, and map the growth of supermassive black holes and their environments. It will resolve the hard X-ray emission from dense regions of our Galaxy to understand the high-energy source populations, and investigate dark matter candidate particles through their decay channel signatures. *HEX-P* could be built and launched in the next decade, resulting in overlap with ESA's *Athena* (Section 11.1.4). Simultaneous observations with both observatories, which have similar ~5″ spatial resolution, would greatly enhance their scientific reach, for example increasing the ability to probe a range of phenomena from the detailed physics of black hole accretion to hot, merger-driven shocks in clusters—both of which have continua extending to high energy.

### 11.2.5 The *Spectroscopic Time-Resolving Observatory for Broadband Energy X-Rays*

The *Spectroscopic Time-Resolving Observatory for Broadband Energy X-Rays* (*STROBE-X*)[25] is a probe-class mission concept study funded by NASA in advance of Astro2020. *STROBE-X* (Figure 11.9) is designed to perform timing and spectroscopy over both a broad energy band (0.2–30 keV) and a wide range of timescales from microseconds to years. *STROBE-X* comprises two narrow-field instruments and a wide-field monitor (WFM).

The X-ray Concentrator Array (XRCA) covers the low-energy band (0.2–12 keV). The lightweight, high-throughput "concentrator" optics, optimized for point-like (<2′) sources with no imaging requirement, are a scaled-up version of those implemented on the NICER[26] Mission of Opportunity currently operating on the *International Space Station*. The optics array design has a 3 m focal length, effective area of 21,760 cm$^2$,

[25] https://gamma-ray.nsstc.nasa.gov/Strobe-X/

[26] https://www.nasa.gov/nicer

**Figure 11.9.** The *STROBE-X* spacecraft reproduced from Ray et al. (2019). (Courtesy of NASA.)

and collected photons will be incident on small silicon-drift detectors with CCD-level (85–175 eV) energy resolution, 100 ns time resolution, and low background rates.

The Large Area Detector (LAD), covering the higher-energy band (2–30 keV), uses large-area, collimated silicon-drift detectors that were developed for the European LOFT[27] mission concept. The effective area will be 51,000 $cm^2$ at 10 keV, with 200–300 eV (FWHM) energy resolution and 10 μs timing resolution.

Each instrument will provide an order-of-magnitude improvement in effective area over its predecessor. *STROBE-X* also includes a sensitive wide-field monitor (WFM), both to act as a trigger for pointed observations of X-ray transients and to provide high duty-cycle, high time-resolution, and high spectral-resolution monitoring of the variable X-ray sky. The WFM design reaches ~20× the sensitivity of the *RXTE* All-Sky Monitor, enabling multiwavelength and multimessenger investigations with a large instantaneous field of view.

## References

Barcons, X., Barret, D., Decourchelle, A., et al. 2017, AN, 338, 153
Gaskin, J. A., Swartz, D. A., Vikhlinin, A., et al. 2019, JATIS, 5, 021001
Guainazzi, M., & Tashiro, M. S. 2018, arXiv:1807.06903
Harrison, F. A., Craig, W. W., Christensen, F. E., et al. 2013, ApJ, 770, 103
Madsen, K. K., Harrison, F., Broadway, D., et al. 2018, Proc. SPIE, 10699, 106996M
Mushotzky, R. F., Aird, J., Barger, A. J., et al. 2019, arXiv 1903.04083
Ray, P. S., Arzoumanian, Z., Ballantyne, D., et al. 2019, arXiv 1903.03035
Smith, R. K., Abraham, M., Allured, R., et al. 2017, Proc. SPIE, 10397, 103970Q
Weisskopf, M. C., Ramsey, B., O'Dell, S., et al. 2016, Proc. SPIE, 9905, 990517
Zhang, W. W., Allgood, K. D., Biskach, M. P., et al. 2019, JATIS, 5, 021012

[27] https://www.isdc.unige.ch/loft/

www.ingramcontent.com/pod-product-compliance
Ingram Content Group UK Ltd.
Pitfield, Milton Keynes, MK11 3LW, UK
UKHW051042120926
8556IPUK00030B/821

9 780750 321617